MEDIA AND COMMUNICATIONS - TECHNOLOGIES, POLICIES AND CHALLENGES

COMMUNICATION SYSTEMS

NEW RESEARCH

MEDIA AND COMMUNICATIONS - TECHNOLOGIES, POLICIES AND CHALLENGES

Additional books in this series can be found on Nova's website under the Series tab.

Additional e-books in this series can be found on Nova's website under the e-book tab.

MEDIA AND COMMUNICATIONS - TECHNOLOGIES, POLICIES AND CHALLENGES

COMMUNICATION SYSTEMS

NEW RESEARCH

VYACHESLAV TUZLUKOV
EDITOR

nova publishers
New York

Library of Congress Cataloging-in-Publication Data

Communication systems (Nova Science Publishers)
Communication systems : new research / editor, Vyacheslav Tuzlukov (School of Electrical Engineering & Computer Science, Kyungpook National University, Sankyuk-dong, Buk-gu, Daegu South Korea).
pages cm
Includes bibliographical references and index.
ISBN: 978-1-62618-654-5 (hardcover)
1. Wireless communication systems. I. Tuzlukov, V. P. (Viacheslav Petrovich), editor of compilation. II. Title.
TK5103.2.C6435 2013
621.384--dc23

2013015684

Published by Nova Science Publishers, Inc. † New York

CONTENTS

PREFACE

Communication systems including wireless communications and wireless sensor networks, mobile communication, and satellite communications have been one of the major revolutions of the last three decades. We are witnessing a very fast growth in these technologies where communication systems have become so ubiquitous in our society and indispensable our daily lives. The demand for new services to support high speed wideband Internet access and advanced high quality real-time video applications push the researchers to investigate new technologies in communication systems, namely, wireless communications and wireless sensor networks, mobile communication, and satellite communications.

Progress in communication systems continues as this book is being written. Although there have been many journals and conference publications regarding the wireless communications and wireless sensor networks, mobile communication, and satellite communications, they are often in the context of academic research or theoretical derivations and sometimes omit practical considerations. Although the literature has many conference and journal papers, technical reports, and standard contributions, they are often fragmental engineering works and thus are not easy follow up. The objective of this book is to present new research avenues in wireless communications and wireless sensor networks, mobile communication, and satellite communications and to accelerate research and development by serving as a forum in which both academia and industry can share experiences and report original studies and works regarding all aspects of wireless communications and wireless sensor networks, mobile communication, and satellite communications. In addition, this book has a great educational value because it aims to serve as a virtual, but nonetheless effective bridge between academic research in theory and engineering development in practice, and as a messenger between the technical pioneers and the researchers who followed in their footstep.

The book that is titled "Communication Systems: New Research", focuses on the current research topics from wide range of communications including wireless communications and wireless sensor networks, mobile communication, and satellite communications and provides "on-going" research progress on these issues. During the preparation of this book, I emphasizes to the authors to add recent research findings and future works in this area and to cite latest references in the chapter. For this reason, a variety of novel technologies in wireless communications and wireless sensor networks, mobile communication, and satellite communications are investigated in this book. The authors attempt to present these topics in detail. Insightful and reader-friendly descriptions are presented to nourish readers of any level, from practicing and knowledgeable communication engineers to beginning or professional researchers. All interested readers can easily find noteworthy materials in much greater detail than in previous publications and in the references cited in these chapters.

This book includes twelve chapters that were authored by the well-known researchers in the world. Each chapter was written in an introductory style beginning with the fundamentals, describing approaches to the hottest issues and concluding with a comprehensive discussion. The content in each chapter is taken from many publications in prestigious journals and conferences and followed by fruitful insights. The chapters in this book also provide many recent references for relevant topics, and interested readers will find these references helpful when they explore these topics further.

Chapter 1 - addresses to the demands for wireless broadband applications that are continuously increasing and it is becoming extremely hard to find vacant frequency bands, to deploy new communication services or enhance the existing ones. Measurements of the spectrum occupancy show that spectrum shortage results from the spectrum management policy rather than from the physical scarcity of usable frequencies. In order to overcome this problem and improve spectrum utilization, cognitive radio concept has been proposed. This chapter provides an overview of cognitive radio for opportunistic spectrum access and related research topics. Cognitive Radio is the key technology allowing radio equipment to obtain knowledge of its radio environment and to dynamically adjust its operational parameters in order to improve its performance. The number of promising concepts for cognitive radio is introduced and discussed in this chapter, starting from the rationale behind the cognitive radio technology, passing through the area of spectrum awareness, spectrum management, and finally focusing on spectrum sensing methodologies.

Chapter 2 - highlights that over the last two decades, wireless communications have gained more and more importance in our everyday life. The rapidly growing of mobile wireless access services, together with an outdated fixed spectrum access policy is speeding up the need for more radio spectrum, a finite resource that is the basis of wireless communications. Moreover, the unlicensed portions of the spectrum, like Industrial, Scientific and Medical (ISM) bands, have become increasingly crowded. To increase the spectrum utilization, and therefore its usage efficiency, a big effort has been employed to develop and improve novel and alternative spectrum management policies, often referred to as Dynamic Spectrum Access (DSA). In DSA schemes, one piece of spectrum can be allocated to one or more users, which are called Primary Users (PUs) or licensed users, using the traditional fixed spectrum access policy. However, these users do not own the exclusivity of accessing the licensed channel, although they have higher priority in using it. But when the licensed channel is not being used by the PUs, the channel is vacant, leading to a spectrum underutilization. In this case, other users, denominated as Secondary Users (SUs) or unlicensed users, can dynamically access to the temporary vacant channel without interfering with PUs, hence improving the spectrum efficiency. To implement the DSA technique, SUs have the responsibility to sense the spectrum, searching for a vacant band, and use it as long as it is shown to be free from licensed user: this set of actions is also denoted as a cognitive capability, and SUs with this ability are usually called as Cognitive Radio (CR) users. The aforementioned characteristics of SUs, and therefore the peculiar characteristics of Cognitive Radio Networks (CRNs) distinguish them from the traditional wireless networks. Thus, traditional cross-layered Physical (PHY)-Medium Access Control (MAC) architectures become unsuitable for this new type of networks. In CRNs, the PHY layer should be responsible to perform spectrum sensing tasks, enabling Cognitive Radio (CR) users to identify the spectrum holes, and also to reconfigure and optimize the transceiver based on the results acquired from the environment. Regarding the MAC sub layer, it should be capable to choose between the spectrum sensing and the spectrum access. The sensing/access coordination should be based on a tradeoff between sensing opportunities requirement and spectrum access requirements. This chapter reviews the most popular decentralized CR MAC protocols already proposed for CR networks. We start with a brief overview describing the basics of the CR principles. Then, we describe the main features that characterize the different CR MAC protocols, emphasizing their strengths and weaknesses. A new taxonomy that classifies the described protocols in several dimensions is also proposed. Finally, the open research problems regarding the design of distributed CR MAC protocols are pointed out.

Chapter 3 - is devoted to discussion of the physical radio resource block (PRB) scheduling under femtocell employment in LTE networks. A simple feedback scheme and collaboration approach based on initial vision for femtocell-to-femtocell interface are proposed to speed up the handover procedure between femtocells, to reduce the signaling overhead, and to balance the traffic load among femtocells. The collaboration approach is compatible with different femtocell access modes. Other important aspects such as the power control, radio channel condition reporting, path loss models, and throughput definition are also discussed.

Chapter 4 - discusses vehicular ad-hoc networks (VANETs) that are a critical component of the intelligent transportation and traffic management systems (ITTSs) which involve the application of advanced information processing, communications, sensors, and control technologies in an integrated manner to improve the functionality and the safety of both public and private transportation systems, providing drivers with timely information on road and traffic conditions including collisions, incidents, congestion, surface condition, etc., and achieving smooth traffic flow on the roads. In order to provide reliable and secure communications, intrusion detection systems (IDSs) can serve as a second defense wall after prevention-based approaches, such as encryption. This Chapter addresses the problem of detecting illicit wireless transmissions from the physical layer perspective, assuming the presence of regular ongoing legitimate transmissions. We consider energy detection-based spectrum sensing, which is a widely used spectrum sensing technique due to its simplicity. In particular, we assume all monitoring nodes perform local energy detection by collecting M samples of the received signals and, then, compare against a local threshold λ. A decision is made to activate an intrusion alarm, when the accumulated energy is greater than λ.

Chapter 5 - emphasizes that a key challenge for high-speed broadband applications is the dispersive nature of frequency-selective fading channels, which causes the so-called intersymbol interference (ISI), leading to an inevitable performance degradation. More specifically, in the time domain equalization, when the channel tie spread is larger than one symbol period, interference among consecutive transmitted symbols happen, which leads to ISI and, consequently, a distortion in the received signal. Single-carrier (SC) frequency domain equalization (FDE) was proposed as an alternative promising approach to OFDM for ISI mitigation. SC-FDE avoids the drawbacks associated with OFDM, including a large peak-to-average power ratio (PARP), intolerance to amplifier nonlinearities, and high sensitivity to carrier frequency offset (CFO), while enjoying a comparable complexity and performance. Unlike time-domain equalization whose complexity grows exponentially with channel memory and spectral efficiency, SC-PDE enjoys low complexity due to its efficient FFT/IFFT operation-based transceiver structure, which is very similar to that of OFDM. In this chapter, we review the recent research on SC-FDE (conventional as well as distributed) and its applications. First, we review the basics of SC transmissions coupled with FDE, and discuss its advantages over OFDM. Then we summarize recent studies on SC-FDE integration with space-time block coding (STBC), also known as STBC-SC-FDE. We next summarize recent studies on the performance of SC-FDE for cooperative STBC systems, also known as distributed (D-STBC-SC-FDE, as extension of conventional STBC to cooperative systems. Finally, the chapter is concluded by providing novel results on performance of linear equalizers in two-way relaying cooperative systems with SC-FDE scheme.

Chapter 6 - focuses on the use the generalized approach to signal processing (GASP) in noise in wireless communication systems. The GASP is based on a seemingly abstract idea: the intorduction of an additional noise source that does not carry any information about the signal and signal parameters in order to improve the qualitative performance of wireless communication systems. In the present chapter we would like to discuss the following aspects of GASP in wireless communication systems:

- Main theoretical statements and brief descripotion;
- Array signal processing;
- Multichannel adaptive signal processing;
- Quadriphase DS-CDMA;
- Optimal waveforms for MIMO wireless communication systems

Chapter 7 - deals with the deficiency of wide frequency bands within the conventional spectrum that poses limits on the realization of broadband wireless communication links. The fact that the Extremely High Frequencies (EHF) band is relatively free of users, combined with recent developments of millimeter wave technologies, enables the utilization of frequencies above 30GHz for ultra-wide band communication links and radar applications. One of the principal challenges in realizing modern wireless communication links in the EHF band are absorptive and dispersive effects emerging while millimeter waves propagate in the atmospheric medium. When millimeter-wave radiation passes through the atmosphere, it suffers from selective molecular absorptions peaking at 22GHz and 183GHz, where resonance absorption of water occurs, as well as at 60GHz and 119GHz, due to absorption resonances of oxygen. Between these frequencies, minimum attenuation is obtained at 35GHz (Ka-band), 94GHz (W-band), 130GHz and 220GHz, which are known as atmospheric transmission 'windows'. The transmission characteristics are also determined by weather conditions such as temperature, pressure and humidity. Attenuation due to fog, haze, clouds, rain and snow is one of the dominant causes of fading in wireless communication links operating in the EHF band. In this chapter, we present a space-frequency theory for the propagation of an ultra-wide band radiation in a lossy dielectric media. Characterization of the atmospheric medium is via its refractivity leading to a transfer function, which describes the response of the medium in the frequency domain. This description enables the consideration of broadband signals taking into account inhomogeneous absorptive and dispersive effects of the medium.

Chapter 8 - addresses to communication systems employing linear modulation for that an expression for the probability of symbol error is derived under the assumption that the matched filter at receiver is sampled with a timing offset. This expression is used as an objective to design filters for digital data transmission that are robust in the sense that they attain low symbol error rates in the presence of sampling jitter. Several simplifications lead to a convex objective function which is easily optimized subject to linear constraints for controlling attributes such as stop band attenuation. The simplified objective measures inner boundary of the eye pattern at points away from the ideal sample times. Filters designed using this technique satisfy the Nyquist property and have equiripple behavior in the stop band. Comparisons with existing filter design techniques are presented.

Chapter 9 - discusses that in a point-to-point communication, the use of multiple transmitter and receiver antennas enables an increased data throughput through spatial multiplexing and an increased range by exploiting the spatial diversity. The design of space time coding schemes that fully achieve the available diversity and the multiplexing gain in a MIMO system has been extensively addressed in literature yielding to the design of the optimal family of codes called perfect space time codes constructed from cyclic division algebra. These codes, originally designed for flat fading channels, received a lot of attention in industry in the last few years. However, the recent standards that use multiple antenna terminals are based on more realistic assumptions involving the use of outer codes and multi-taps channels. This chapter will give a literature overview on the design criteria of space time coding technique and their application in industrial systems.

Chapter 10 - introduces an overview of recent results on unequal power allocation for the transmission of JPEG2000 coded images over frequency selective channels. During transmission of the JPEG2000 codestream over frequency selective channels, the negative effects of intersymbol interference and intercarrier interference are eliminated by designing two separate systems for downlink and uplink data transmission: an orthogonal frequency division multiplexing system (downlink) and a single carrier frequency domain equalization system (uplink). The performance of these two systems is analyzed under deployment of multiple transmitter antennas, to achieve spatial diversity by incorporating space time block coding technique into the transmitter model. In addition, we explore addition of a linear constellation precoder matrix to the OFDM system to achieve multipath diversity during transmission of JPEG2000 images. During the transmission, the channel state information is assumed to be available at the receiver and can be feedback to the transmitter. Also, transmission of the header information of the JPEG2000 image codestream is assumed to be error free. To demonstrate the simulation results, the peak signal to noise ratio and the bit error rate profiles of the received images are used as the measure of metric. The results demonstrate a performance evaluation of our proposed systems for JPEG2000 image transmission with competing strategies, and measure transmission bandwidth efficiency and peak signal to noise ratio profile of the received images in our proposed systems.

Chapter 11 - addresses to the fundamental performance metrics of wireless communications systems, i.e. capacity, error probability and outage probability that can be frequently obtained in closed-form by the common classical special functions (for example, Bessel's functions, Hypergeometric's functions or Meijer's functions), the Gaussian *Q*-function, the Marcum *Q*-function, and the Nuttall *Q*-function. In order to obtain new analytical closed-form results from the basic performance metrics, this set of special functions have been recently extended to include either less common classical special functions, e.g. Lauricella functions of the Fox's *H* function, or new generalizations of classical functions, for example, the upper Fox's *H* function or the generalized Fox's *H* function. From a computational point of view, certain Lauricella functions, e.g. F_D, Φ_1, Φ_2 or Φ_3, are easier to compute than Fox type functions because either they have Euler type single integral representations with elementary integrands or their univariate Laplace transforms are elementary being easily inverted numerically. Other Lauricella functions as F_A are more difficult to compute than the F_D function or its confluent forms. However, certain important metrics for wireless communication systems are obtained in closed-form by the function F_A. In this chapter, general results on the applications of the classical Lauricella functions F_A, F_D, Φ_1, Φ_2 and Φ_3 to the performance of wireless communication systems are obtained. Then, new special functions which include these classical Lauricella functions as particular cases are proposed. This new type of functions, named as $\mathcal{A}, \overline{\mathcal{A}}, \mathcal{B}, \mathcal{B}^+$ and $\mathcal{B}^-$, is nearly as easy to compute as the Lauricella functions listed above. At the same time, this new family of special functions will allow us to express in exact closed-form a variety of important analytically open problems in wireless communications.

Chapter 12 - deals with a new systems approach to the modelling of communicative processes. Systemic modelling makes it possible to arrive at generalizations about social phenomena which have the force of the law-like propositions of the natural sciences. In the process documented in Robert Franck's seminal work on systemic modelling, which uses a classical induction method similar to reverse engineering, the necessary functions required to affect a social phenomena or process are first identified. These constitute the formal or theoretical aspect of the social mechanism involved. A practical or "empirical" model is then formulated to describe how the functions are carried out in actual social contexts. Once this model has been validated by its match with observable events, the theoretical model, or "architecture of functions", can then be validated against the practical model. The theoretical aspects of the model make it possible to generalize about the same social process in very different contexts, as well as the possibly of its being transferable to other areas or fields. This chapter will show how data from over 54 video protocol analyses gathered to formulate a model of written communication were used to establish the architecture of functions underpinning communicative interactions. The system of functions thus identified was considered to constitute a generalized principle underpinning design and inquiry processes. It also offered insights into the functions language is thought to perform, suggesting that these are in fact communicative functions rather than linguistic per se. This account offers an exemplar of a systems approach to modelling in the social sciences, while at the same time developing a deep structure which might be used to explain the varied and highly idiosyncratic surface manifestations of spoken, written and computer-mediated interactions.

Briefly, this book will provide a comprehensive technical guide covering fundamentals, recent advances and open issues in communication systems, namely, in wireless communications and wireless sensor networks, mobile communication, and satellite communications to the readers. Objective of the book is to serve as a valuable reference for students, educators, scientists, faculty members, researchers, engineers and research strategists in these rapidly evolving fields and to encourage them to actively explore these broad, exciting and rapidly-evolving research areas.

Professor Vyacheslav Tuzlukov, Ph.D
Editor
Director of Signal Processing Lab
Department of Communications and
Information Technologies
School of Electronics Engineering
College of IT Engineering
Kyungpook National University
Daegu, South Korea

In: Communication Systems: New Research
Editor: Vyacheslav Tuzlukov

ISBN: 978-1-62618-654-5

Chapter 1

THE COGNITIVE RADIO TECHNOLOGY: FUTURE TRENDS IN THE SPECTRUM ACCESS OF NEXT GENERATION COMMUNICATION SYSTEMS

Francesco Benedetto and Antonio Tedeschi
Signal Processing for Telecommunications and Economics (SP4TE) Lab.
Department of Applied Electronics, University of ROMA TRE, Rome, Italy

ABSTRACT

The demand for wireless broadband applications is continuously increasing and it is becoming extremely hard to find vacant frequency bands, to deploy new communication services or enhance the existing ones. Measurements of the spectrum occupancy show that spectrum shortage results from the spectrum management policy rather than from the physical scarcity of usable frequencies. In order to overcome this problem and improve spectrum utilization, cognitive radio concept has been proposed. This chapter provides an overview of cognitive radio for opportunistic spectrum access and related research topics. Cognitive Radio is the key technology allowing radio equipment to obtain knowledge of its radio environment and to dynamically adjust its operational parameters in order to improve its performance. A number of promising concepts for cognitive radio are introduced and discussed in this chapter, starting from the rationale behind the cognitive radio technology, passing through the area of spectrum awareness, spectrum management, and finally focusing on spectrum sensing methodologies.

Keywords: Cognitive Radio, Spectrum Access, Spectrum Sensing, White Spaces, Wireless Communication Systems

1. INTRODUCTION

Spectrum scarcity is becoming a relevant issue in the field of broadband communication. The demand for wireless broadband applications is continuously increasing and applications with quality of service (QoS) requirements are becoming more and more popular [1]-[3]. However, it is becoming extremely hard to find vacant frequency bands, to deploy new communication services or enhance the existing ones. Measurements of the spectrum occupancy show that spectrum shortage results from the spectrum management policy rather than from the physical scarcity of usable frequencies. Recently, there are a lot of researches focusing on the Cognitive Radio (CR) technology [4]. In fact, CR is the key technology allowing radio equipment to obtain knowledge of its radio environment and to dynamically adjust its operational parameters in order to improve its performance. Implementing cognitive radio networks requires knowledge in different fields of expertise, regarding not only scientific and technological capacity (for the physical deployment of such networks), but also marketing and management knowledge for the regulation of primary and secondary users consistence. However, CR systems promise many new and exciting benefits for radio users. The list of CR benefits is long, but some of the most important are interoperability with new and legacy radio systems; ability to squeeze a staggering amount of radio capability in a very small package; and the ability to implement networking tasks on each radio that run transparently to the user. Such a technology opens the possibility to implement reconfigurable and cost-effective architectures for wireless devices. Such innovation will enable portability among different technologies and stimulate the harmonization of wireless networks, through the realization of multi-task, multi-vendor wireless devices. Moreover, CR will open the door for broadband usage in secure communications, which is currently restricted in few available technologies. The capabilities of CR will allow harmonizing the needs of commercial, public, safety and military users. Creating a set of standards acting as common framework to address the needs of different stakeholders would therefore allow for huge economies of scale for future CR products. Standardization in this field has great potential for stimulating innovation and competitiveness on a worldwide scale.

This chapter focuses on the CR technologies and how they would bring many substantial benefits in the next generation wireless systems. CR networks, thanks to the Spectrum Sensing functionality, promise a better utilization of the radio spectrum, which is expected to mitigate the ever increasing "spectrum scarcity" problem. A more rational exploitation of radio spectrum would open the opportunity for new economies, allowing for the possibility to deploy new communication services or enhance the existing ones. Hence, after a brief introduction, the second paragraph of this chapter highlights the motivations behind CR technology, starting from radio spectrum regulations to the definition in CR systems and discussing the operating deployment scenarios of such a technology and the obtainable benefits. Then, the third paragraph is about spectrum management. The active coexistence of primary (licensed) and secondary (unlicensed) users in space, time and frequency domain imposes, indeed, unique challenges to the spectrum management in CR systems. A brief overview of the principle techniques of spectrum access is shown, along with a detailed discussion of the application of the game theory to the problem of rate and power control, aiming at implementing an efficient use of the radio spectrum. The cognitive cycle is discussed in the fourth paragraph. In particular, the techniques required by the CR system to adaptively modify its characteristics and to access the radio spectrum without causing excessive interference to primary licensed users are illustrated. In the fifth paragraph, the essential functionality (i.e. spectrum sensing), used to identify spectral bands not employed (spectrum holes) by primary licensed users, is depicted in details. Several strategies have been proposed in literature for spectrum sensing. The conventional solution relies on the energy (or power) detector, where the energy (or power) of the received signal in a certain band, is compared with a pre-selected threshold to decide whether a licensed user is present or not. However, in strongly noisy channels more sophisticated spectrum sensing strategies have been proposed, such as detection techniques based on cyclostationarity or on autocorrelation functions. Such approaches are particularly suitable for Orthogonal Frequency Division Multiplexing (OFDM) based system, which is very popular in wireless communications (i.e. WiFi, Long Term Evolution (LTE), WiMax). The chapter ends with a summary of the most important and practical aspects of the CR technology.

2. Motivations of the CR Technology

Radio spectrum is an important and natural resource shared by various types of wireless services. It can be repeatedly re-used, if certain technical conditions are met. In particular, radio spectrum can simulataneously accommodate a limited number of users. Therefore, radio spectrum requires careful planning and management to maximize its value for all users.

Currently, spectrum regulatory framework is based on static spectrum allocation and assignment policy. Radio spectrum is globally allocated to the radio services on a primary or secondary basis. This is very well depicted and regulated in the Radio Regulations published by the International Telecommunication Union (ITU) [5], which contains definitions of these services and a table of allocations for each of three ITU geographic world regions. On the European level, radio spectrum is governed in the European Union (EU) by the Radio Spectrum Policy Group (RSPG) and Radio Spectrum Committee (RSC) and by European Conference of Postal and Telecommunications Administrations (CEPT). Additionally, national regulatory agencies define national allocation table and assign radio spectrum to license holders on a long term for large geographical regions on exclusive basis. Generally, users can use radio spectrum only after obtaining an individual license issued by the National regulatory agency. This approach helps in designing the system, since it is easier to make a system that operates in a dedicated band than a system that can use many different bands over a large frequency range. In addition, spectrum licensing offers an effective way to guarantee adequate quality of service and to prevent interference, but, unfortunately, leads to a highly inefficient use of the radio spectrum resource. Analyzing the Article 5 of Radio Regulations [5] and national allocation tables, it can be said that usage of radio spectrum bands is already determined. Furthermore, in national spectrum assignment databases almost all frequency bands of commercial or public interest are already licensed. Current predictions of further growth of demand for wireless communication services show substantial increase in the demand of radio spectrum. These circumstances raise serious concerns about future radio spectrum shortages.

Nevertheless, related radio spectrum observation surveys have proved that most of the allocated spectrum is not fully used [6]-[12]. The measurements of the Federal Communications Commission (FCC) in several cities (e.g. Atlanta, New Orleans, and San Diego) in 2002 revealed that there are large variations in the intensity of spectrum use below 1 GHz [6], [7]. By observing two non-adjacent 7 MHz spectrum bands with a sliding 30 seconds window, the measurements showed that a fraction of 55-95 % of the observed frequencies were idle during the observation period on one band, while on the other band frequencies were almost fully idle. Then, a lot of spectrum occupancy measurements were conducted on the bands between 30 MHz and 3 GHz at six locations in the USA [8]. The average occupancy in those locations was found to be only 5.2% with a maximum occupancy of 13.1% in New York City and a minimum occupancy of 1% in rural areas. Similar spectrum measurements conducted in Europe [9]-[12] (i.e. Germany, Spain, Netherlands, Ireland, France, Czech Republic) show higher spectrum occupancy if compared to USA, but still rather low (e.g. 32% for the band 20-3000 MHz in Aachen area, Germany). Generally, it can be said that spectrum occupancy is moderate below 1 GHz and is very low above 1 GHz. Large discrepancies in radio spectrum allocation assignment and actual radio spectrum usage show that spectrum shortages result from the outdated spectrum management policy rather than the physical scarcity of usable radio spectrum. In order to satisfy the EU Digital Agenda goals and the future market demand for mobile and broadband services, we can envisage deployment of next generation broadband wireless networks and services which will need a rapid and more flexible access to the radio spectrum in the ultra high frequency (UHF) band. The general trend towards more flexible and efficient spectrum management is further driven by the continuous development of new technologies. To deal with increasing conflict of spectrum allocation congestion and spectrum usage under-utilization, cognitive radio approach (see for example [13]-[18]) has been recently proposed as a method which allows secondary users to opportunistically utilize already licensed bands. Cognitive radio using opportunistic spectrum access has the possibility to improve spectrum utilization efficiency and in perspective to allow next generation mobile networks access radio spectrum bands. Coexistence with licensed legacy users in the shared spectrum is a pre-requisite. Although cognitive radio is an interesting and disruptive concept promising significant improvements in radio spectrum usage efficiency, it still faces a lot of research challenges in its way from concept to practical implementation in everyday use.

2.1. Radio Spectrum Regulation

Regulation of radio spectrum has its origin in the economic regulation of railroads. Since the beginning of the 19th century, communication services of telephony, radio and television have been regulated, according to this model, to provide a service to the public on a non-discriminatory basis to fair and reasonable prices and conditions. Regulators approach spectrum regulation in determining how particular bands of spectrum can be used, make rights available to licenced or unlicensed users and define rules constraining the accesses to this spectrum. Ideally, the regulators' decision making aims thereby at increasing of public welfare and reflects the public interest. In the context of radio spectrum regulation, a distinction has to be made between *trading*, as a transfer of spectrum usage rights and *liberalization*, as a weakening of restrictions and limitations associated with spectrum usage rights, related to technologies and services.

The report of the FCC's Spectrum Policy Task Force [19] defines spectrum regulatory mechanisms evidencing an *exclusive use* model, a *command-and-control* model and a *commons* or *open access* model. The command-and-control is currently the most often used regulation model and refers to the licensed spectrum for shared usage and unlicensed spectrum. Radio spectrum regulation has to take into account the development of access protocols and standards to balance the following goals [20]:

- An adequate QoS should be possible to all radios, depending on the supported applications.
- No radio should be blocked from spectrum access and transmission for extended durations.
- Spectrum management policies and standards should not slow down innovations in the economically significant, but rapidly changing, communication sector.
- The limited available spectrum should be used efficiently, including special re-use of spectrum.
- Spectrum can be used in a dynamically adaptive way, taking the local communication environment like spectrum usage policies into account.
- The costs of devices should be not increased significantly through techniques prescribed by regulation.

2.1.1. Licensed Spectrum

Large parts of the radio spectrum are allocated to licensed radio services in a way that is often referred to as command-and-control. Licensing spectrum covers the exclusive access to the spectrum and spectrum sharing of the licensed spectrum through strictly regulated devices. In case of exclusive spectrum usage, a license holder pays a fee to have this privilege. Exclusive access rights have the advantage of preventing potential interference which implies dangers to a reliable and thus chargeable communication. In case of spectrum scarcity, licensed spectrum is highly valuable leading to economic profits, as consumers need to pay for using it. Having an immense commercial impact, spectrum licenses can be bound to requirements which are to be fulfilled like a concrete transmission technology allowed in this spectrum or a certain percentage of population to be reached by the network when purchasing the spectrum for wireless communication. The UMTS auctions in Europe are an example of this.

Today's most often used licensing model is to license spectrum for shared usage restricted to a specific technology. Emission parameters like transmission power and interference to neighboring frequencies like out of band emissions are restricted. Regulation takes care of protection against interference and for a limited support of coexistence capabilities like Dynamic Channel Selection (DCS) in Digital Enhanced Cordless Telecommunication (DECT) that are mandatory and part of the standard.

2.1.2. Unlicensed Spectrum

The access to unlicensed spectrum is open but its utilization is strictly regulated. Actually, a lot of users are sharing the same unlicensed spectrum. Spectrum usage is allowed to all devices that satisfy certain technical rules or standards in order to mitigate potential interference. Examples for these technical rules are limitation of transmission power or advanced coexistence capabilities. The usage rights at unlicensed spectrum are flexible and no concrete methods to access spectrum are specified [20].

As TV bands in the US are often under-utilized, the FCC proposed in 2004 to allow unlicensed systems the secondary usage of this spectrum [23]. In 2004, the FCC also initiated the opening of new spectrum for wireless broadband communication in the 3.65-3.7 GHz band for fixed and mobile devices transmitting at higher power [23]. It is envisaged that multiple users share this spectrum through the use of contention-based protocols, minimizing interference between fixed and mobile operation. These contention-based protocols will help to reduce the possibility of interference from co-frequency operation by managing each station's access to spectrum.

2.2. Definition of CR Systems

The term cognitive radio is derived from "cognition". According to Wikipedia [24] cognition is referred to as:

- Mental processes of an individual, with particular relation.
- Mental states such as beliefs, desires and intentions.
- Information processing involving learning and knowledge.
- Description of the emergent development of knowledge and concepts within a group.

Resulting from this definition, the cognitive radio is a self-aware communication system that efficiently uses spectrum in an intelligent way. It autonomously coordinates the usage of spectrum in identifying unused radio spectrum on the basis of observing spectrum usage. The classification of spectrum as being unused and the way it is used involves regulation, as this spectrum might be originally assigned to a licensed communication system.

Cognitive radio as a new concept was firstly introduced by Joseph Mitola and Gerald Maguire in [16] where cognitive radio is presented as an extension of software defined radio (SDR) enhancing flexibility of personal wireless services. Cognitive radio architecture as an integrated agent for SDR in the intersection of personal wireless technology and computational intelligence is further developed in Mitola's Doctoral Dissertation [17]. The definitions of cognitive radio are still being developed by industry, academia and standardization bodies. A cognitive radio is assumed to be a fully re-configurable radio device that can *cognitively* adapt itself (*i*) to the communications requirements of its user, (*ii*) to the radio frequency environment in which it is operating and (*iii*) to the various network and regulatory policies [14]. Cognitive radio is an intelligent wireless communication system that is aware of its surrounding environment and uses the methodology of understanding by building to learn from the environment and adapt its internal states to statistical variations in the incoming radio frequency stimuli by making corresponding changes in certain operating parameters in real time, with two primary objectives: highly reliable communications whenever and wherever needed and efficient utilization of radio spectrum [25]. Fully capable cognitive radio is unlikely to be achieved in the next years, but certain cognitive radio features will be gradually implemented in radio equipment in the next future.

The FCC has identified in [26] the following features that cognitive radios can incorporate to enable a more efficient and flexible usage of spectrum:

- *Frequency Agility* – The radio is able to change its operating frequency to optimize its use in adapting to the environment.
- *Dynamic Frequency Selection* (DFS) – The radio senses signals from nearby transmitters to choose an optimal operation environment.
- *Adaptive Modulation* – The transmission characteristics and waveforms can be reconfigured to exploit all opportunities for the usage of spectrum
- *Transmit Power Control* (TPC) – The transmission power is adapted to full power limits when necessary and/or to lower levels to allow greater sharing of spectrum.
- *Location Awareness* – The radio is able to determine its location and the location of other devices operating in the same spectrum to optimize transmission parameters.
- *Negotiated Use* – The cognitive radio may have algorithms enabling the sharing of spectrum in terms of prearranged agreements.

Strictly following this definition, modern Wireless Local Area Networks (WLANs) can already be regarded as cognitive radios: IEEE 802.11 devices operate with a listen-before-talk spectrum access, dynamically change the operation frequencies and control their transmission power. In late research, cognitive radios are also referred to as *spectrum agile radios* [27], [28] to indicate an emphasis on dynamic spectrum usage. Mangold *et al.*, in [27], focus thereby on IEEE 802.11*k* for radio resource measurements as a technique to facilitate the development of spectrum agile radios, while Mangold *et al.* in [28] introduce spectrum agile radios as a society of value oriented machines. Basic concepts are taken there from social science to classify the social action of independent decision-makers. This understanding of cognitive radios is summarized in the following definition of cognitive radio from Haykin [14]:

> "Cognitive radio is an intelligent wireless communication system that is aware of its surrounding environment (i.e., outside world), and uses the methodology of understanding-by-building to learn from the environment and adapt its internal states to statistical variations in the incoming radio frequency stimuli by making corresponding changes in certain operating parameters (e.g., transmit power; carrierfrequency, and modulation strategy) in real-time, with two primary objectives in mind: (i.) highly reliable communication whenever and wherever needed and (ii.) efficient utilization of the radio spectrum".

2.3. Benefits of the CR Technology

If the Cognitive Radio Technology (CRT) is going to be attractive for most actors in the wireless industry, it has to provide a significant benefit compared to what is possible with today's and tomorrow's mainstream wireless technology. Mainstream technology like 3GPP's Long Term Evolution (LTE), with the evolution towards LTE-Advanced, and Wi-Fi, has a great momentum in the market, and will also provide significant improvements in performance as well as cost in the years to come. Three top criteria have been defined in order to select feasible deployment scenarios for a CRT providing both managed QoS and high mobility:

- *Benefit from CRT technology.* The CRT solution should be able to provide a significantly better performance than existing (conventional) systems.
- *Benefit for actors.* Deploying CRT for a particular scenario should provide a significant potential benefit for the actors. It should have a joint maximized benefit both for end users and industrial actors. A successful CRT should be commercially attractive. This criterion addresses the commercial side of the CRT, and the selected scenarios must be likely to provide a better business case than conventional systems.
- *Managed QoS and mobility.* The scenario should cover a range of QoS and/or mobility demands. A scenario's QoS requirements depend on the traffic classes that it will serve and how demanding these traffic classes are.

Further, seven criteria have been used for targeting the most interesting and promising scenarios for business case studies:

- *Market Potential.* The scenario should have a large market potential, e.g. with respect to the number of user terminals or expected revenue for the service. This potential could actually come from reduced costs, e.g. reduced spectrum costs or lower power requirements.
- *Best Solution.* No other solution should appear as a better solution for the given scenario.
- *Technical Feasibility.* It must be probable that this system can be implemented with current state of the art technology, or beyond state of the art technology achievable within a reasonable time frame.
- *Economic Feasibility.* It must be probable that within a period of 3-10 years it will be possible to produce equipment and services to a cost that match the users' willingness to pay. The scenario must offer profitability for all major actors in its ecosystem.
- *Regulatory Feasibility.* If the solution requires regulatory changes in order to be deployed, the changes should be such that it is reasonable to expect that they can be realized within a reasonable time frame.
- *Ecosystem Feasibility.* The ecosystem may consist of customers, partners, suppliers, competitors and local and national authorities. If the scenario imposes great changes in the ecosystem (e.g. roles that disappear), it will be much harder to get acceptance for the solution in the industry.
- *Benefits for the society.* Local or national authorities may be willing to support deployment of a system if the social benefits it represents are large. Political support can also make it much easier to get acceptance for regulatory changes.

2.4. Operating Scenarios

2.4.1. Scenario Femto-cells

The femtocell scenario, depicted in Figure 1, describes a user situation with low mobility, but high demands on throughput and QoS. It may also be described as a *hot spot* scenario. Femtocells are always connected to an infrastructure. Both indoor and outdoor deployment is possible. The stakeholders in this scenario are both mobile and fixed operators as well as private and enterprise users. Examples of use cases for this scenario are:

- Private wireless access solution of the same type as Wi- Fi is used today.
- Public hot spots, where several femtocells comprise a larger coverage area.
- The use of indoor femtocells to provide outdoor coverage in e.g. urban/suburban streets.

The main benefits of using cognitive radio for femtocells are:

- Better interference control than current 3G/LTE femtocell technology which can improve capacity and coverage.
- Better user experience due to more frequencies being available and potentially larger coverage.

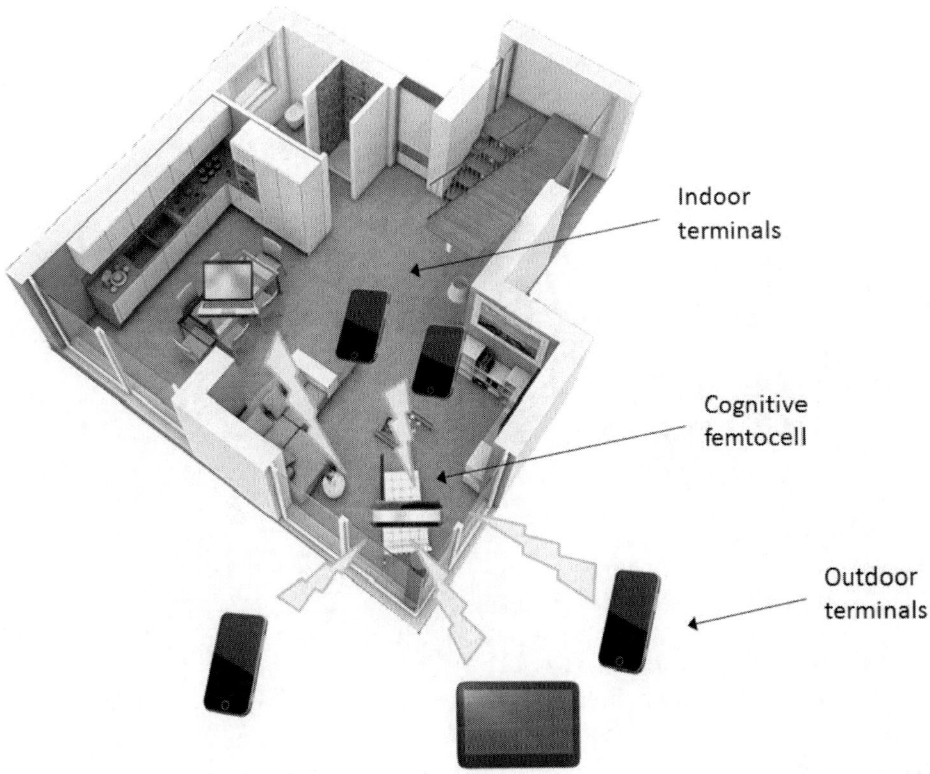

Figure 1. Hot-spot scenario.

2.4.2. Scenario Cellular Extension in Whitespace

Cellular extension in white space, depicted in Figure 2, is where mobile network operators (e.g. LTE-operators) will utilise white space spectrum in addition to their own licensed spectrum. The suitability of a spectrum band for this scenario depends on whether it is to be used for coverage or capacity enhancements. The stakeholders in this scenario are mainly network operators and service providers. Examples of use cases for this scenario are:

- Increased mobile broadband coverage in rural areas with low traffic demand.
- Peak hour traffic offloading.
- Rural broadband involving the provision of wireless Internet connectivity to homes in rural locations through a base station.

The main benefits of using cognitive radio in this scenario are:

- Better user experience due to more frequencies being available and potentially larger coverage.
- Increased operational bandwidths, resulting in improved load balancing, improved link quality and more flexible services.
- The use of low frequencies increases range and the transmit power can be kept low. This reduces power consumption and reduces health risk concerns (especially for uplink transmissions).

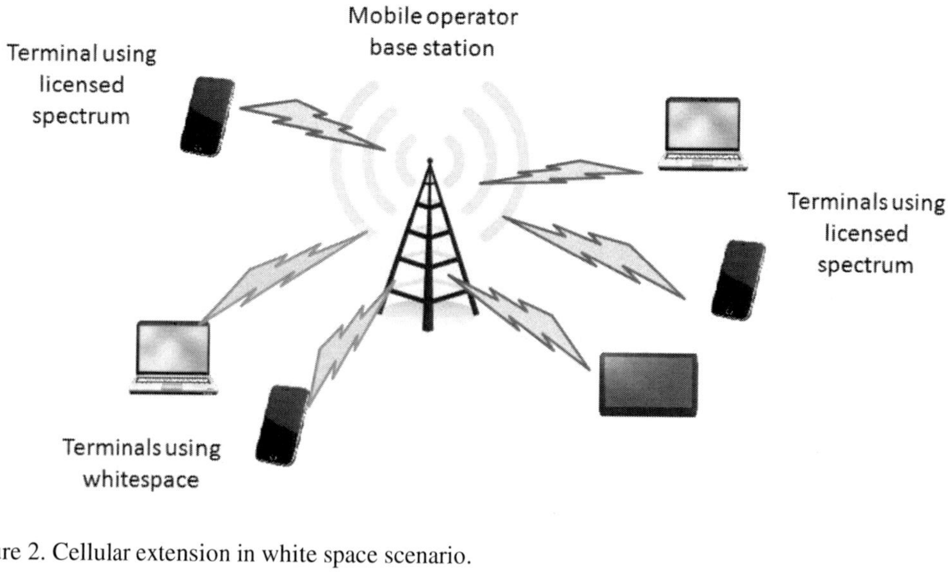

Figure 2. Cellular extension in white space scenario.

2.4.3. Scenario Cognitive ad hoc Network

The cognitive ad hoc network scenario, depicted in Figure 3, typically includes properties of high dynamics and different nodes and terminals. Ad hoc networks are typically limited in space and time. The stakeholders in such a scenario are, among others, end users (both private and enterprise), equipment vendors and the public sector. Examples of use cases for this scenario are:

- Emergency ad hoc networks with several actors (police, paramedics and fire fighters) who will typically have two needs: One is to communicate efficiently between one another; the other is to establish a connection to a rescue co-ordination centre.
- A network established for a business meeting to exchange documents and other information. Dependent on the type of event, such a network may be partly preplanned before the actual event.

The benefits of using cognitive radio for ad hoc networks are:

- The capacity can be increased to serve peak demands without the need for such bandwidth to be allocated during off-peak times.
- The use of low frequency bands is beneficial especially in emergency scenarios due the improved propagation through walls.

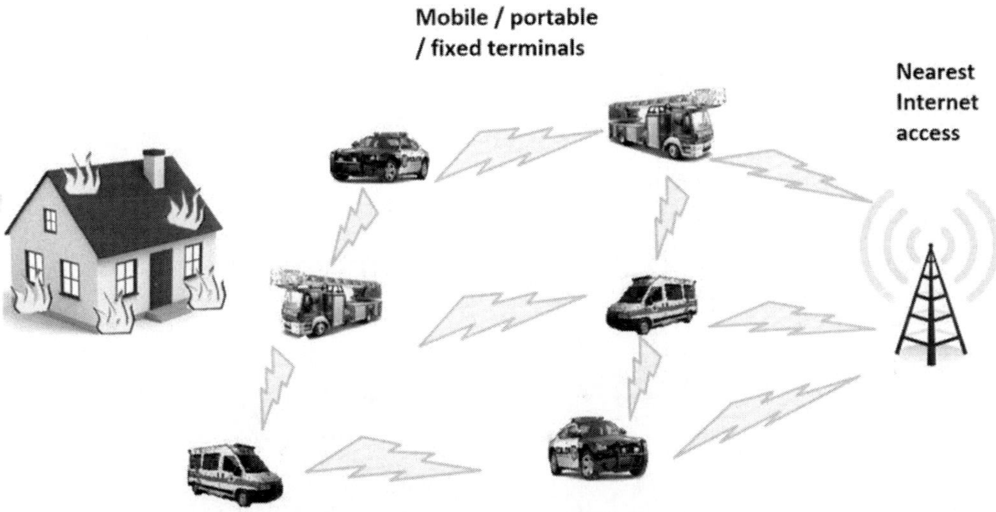

Figure 3. The cognitive ad hoc network scenario.

3. Spectrum Management

3.1. Spectrum Access

Active coexistence of primary licensed and secondary users in space, time and frequency domain impose unique challenge to the spectrum management in cognitive radio systems. Basic spectrum management functions are spectrum decision, spectrum sharing and spectrum mobility [29], [30]. All these functions will be detailed in par. 4 while now let us focus on spectrum access models. Spectrum access models can be categorized as exclusive use and shared use models [31, 32] as shown in Fig 4. Command and control model and long-term exclusive use are traditional spectrum access models used in conventional radio systems [32]. Characteristic of dynamic exclusive model is that radio spectrum is used exclusively by one system in a determined spectrum hole. In order to improve spectrum efficiency some level of flexibility is introduced. At different points in time, the cognitive users can access the radio spectrum under defined rules. Flexibility helps licensees to put spectrum to its most valuable use with the most effective technology, without waiting for a regulator's permission. Two approaches have been proposed under this model: spectrum property rights and exclusive dynamic spectrum allocation [31]-[33].

Primary users having spectrum property rights can have various levels of flexibility. They can use assigned radio spectrum however they wish, or they could be restricted to specific radio service or technology. A licence is assigned for temporary basis with long duration or for permanent usage. Using spectrum property right licensee of spectrum can trade, lease or borrow parts of spectrum on secondary spectrum markets to cognitive radio user. Economy and market forces will therefore play an important role in driving toward the most profitable and efficient use of this limited resource. The second approach, exclusive dynamic spectrum allocation aims to improve spectrum efficiency exploiting the spatial and temporal traffic statistics of different services. Based on observed traffic statistics, spectrum is shared between different services. In a given region and at the given time, spectrum is assigned to services on exclusive use, but this allocation varies at a much faster scale than the static policy. Dynamic spectrum allocation can take advantage of daily user's migration from residential to business areas, or day and night variations of usage statistics. Furthermore, governmental and emergency applications have exclusive access to large portions of radio spectrum which are rarely used. This radio spectrum can be also used for some commercial application under dynamic spectrum allocation model.

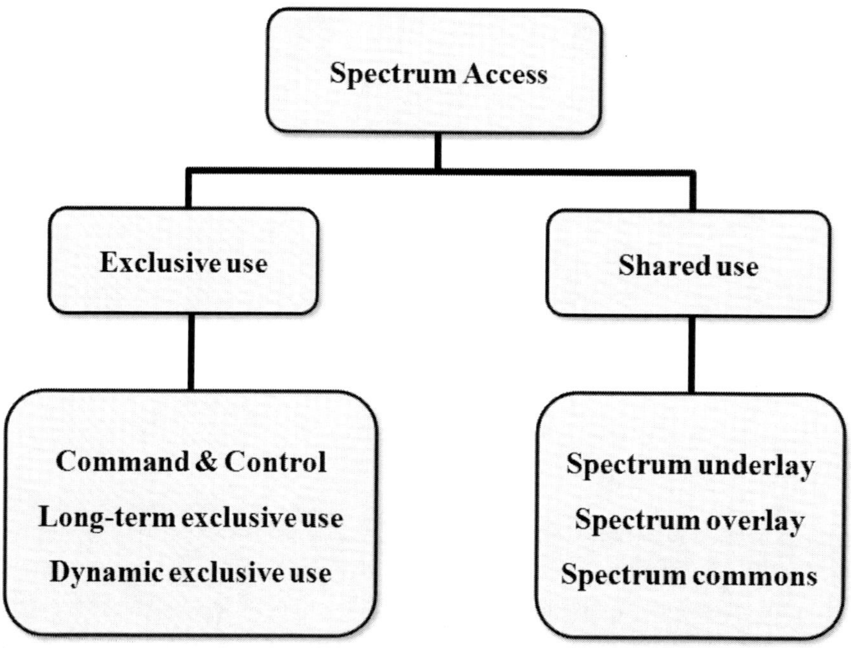

Figure 4. Spectrum access models.

In shared use spectrum access model [30]-[35], the radio spectrum can be simultaneously used by a primary user and a secondary user, if satisfying interference constraint. In this model, unlicensed users can opportunistically access the radio spectrum if it is not occupied or fully used by primary users. Licensed spectrum is consequently opened to secondary users, while limiting the interference observed by primary users. Interference constraints for secondary users have to be defined carefully in order to allow primary users to operate without noticeable reduction of service quality. Underlay or interference avoidance model [31] allows concurrent transmission of primary and secondary users in ultra wideband fashion. The transmit power of the secondary user is limited so that the generated interference is below the noise floor for the primary user. Due to power constraints systems using underlay model can be used only for short-range communication. By spreading transmitted power over ultra-wide frequency band (UWB), secondary users can achieve high data rates on short distances. In the spectrum overlay model [31], a primary user receives an exclusive right to spectrum access.

However, if the spectrum is not utilized at a particular time or frequency, it can be opportunistically accessed by secondary user. Cognitive radio opportunistically communicates in a non-intrusive manner over the spectrum holes. Spectrum commons mode employs open sharing among peer users as the basis for managing radio spectrum. Spectrum commons model [32, 33] requires radio spectrum sharing without priority allocation to service or class of users. In a shared radio spectrum band, devices might cooperate or merely co-exist. When devices cooperatively share radio spectrum band they have to use common inter-networking protocol and communicate with each other. Cooperative approach is more technologically demanding, but most of the possible time and spectrum collisions can be avoided.

3.2. Joint Rate and Power Control

In such a scenario, joint rate and transmitter power control in cognitive radio network assures interference free operation while satisfying quality and capacity goals. In fact, the available wireless spectrum is a limited resource and the need for spectrally efficient systems has motivated the development of adaptive transmission techniques, several of which are yet to be standardized. These procedures adapt some of the users' parameters according to the communication channel conditions, signal-to-interference-plus-noise ratio (SINR), rate requirements, bit error rate (BER) needs, and battery constraints. In fact, while varying the data rates has the goal to improve the QoS of the real-time transmission, the power control contributes to extend the battery life of the mobile terminals [36].

In the context of (wide-band) direct-sequence code division multiple access (DS-CDMA) wireless networks, power control has traditionally been the single most important adaptation parameter and has been thoroughly studied (see [36] and the references therein). Recent efforts on adaptation in CDMA networks have also focused on adapting the transmission rate [38]. In fact, rate adaption and power control in wireless networks are strictly connected to each other: rate control regulates the source rates to avoid overwhelming any link capacity which depends on interference levels, which are in turn adjusted by power control policy [37]. Providing flexible transmission rates for each transmitter/receiver pair, as well as an efficient use of the shared radio spectrum, require joint power and rate control optimization algorithms [38 and reference therein]. Game theory was shown to be an appropriate tool for finding both power and rate flow control algorithms [39]. Different game models (e.g., non-cooperative/cooperative, static/dynamic, and complete/incomplete information games) have been developed to study the behavior of transmitting nodes to access the wireless channel(s) and obtain the multiple access solution (i.e. the equilibrium) [40]. The common aim of these models is to improve network performance (e.g., throughput maximization, resource consumption minimization, and QoS guarantee) given self-interest or group rationality of transmitting nodes [40]. The idea is to define for each user a function, namely the utility function, to be maximized according to some networks characteristics. For power control, the utility function usually depends on both the signal-to-interference ratio (SIR) and the transmission power of the terminal [41]-[44]. For joint rate and power control, the utility function must take into account also that each user is capable of variable transmission rates [45]. In particular, the authors in [45] formulate a non-cooperative joint transmission rate and power control game (NRPG) to determine the optimal rate and power of the transmission, maximizing the utility function of each user.

As system model, let us consider the uplink communication of a single cell CDMA system with N mobile users that transmit data to the base station. It is well known (see [46]) that the SINR corresponding to the j-th user in the cell can be expressed as follows:

$$\gamma_j = \frac{B_W}{r_j} \cdot \frac{h_j \cdot p_j}{\sum_{\substack{k=1 \\ k \neq j}}^{N} h_k \cdot p_k + \sigma^2} \tag{3.2.1}$$

where B_W is the available spread-spectrum bandwidth, h_j is the set of path gains from the j-th mobile user to the base station, p_j is the power and r_j the transmission rate respectively of the j-th user, while σ^2 is the power spectral density of the additive white Gaussian noise (AWGN). The SINR expression of (1) assumes that, in the CDMA system, users are assigned pseudorandom signature sequences and perform conventional matched filtering operations [46]. Since the available bandwidth B_w is shared among all the N users of the same cell, the transmission of the j-th user causes interference to the transmission of the other N-1 users. As a consequence, eq. (3.2.1) represents an indicator of the QoS (i.e. an indicator of the satisfaction of j-th user in transmitting at a given power and rate). Since most of the terminals in a wireless network are battery-powered, satisfying the QoS requirements while maintaining low energy consumption is very critical for the system performance, in fact energy efficiency has a direct impact in prolonging the life of the terminals. Users should achieve the satisfactory QoS level transmitting at the minimum power, given rate and SINR values. Clearly, higher SINR levels at the output of the receiver correspond to lower BER and obviously higher throughputs. Conversely, achieving a high SINR level often requires the user to transmit at a high power which in turn results in low battery life. This tradeoff can be captured by defining the utility (or payoff) function of a user as the ratio of its throughput (T_j) to the transmit power as follows:

$$u_j = \frac{T_j}{p_j} \tag{3.2.2}$$

In a NRPG-theoretic approach, the utility function introduced in (3.2.2) can be explicitly expressed (in bits/Joule) in terms of rate and power of the j-th user as follows [45]:

$$u_j\left(r_j, p_j\right) = \frac{L \cdot r_j \cdot f\left(\gamma_j\right)}{M \cdot p_j} \tag{3.2.3}$$

where L is the number of information bits transmitted in a packet of length M and $f(\gamma_i)$ is the efficiency function, or frame success rate (FSR), expressed by:

$$f\left(\gamma_j\right) = \left(1 - 2 \cdot P_e\right)^M \tag{3.2.4}$$

and evaluated for a fixed BER equal to P_e. In [45], the authors consider only non-coherent frequency shift keying (NFSK) modulation that has the advantages of both low implementation complexity and low energy consumption. The BER of this kind of modulation is expressed by:

$$P_e = \frac{1}{2} e^{-\gamma_j/2} \tag{3.2.5}$$

Then, the SINR of the j-th user γ_j can be rewritten in terms of rate and power by means of the following:

$$\gamma_j = c_j \frac{p_j}{r_j} \tag{3.2.6}$$

where the quantity:

$$c_j = \frac{B_W \cdot h_j}{\sum_{\substack{k=1 \\ k \neq j}}^{N} h_k \cdot p_k + \sigma^2} \tag{3.2.7}$$

does not depend on the rate and power of the j-th user. The goal of each user is hence to modify and adaptively update rate and transmission power in a distributed fashion in order to obtain the maximum payoff (i.e. so that the utility function of each user is maximized). The utility function of the conventional NRPG approach can be finally expressed as:

$$u_j\left(r_j, p_j\right) = \frac{L}{M} \frac{r_j}{p_j} f\left(\gamma_j\right) \tag{3.2.8}$$

Spectrum access and transmit power control functions are basic spectrum management functions aiming at implementing efficient use of radio spectrum. Different methods of exclusive and dynamic spectrum access are investigated in literature. A great deal of attention have been paid to spectrum overlay model of shared use dynamic spectrum access since there is no need to make changes to primary legacy systems and therefore most appropriate for cognitive radio systems in near future.

4. Spectrum Awareness

4.1. Spectrum Access

One of the most important features of the cognitive radio is the ability to acquire, measure, sense, learn and be aware of the radio's operating environment in order to recognize spectrum space opportunities and efficiently use them for adaptive transmission. In order to achieve these objectives, cognitive radio is required to adaptively modify its characteristics and to access radio spectrum without causing excessive interference to the primary licensed users. Cognitive cycle of cognitive radio operation as secondary radio system is shown in Figure 5.

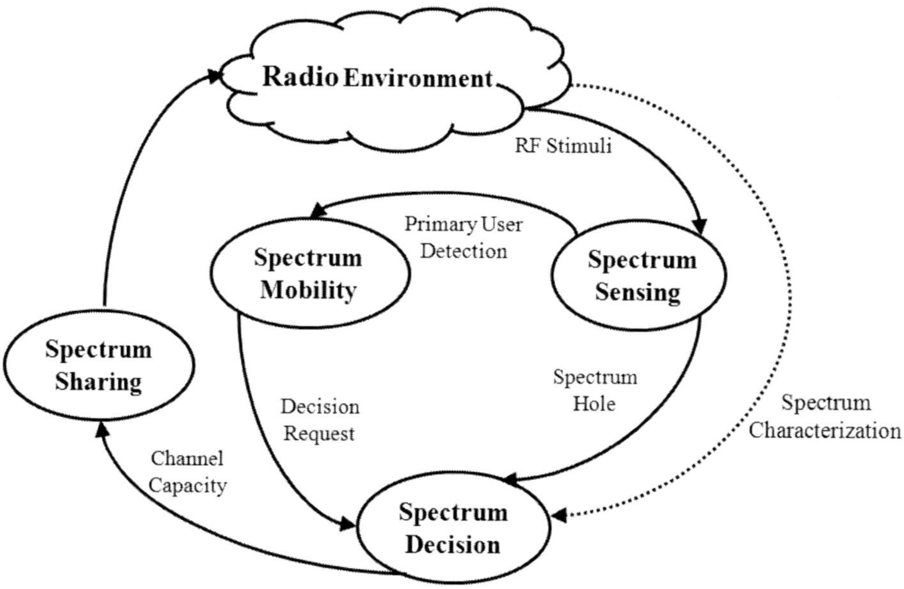

Figure 5. Cognitive cycle of cognitive radio operation.

The building blocks of the cognitive cycle and of the spectrum awareness process are represented by four major steps [47]:

- *Spectrum sharing*: There may be multiple CR users trying to access the spectrum, and so CRN access should be coordinated to prevent multiple users colliding in overlapping portions of the spectrum (see par. 4.1).
- *Spectrum mobility*: CR nodes are regarded as roaming mobile users to the spectrum. Hence, if the specific portion of the spectrum in use is required by a primary user, the communication must be continued in another vacant (or available under appropriate incentives) portion of the spectrum (see par. 4.2).

- *Spectrum decision*: Based on the spectrum availability, CR users can allocate a channel. This allocation not only depends on spectrum availability, but is also determined based on internal (and possibly external) policies (see par. 4.3).
- *Spectrum sensing*: CR users can allocate only an unused portion of the spectrum. Therefore, these users should monitor the available spectrum bands, capture their information and then detect spectrum holes (see par. 4.4).

4.2. Spectrum Sharing

Since there is a number of secondary users participating in usage of available spectrum holes, cognitive radio has to achieve balance between its self-goal of transferring information in an efficient way and altruistic goal to share the available resources with other cognitive and non-cognitive users. This is done with policy rules determining cognitive radio behavior in radio environment. This topic has generated a lot of research interest in the recent past, see for instances [48] - [51].

4.3. Spectrum Mobility

If primary user starts to operate, cognitive radio has to stop its operation or to vacate currently used radio spectrum and change radio frequency. In order to avoid interference to primary licensed user this function has to be performed in real time, therefore cognitive radio has to constantly investigate possible alternative spectrum holes. For details, readers can refer to [52], [53].

4.4. Spectrum Decision

Based on spectrum sensing information cognitive radio selects when to start its operation, operating frequency and its corresponding technical parameters. Cognitive radio primary objective is to transfer as much information as possible and to satisfy the required quality of service, without causing excessive interference to primary users. Additionally, cognitive radio may use data from regulatory database and policy database in order to improve its operation and outage statistics [54], [55].

4.5. Spectrum Sensing

Spectrum awareness can be classified as passive and active awareness (or also called spectrum sensing) [56] – [60]. Fig 6 shows basic classification of spectrum awareness methodologies for cognitive radio. In the passive awareness radio spectrum information is received from outside world like from primary communications system, from server, from centralized databases, or predefined policy set [57], [60]. In this approach relevance of data in space and time is critical and an additional communication channel is needed for acquiring information. While leading to simplified secondary transceivers, these methods require some modifications to the legacy primary system, additional data acquisition, data storage resources, data management system, and additional network capacity. Generally, passive awareness results in rather static secondary usage without optimally exploiting spectrum space opportunities. On the other side active awareness is based on spectrum sensing which can be performed in cooperative and non-cooperative manner. Cooperative spectrum sensing significantly improves error statistics even with a small number of non-correlated sensing entities [61], [62].

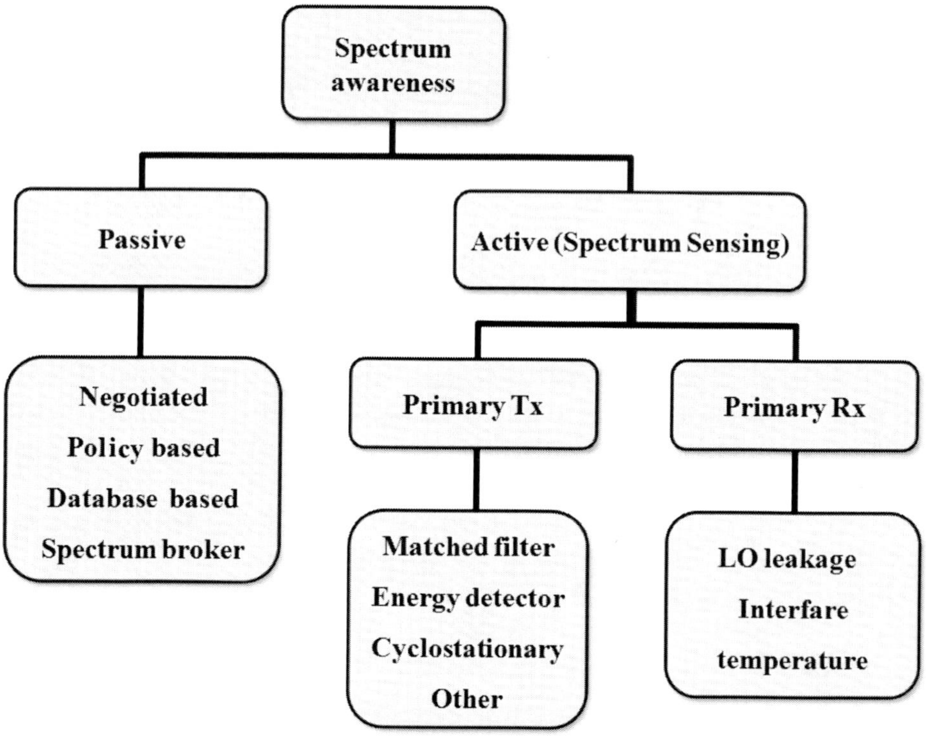

Figure 6. Spectrum awareness classification for cognitive radio.

In active awareness, secondary users actively sense the radio environment and adapt their transmissions based on the measurements. In a non-cooperative model, secondary users make their decisions independently based on the observations about the spectrum environment. In a cooperative situation, local measurements will be combined and signalled to all secondary users before decisions about spectrum use are made. Active awareness is more complex and hardware demanding, but generally it results in improved secondary users spectrum usage statistics compared to spectrum usage based on passive awareness [57] – [59]. Cognitive radio systems may employ either one or both forms of awareness, thus the discussed approaches should not be viewed as mutually exclusive. In systems based on negotiated spectrum use [57], [60] the primary system informs secondary users about the allocated frequencies and the available spectrum opportunities using spectrum beacons. Negotiation is usually centralized and includes determination of geographical, temporal, technical, financial, service quality and interference constraints and conditions. In the policy based approach [57],

[60], the national regulatory agency identifies a licensed band of the radio spectrum where use is low or the band is used with a deterministic pattern. The regulatory agency assigns a set of policies that provide rules and constraints concerning how to use this radio spectrum for secondary use. Secondary devices repeatedly seek for updates of policies that are relevant for their regulatory domain and update their information bases. After updating information, secondary users adapt their transmission parameters like frequency and power to meet predefined policies. A primary system or the regulatory authority can maintain a database of frequency resources which includes location information and an estimate of the interference range of the secondary user. Frequencies used by the licensed system can be seen and checked from this table. When a secondary user needs to transmit, it checks the table, chooses an available band and reserves it to its use. Other secondary users can then see that this particular band is occupied by a secondary user and can choose other resources for their use. When a primary or secondary user stops transmitting the associated band is released from the table and made available to other users.

In spectrum broker approach [57], [60], a server can be used to enable coexistence of primary and secondary radios in a shared environment in a centralized fashion. The centralized spectrum server obtains information about surrounding and interference through local measurements from different terminals and then offers suggestions to the efficient spectrum use. Service providers and users of the networks do not a priori own any spectrum; instead they obtain time bound rights from a regional spectrum broker to a part of the spectrum and configure it to offer the network service. General design constraint of cognitive radio is limiting cumulative interference to licensed primary users below a prescribed level determined by regulatory policy or primary spectrum user. Since interference actually takes place at the receiver location, active spectrum awareness should be focused on detecting the receiving activities of primary users. Without assuming cooperation between primary and secondary system, primary receivers are much harder to detect than primary transmitters using active awareness techniques. In [63] authors proposed method where primary receivers are detected by measuring local oscillator leakage power by the RF front end. In some countries, this method was historically used for the detection of TV viewers not paying for TV subscription. The difficulty of this method is in the short detection range and the long detection time to achieve accuracy, which implies building a large network of passive sensors assisting secondary users in spectrum decision.

Another receiver centric method of active spectrum awareness is based on measuring interference power in the secondary user surroundings and converting it to interference temperature [56]. Interference temperature constraint is calculated as the sum of receiver noise floor level at the primary system service range plus the interference cap determined as maximum positive variation of noise level as depicted in Figure 7. Comparing measured interference temperature and interference temperature constraint determines interference temperature gap which can be additionally filled by cognitive radio system. Interference temperature gap sets up accurate measure of allowed cumulative interference level that primary receiver could tolerate without reduction in service area. This additional interference gap can be used for short range secondary transmissions. Another approach of active spectrum awareness is to transform the problem of detecting primary receivers into detecting primary transmitters. In this approach, the most popular methods are matched filter detection, energy detection, cyclostationary feature detection and others (e.g. waveform based sensing, radio identification, multitaper spectral estimation) [56] – [62].

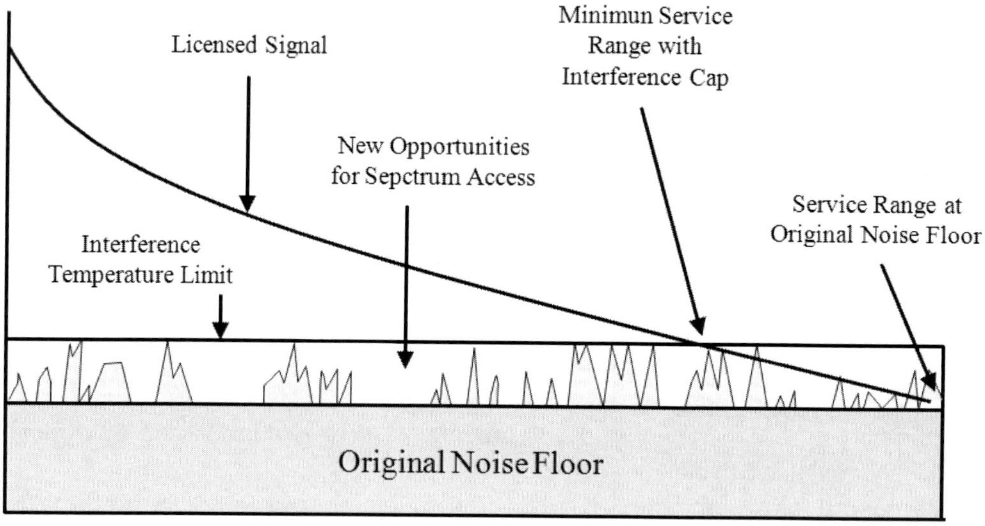

Figure 7. Temperature of interference.

Matched filter detection [56], [58] – [59] is the optimum method for primary user detection in stationary Gaussian noise channel since it maximizes received signal-to-noise ratio comparing to other detection methods. An advantage of this method is that it requires short time and a low number of samples to achieve required level of false alarm or missed detection. Matched filter detection has high processing gain, but sensing devices have to achieve coherency and demodulate primary user signal. Therefore, a secondary user needs *a priori* knowledge of the primary user signal at both physical and medium access control layers. This is possible since most of wireless networks have pilot, preambles, synchronization word or spreading codes that can be used for coherent detection. On the other side, matched filter detection requires a dedicated receiver for every primary user signal class and has large power consumption which makes it impracticable for wider implementation in spectrum awareness for cognitive radio. Energy detection method [56], [58] – [59] is the most used for primary user transmitter detection because of its low computational and implementation complexities. In this method, widely used in radiometry, the detecting receiver performs non-coherent detection and do not need any knowledge of the primary user signal or its statistics. The energy of the received signal is obtained by integrating the received signal with the observation time interval and the receiver bandwidth. Processing gain of energy detector is proportional to the fast Fourier transform (FFT) size and observation time. The signal is detected by comparing the energy of the received signal to the threshold level. Threshold level depends on the noise floor and required level of false alarm and is highly susceptible to uncertainty in noise power or in-band interference. Energy detector cannot differentiate between primary signals, noise or interference and has poor performance under low signal-to-noise ratio. Energy detection method is not appropriate for detecting spread spectrum signals for which more sophisticated signal processing algorithms need to be used.

Cyclostationary detection [56], [58] – [59] is based on the inherent redundancy in the primary user transmitted signals. Modulated signals in most of the modern communication systems are associated with sine wave carriers, pulse trains, repeated digital spreading or frequency hopping sequences. Signal can be modelled as a cyclostationary random process, due to built-in periodicity in the signal or in its statistics. Distinctive feature of cyclostationary signals is that they exhibit a correlation between widely separated spectral components. Spectral cyclic correlation function is used for detecting primary users since modulated signal have nonzero correlation components. One of the advantages of this detector is that it can differentiate wanted signal from the noise because modulated signals are cyclostationary with spectral correlation and noise is in wide-sense stationary process with no correlation. Primary user transmissions can be detected using cyclostationary detection even with negative signal-to-noise ratio. A drawback of this method is that it is computationally complex and requires a significantly long observation time. Passive and active spectrum awareness techniques are used for obtaining information about radio spectrum use. Several methods or cooperative spectrum awareness can be combined in order to improve spectrum decision accuracy. Radio spectrum statistics based on historical sensing can also provide useful insight for spectrum awareness. Generally, energy detection methodology is regularly used because of its low complexity, but it has major drawbacks. Match filtering methodology is much more accurate, but it is most complex since detailed information about primary transmission is necessary. Information about past and current spectrum use offers foundation for opportunistic spectrum space access in cognitive radio. Functions of spectrum management in cognitive radio system are to access, assign and use radio spectrum efficiently without causing excessive interference to primary users.

4.6. Spectrum Sensing for OFDM Systems

Orthogonal Frequency Division Multiplexing (OFDM) is a technique commonly used for high-data-rate wireless communications due to its inherent error resistance in a multi-path environment and has been chosen for several broadband wireless local area network (WLAN) standards, such as IEEE 802.11, IEEE 802.16, European HIPERLAN/2, and Japanese MMAC [65], [66]. Recently, OFDM has been also selected as the modulation scheme for both the 3GPP Long Term Evolution (LTE) the DVB-T (Digital Video Broadcasting-Terrestrial) systems [68]. The OFDM modulation is going to be a key technology for the future of broadband wireless telecommunications. The main idea of OFDM is to send data in parallel over a number of spectrally overlapping temporally orthogonal sub-carriers. It reduces the ISI (Inter-Symbol Interference) effect by making the signal period of sub-carriers much larger than the delay spread of the channel. In terms of frequency domain characteristics, this technique introduces frequency diversity to combat fading and thus enhances the system performance. Input data bits are first divided into data blocks and then mapped to a constellation (e.g. binary phase shift keying BPSK, quadrature amplitude modulation QAM, 16-QAM, etc…) by the digital modulator. Then the symbol sequence in frequency is modulated by inverse FFT (IFFT) to an OFDM symbol sequence. A cycle prefix (CP) is attached to every OFDM symbol to mitigate the effect of channel delay spread, and a preamble is inserted in every slot for timing. Finally, the constructed data frame is ready to be transmitted on the communication link [64]. Since OFDM systems use the Fast Fourier transform for multi-carrier modulation of the data to be transmitted, the user information after applying the FFT transformation follows the cyclic shift properties of the FFT matrix. By applying a cyclic prefix, which merely involves pre-pending some number of the ending data vector entries to the beginning of the OFDM symbol to be transmitted, then the interference per sub-carrier will resemble a flat fading channel (as long as the maximum delay spread of the channel is less than the length of the cyclic prefix) [67].

At the receiving side, the received signal is serial-to-parallel converted, the preambles and CP are extracted from the received symbol stream, and the remaining OFDM symbol is passed through the FFT demodulator. The estimated transmit symbols are then demodulated and decoded. Finally, the decoded source bit-streams are obtained as the system outputs. The cyclic guard interval is introduced during the transmission process in order to preserve the orthogonality of the sub-carriers and the independence of subsequent OFDM symbols. The guard interval is established as a copy of the last part of the OFDM symbol and the duration T_g is just made larger than the maximum excess delay of the radio channel. The guard interval is then transmitted just before the effective part of the symbol that means the demodulation by FFT at the receiving side will not be initiated before T_g.

The presence of the CP gives to the OFDM signals the following well-known characteristic: the autocorrelation coefficients:

$$\rho = E\left[x(t)x(t+\tau)\right] / E\left[x(t)x(t)\right] \tag{4.6.1}$$

are non-zero at delays equal to $\tau = \pm T_d$, where T_d is the symbol length of an OFDM block with a CP of length Tc and E[·] means expectation operator. The authors in [69] propose a method for the detection of an OFDM signal which is based on this convenient property. Detection is based on the auto-correlation coefficients of the received signal, which should provide a non-zero value in the H_1 hypothesis (i.e. presence of primary signal), and a null value otherwise:

$$\begin{aligned} H_0 &: \rho\left(\pm T_d\right) = 0 \\ H_1 &: \rho\left(\pm T_d\right) = \rho_1 \end{aligned} \tag{4.6.2}$$

where, it can be shown that:

$$\rho_1 = \frac{T_c}{T_c + T_d} \frac{\sigma_s^2}{\sigma_s^2 + \sigma_w^2} \tag{4.6.3}$$

The decision is based on an estimation of the correlation coefficient from a set of M observations (with M >> T_d) of the received signal [x(0), x(1)....x(T_d +M-1)]. In other terms, the testing variable is expressed as the maximum likelihood estimate of the correlation coefficient [69]:

$$\hat{\rho}_{ML} = \frac{\frac{1}{M}\sum_{t=0}^{M-1} x(t)x(t+T_d)}{\frac{1}{2(M+T_d)}\sum_{t=0}^{M+T_d-1} |x(t)|^2} \tag{4.6.4}$$

For large values of M, the testing variable in (4.6.4) is asymptotically Gaussian under the two assumptions, because of the central limit theorem. Hence, the hypothesis testing can be written as follows:

$$H_0 : \hat{\rho}_{ML} \sim N\left(0, \frac{1}{2M}\right)$$
$$H_1 : \hat{\rho}_{ML} \sim N\left(\rho_1, \frac{\left(1-\rho_1^2\right)^2}{2M}\right) \quad (4.6.5)$$

The test threshold can be now tuned from a straightforward evaluation of the Gaussian integral for a fixed probability of false alarm.

$$v = \frac{1}{\sqrt{M}} erfc^{-1}\left(2\, P_{FA}\right) \quad (4.6.6)$$

where P_{FA} is the false alarm probability and $erfc^{-1}(\cdot)$ is the well-known complementary error function. Moreover, the detection probability (P_D) can be determined by means of the error function under the H_1 hypothesis:

$$P_D = \frac{1}{2} erfc\left(\sqrt{M}\,\frac{v-\rho_1}{1-\rho_1}\right) \quad (4.6.7)$$

In the case that the CP is known to the receiver, the performances of the overall system can be strongly improved. In fact, in this case, and assuming perfect synchronization between the receiver and the transmitter, the test statistics in (4.6.7) can be expressed as:

$$H_0 : \rho_{cML} \sim N\left(0, \frac{1}{2M_c}\right)$$
$$H_1 : \rho_{cML} \sim N\left(\rho_{cl}, \frac{\left(1-\rho_{cl}^2\right)^2}{2M_c}\right) \quad (4.6.8)$$

with:

$$\rho_{cl} = \frac{\sigma_s^2}{\sigma_s^2 + \sigma_w^2} \quad (4.6.9)$$

and $M_c = T_c\, N_s$, where N_s is the number of OFDM blocks over which the autocorrelation coefficient is estimated. As a consequence, the mean of the estimate has increased by a factor T_s/T_c under the hypothesis H_1, thus increasing the detection probability. It has to be noted that, even if the CP length may be known to the receiver, since it is related to the communications standard, perfect synchronization between the transmitter and the receiver is not a typical operating case. Hence, it can be considered as an upper bound of the system performances.

It is interesting to underline that a lot of spectrum, sensing methods recently proposed in the literature takes their rationale form previously published works in other fields. For example, searching for the presence of a primary user in a spectrum sensing approach is equivalent to search for the presence of the correct code in the code acquisition stage typical of code division multiple access (CDMA) techniques [71] - [77]. Following this approach, a blind method for spectrum sensing of OFDM systems has been introduced in [78]. In particular, the authors of [78] propose a strategy for the detection of an OFDM-based primary user, based on the recently introduced "Rayleigh-ness test" [70]. This test was originally proposed for code acquisition in DS-CDMA communications to determine the presence of a useful signal, with the correct code offset. The Rayleigh-ness test is used to determine the presence (hypothesis H_1), or absence (hypothesis H_0) of such signal from the observation of the (magnitude of the) cross-correlation samples. The statistical distribution of these samples is the Rice probability density function (PDF) in the H_1 hypothesis while it reduces to the Rayleigh PDF otherwise [70]. The Rayleigh-ness test can discriminate between the two alternate hypothesis by deciding on the presence of a (statistically relevant) mean value μ of a complex Gaussian model generating both a Rayleigh ($\mu = 0$) or Rice ($\mu \neq 0$) distribution. The decision is based on the magnitude of the cross-correlation samples detected at the matched filter's output, which can be described as complex Gaussian variable. The rationale behind the modified Rayleigh-ness detection method is as follows. The correlation method discussed in the previous section states the presence of a primary user (i.e. H_1 hypothesis) when the testing variable ρ_{ML} has non-zero mean (i.e. $E[\rho_{ML}]= \rho_1$). This is also equivalent to state whether the observed series is a portion of one realization of a Rice (or Rayleigh) distributed random process. Thus, the Rayleigh-ness test can be effectively applied for spectrum sensing in cognitive radio networks. The original Rayleigh-ness procedure is based on the cross-correlation samples detected at the output of the matched filter. Matched filter is the optimal detector when the transmitted signal is known. However, in a cognitive radio network the a priori knowledge about the signals of interest is generally very low, therefore matched filter detector are rarely used in practice. Moreover, sensing strategies should be as blind as possible to detect signals from different communication standards. For such reason, we have accordingly modified the Rayleigh-ness test by substituting the cross-correlation samples with the auto-correlation ones.

The fourth order moment of a real and positive Rice random variable $|\eta|$ generated as the magnitude of a random complex Gaussian variable η with mean value λ, depends on the second-order moment according to the following relationship:

$$|\lambda|^4 = 2E\left(|\eta|^2\right)^2 - E\left(|\eta|^4\right) \tag{4.6.10}$$

The observed series follows a Rayleigh distribution if λ=**0**, and a Rice distribution otherwise. Exploiting such property, the decision whether a primary user is present or not can be made from an estimation of the forth-order moment of the auto-correlation series. The testing variable is defined as follows:

$$X_c = 2\left[\frac{1}{M}\sum_{t=0}^{M-1}|x(t)|^2\right]^2 - \frac{1}{M}\sum_{t=0}^{M-1}|x(t)|^4 \tag{4.6.11}$$

where M is the number of observations of the received signal over which the estimation is made. The testing variable is asymptotically Gaussian because of the central limit theorem. The modified Rayleighness test has been proved to perform more effectively than the conventional autocorrelation-based method. In particular, we report here the probability of detection evaluated under the constant false alarm rate CFAR procedure [71] – [77] versus values of the Signal-to-Noise-Ratio (SNR) of practical interest. In the reported analysis, a $P_{FA}=10^{-3}$ and white Gaussian noise have been used. Moreover, several Monte-Carlo simulation trials (10^6 independent runs) have been implemented to numerically evaluate the detection probabilities of the two methods. In all the addressed cases an OFDM signal with a 16-QAM modulation and a symbol length of T_d=256 have been used. Figure 8 shows the probability of detection of both the conventional and the new approach, for a CP length of $T_c = T_d/4$ (i.e. the length of the CP for the WLAN system). In particular, we report here the curves (by theory and simulation) referring to the following cases: (*i*) the CP is known (i.e. the upper bound of the system's performance); (*ii*) the autocorrelation-based method (T_d known); (*iii*) modified Rayleighness test (T_d known); (*iv*) modified Rayleighness test (T_d unknown).

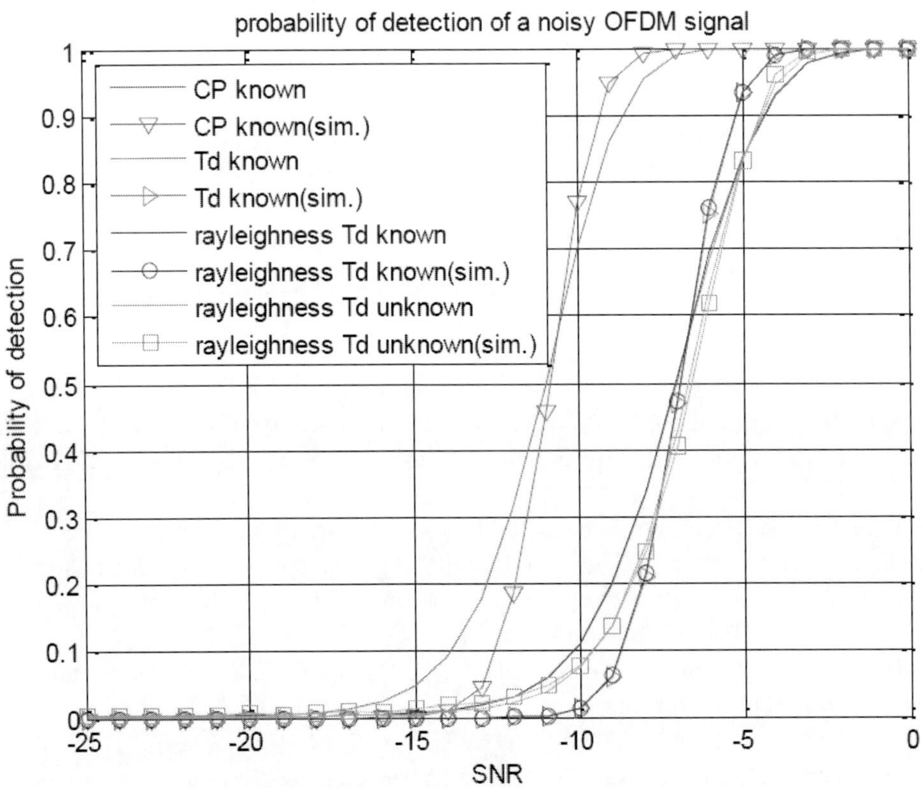

Figure 8. P_D for $T_c=T_d/4$ (i.e. the length of the CP for the WLAN system).

As it can be seen, the conventional approach slightly outperforms the modified Rayleighness method. Anyways, the small loss in performance is greatly compensated by the "blindness" of the strategy that requires no additional information on the parameters of the signal of interest. If the CP length is known to the receiver the performance of the conventional test can be obviously even higher. This is the best possible case, where all the signal's parameters are known to the detector. Nevertheless, this is not the typically operating case and we refer to this situation as the upper limit to the system performance improvement.

CONCLUSION

This chapter has provided an overview of the cognitive radio technlogy for opportunistic spectrum access and related research topics. Implementing cognitive radio networks requires knowledge in different fields of expertise, regarding not only scientific and technological capacity (for the physical deployment of such networks), but also marketing and management knowledge for the regulation of primary and secondary users consistence. A number of promising concepts for cognitive radio have been introduced and discussed in this chapter, starting from the rationale behind the cognitive radio technology, passing through the area of spectrum awareness, spectrum management, and finally focusing on spectrum sensing methodologies.

We have highlighted the motivations behind CRT, starting from the radio spectrum regulations to the principle definition in CR systems and discussed the operating deployment scenarios of such a technology and the obtainable benefits. Then, a brief overview of the principle techniques of spectrum access has been proposed, along with a detailed discussion of the application of the game theory to the problem of rate and power control, aiming at implementing an efficient use of the radio spectrum. The cognitive cycle has been introduced and its essential functionality (i.e. the spectrum sensing), used to identify the spectral bands that are unused by primary licensed users, has been depicted in details.

REFERENCES

[1] F. Benedetto, Editorial, *Recent Patents on Computer Science* (Special Issue on Recent Advances in Cognitive Radio Communications), vol. 5, no. 2, pp.81-82, 2012. ISSN: 1874- 4796.

[2] F. Benedetto, G. Giunta, A. Tedeschi - "Performance Analyis of Tracing Watermarking in the YST Domain for 3G Video-on-Demand Applications" - *International Journal of Sensor Networks and Data Communications,* vol. 1, 7 pages, 2012. DOI: 10.4303/ijsndc/X110504.

[3] F. Benedetto, G. Giunta, A. Neri - "A Bayesian Business Model for Video-Call Billing for End-to-End QoS Provision" - *IEEE Trans. on Vehicular Technology* , vol. 58, no. 2, pp. 836 - 842, Feb. 2009. DOI: 10.1109/TVT.2008.925316.

[4] F. Benedetto, G. Giunta, A. Neri - "QoS Assessment of 3G Video-Phone Calls by Tracing Watermarking Exploiting the New Colour Space "YST"" - *IET Communications* (formerly IEE Proceedings on Communications), vol. 1, no. 4, pp. 696 - 704, Aug. 2007. DOI: 10.1049/iet-com:20060331.

[5] ITU Radio Regulations, *International Telecommunication Union,* Genève, 2008.

[6] Federal Communications Commission, *"Facilitating Opportunities for Flexible, Efficient and Reliable Spectrum Use Employing Cognitive Radio Technologies"*, notice of proposed rulemaking and order, FCC 03- 322, December 2003.

[7] Federal Communications Commission Spectrum Policy Task Force, *"Report of the Spectrum Efficiency Working Group"*, November 2002.

[8] Shared Spectrum Company, *"Spectrum Occupancy Measurements"*, 2005, retrieved from www.sharedspectrum.com/measurements on 27/10/2009.

[9] K. N. Steadman, A. D. Rose, and T. T. N. Nguyen, "Dynamic Spectrum Sharing Detectors", *In Proc. of IEEE International Symposium on New Frontiers in Dynamic Spectrum Access Networks* (DySPAN 2007), Dublin, Ireland, April 2007, pp. 276-282.

[10] M. Wellens and P. Mähönen, "Lessons Learned from an Extensive Spectrum Occupancy Measurement Campaign and a Stochastic Duty Cycle Model", *Proc. of TridentCom 2009*, Washington D.C., USA, April 2009, pp. 1-9.

[11] M. Lopez-Benitez and F. Casadevall, *"On the Spectrum Occupancy Perception of Cognitive Radio Terminals in Realistic Scenarios"*, International Workshop on Cognitive Information Processing, Elba, June 2010, pp. 99-104.

[12] V. Valenta, R. Maršalek, G. Baudoin, M. Villegas, M. Suarez and F. Robert, "Survey on Spectrum Utilization in Europe: Measurements, Analysis and Observations", in *Proc. of ICST Conference on Cognitive Radio Oriented Wireless Networks and Communications* (CROWNCOM), Cannes, France, June 2010, pp. 1-5.

[13] I. F. Akyildiz, W. Y. Lee, M. C. Vuran and S. Mohanty, "NeXt Generation / Dynamic Spectrum Access / *Cognitive Radio Wireless Networks: A Survey"*, Computer Networks, 2006, pp. 2127-2159.

[14] S. Haykin, "Cognitive Radio: Brain-empowered Wireless Communications", *IEEE Journal on Selected Areas in Commun.*, Vol. 23, No. 2, February 2005, pp. 201-220.

[15] A. L. Drozd, I. P. Kasperovich, C. E. Carroll and A. C. Blackburn, "Computational Electromagnetics Applied to Analyzing the Efficient Utilization of the RF Transmission Hyperspace", In *Proc. of IEEE/ACES Conf. on Wireless Comm. and Applied Computational Electromagnetics*, Hawaii, USA, April 2000, pp. 1077-1085.

[16] J. Mitola and G. Q. Maguire, "Cognitive radios: Making Software Radios More Personal", *IEEE Pers. Commun.*, Vol. 6, No. 4, August 1999, pp. 13-18.

[17] J. Mitola, "Cognitive Radio: An Integrated Agent Architecture for Software Defined Radio", *PhD thesis*, KTH Royal Institute of Technology, Stockholm, Sweden, 2000.

[18] L. Berlemann, G. Dimitrakopoulos, K. Moessner, J. Hoffmeyer, *"Cognitive Radio and Management of Spectrum and Radio Resources in Reconfigurable Networks"*, Wireless World Research Forum, Working Group 6 White Paper Cognitive Radio and Management of Spectrum and Radio Resources, 2005.

[19] Federal Communications Commission Spectrum Policy Task Force, *"Report of the Spectrum Rights and Responsibilities Working Group,"* ET Docket No. 02-135, 15 November 2002.

[20] J. M. Peha, "Wireless Communications and Coexistence for Smart Environments," *IEEE Personal Communications*, pp. 66-68, October 2000.

[21] Federal Communications Commission, *"Report and Order (FCC 03-287): Revision of Parts 2 and 15 of the Commission's Rules to Permit Unlicensed National Information Infrastructure (U-NII) devices in the 5 GHz band*," ET Docket No. 03-122, 18 November 2003.

[22] Federal Communications Commission, *"Notice of Proposed Rulemaking (FCC 04-113): Unlicensed Operation in the TV Broadcast Bands,"* ET Docket No. 04-186, 25 May 2004.

[23] Federal Communications Commission, *"Notice of Proposed Rulemaking (FCC 04-100): Unlicensed Operation in the Band 3650 – 3700 MHz,"* ET Docket No. 04-151, 23 April 2004.

[24] Wikipedia, http://www.wikipedia.org, June 2005.

[25] M. Matinmikko, M. Mustonen, H. Sarvanko, M. Höyhtyä, A. Hekkala, A. Mämmelä, M. Katz and M. Kiviranta, *"A Motivating Overview of Cognitive Radio: Foundations, Regulatory Issues and Key Concepts",* First International Workshop CogART, Aalborg, February 2008, pp. 1-5.

[26] Federal Communications Commission, *"Report and Order (FCC 05-57): Facilitating Opportunities for Flexible, Efficient, and Reliable Spectrum Use Employing Cognitive Radio Technologies,"* ET Docket No. 03-108, 11 March 2005.

[27] S. Mangold, Z. Zhong, K. Challapali, and C. T. Chou, "Spectrum Agile Radio: Radio Resource Measurements for Opportunistic Spectrum Usage," in *Proc. of 47th annual IEEE Global Telecommunications Conference,* Globecom 2004, Dallas TX, USA, 29 November - 3 December 2004.

[28] S. Mangold, S. Shankar, and L. Berlemann, "Spectrum Agile Radio: A Society of Machines with Value-Orientation (invited paper)," in *Proc. of European Wireless Conference* 2005, EW'05, Nicosia, Cyprus, 10-13 April 2005.

[29] I. F. Akyildiz, W.-Y. Lee, K. R. Chowdhury: *"CRAHNs: Cognitive Radio Ad Hoc Networks",* Ad Hoc Networks, Elsevier, Vol. 7, No. 5, July 2009, pp. 810-836.

[30] I. F. Akyildiz, W. Y. Lee, M. C. Vuran and S. Mohanty, *"A Survey on Spectrum Management in Cognitive Radio Networks",* IEEE Communications Magazine, April 2008, pp. 40-48.

[31] Q. Zhao and B. M. Sadler, "A Survey of Dynamic Spectrum Access", *IEEE Signal Processing Magazine,* May 2007, pp. 79-89.

[32] J. M. Peha, "Approaches to Spectrum Sharing", *IEEE Communications Magazine,* February 2005, pp. 10-12.

[33] T. A. Weiss and F. Jondral, "Spectrum Pooling: An Innovative Strategy for the Enhancement of Spectrum Efficiency", *IEEE Radio Communications Magazine,* March 2004, pp. S8-S14.

[34] S. Geirhofer, L. Tong and B. M. Sadler, "Dynamic Spectrum Access in the Time Domain: Modelling and Exploiting White Space", *IEEE Communications Magazine,* May 2007, pp. 66-72.

[35] C. Politis, "Managing the radio spectrum", *IEEE Vehicular Technology Magazine,* March 2009, pp. 20-26.

[36] Muqattash, A.; Shu, T.; Krunz, M., "On the Performance of Joint Rate/Power Control with Adaptive Modulation in Wireless CDMA Networks", *Proc. of INFOCOM 2006. 25th IEEE Int. Conf. on Computer Commun.*, pp. 1 – 12, 2006.

[37] Tran, N.H.; Choong Seon Hong, "Joint Rate and Power Control in Wireless Network: A Novel Successive Approximations Method", *IEEE Commun.* Letters, vol. 14, no. 9, pp. 872-874, 2010.

[38] Hayajneh, M.; Abdallah, C.T., "Distributed joint rate and power control game-theoretic algorithms for wireless data", *IEEE Commun. Letters,* vol. 8, no. 8, pp. 511-513, 2004.

[39] V. Srivastava, J. A. Neel, A. B. MacKenzie, J. E. Hicks, L. A. DaSilva, J. H. Reed, and R. P. Gilles, "Using game theory to analyze wireless ad hoc networks," *IEEE Communications Surveys and Tutorials,* vol. 7, no. 5, Fourth Quarter 2005, pp. 46–56.

[40] Akkarajitsakul, K.; Hossain, E.; Niyato, D.; Dong In Kim, "Game Theoretic Approaches for Multiple Access in Wireless Networks: A Survey", *IEEE Commun. Surveys & Tutorials,* vol. 13, no. 3, pp. 372-395, 2011.

[41] T. Alpcan, T. Basar, and S. Dey, "A power control game based on outage probabilities for multicell wireless data networks," *IEEE Trans. Wireless Commun.,* vol. 5, no. 4, pp. 890-899, Apr. 2006.

[42] D. J. Goodman and N. B. Mandayam, "Power control for wireless data," *IEEE Personal Commun. Mag.*, vol. 7, no. 2, pp. 48-54, Apr. 2000.

[43] S. Gunturi and F. Paganini, "A game theoretic approach to power control in cellular CDMA," in *Proc. 58th IEEE Veh. Technol.* Conf. - VTC 2003 Fall, vol. 3, Orlando, FL, Oct. 2003, pp. 2362-2366.

[44] C. U. Saraydar, N. B. Mandayam, and D. J. Goodman, "Efficient power control via pricing in wireless data networks," *IEEE Trans. Commun.,* vol. 50, no. 2, pp. 291-303, Feb. 2002.

[45] M. R. Musku, A. T. Chronopoulos, D. C. Popescu, A. Stefanescu, "A Game-Theoretic Approach to Joint Rate and Power Control for Uplink CDMA Communications*", IEEE Trans. Commun,* vol. 58, no. 3, pp. 923-932, March 2010.

[46] C. U. Saraydar, N. B. Mandayam, and D. J. Goodman, "Efficient power control via pricing in wireless data networks," *IEEE Trans. Commun.,* vol. 50, no. 2, pp. 291-303, Feb. 2002.

[47] Akyildiz, I.F., "Spectrum, network and operations management in cognitive radio networks", *IEEE Network Operations and Management Symposium,* 2008. NOMS 2008.

[48] Nie Nie, Cristina Comaniciu,"*Adaptive Channel Allocation Spectrum Etiquette for Cognitive Radio Networks",* Springer Science Business Media, 2006.

[49] Raul Etkin, Abhay K. Parekh, David Tse, "Spectrum Sharing for Unlicensed Bands," *IEEE Journal on Selected Areas in Communications,* vol. 25, pp. 517–528, April 2007.

[50] Bouida, Z.; Qaraqe, K.; Abdallah, M.; Alouini, M., "Performance Analysis of Joint Multi-Branch Switched Diversity and Adaptive Modulation Schemes for Spectrum Sharing Systems," *IEEE Trans. on Commun.,* early access, pp. 1 – 11, 2012.

[51] Kalathil, D.; Jain, R., "Spectrum Sharing through Contracts for Cognitive Radios", *IEEE Trans. on Mobile Computing,* early access, pp. 1-14, 2012.

[52] Won-Yeol Lee; Akyildiz, I.F., "Spectrum-Aware Mobility Management in Cognitive Radio Cellular Networks" , *IEEE Trans. on Mobile Computing,* vol. 11, no. 4, pp. 529 – 542, 2012.

[53] Christian, I.; Sangman Moh; Ilyong Chung; Jinyi Lee, "Spectrum mobility in cognitive radio networks", *IEEE Commun. Mag.,* vol. 50, no. 6, pp. 114 – 121, 2012.

[54] Won-Yeol Lee; Akyldiz, I.F., "A Spectrum Decision Framework for Cognitive Radio Networks", *IEEE Trans. on Mobile Computing,* vol. 10, no. 2, pp. 161 – 174, 2011.

[55] Li-Chun Wang; Chung-Wei Wang; Adachi, F., "Load-Balancing Spectrum Decision for Cognitive Radio Networks", *IEEE J. on Selected Areas in Commun.,* vol. 29, no. 4, pp. 757 – 769, 2011.

[56] I. F. Akyildiz, W. Y. Lee, M. C. Vuran and S. Mohanty, *"NeXt Generation / Dynamic Spectrum Access / Cognitive Radio Wireless Networks: A Survey",* Computer Networks, 2006, pp. 2127-2159.

[57] M. Matinmikko, M. Höyhtyä, M. Mustonen, H. Sarvanko, A. Hekkala, M. Katz, A. Mämmelä, M. Kiviranta and A. Kautio *"Channel State Estimation and Spectrum Management for Cognitive Radios, Cognitive Radio: An Intelligent Wireless Communication System",* VTT Research report, No. VTT-R-02219-08, March 2008.

[58] T. Yücek and H. Arslan, "A Survey of Spectrum Sensing Algorithms for Cognitive Radio Applications*", IEEE Communications Surveys and Tutorials,* Vol. 11, No.1, 1Q2009, pp. 116-130.

[59] D. Čabrić, S. M. Mishra and R. Brodersen, "Implementation Issues in Spectrum Sensing for Cognitive Radios", *Conference Record of the 38th Conference on Signals,* Systems and Computers, November 2004, pp. 772-776.

[60] A. Ghasemi and A. S. Sousa, "Spectrum Sensing in Cognitive Radio Networks: Requirements, Challenges and Design Trade-offs", *IEEE Communications Magazine,* April 2008, pp. 32-39.

[61] S. M. Mishra, A. Sahai and R. Brodersen, "Cooperative Sensing Among Cognitive Radios", in *Proc. of IEEE International Conference (ICC 2006),* June 2006, pp. 1658-1663.

[62] A. Ghasemi and E. S. Sousa, "Collaborative Spectrum Sensing for Opportunistic Access in Fading Environments", *in Proc. of IEEE International Symposium on New Frontiers in Dynamic Spectrum Access Networks* (DySPAN 2005), pp. 131-136.

[63] B. Wild, K. Ramchandran, "Detecting Primary Receivers for Cognitive Radio Applications", in *Proc. of IEEE International Symposium on New Frontiers in Dynamic Spectrum Access Networks* (DySPAN 2005), pp. 124-130.

[64] H. Yang, "A Road to Future Broadband Wireless Access: MIMO-OFDM-based Air Interface", *IEEE Commun. Magazine,* vol. 43, no. 1, pp. 53-60, Jan. 2005.

[65] Y.-H. You, W.-G. Jeon, J.-H. Paik, M.-J. Kim, D.-S. Kim, H.-K. Song, "Performance Evaluation of OFDM-CDMA with Multiple Antennas for Broadband Wireless Access Networks", *IEEE Trans. on Vehic. Techn., vol.* 54, no. 1, pp. 385-398, Jan. 2005.

[66] A. Doufexi, S. Armour, M. Butler, A. Nix, D. Bull, J. McGeehan, and P. Karlsson, "A comparison of the HIPERLAN/2 and IEEE802.11a wireless LAN standards," *IEEE Commun. Mag.,* vol. 40, no. 5, pp. 172–180, 2002.

[67] G.D. Mandyam, "Interspersed Sinusoidal Transforms for OFDM Systems", *38th Asilomar Conf. on Signals, Systems and Computers,* vol. 1, pp. 1025-1029, Nov. 2004.

[68] Digital video broadcasting (DVB); *Framing structure, channel codingand modulation for terrestrial television, European Standard* (EN) 300 744 V1.6.1, European Telecommunications Standards Institute (ETSI),Jan. 2004.

[69] S. Chaudhari, V. Koivunen, V. Poor, "Autocorrelation-Based Decentralized Sequential Detection of OFDM Signals in Cognitive Radio", *IEEE Trans. on Sign. Proc.,* vol.57, pp 2690-2700, Jul. 2009.

[70] G. Giunta, L. Vandendorpe, "A "Rayleigh-ness" Test for DS/SS Code Acquisition", *IEEE Trans. on Commun.,* vol. 51, no. 9, pp.1492-1501, Sept. 2003.

[71] F. Benedetto, G. Giunta, S. Bucci - "A Unified Approach for Time Delay Estimation in Spread Spectrum Communications" - *IEEE Trans. on Communications* ,vol. 59, no. 12, pp. 3421-3429, Dec. 2011. DOI: 10.1109/TCOMM.2011.100411.090422.

[72] F. Benedetto, G. Giunta - “A Fast Time-Delay Estimator of PN Signals” - *IEEE Trans. on Communications,* vol. 59, no. 8, pp. 2057-2062, Aug. 2011. DOI: 10.1109/TCOMM. 2011.051311.100025.

[73] F. Benedetto, G. Giunta - “On Efficient Code Acquisition of Optical Orthogonal Codes in Optical CDMA Systems” - *IEEE Trans. on Communications* , Vol. 58, no. 2, pp. 438-441, Feb. 2010. DOI: 10.1109/TCOMM.2010.02.080043.

[74] F. Benedetto, G. Giunta - “A Self-Synchronizing Method for Asynchronous Code Acquisition in Band-Limited Spread Spectrum Communications” - *IEEE Trans. on Communications,* vol. 57, no. 8, pp. 2410-2419, August 2009. DOI: 10.1109/TCOMM. 2008.08.070030.

[75] F. Benedetto, G. Giunta, L. Vandendorpe - “A Blind Equalization Algorithm Based on Minimization of Normalized Variance for DS/CDMA Communications” - *IEEE Trans. on Vehicular Technology,* vol. 57, no. 6, pp. 3453-3461, Nov. 2008. DOI: 10.1109 /TVT.2008.918693.

[76] G. Giunta, F. Benedetto - “Spread-Spectrum Code Acquisition in the Presence of Cell Correlation” - *IEEE Transactions on Communications,* vol. 55, no. 2, pp. 257-261, Feb. 2007.

[77] F. Benedetto, G. Giunta, A. Toscano, L. Vegni - "Dynamic LOS/NLOS Statistical Discrimination of Wireless Mobile Channels ", *2007 IEEE 65th Vehicular Technology Conf. VTC2007 Spring,* 23-25 April 2007, Dublin, Ireland.

[78] E. Guzzon, F. Benedetto, G. Giunta - "Performance Improvements of OFDM Signals Spectrum Sensing in Cognitive Radio" - *76th IEEE Int. Conf. on Vehicular Technology, IEEE VTC-Fall 2012,* Quebec City, Canada, 3-6 Sept. 2012.

In: Communication Systems: New Research
Editor: Vyacheslav Tuzlukov

ISBN: 978-1-62618-654-5

Chapter 2

ADVANCES ON MEDIUM ACCESS CONTROL FOR COGNITIVE RADIO NETWORKS

M. Luís[1,2,3]*, ***R. Oliveira***[1,2,3], ***R. Dinis***[1,2,3], ***L. Bernardo***[1,2]

[1] Departamento de Engenharia Electrotécnica, Faculdade de Ciências e Tecnologia, Universidade Nova de Lisboa, Caparica, Portugal
[2] UNINOVA, Caparica, Portugal
[3] IT, Instituto de Telecomunicações, Portugal

Abstract

Over the last two decades, wireless communications have gained more and more importance in our everyday life. The rapidly growing of mobile wireless access services, together with an outdated fixed spectrum access policy is speeding up the need for more radio spectrum, a finite resource that is the basis of wireless communications.

To increase the spectrum utilization, and therefore its usage efficiency, a big effort has been employed to develop and improve novel and alternative spectrum management policies, often referred to as Dynamic Spectrum Access (DSA). In DSA schemes, one piece of spectrum can be allocated to one or more users, which are called Primary Users (PUs) or licensed users, using the traditional fixed spectrum access policy. However, these users do not own the exclusivity of accessing the licensed channel, although they have higher priority in using it. But when the licensed channel is not being used by the PUs, the channel is vacant, leading to a spectrum underutilization. In this case, other users, denominated as Secondary Users (SUs) or unlicensed users, can dynamically access to the temporary vacant channel without interfering with PUs, hence improving the spectrum efficiency.

The aforementioned characteristics of SUs, and therefore the peculiar characteristics of Cognitive Radio Networks (CRNs) distinguish them from the traditional wireless networks. Thus, traditional cross-layered Physical (PHY)-Medium Access Control (MAC) architectures become unsuitable for this new type of networks. In CRNs, the PHY layer should be responsible to perform spectrum sensing tasks, enabling Cognitive Radio (CR) users to identify the spectrum holes, and also to reconfigure and optimize the transceiver based on the results acquired from the environment. Regarding the MAC sublayer, it should be capable to choose between the spectrum sensing and

*E-mail address: nmal@campus.fct.unl.pt

the spectrum access. The sensing/access coordination should be based on a tradeoff between sensing opportunities requirement and spectrum access requirements.

This chapter reviews the most popular decentralized CR MAC protocols already proposed for CR networks. We start with a brief overview describing the basics of the CR principles. Then, we describe the main features that characterize the different CR MAC protocols, emphasizing their strengths and weaknesses. A new taxonomy that classifies the described protocols in several dimensions is also proposed. Finally, we point out open research problems regarding the design of distributed CR MAC protocols.

Keywords: Cognitive Radio Networks, Medium Access Control, CR MAC Protocols

1. Introduction

Cognitive Radios (CRs) are characterized by their cognitive capability and reconfigurability, supporting the design and the implementation of new and *valuable* Dynamic Spectrum Access (DSA) techniques. The cognitive capability provides spectrum awareness, while reconfigurability enables a CR device to dynamically adapt its operating parameters to the surrounding environment [1]. More specifically, a CR device can be parameterized to i) dynamically transmit and receive over widely separated frequency bands and adapt its transmission power, modulation type and frequency range, through Software-Defined Radio (SDR) technology [2], and ii) determine its optimal spectrum access scheme [3].

A typical Cognitive Radio Network (CRN) environment consists of a number of licensed users, or Primary Users (PUs), forming a Primary Radio Network (PRN), and a set of unlicensed users, or Secondary Users (SUs). The PRN is referred to an existing network, where PUs have the permission to operate over pre-defined licensed spectrum bands. PUs may be controlled by Base Stations (BSs) and, since they have the priority to access the channel, their operations should not be affected by SUs' transmissions. On the other hand, SUs do not have the license to communicate in licensed frequency bands but they will try to access them, by opportunistically exploiting the unused portions of the spectrum, always avoiding causing interference to licensed users.

Regarding the CRN architecture, CRNs can be classified as infrastructure-based CRNs or as Cognitive Radio *ad hoc* Networks (CRahNs) (also denoted as Distributed CRNs or Infrastructureless CRNs). Infrastructure-based CRNs are characterized by having a central entity, usually a BS or an Access Point (AP), while the CRahNs are not coordinated by a central unit, and the communication between SUs is made through *ad hoc* connections. However, the difference between these two architectures does not rely only in the communication method. Regarding the spectrum sensing and spectrum access decisions, in infrastructure-based CRNs the local spectrum observations and analysis are performed by each SU and then transmitted to the central unit. By this way a central unit can make decisions on how to avoid interfering with PUs. By the other hand, in CRahNs each SU is responsible for independently sensing the spectrum and determine when and which part of the spectrum should be opportunistically used, which requires a more complex and robust Medium Access Control (MAC) scheme.

In order to adapt to dynamic spectrum environment, CR devices must perform spectrum-aware/access operations that all together form the *cognitive cycle* [4, 5]. This cycle, illustrated in Figure 1, consists of four spectrum management functions: spectrum sensing, spectrum decision, spectrum sharing and spectrum mobility. The main features of these spectrum management functions are:

- **Spectrum Sensing:** In DSA context, SUs are able to access to vacant licensed channels. For that, each SU should monitor the available spectrum, searching for spectrum access opportunities. Spectrum sensing is the management function capable of realize such task, and it is usually considered as the most important function in the cognitive cycle because it provides the information to the remaining cycle.
- **Spectrum Decision:** After sensing the spectrum, SUs select the most appropriate frequency band to transmit, not only in terms of availability, but also in terms of their Quality of Service (QoS) requirements. In order to design a spectrum decision algorithm that incorporates dynamic spectrum characteristics, SUs receive the information about the PUs' activity through the spectrum sensing management function.
- **Spectrum Sharing:** In CRahNs, multiple SUs will make their access to the available spectrum without a central coordinator. Then, each SU's spectrum access must be carefully managed to prevent collisions. This management function provides the capability to share the spectrum with other SUs by adopting optimal resource allocation schemes.
- **Spectrum Mobility:** When a PU transmission is detected by SUs, they should stop their activity and immediately vacant the channel. The spectrum mobility management block must be constantly aware of the spectrum changes during the sensing and access periods, to detect link failures, and to switch the current transmissions to a new and vacant frequency band.

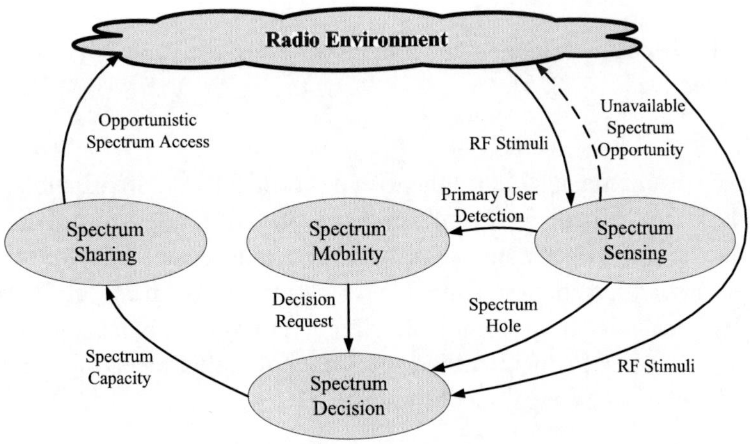

Figure 1. The cognitive radio cycle.

Traditional Physical (PHY)-MAC cross-layer architectures are not prepared to support the CR characteristics and demands. As an example, the spectrum mobility management function usually does not make part of classical MAC schemes, because all nodes have the same priority to use the spectrum. Also, the new spectrum sensing management function has to search the spectrum looking for vacant channels, using sensing techniques that usually are not used by the existing PHY layer schemes (these sensing techniques will be summarized in Subsection 2.2.).

Several PHY-MAC cross-layer frameworks have been designed and adopted by the community to incorporate all the spectrum management functions needed to dynamically access the spectrum. Figure 2 illustrates again the spectrum management functions, according to its role and functionality inside the PHY-MAC cross-layer framework. In the PHY layer, the spectrum sensing management function is an essential essential component that enables SUs to identify spectrum holes. The spectrum mobility management function should operate in both PHY and and MAC layers because it must provide information about the spectrum changes, not only during the sensing periods (PHY responsibility), but also during the access periods (MAC responsibility).

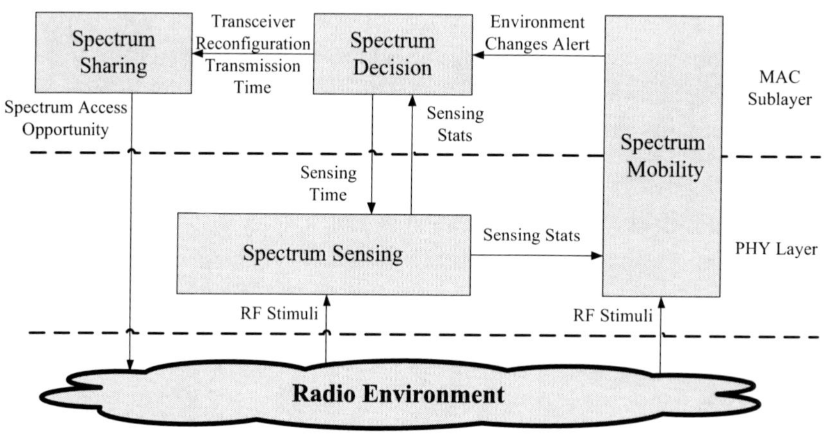

Figure 2. The CR spectrum management framework: an Open Systems Interconnection (OSI) model perspective [6, 7, 8, 9].

The specific spectrum management functions performed by the MAC sublayer of a CR device must include spectrum decision and spectrum sharing. The spectrum decision management function controls the sensing and access operations, establishing the transmission and sensing durations, which are based on the information sent by the spectrum sensing and spectrum mobility management blocks. Finally, the spectrum sharing management function governs the spectrum access to the identified spectrum holes after receiving the information from the spectrum decision management block.

The rest of the chapter is organized as follows: Section 2. tackles the spectrum sensing problem, presenting its challenges and describing most of the sensing techniques already proposed. Section 3. overviews the importance of having an appropriate MAC sublayer scheme in CRNs. It also provides a considerable and complete list of the most studied and successful MAC schemes for CRN and briefly describe some standards already proposed for CRNs. Finally, Section 4. overviews the main conclusions of this chapter.

2. Spectrum Sensing

Cognitive radio devices are designed to be aware of and sensitive to the changes in their surrounding, which makes spectrum sensing one of the key enabling functions in CRNs. To comply with these requirements, the spectrum sensing framework can be divided into two major tasks: i) spectrum exploration in search for spectrum hole opportunities, during the spectrum sensing period, and ii) frequent spectrum monitoring, during the spectrum access periods, to detect the presence of PUs so as to avoid interference with them.

Although spectrum sensing is usually referred as measuring the spectrum content, or measuring the radio frequency energy over the spectrum, in CR context this term evolves to a more general term that includes acquiring the spectrum usage characteristics such as signals' modulations, waveforms, bandwidths, carrier frequencies, etc. [10]. However, this requires more powerful signal analysis techniques with additional computational complexity.

Spectrum awareness is achieved by employing multiple techniques, including: geolocation and database [11, 12, 13, 14]; using periodic beacons [15]; local radio observations performed by the CR devices [16, 17, 18, 19, 20, 21]. According to the classical definition of cognitive radio by *Simon Haykin* and *Joseph Mitola* [22], a CR should be able to *be autonomous* and *sensitive to the environment*, which demands for real-time radio spectrum monitoring.

Using the geolocation and database spectrum awareness techniques all PUs must be equipped with a geolocation/position technology and update their location into a database. Hence, SUs can access to this database and be aware about the spectrum band that might be available in a given location. Due to storage limitations, this technique is only used to tag fixed PUs (e.g. UHF/VHF TV antennas). An alternative method to perform spectrum sensing is to deploy beacon transmitters on each PU, which periodically transmit *disabling beacons* with sufficiently strong power that any SU detecting the beacon will avoid using the channel. However, this method is not economically attractive because it requires that each PU must be equipped with at least two transceivers, incurring excessive costs.

However, the most used spectrum awareness technique is the local radio observation. Here, each SU periodically observes the spectrum searching for PU's activity. Spectrum sensing techniques based on local radio observations can be classified into three groups [23]: primary receiver detection, interference temperature management and primary transmitter detection. The primary receiver detection is based on the detection of PUs that are receiving data within the communication range of a SU. When operating in receiving mode, some PUs emit the Local Oscilator leakage power from its Radio Frequency front-end, which can be used by SUs to detect its presence [24]. However, this signal is typically weak and with very short range, which demands for a very accurate detector. This technique is frequently adopted in the detection of TV receivers [25].

The idea behind the interference temperature management, proposed by the Federal Communications Commission (FCC) in 2003 [16], is to set up an upper interference limit for a given frequency band, such that the SUs are not allowed to cause harmful interference to PUs. As long as SUs do not exceed this limit by their transmissions, they can use this frequency band. However, the difficulty of the interference model lies in accurately determining the interference temperature limit, because SUs cannot distinguish between the

PUs' signals and noise. In 2007 the FCC decided to terminate the proceeding related to the interference temperature management model, mainly because it exhibits a low performance when compared to the amount of interference it can cause to primary users [26].

Due to the challenges imposed by the previous spectrum sensing approaches, most of the research in spectrum sensing focuses on the primary transmitter detection technique. This technique is similar to the primary receiver detection but, in this case, the objective is to detect PUs' transmitters. In order to distinguish between used and unused spectrum bands, SUs observe the spectrum during the sensing period T and make their decisions about the spectrum occupancy based on the following hypothesis model:

$$\begin{aligned} \mathcal{H}_0 &: r(t) = w(t) \\ \mathcal{H}_1 &: r(t) = s(t) + w(t), \end{aligned} \tag{1}$$

where $r(t)$ is the received signal, $w(t)$ is a zero-mean Additive White Gaussian Noise (AWGN) and $s(t)$ is the PU transmitted signal. The first condition, $\mathcal{H}_0$, states that there are no PUs in a certain spectrum frequency, while the second condition, $\mathcal{H}_1$, indicates that there exists some PU activity. Subsection 2.2. describes three of the most studied primary transmitter detection schemes in the literature, and how they adapt the previous hypothesis test to decide if the sensed spectrum is occupied.

The primary transmitter detection, similarly with the others spectrum sensing techniques, presents several implementation challenges. The most crucial one is to determine the decision threshold λ, that will be used to distinguish between the hypothesis $\mathcal{H}_0$ and $\mathcal{H}_1$. More details about the difficulties associated with the decision threshold computation will be given in Subsection 2.2..

The performance of the primary transmitter detector is related with two probabilities: i) the Probability of Detection P_D, that states the probability of detecting a signal and, ii) the Probability of False Alarm P_{FA}, which denotes the probability of a SU to erroneously classify the spectrum frequency as occupied, when there is no PU activity. These probabilities can be formulated as

$$P_D = Pr(Y > \lambda|\mathcal{H}_1) \tag{2}$$

$$P_{FA} = Pr(Y > \lambda|\mathcal{H}_0), \tag{3}$$

where Y is the detector output, that depends on which primary transmitter detection scheme is being used. In order to optimize the performance of the primary transmitter detector we have to increase the probability of detection, to avoid causing interference to PUs, and decrease the probability of false alarm, to prevent underutilization of transmission opportunities. But, as we can see in Equations (2) and (3), both probabilities are directly related with the decision threshold, and so λ should be selected by finding an optimum balance between P_D and P_{FA}. In practice, the decision threshold is chosen to obtain an acceptable predefined false alarm probability [27], but different criteria can be adopted [28].

Subsection 2.1. overviews the challenges in performing spectrum sensing operations and Subsection 2.2. describes three of the most used spectrum sensing techniques.

2.1. Design and Implementation Challenges

Before detailing some of the most used spectrum sensing techniques in CR, we briefly enumerate several challenges associated with the design and implementation of the spectrum sensing framework.

2.1.1. Hardware Requirements

Traditional receivers are capable of processing narrow baseband signals with moderately low complexity and low power processors. However, and due to the peculiar characteristics of the spectrum sensing concept, CR terminals are required to process transmissions over a much wider band, looking for an access opportunity. Hence, CR devices should be able to capture and analyze a relatively larger band in a shorter period of time in order to identify vacant channels in the licensed radio spectrum.

Spectrum sensing can be performed via two different architectures: single-radio and dual-radio [17]. In the single-radio architecture only a specific time slot is allocated for spectrum sensing. As a result of using single-radio architecture, the SU gets a limited sensing duration and a decrease in the spectral efficiency: the single-radio has to perform spectrum sensing and spectrum access techniques in an alternate way. The obvious advantage of single-radio architecture is its simplicity and lower cost [29]. Using the dual-radio sensing architecture, one radio is dedicated for spectrum monitoring while the other is dedicated for data transmission, which increases the spectral efficiency. However, this architecture increases the power consumption and hardware costs [18].

Section 3. will address the limitations of using a single or dual-radio architectures in a MAC framework perspective.

2.1.2. Hidden Primary User Problem

The hidden Primary User problem is similar to the well known hidden node problem in the Carrier Sensing Multiple Access (CSMA) scheme [30]. Basically it occurs when a node C is within the range of a node B, but out of the range of a node A. When the node A transmits to node B, the node C can not detect the transmission, assuming that the channel is idle. The hidden Primary User problem can also be caused by other environmental factors, including severe multipath fading or shadowing observed by secondary users while scanning the spectrum for PUs' transmissions. Due to these factors, the PUs signal becomes undetected to SUs, which, in their turn, will access to the channel without permission, causing harmful interference to PUs. This problem is usually solved using cooperative sensing [19].

2.1.3. Detecting Spread Spectrum Primary Users

The reliability of wireless communications is usually improved by the wireless devices by adopting spreading-spectrum techniques such as Frequency-Hopping Spread-Spectrum (FHSS) or Direct-Sequence Spread-Spectrum (DSSS). FHSS devices are constantly changing their operational frequencies to multiple channels using an hopping pattern known by both receiver and transmitter. DSSS devices are similar to FHSS devices, however they use a single band to spread their energy. Using these two techniques, wireless devices behave robustly against natural interference, noise and jamming, increasing the communication reliability.

The spreading-spectrum techniques also introduce some challenges related with PUs' detection. When the signal power of the PUs is distributed over a wide frequency range (DSSS case), or even not present in some channel at a given time instant (FHSS case), SUs may wrongly assume that the channel is free. This problem can be partially avoided by knowing the hopping pattern together with a tight synchronization with the PUs, which is a very complex procedure [10, 20].

2.1.4. Sensing Duration and Frequency

One of the most challenging tasks when designing spectrum sensing frameworks for single-radio architectures, is the choice of the spectrum sensing duration and frequency. The selection of these sensing parameters establish a tradeoff between the speed and the reliability of the sensing. Sensing frequency depends on the capabilities of the SUs itself and the temporal characteristics of PUs in the environment [31]. For example, when detecting TV channels, the sensing frequency should be relaxed because it is known that a TV station does not change its frequency inside the same geographical area. Another factor that affects the sensing frequency is the interference tolerance of PUs when accessing to some licensed channel: in this case, SUs must be aware if some PU accesses the licensed channel, in order to release it immediately.

Regarding the spectrum sensing duration, this value is usually obtained through optimization problems where the goal is to maximize the average throughput of SUs while protecting PUs from causing interference [29, 32]. Although *higher* sensing times ensure the correct detection of the spectrum, this may result in a comparatively smaller duration for accessing the channel, lowering the throughput. Similarly, by choosing a *shorter* sensing duration, SUs will have more time to access the channel, diminishing the sensing accuracy.

2.1.5. Cooperative Sensing

As mentioned before, cooperative sensing can be used to solve the hidden PU problem, but also to improve the sensing accuracy in a general manner. However, this solution can also become a very challenging task. When a SU detects some PU activity it should notify its observations, or decisions, to its neighbors. To this end, a reliable control channel is needed for discovering neighbors of a SU as well as exchanging sensing information (more details about control channels will be given in Section 3.). Additionally, asynchronism between SUs make it difficult to exchange information between them.

Considering that the sensing information is successfully exchanged between SUs, each one have to combine all the information and make a decision about the spectrum occupancy. For that, SUs can use one of the two methods: i) soft information combining, where the information send by each SU includes the reliability level of each measure, and the decision is usually based on the likelihood ratio test [33, 34]; ii) or using hard information combining, where no information about each measure's reliability is send, and the decision is based on AND, OR or M-out-of-N methods [35]. The likelihood ratio test, or its equivalent the log-likelihood ratio test [36], is a statistical test used to make a decision between two hypotheses (i.e. $\mathcal{H}_0$ or $\mathcal{H}_1$) based on the value of this ratio, usually related with the

Probability of Detection and the Probability of False Alarm. As for the hard information combining methods: using the AND rule the spectrum will be considered occupied if all the cooperative SUs considered occupied; using the OR rule, the spectrum is classified occupied if at least one of the cooperative SUs have considered it like that; and finally, the M-out-of-N rule states that at least M SUs must consider the spectrum occupied to be also considered occupied by the computing SU. Results presented in [33] show that soft information combining performs better than hard information combining methods in terms of the probability of missed opportunity. On the other hand, hard information combining methods are found to be as good as soft methods when the number of cooperating users is *high* [37].

2.1.6. Asynchronous Sensing/Transmission

In distributed CRNs, SUs usually have independent and asynchronous sensing and transmissions schedules. Then, each SU will possibly detect the transmission of other SUs as well as PUs during the sensing period. Assuming that each SU is using the most commonly used spectrum sensing technique, the energy detector (more details about spectrum sensing techniques will be given in Subsection 2.2.), SUs won't be able to distinguish between the transmission of other SUs and PUs. As a result, the SUs detected during the sensing period will increase the number of false alarm situations, ruining the spectrum sensing accuracy.

2.2. Spectrum Sensing Techniques for Cognitive Radio

A number of different techniques have been proposed for identifying the presence of signal transmissions. However, this document only focuses on the three most studied spectrum sensing methods in the literature: the energy detector, the matched-filtering and the cyclostationarity-based sensing detector. More information about other spectrum sensing methods can be found in [7, 8, 10, 24, 38].

2.2.1. Energy Detector Based Sensing

The energy detector based sensing is the most popular spectrum sensing due to its low computational and implementation complexities [17, 18, 19, 20]. In addition, it is more generic compared to other methods as receivers do not need any knowledge on the PUs' signal. In the energy detection, SUs sense the presence or absence of the PUs based on the energy of the received signals. Figure 3 shows the operation mode of this method. The measured signal $r(t)$ is squared and integrated over the sensing period T. Then, the output of the integrator is compared with a threshold λ to decide if a PU is present [39].

Some of the challenges with the energy detector based sensing include failure to differentiate interference from PUs and SUs, already referred in the previous Section, poor performance under low Signal to Noise Ratio (SNR) [40] and the selection of the decision threshold, probably the most challenging task. The interference threshold depends on the noise variance, and a small noise power estimation error will cause a significant performance loss [41]. Several proposals about the interference threshold estimation are given in [39, 42, 43].

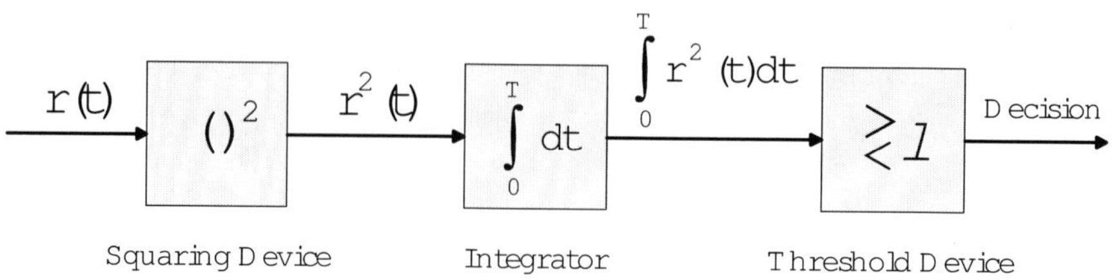

Figure 3. Block diagram of energy detection.

2.2.2. Matched-Filtering

The matched filter is known as the optimum method for detection of PUs when the transmitted signal is known and the noise is AWGN [21]. As shown in Figure 4 the matched filter detection consists in correlating the known original PU signal $s(t)$ with the measured signal $r(t)$ during the PUs' symbol duration. Then, the output of the correlation is compared with a threshold λ, as in the energy detector technique.

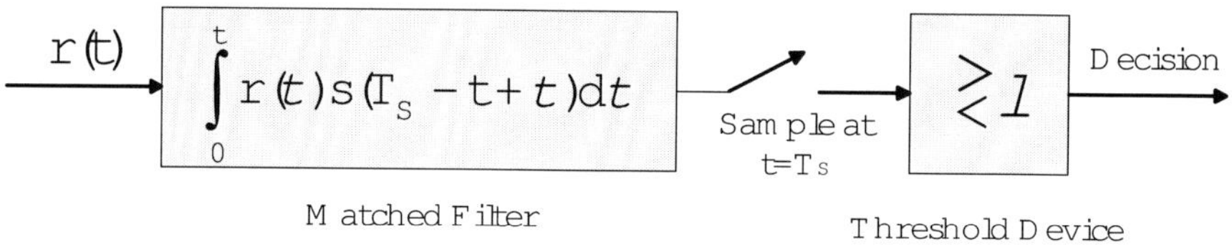

Figure 4. Block diagram of matched filter.

The main advantage of matched filtering, besides its accuracy, is the short time to achieve a certain probability of false alarm as compared to other methods [44]. However, matched-filtering requires not only a priori knowledge of the characteristics of the PU signal, but also the synchronization between the PU transmitter and the SU receiver. Moreover, SUs also need to have different multiple matched filters, each one for each type of PUs' signal, increasing the implementation complexity of the sensing unit.

2.2.3. Cyclostationarity-Based Sensing

Cyclostationarity-based sensing determines the presence of PUs by exploiting the cyclostationarity features of the received signals [17, 20, 45, 46, 47]. These cyclostationarity features are caused by the periodicity in the signal or in its statistics like mean and autocorrelation [48], or they can be intentionally induced to assist spectrum sensing [49, 50]. Figure 5 illustrates the operating mode of this detection technique, where the Power Spectral Density function is replaced by the Cyclic Spectral Density (CSD) function: the CSD function outputs peak values when the cyclic frequency is equal to the fundamental frequencies of the received signal $r(t)$.

The main advantage of this detection technique is its robustness to the uncertainty in noise power and the capability to distinguish different PUs' signals. This is a result of

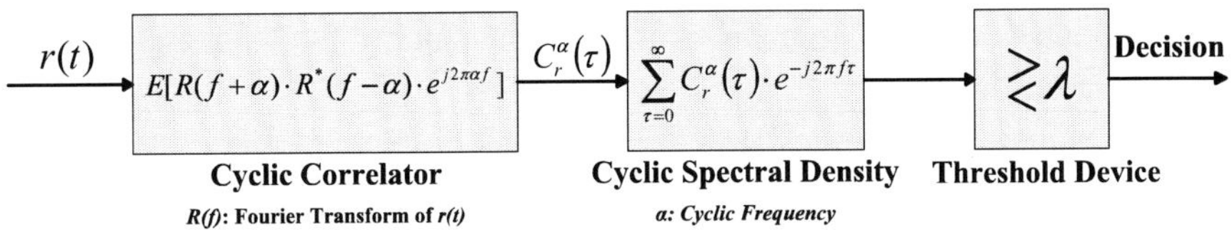

Figure 5. Block diagram of cyclostationarity-based detector.

the fact that noise is wide-sense stationary with no correlation while modulated signals are cyclostationary with spectral correlation due to the redundancy of signal periodicities [47]. Still, cyclostationarity-based detection is computationally very complex and requires significantly long sensing time and power consumption [51].

3. Medium Access Control Schemes for CRNs

Medium Access Control schemes are responsible for handling the medium access control and coordination over wireless channels. Conventionally, control and coordination is treated in a fixed manner, using well-known multiple access techniques such as Time Division Multiple Access (TDMA) or Frequency-Division Multiple Access (FDMA), or through random access techniques like CSMA [52]. In wireless networks several users compete for the wireless spectrum, increasing the number of access collisions, and consequently increasing the interference between them. Thus, MAC schemes should be carefully designed to reduce the number of collisions, therefore improving the wireless network throughput. More information about the basics of MAC schemes are available in [52, 53, 54].

The already aforementioned peculiar characteristics of CRNs distinguish them from traditional wireless networks. Nevertheless, many traditional MAC protocols have been proposed for CRNs although exhibiting low performance. This is mainly because traditional MAC schemes work well under the current static spectrum policies, but they are not suitable for opportunistic spectrum access: classical MAC schemes cannot vary their operation parameters according to the changes in heterogeneous networks, typically the ones found in CRNs. Furthermore, CR MAC schemes need to consider practical constraints like the availability of spectrum opportunities and the number of transceivers available for sensing and communication operations.

As a first step in the development of MAC schemes for CRNs, multichannel MAC protocols for *traditional* multichannel *ad hoc* wireless networks have been adapted to operate in CR scenarios, because both address the following similar problems [55, 56, 57]: i) multichannel MAC protocols were proposed to increase the spectrum utilization of single-channel *ad hoc* wireless networks; ii) they operate in a multichannel context and iii) they face the multiple channel terminal problem [58]. However, due to their sensing functionalities, not required in traditional multichannel MAC schemes, CRs become more spectrum agile providing protection to PUs. Also, in a multichannel network the number of channels available at each user is fixed, while it varies in a cognitive scenario. Finally, the time-scale in which a CR user operates is quite different from that of a user in a multichannel *ad hoc* network: SUs must periodically sense the wireless environment and then quickly adapt their behavior to comply with interference constraints. Regarding all these differences between multichannel wireless networks and CRNs, the multichannel MAC schemes established a good starting point to the development of CR MAC protocols.

Below are summarized specific attributes that must integrate a generic MAC protocol for CRNs [59, 60]:

- The protocol should be *transparent* to PUs;
- The protocol should provide guarantees on PRN's performance;
- The protocol should allow cooperation among SUs to improve spectrum efficiency and fairness among them;
- The protocol should make efficient sensing and spectrum assignment decisions to explore the spectrum access opportunities. These decisions should take into account the dynamic of PUs;
- The protocol should provide an effective distributed coordination scheme for exchanging control information without assuming a predefined dedicated control channel.

3.1. Design and Implementation Issues

In this Section the main challenges in designing CR MAC protocols are discussed, such as the configuration of the common control channel or how to handle with the multichannel hidden terminal problem, among others.

3.1.1. Control Channel Configuration

The amount of signaling exchanged in a CRN is substantially larger than in a classical wireless network [55, 61]. As an example, imagine that in a CRahN two SUs want to communicate but, at that moment, they are *seeing* a different set of channels. In this case, and since we are in the absence of a central coordinating unit, both SUs will have to negotiate which channels will be used to perform the communication. Thus, several CR MAC protocols exploit a Control Channel (CC)[1] to perform the transmitter-receiver negotiation about the communication channel. Moreover, the CC has also another important role which is the exchange of spectrum sensing information, in case of cooperative sensing. Then, a reliable CC should be settled to ensure the proper functioning of the CRN.

Nowadays, it is possible to identify four implementations of CCs that are normally used as a classification parameter of CR MAC protocols (as we will see in the next Subsection). These CCs are classified according to their design and implementation methods: unfortunately none of them are totally satisfactory, and the community is still discussing which solution brings more guarantees to CR MAC protocols. Below are described the main characteristics of each implementation, together with the respective advantages and drawbacks:

[1]The CC is usually referred in the literature as a Common Control Channel (CCC), however it only makes sense to be treated as *common* when it is used by the entire CRN. From now on, CCC is only used when all the SUs are using the same CC, *i.e.*, a CCC is equivalent to a global CC.

- **Dedicated Common Control Channel:** In this implementation, all the SUs in the same CRN share the same dedicated channel to exchange signaling information, sensing outcome and perform channel selection. Then, all SUs should overhear the control message exchanges, even during data exchanges, which demands for an exclusive transceiver dedicated to the CCC. In single-radio architectures, the transmission of control and data messages becomes time multiplexed, turning the MAC protocol operation more complex. Using a dedicated CCC can cause a single point of failure, become a performance bottleneck or even raise security issues. However, the major drawback of using this CCC implementation is the possibility of a PU to become active on the CCC. For this reason, existing MAC protocols often assume that the dedicated CCC is always free from the PUs, being implemented in an unlicensed band (e.g. Industrial, Scientific and Medical (ISM) band) [62, 63, 64, 65] or in an underlay Ultra-Wide Band (UWB) [66]. The dedicated CCC requires synchronization among SUs to make sure of when they have to tune to the CC for control message exchanges, and when they can resume data transmission. This synchronization can be achieved through either local or global time synchronization: local time synchronization is usually preferred in CRNs where coordination is limited to neighboring nodes (e.g. [64]); on the other hand, with global time synchronization periods of data transmission, channel negotiation and channel sensing are predefined and known to all SUs (e.g. [67]).

- **Dynamically Configurable Control Channel:** To overcome the drawbacks of the dedicated CCC, some CR MAC protocols have been using a CC that is dynamically configurable. In a matter of fact, this implementation of CCs introduces a simple but important modification when compared to the dedicated CCs: instead of waiting for the CC to be free from PUs, SUs will renegotiate a new CC. The dynamically configurable CC has two variants: it can be a local configurable CC (single-hop awareness) or a global configurable CC (multi-hop awareness). The former one is used when the CRN is divided into cells or groups of SUs, and each group share the same local CC [68], while the last one is used for the entire CRN. Applying the dynamically configurable local CC in a multi-hop network scenario, the entire connectivity can not be assured, because different groups adopt different local CCs; using the dynamically configurable global CC, the network will spend a lot of time and control overhead converging to the same CCC available to all SUs.

- **Split Phase Control Channel:** The split phase CC method permits to exploit only single-radio architectures, but with a cost in terms of synchronization overheard. This implementation divides the time frame into alternating control and data periods (the main difference when compared to both previous approaches): during the control period all SUs switch their transceivers to overhear control messages and to decide which channels will be used for data transfers; in the data period, transmissions are performed [61, 69]. The split phase CC requires stronger synchronization to identify control and data periods. The advantage is that the CC is not dedicated for control messages and it can be used for data transmissions, when only one transceiver is required.

- **Hopping-Based Control Channel:** The hopping-based CC is being used as a promising technique to overcome the need for a CC, and alleviate the aforementioned CC bottleneck problem [70, 71]. In this implementation, also denominated as blind rendezvous [72], all idle SUs will hop across a set of channels, following the same hopping pattern. When both CR sender and receiver successfully exchange control messages about the data channel to use, they stop hopping and start the data transmission. Once done, both SUs resynchronize back with the hopping sequence. The hopping-based CC has the advantage of using all channels for transmission and control, and does not requires a single channel to be free from PU activity. The most challenging task in this implementation is the stringent time/channel synchronization required between SUs.

It is important to clarify the following: at a first glance, it can be misinterpreted that all single-transceiver CR MAC protocols belongs to the class of split phase CCs. This assumption is in part correct, because it is obvious that all SUs with only one transceiver will have to *split* their transceiver between control and data operations. The difference remains in the use of the CC: in the split phase approach the CC is used for both control signaling and data transmissions; on the other hand, if the CC is used only for control signaling exchange, then it is considered a dedicated or dynamically configurable CC, instead of a split phase CC.

Then, we can further group the previous CC implementations into two major categories: i) non-sequenced assigned CCs, represented by the first three implementations, the dedicated CCC, dynamically configurable CC and split phase control channel; and ii) sequenced assigned CCs, characterized by the hopping-based control channel implementation. The non-sequenced assigned term refers to the fact that, from the moment that the CC is established, the following control packets will be exchanged without being necessary to negotiate a new CC[2]. On the other hand, CR MAC protocols with sequenced assigned CCs do not have a predefined CC, and the control packets are exchanged while both sender and receiver hop according to the same hopping pattern, *i.e.*, the CC is given by the set of the hopping channels used for negotiation, which can be different from the following set.

The implementation of a CC is not mandatory [69, 73] however in this case the MAC design becomes more challenging because there is no predefined channel to start the negotiation with the intended receiver. Then, the CR MAC protocol must find a way to guarantee that sender and receiver will, somehow, meet in some channel. In Subsection 3.3., two distributed CR MAC protocols without CC are analyzed. The analysis focuses on the method developed by the authors to guarantee that two SUs will meet in a channel to perform the data channel negotiation. Regarding the other protocol, the authors assume that the CRN is sharing a single transmission channel with PUs, and then, there is no need for channel negotiation.

3.1.2. Multichannel Hidden Terminal Problem

In CRNs, each SU has to sense the spectrum searching for spectrum opportunities but, assuming that each SU is equipped with only one transceiver, it can only sense one channel

[2]In the dynamically configurable CC implementation, the CC can be renegotiated when PUs are detected in the frequency band used by the CC. Otherwise the CC will always be the same.

at a time with a finite duration. So, when a SU is using its single transceiver to sense a single channel, important information is being exchanged in another channel (e.g. a channel negotiation between SUs, or even the return of a PU). Since the SU is not able to hear this information, there is an higher probability it will cause unintentionally interference to PUs, or even to other SUs [58].

In fact this problem also exists in multi-radio architectures. The unique solution to solve this problem is to have the same number of sensing transceivers as channels to sense, which is quite intolerable. However, increasing the number of transceivers will decrease the number of multichannel hidden terminal problem occurrences, but will also increase the complexity and the cost of the CR device. Other solution to handle this problem is to ensure the synchronization between SUs, hence increasing the signaling mechanism [70, 74]. Again, this solution is also very difficult to implement in a single-radio architecture, because it will increase the control overhead and does not offer any protection to PUs.

3.1.3. Quality of Service

Provisioning QoS in a CRN is a very challenging task, and it is impossible to achieve without the necessary support from the CR MAC protocol: the dynamic nature of the heterogenous spectrum can cause rapid variations in the available communications channels. When in a CRN scenario, we have to preserve the QoS for PUs while offering a fair, but always dependent on the available spectrum, level of service to SUs. To protect the PUs' QoS, SUs have to accurately detect the presence of PUs, which depends, not only in the propagation conditions and noise, but also in the duration of the sensing and access periods. Regarding the SUs' QoS, they should access the channel whenever possible [22], which directly depends on the channel utilization intensity and traffic pattern of the PUs [75]. Then, it is necessary to find a tradeoff between the sensing and access duration to guarantee an acceptable level of QoS to both PUs and SUs.

3.1.4. Time Synchronization

Time synchronization is a very important feature in CRNs, especially in CRahNs. Due to the lack of a cental unit, SUs have to negotiate a data channel, which becomes very difficult without time synchronization. Moreover, time synchronization is also crucial to avoid spectrum sensing mistakes: all SUs should perform spectrum sensing tasks at the same time to guarantee that all the detected spectrum activity is caused by PUs. Then, time synchronization is needed for network establishment and for coordination among SUs. However, achieving network-wide time synchronization without a central entity is a very challenging research problem. Several groups have already proposed the use of the Global Positioning System (GPS) for time synchronization [31, 76, 77, 78].

3.1.5. Adapting to PU transmission

There are several licensed devices that have specific transmission patterns, e.g. television broadcast stations which have pre-determined spectrum usage times and duration [9, 79, 80, 81], or may have occasional random access to the channel, like public service agencies [6] or radars. In these situations, the CR MAC protocol should be able to infer the nature of PUs and adopt new dynamic power control and transmission scheduling techniques to minimize the level of interference caused to PUs, increasing the spectrum usage.

3.1.6. Multi-hop Network Design

The design of multi-hop wireless networks is a very challenging task, not only in CRahNs but in any *ad hoc* wireless network, mainly due to their dynamic and heterogeneous nature [82]. In multi-hop *ad hoc* wireless networks the wireless channel is highly unreliable and its capacity may vary dramatically from hop to hop. The hidden terminal problem in a multihop network scenario is responsible for a high degradation in terms of throughput. Nodes' mobility causes rapid changes in terms of network topology which demands for a frequent update of the network topology, increasing the amount of control traffic. Finally, a decentralized multi-hop wireless network only uses the local information available to any node to adapt its parameters to optimal points of operation. Therefore, it is preferable to use distributed algorithms instead of gathering all the network information in a single node, which avoids a single point of failure and decreases the amount of control traffic.

To this classical multi-hop network issues, CRahNs add one more: the choice of the CC. The global CC addresses the multi-hop communication easily, however it is very difficult to configure a CCC for all SUs, mainly due to the availability of the same spectrum band for all the nodes in the network [78, 83]. The local CC is a better solution to address the multi-hop when compared to the CCC, however it still introduces moderate control overheads, because it is necessary to settle CCs between groups [84, 85]. Dynamic configurable CCs, global or local, involve higher overheads due to their repeatedly reconfiguration, which are justified by the frequent changes occurring in the network topology and in the occupancy of the bands by PUs.

3.2. Classification of CR MAC Protocols

CR MAC protocols, just like the traditional wireless MAC protocols, can been classified according to several features: sometimes these features are directly related with the characteristics of the MAC protocol (e.g. CC design, Medium Access Schemes and Network Optimization); sometimes they are related with physical characteristics (e.g. number of existing transceivers); or even related with the network topology. Below we briefly describe some of the features that are usually used to classify CR MAC protocols [6, 7, 9, 55, 59, 60, 86, 87, 88]:

- **Network Topology:** CR MAC protocols are broadly divided into two major classes according to the architecture of the wireless network: they can be infrastructured or infrastructureless. In infrastructured CRNs, CR MAC protocols need a central entity, such as a BS or an AP, that manages network activities, synchronizes and coordinates operations among SUs. Usually, the central unit is fixed and forms a single hop link with the SUs that are within its coverage area, being coordinated by the central unit. This architecture helps in the coordination among the SUs for collecting the spectrum sensing information from each SU, and allows the spectrum decisions to be localized. In infrastructureless CRNs (here denoted as CRahNs), SUs are responsible for making spectrum sensing and spectrum access tasks without the help of a central unit. Though this topology is scalable and has flexible deployment, the distributed spectrum sensing, sharing and access demands for an increased cooperation between SUs. Therefore, maintaining time synchronization throughout the entire network is the most challenging task to be solved in a infrastructureless CR MAC protocol.

- **Number of Transceivers:** CR MAC protocols can also be classified according to the number of existing transceivers. If SUs have a single transceiver, the protocols are denoted as single-transceiver architectures. If SUs are equipped with two or more transceivers, the architecture is denoted as a multi-transceiver architecture. The number of transceivers has a big influence in the performance of CR MAC protocols. In single-radio architectures, a single transceiver perform spectrum sensing and spectrum access. In multi-radio architectures, each SU usually has two or three transceivers: one of them is always listening the CC; the second transceiver is used for data exchange; and in some cases a third transceiver is used for advertisement tasks, e.g. busy tones.

- **Control Channel Design:** As presented in the previous Subsection, the CCs can be divided into two groups. Similarly, CR MAC protocols adopt the same classification, depending on which implementation of CC they are using. CR MAC protocols can be classified as i) having a non-sequenced assigned CC, using a dedicated CCC, a dynamically configurable CC or a split phase CC, or ii) adopting a sequenced assigned CC, if the hopping-based CC implementation is used. Regarding the single-transceiver CR MAC protocols, it is presented two examples of CR MAC protocols that avoid the adoption of a CC.

- **Medium Access Schemes:** CR MAC protocols are also classified based on the access scheme adopted by the MAC protocol: random access, time slotted or hybrid. Random access CR MAC protocols do not need time synchronization, and are generally based on the Carrier Sensing Multiple Access with Collision Avoidance (CSMA/CA) principle [52]. In this case, SUs monitor the spectrum searching for others SU transmissions, and transmit after a backoff period to prevent simultaneous collisions. Time slotted CR MAC protocols need network synchronization and the time is divided into slots for both control and data channels. Time synchronization is considered i) global, if the periods of data transmission, channel negotiation and channel sensing are predefined and known to all SUs in the network, or ii) local if the synchronization is just limited to neighboring SUs. Finally, hybrid CR MAC protocols apply a mix of the previous approaches: the CC is usually slotted, demanding for a synchronized access, while the data exchange does not requires time synchronization, because random access schemes are used.

- **Network Optimization:** CR MAC protocols can also be classified according to their concern about the network optimization. They can be classified as Dynamic Spectrum Allocation MAC protocols if complex optimization algorithms (e.g. graph theory, stochastic, game theoretic, genetic, etc.) are used to achieve a global efficient exploration of the available resources; or they can be classified as Direct Access Based (DAB) MAC protocols if there are no global network optimization, since each sender-receiver pair tries to maximize its own optimization goal. DAB MAC protocols can also be divided into two general groups: i) contention based protocols, where the sensing information is only shared between the sender and the receiver and ii) coordination based protocols, where each SU always shares its sensing information with their neighbors, improving the system overall performance.

- **Channel Occupancy Model:** The channel occupancy model can also be used a classification feature for CR MAC protocols. Depending on the interference relationships between SUs, CR MAC protocols can be classified into two categories: protocol and physical. In the protocol model, there are no interference between SUs because it is eliminated by enforcing the exclusive channel occupancy between SUs, *i.e.*, a channel already occupied by a SU can not be simultaneously occupied by another SU, inside the same interference area. On the other hand, in the physical model, SUs are allowed to overlap their data channels if each SU transmitter selects its transmission power in order to protect the unintended receivers from its interference. This model is traditionally denominated as interference channel model, and it implies that each SU transmitter should be aware about the environment conditions and SUs' location, in order to accurately estimate the interference that he is causing to other SUs. Sometimes the channel occupancy model is not considered as a classification feature for CR MAC protocols because the classification principle is more related with PHY issues. However, since both approaches (protocol and physical) demand for distinguished PHY-MAC cross-layer behaviors, we consider the channel occupancy as a valid classification feature for CR MAC protocols.

Figure 6 illustrates a CR MAC protocol classification tree based on some of the features presented before. This classification tree comes with an extensive list of the most studied CR MAC protocols in the literature. Each CR MAC protocol mentioned in the tree will be properly described in the next Subsection.

3.3. CR MAC Protocols: Analysis and Discussion

This Subsection presents an extensive analysis of the most relevant CR MAC protocols proposed in the literature. This review focuses on the operating mode of each CR MAC protocol, based on the features previously described. Advantages and drawbacks are enumerated for each CR MAC protocol. Regarding the protocols already standardized or involved in a standardization process will be discussed in the next Subsection.

In terms of network architecture, CR MAC protocols are broadly divided into infrastructured CR MAC and infrastructureless CR MAC. The analysis of infrastructureless CR MAC protocols is the main objective of this Chapter. However, because some common protocol's behaviors are also present in several infrastructured CR MAC protocols, we include three classical infrastructured CR MAC protocols in the review.

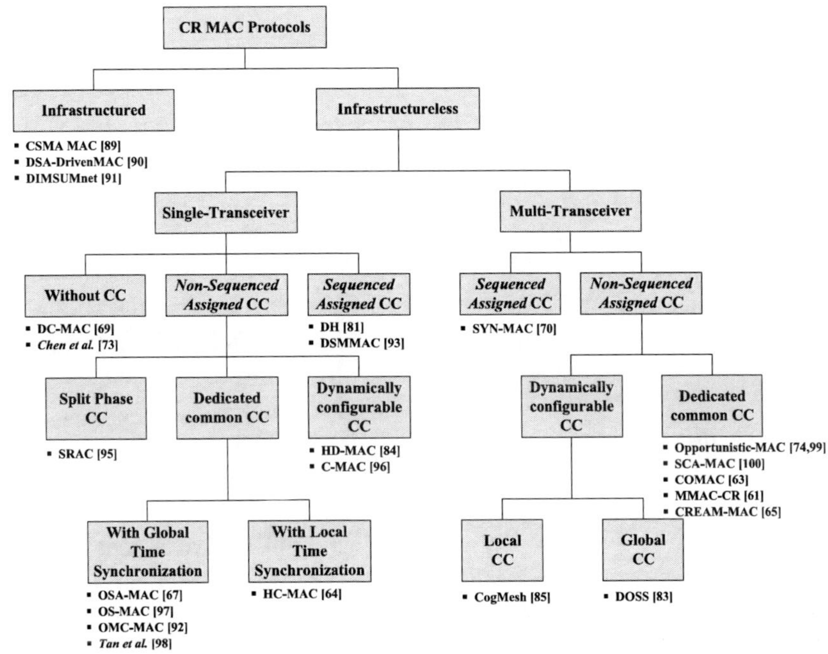

Figure 6. A CR MAC protocol classification tree.

3.3.1. Infrastructured CR MAC Protocols

CSMA MAC

Lien et al.[89] proposed a CSMA based protocol for single-transceiver architectures that uses in-band signaling. CSMA MAC protocol considers that the CRN is composed by a BS and multiple SUs that attempt to transmit packets to the BS. The proposed protocol ensures the coexistence among SUs and the PUs by adapting the transmission power and rate of the CRN, i.e., as long as the interference caused to PUs is contained within a predefined threshold. This adaptation is based on the distance of each SU from the CR BS and the noise power.

CSMA MAC protocol assumes that the PRN follows a classical CSMA scheme with a four-way handshake (Request To Send (RTS)-Clear To Send (CTS)-DATA-ACKnowledge (ACK)), in which the PU undertakes carrier sensing for a period τ_{PU}. To guarantee that each PU has the priority over the spectrum, the CR BS establishes for each SU a carrier sensing period τ_{SU} bigger than the carrier sensing period used for each PU, *i.e.* $\tau_{SU} >> \tau_{PU}$. Besides, in order to minimize the risk of interference to PUs, each SU is allowed to send just one packet after each successful channel contention. The proposed CR MAC protocol has the advantage of considering coexistence between PRNs and CRNs. However, the coexistence is based on some interaction between both networks, because SUs have to know the correct value of PUs' carrier sensing period. Furthermore, the transmission power for the SUs is only partitioned into two levels (high and low), which may not be sufficient to protect all PUs due to possible changes in the PRN topology.

DSA-Driven MAC

Zou and Chigan [90] proposed a centralized game theoretic DSA based CR MAC protocol denominated as DSA-Driven MAC. The proposed protocol adopts an hybrid medium access scheme because the data exchange occurs in pre-determined time slots, while the control signalling is transmitted by a random access scheme. It is considered infrastructured because the CRN is organized in different clusters and the game policy in each cluster is managed by a unique central unit that belongs to the cluster, solving the eventual scalability and time synchronization problems that appear when the game is played flat (by the all CRN in a non-clustered way). The structure of the DSA-Driven MAC framework is divided into four components, as shown in Figure 7:

- **DSA Algorithm:** The game theoretic DSA algorithm aims at finding an optimization solution for the medium access by reaching the Nash equilibrium. Each SU keeps updating its strategy to maximize its own local utility function, taking into account QoS and fairness requirement. When the Nash equilibrium is reached all the SUs transmit over the agreed spectrum with appropriate transmission power.

- **Clustering Algorithm:** The clustering algorithm is responsible for the formation of the clusters. In this case the clustering algorithm satisfies the following properties: a) only SUs with non-negligible mutual impacts should be within the same cluster running the same game; b) the clustering algorithm should be scalable to fit any size/topology of networks; c) the cluster should be stable to avoid frequent reconstruction; d) the cluster formation process should be as simple as possible to reduce the overhead for cluster construction.

- **Negotiation Mechanism:** The negotiation mechanism deals with the control message exchange and coordination of the actions of the SUs. This negotiation occurs over a local CC and is composed of two stages: i) an inquiry stage, responsible for identifying the SUs that wish to start data communication and, ii) a formal negotiation stage, where those SUs that have packets to transmit will start to play the game theoretic DSA algorithm.

- **Collision Avoidance Mechanism:** The collision avoidance mechanism aims at avoiding collisions among negotiations in different clusters, relying on out-of-band busy tones. The mechanism uses two types of busy tones: a) the inside-cluster busy tone, activated by a receiving SU, is used to prevent other nodes external to each cluster from interfering with the ongoing negotiation and, b) the outside-cluster busy tone which is emitted by a SU overhearing messages from other clusters, in order to avoid initiating a new round of negotiation within the cluster.

However, this infrastructured CR MAC proposal does not take into account several details. The first one is that the time and the number of negotiation iterations taken to converge to the Nash equilibrium may become too large, depending on the size of the cluster and on the payoff of each utility function. A too long negotiation mechanism could not be reliable in CRNs since the spectrum may become occupied at anytime. Also, the proposal faces several scalable issues since the negotiation delay increases with the number

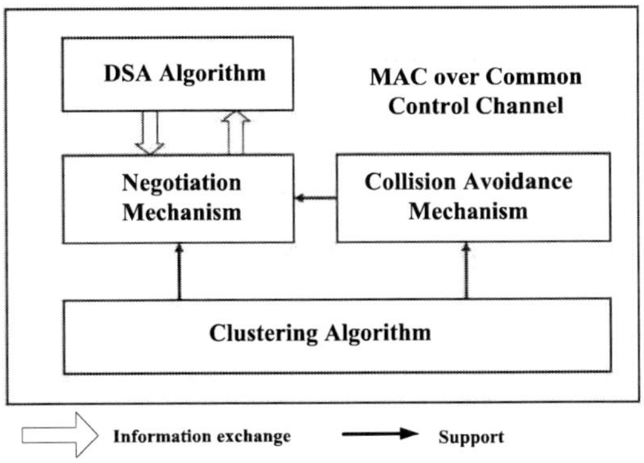

Figure 7. DSA-driven MAC framework.

of SUs. This situation could be fixed by decreasing the number of SUs by cluster, but then, the number of clusters will increase, increasing the amount of busy tones in the CRN. The other drawback of this protocol is the absence of sensing support, which is assumed to be readily available.

DIMSUMnet

Buddhikot et al. [91] proposed in 2005 a new centralized spectrum leasing scheme, the Dynamic Intelligent Management Spectrum for Ubiquitous Mobile-access Network (DIMSUMnet) architecture. This scheme leases the channels from the range of the PUs' spectrum, here denominated as Coordination Access Band (CAB), which may include idle channels allocated to cellular communication systems or TV spectrum.

The architecture of DIMSUMnet is composed by four components: a) the SPectrum Information and Management (SPIM) broker, responsible for constructing a geographical map of the spectrum allocation (denoted as Spectrum Allocation Map (SAM)); b) the Radio Access Network (RAN), consisting of a set of BSs; c) the RAN MANager (RANMAN), responsible for authenticating the spectrum access requests made by the SUs, and also for controlling the power, load, price and lease stipulations of the SUs on network operations; and d) the SU devices, that may optionally contain a spectrum-sensing component which periodically measures the CAB activity and communicates this information to BSs. The DIMSUMnet architecture sets up two sub-MAC protocols to be used during the negotiation phase. The first one, entitled SPEctrum Leasing (SPEL) protocol, operates between the SPIM broker, the RANMAN and BSs, while the second is used between the BS and the SU devices, called SPectrum Information Channel (SPIC) protocol. Therefore, the BS work as a mediator between the SU devices and the SPIM, and it can also apply for spectrum lease and may in turn lease spectrum to the SUs.

One of the problems of the DIMSUMnet is that the authors do not provide any simulation results to be used as a comparison. Moreover, the architecture seems to be very complex, and some instances could be avoided.

3.3.2. Infrastructureless CR MAC Protocols

According to our CR MAC classification (Figure 6), infrastructureless CR MAC protocols can be divided according to the number of transceivers per device. Single-transceiver CR MAC protocols are more vulnerable to the Multi Hidden Channel Problem since a SU equipped with a single transceiver can only listen a channel at any time. Therefore, the SU can miss control messages when its transceiver is busy transmitting or receiving data. Moreover, single-transceiver CR MAC protocols present a smaller theoretical channel utilization when compared to multi-transceiver CR MAC protocols [92]. On the other hand, multi-transceiver CR MAC protocols can listen to both data and control messages simultaneously. However, the multi-radio architecture is more expensive and computationally complex. Let us start with single-transceiver CR MAC protocols.

3.3.2.1 Single-transceiver CR MAC Protocols

Besides the amount of existing transceivers, CR MAC protocols can also be classified depending on how SUs exchange control messages for channel negotiation and coordination. Distributed CR MAC protocols using a CC can be divided into two major groups: sequenced assigned CC and non-sequenced assigned CC. The protocols with a sequenced assigned CC, as is the case in [81, 93], adopt pre-defined channel hopping patterns, solving some of the challenges imposed by the non-sequenced assigned CC approaches: CC saturation and higher probability of CC interruption. However, it is also possible to develop CR MAC protocols without using a CC (e.g. [69, 73]), which are described below.

DC-MAC

Zhao et. al [69] proposed in 2007 the Decentralized Cognitive MAC (DC-MAC) protocol, a CR MAC protocol without requiring a CC. As previously discussed, SUs with single transceiver have a limited observation of the radio spectrum. Based on this fact, the authors proposed a channel sensing/access policy which considers the partial knowledge of the licensed channels state at SUs. DC-MAC's spectrum sensing/access policy is based on the theory of Partially Observed Markov Decision Process, which is a generalization of a Markov decision process [94] where traffic characteristics of the PUs are represented as a Markov chain. DC-MAC integrates the design of spectrum access protocols at the MAC sublayer, with spectrum sensing at the PHY layer and traffic statistics determined by the application layer. Two main issues are addressed: a) joint consideration of the spectrum sensing and spectrum access issues, like sensing or access errors, and b) transmitter-receiver synchronization without requiring a CC.

DC-MAC is one of the most studied CR MAC protocols in the literature because it does not requires a CC. Hence, the main question is: how is it possible to guarantee that sender and receiver will meet in the same channel to perform data communication? In DC-MAC the time frame is divided into slots and, at the beginning of each slot, each SU chooses a set of spectrum bands to sense and another set of bands for transmission. Those lists are formed throughout an optimization process based on the following objective: maximize the throughput of SUs while i) limiting the interference to the PUs, and ii) exploiting the past history of the spectrum band. For that, each SU maintains a list of all channels already sensed, denoted as a *belief* vector. In the end of each slot, the *belief* vector is updated with the probability of each channel to become idle or busy in the next slot, based on the previous observations. At this point, the authors assume that all the neighboring SUs have the same view of the channel, and then, they will have a similar *belief* vector. Based on this assumption, the sender will continuously visit the channels where the intended receiver is expected to be tuned with higher probability, *i.e.*, the channels listed in the *belief* vector with higher idle state probability. The rendezvous between transmitter and receiver is detailed as follows:

- The transmitter senses the channel with highest idle state probability contained in the belief vector;
- Depending on the sensing outcome, the transmitter assumes that the channel is free and send a RTS message to the receiver;
- The receiver, upon receiving the RTS, replies with a CTS message if the channel is also available at the receiver side;
- After receiving the CTS message, the sender starts the transmission. In case of transmission errors or collisions, sender and receiver will hop to a different communication channel, based on the *belief* vector.
- When the transmission is complete the receiver transmits an ACK message signaling the success of the operation;
- Finally, both sender and receiver update the respective *belief* vectors.

DC-MAC protocol is one of the few opportunistic MAC protocols that include sensing and access errors in its design. However, the implementation of the DC-MAC protocol is limited by the assumption that the transition probabilities in the Markov channel model, representing the channel state variations, are known. In practice, this assumption may not be available. Furthermore, the results presented by the authors also assume that the PUs' transition probability remains unchanged during the evaluation simulations, which simplifies the MAC protocol design. DC-MAC also requires that the signal of PUs may be distinguishable from those of SUs in order to build a consistent *belief* vector.

Slotted CR-ALOHA and CR-CSMA

Chen et al. [73] proposed in 2011 a two-level opportunistic spectrum access strategy to optimize the system performance of a CRN and to adequately protect the operation of the PRN. In this CR MAC protocol no additional CC is needed because the CRN and the PRN share a single transmission channel, and then there is no need for a channel negotiation. In the first level, SUs maintain a sufficient detection probability to avoid interference with PUs while the spectrum sensing time is optimized to control the total traffic rate of the

CRN allowed for random access when the channel is detected available. Spectrum sensing is periodically executed by each SU before its packet transmission in order to detect PU's activity. For that, energy detection is assumed, and the probability of detection is set according to the protection requirement: if PUs are detected to be active, the SU will be blocked for transmission in the current frame; otherwise it will attempt to transmit. If a SU qualifies to transmit, the packet transmission probability is determined by using both probability of detection and probability of false alarm.

At the second level, the authors proposed two MAC with distinct medium access schemes to deal with the packet scheduling of the CRN: the slotted Cognitive Radio ALOHA (CR-ALOHA) and the Cognitive-Radio-based Carrier Sensing Multiple Access (CR-CSMA). The slotted CR-ALOHA is developed from the conventional slotted ALOHA [52], which differs in the discrete channel access time and the constraint of protecting the primary network. The authors assume that, for each frame, the data transmission duration is slotted, and the slot size is equal to the length of a fixed packet transmission time together with the propagation delay. By this way, in each frame duration slot, more than one SU can compete for channel access which reduces the waiting time. The slotted CR-ALOHA operates as follows:

- If a SU detects that the channel is available in the current frame, any packet arriving in the last slot of the previous frame will be transmitted in the first slot of the current frame; otherwise, if the packet arrives in the nth remaining slots, it will start to transmit the packet at the beginning of the $(n + 1)$th slot;

- If the channel is unavailable, any packet arrival within this frame up to the next-to-last slot will be blocked to the end of this frame, and then uniformly retransmitted with a backoff window;

- The current transmission is successful when there is only one packet that was transmitted; otherwise a collision occurs, and the packets involved will be retransmitted using non-related random delays to avoid continuously repeated conflicts;

- Finally, any arrival in the last slot of one frame will be processed in the next frame.

It is know that the conventional slotted ALOHA protocol presents a maximum throughput of around 0.368. However, with this approach the authors show that the maximum throughput achieved by the CR-ALOHA underperforms the traditional one, since PUs increase the number of collision.

In CR-CSMA the authors assume that the data transmission duration of each frame is now divided into minislots having a similar duration to the slot time used in the Institute of Electrical and Electronics Engineering (IEEE) 802.11 Distributed Coordination Function (DCF) (20 μs). The concepts of slot and minislot are different due to the different schemes's behavior. CR-CSMA requires that each packet must be transmitted at the beginning of the next minislot. The details of CR-CSMA can be described as follows:

- If a SU detects that the spectrum is free from PUs during the current frame, any arrival during the last mininslot of the previous frame will be transmitted in the first minislot; otherwise, if it arrives in the nth remaining minislots, the following conditions hold:

a) if the channel is idle, it will be transmitted at the beginning of the next minislot;
b) if the channel is busy, SU keeps sensing until the channel becomes idle again and then tries to transmit;

- If a SU detects that a PU is active, any arrival within the current frame up to the next-to-last minislot will be blocked until the end of this frame, and then, the SU chooses a uniformly distributed backoff time to retransmit;

- Again, the current transmissions will be successful if there are no other packets being transmitted at the same time; otherwise, the transmission fails and the packet will be retransmitted after a random delay to avoid repeated conflicts;

- Finally, any packet arriving during the last frame's minislot must wait to be processed in the next frame.

The difference between slotted CR-ALOHA and CR-CSMA is that: in slotted CR-ALOHA SUs transmit only from the beginning of a slot, and so they do not have to keep sensing the channel until the end of the frame. On the other hand, under CR-CSMA SUs become more aggressive because they keep sensing until the channel becomes idle. The performance comparison presented by the authors for the same sensing time conclude that CR-CSMA always outperforms CR-ALOHA in terms of throughput and delay. The performance gain is due to a more efficient packet scheduling scheme where the packets suffer a lower transmitting service time.

DSMMAC

Hou et al. [93] have recently proposed a new asynchronous multichannel MAC protocol for distributed CRNs entitled Difference-Set-based asynchronous Multichannel MAC (DSMMAC). The DSMMAC protocol take advantage of the unique rotation closure property of difference sets, which enables any two SUs to communicate with a nonzero probability without global synchronization. By combining multiple disjoint difference sets and adopting the same difference-set-based hopping sequence, two SUs will be able to meet within a certain probability, even without synchronization. When a SU joins the network, it randomly selects an initial channel and attempts to access the channel by starting a handshake process. If the transmitter-receiver pair successfully exchanges the handshake information, the pair stops hopping and starts the transmission over the same channel. DSMMAC allows multiple sender-receiver SUs to simultaneously rendezvous and communicate over different channels[3] without a centralized coordinator, which significantly improves the channel utilization. By using the same hopping sequence, DSMMAC reduces the signaling overhead, because there is no need for exchanging the hopping sequence, and all the available channels can be used for data transmission. However, the DSMMAC protocol assumes that all SUs know the hopping pattern, given by the manufacturer or the service provider.

[3]Channel orthogonality is assumed.

DH

The key feature of the sequenced assigned CR MAC protocols is the hopping pattern adopted by the SUs: it allows continuous control/data transmission between PUs and simultaneously tries to assure unimpaired operation of PUs. *Willkomm et al.* [81] proposed a new hopping approach for Dynamic Frequency Hopping (DFH) in CRNs. Double Hopping (DH) is a clustering DFH scheme that proposes a distributed algorithm to generate a hopping pattern that minimizes the number of used frequencies based on the following premise: maximizing the number of used channels for a transmission would reduce the interference to PUs[4]. However this assumption exposes an important handicap of this DFH scheme: maximizing the number of used channels for a transmission may lead to channel overassignment, and consequently it will become difficult to guarantee a desirable level of QoS to SUs.

The DH scheme divides the network into cells, where each cell consists of a CR clusterhead and associated SUs. Each clusterhead is responsible for the DH pattern generation of all cells in the network. For that, it uses a graph coloring algorithm to perform frequency assignment, minimizing the number of used frequencies and ensuring a collision free assignment of frequencies among neighboring cells. Two graph coloring based alternatives are proposed: a) a centralized approach denominated as Optimal Frequency Assignment (OFA) and b) a distributed heuristic approach, the Distributed Frequency Assignment (DFA). To create the hopping pattern, the frequency channels are chosen based on the interfering degree, *i.e.*, a SU does not choose its channel until it receives the decisions from its neighbors with higher degree. Based on the results presented by the authors, the centralized DH approach slightly outperforms the sub-optimal approach in terms of used channels, while the centralized DH scheme presents a lower computational complexity, resulting in a better scalability.

On the other hand, there are CR MAC protocols adopting a non-sequenced assigned CC. The behavior of these protocols is based on the assumption that the common CC is always in the same band/frequency, sometimes roughly assumed as being always available. There is however an exception: using the dynamically configurable CC approach, SUs have the possibility to negotiate a new CC, which is different from hopping following a predefined sequence. But this change of CC only happens exclusively because the previous one has been occupied by licensed users. Otherwise, SUs will always use the same common CC. As aforementioned, CR MAC protocols with a non-sequenced assigned CC can be divided into three categories: split-phase CC (e.g. [95]), dynamically configurable CC (e.g. [84, 96]) and dedicated common CC (e.g. [64, 67, 92, 97, 98]).

SRAC

Ma et al. [95] designed the Single-Radio Adaptive Channel (SRAC) algorithm that adaptively combines spectrum bands based on the requirements of each SU. For that, the (SRAC) algorithm, a random access based protocol, bases its operation on three key features: a) adaptive discrete channelization, b) cross-channel communication and c) as-

[4]In this case, the expression *maximize the number of used channels* is equivalent to say *choosing the minimum number of channels*.

needed use of spectrum. The adaptive discrete channelization is a tradeoff between performance and practicality because it assumes that any piece of spectrum is a permissible channel, meaning that spectrum can be channelized with essentially arbitrary values for carrier frequency and bandwidth. The idea behind the cross-channel communication, is that a SU can transmit on any channel but it can only receive on a unused channel, called *receive channel*. This way, two SUs are able to communicate when there are multiple jamming sources and there is no common idle spectrum between the sender and the receiver. The information about available *receive channels* is delivered using the Automatic Repeat reQuest (ARQ) mechanism to guarantee that the information is received by all the neighbors. Finally, the use-as-needed spectrum usage policy guarantees that each SU should keep the same receiving channel unless required by its own needs, *i.e.*, to avoid jamming and PUs, or to provide more capacity in response to high channel load. Using this feature, the (SRAC) algorithm keeps the number of used channels low for a better multi-hop support.

However, SRAC MAC protocol does not carefully address the issue of long periods of time during which SUs can not monitor the control packets. It is also unclear how this work detects the jammers. Other disadvantage is related with the absence of a dedicated CC, which forces SUs to send control packets to all the available channels, increasing the control traffic. In the case of a frequent channel switching, the cognitive network may become quickly saturated.

HD-MAC

Zhao et al. [84] proposed a new channel selection metric that jointly considers traffic load, connectivity and interference (latter denominated as *score*). The Heterogenous Distributed MAC (HD-MAC) protocol, which is an improvement of the Multichannel (MMAC) [58], allows distributed coordination of local clusters in a multi-hop CRN. In HD-MAC, neighbor SUs self organize in local groups where coordination is handled through a local CC. This local CC, here denoted as coordination channel, is dynamically configured depending on the channel availability within each group. The MAC structure is organized in super-frames consisting in a beacon period (BP) (used for neighboring discovery), a coordination window (CHWIN) and a data transmission period (DATA). During the CHWIN period, SUs hop to the coordination channel to manage channels negotiation. The access is made according to the CSMA/CA protocol.

The data channel negotiation is as follows: i) transmitter u sends a channel request (CHRTS) message to receiver v including its queue size, related to SU v, and a channel information message related to its available channel list; ii) after receiving the CHRTS message, SU v combines the channel *score* of both nodes and selects the common channel between both SUs with the lower *score* value; iii) the selected channel is sent back to SU u in a CHannel RESpond (CHRES) message, together with the volume of pending packets; iv) upon receiving the CHRES message, SU u sends a CHannel ConFirMation (CHCFM) to SU v, including the length of the pending packets. Neighbor SUs which overhear the CHCFM and CHRES messages obtain traffic information and update the channel *score* accordingly.

HD-MAC allows SUs to use previous data channels without a so complex negotiation: if SUs u and v have marked the channel as "outstanding", for the next data transmissions both SUs will spend less time in the negotiation phase, assuming that the same channel *score* has not been changed. The HD-MAC however presents a moderate reconfiguration, control and communication overheads due to the negotiation mechanism. It also requires tight synchronization among SUs.

C-MAC

The Cognitive-MAC (C-MAC) protocol, presented by *Cordeiro and Challapali* [96] in 2007, uses an half-duplex radio to access multiple channels. The available channels are organized as superframes and one of the channel is configured as a dynamically configurable CC, here denoted as *rendezvous channel* (RC). The RC, like traditional CCs, manages operations such as network wide group coordination, inter-channel synchronization, neighboring discovery and load balancing.

C-MAC's superframe is composed by two periods: a Beacon Period (BP) and a Data Transfer Period (DTP). The BP are kept non-overlapping using an inter-channel synchronization procedure. It is also divided into slots so that nodes operating on the channel can broadcast their beacon frames during their respective slot. The beacon frame is transmitted using the most robust modulation scheme in order to allow SUs to decode the beacon frame even in the presence of PUs or jamming. The first two beacons slots are intended for new SUs. Synchronization among SUs (and not inter-channel synchronization, which is achieved through the RC) is performed using the BP, however SUs have to listen for at least one superframe duration in order to stay synchronized. During the BP the SUs operating on the channel broadcast their beacon frames with information about the previous beacon frames. By this way, every SU has an update of the neighbors's neighbors details like communication schedules, beacon slots, spectrum measurements, location, etc., and such improvements helps to overcome the hidden and exposed node problems. During the DTP, a Quiet Period (QP) is established so that SUs can perform in-band and out-of-band sensing. These measures are then broadcasted in the RC to help other SUs to select a data channel with better quality.

One of the main drawbacks of C-MAC is that all beacons sent by SUs must be accommodated in the BP of a superframe, which has a maximum beacon period length, originating some scalability problems. Besides, the authors do not clarify how the beacon slots are associated to the SUs on a given channel. Furthermore, each SU repeatedly switches to the RC to update its details, resulting in more control message overheads. The control and communication traffic overhead increase with the increase of network size and number of channels.

Time synchronization becomes an important feature for CR MAC protocols with single-transceiver and dedicated common CC. In this situation SUs can not remain tuned to the common CC when they are transmitting. Then, time synchronization among SUs is needed to make sure that SUs know when they have to tune to the common CC for control messages exchange and when they can resume data transmissions. Time synchronization can be achieved through either local (e.g. [64]) or global (e.g. [67, 92, 97, 98]) synchronization.

OSA-MAC

In 2008 *Le and Hossain* [67] proposed the Opportunistic Spectrum Access-MAC (OSA-MAC) protocol for CRN. OSA-MAC assumes that SUs exchange control information in a dedicated common CC. The common CC is assumed as being always available (owned by the secondary service provider), and it demands for a synchronization between all the SUs. For that, OSA-MAC divides the time into slots, and all SUs are synchronized by periodic beacons sent in the beginning of the slot. The OSA-MAC time structure is divided into three phases: a channel selection phase, a sensing phase and a data transmission phase. In the channel selection phase, each sender chooses a potential channel to its receiver by transmitting an *Ad Hoc* Indication Message (ATIM) to its receiver in the common CC. To decrease the probability of collisions, a backoff mechanism similar to the one used in IEEE 802.11 DCF mode is adopted. After agreeing about the desired data channel, both sender and receiver will sense the channel during the sensing phase. If the channel is sensed busy, they will have to wait for a new round. Otherwise they can start transmitting during the data transmission phase using a four-way handshake (RTS/CTS/DATA/ACK).

The channel is selected by the sender by applying one of the two channel selection processes. The sender can use a uniform channel selection, where all the available channels have the same probability of being chosen. This simple channel selection method does not consider different spectrum opportunities in different channels. On the other hand, the sender can choose a channel by using an opportunity-based channel selection taking into account the probabilities of spectrum availability in different channels. The results presented by the authors show that the opportunity-based channel selection performs better when the number of secondary links is small compared to the number of channels.

OS-MAC

Hamdaoui et al. [97] presented the Opportunistic Spectrum-MAC (OS-MAC) protocol described by the authors as being an efficient MAC protocol for spectrum-agile wireless networks. OS-MAC protocol uses pre-determined window periods for coordinating the choice of spectrum channels among the SUs, and for clustering formation. However, within each window period, the spectrum access if random-based, turning the OS-MAC into an hybrid MAC protocol. SUs self organize in groups, denoted as Secondary User Groups (SUGs), to manage the access in both licensed and unlicensed spectrum. Although SUs are organized in clusters, the dedicated common CC adopted by the OS-MAC requires global synchronization because it is used to perform clustering operations and also to realize inter-cluster signalling exchange. Besides the CC, OS-MAC equally divide the radio spectrum into nonoverlapping Data Channels (DCs) used to perform intra-cluster communication, where each DC represents a PU licensed channels.

The OS-MAC operation can be divided into six phases. In the first phase, denominated as network initialization phase, each SU tunes its radio into the common CC to form clusters such that all the members of the same cluster wish to communicate with each other. A new SU is able to create its own cluster or joining the existing ones. At any moment, only one SU is active in a cluster, which is called as Delegate SU (DSU). In the session initiation phase the active DSU chooses a spectrum band for the group and communicates the decision to all group members. Then, during the data communication phase, the SUs of the cluster use a backoff mechanism for accessing the spectrum band, while the DSU monitors the common CC for collecting information of the spectrum environment. During the fourth stage, the update phase, each DSU shares information with other DSUs over the common CC. After updating the information, the DSU may initiate changing of the spectrum used inside the cluster. The DSU will have to try to access the spectrum with smaller periods of backoff in order to win the contention and transmit with higher priority. Finally, in the delegate phase, the delegation role is passed to another SU, more particularly to the first SU that successfully delivered a message during the previously slot.

The main drawbacks in OS-MAC is that spectrum sensing and spectrum mobility functionalities are not carefully addressed, *i.e.* a mechanism to protect licensed transmissions is not implemented, which can increase the interference over the PUs. Furthermore, during the six phases, the common CC is only used during the initialization and update phases, which produces control overhead during those two phases, while it is unused during the other four phases.

OMC-MAC

Jha et al. [92] designed the Opportunistic Multichannel Cognitive-MAC (OMC-MAC) protocol for single-transceiver CRNs where a dedicated global common CC is assumed. The OMC-MAC time frame is divided into Beacon Intervals (BI) used by the SUs for synchronizing purposes. Additionally, the BI is divided into three phases: sensing, contention and data transmission.

During the sensing period, SUs are not allowed to transmit and therefore, the sensing result is free from interference from other SUs. Each SU senses the channel independently and maintains a list of available channels at the end of the sensing phase. During the contention phase, each SU that has data to send contends for the common CC to participate in a channel negotiation. The contention is based on a contention backoff similar to IEEE 802.11 DCF. Using a three-way handshake (RTS/CTS/CRTS) mechanism with the information about the negotiated data channel, OMC-MAC guarantees that other SUs will set some kind of Network Allocation Vectors (NAV) to remind that channel can not be used until the end of the time-frame. This solution effectively addresses the Multi Hidden Channel Problem because all SUs are tuned into the common CC during the negotiation phase. During the data transmission phase, data messages are exchanged without the need for a contention because the channels are already reserved. However, the authors divided the data transmission period in multiple MAC Packet Data Unit (MPDU) plus an ACK message. If one ACK message is missing, SUs assume that the loss is due to the channel error and they re-send the MPDU. If two consecutive ACK messages are missing the sender assumes the presence of a PU and it defers the transmission.

The results presented by the authors show that the OMC-MAC is highly robust in decreasing the effect of sensing error on PUs, when compared to the already studied OSA-MAC. The results also show that OMC-MAC effectively avoids collision among SUs during the data transmission phase in order to efficiently utilize the opportunistically available licensed frequencies.

Tan et al.

Tan and Le [98] proposed a distributed synchronized MAC protocol for CRNs incorporating spectrum-sensing operations. In fact they have proposed two CR MAC protocols, one for single-channel and other for multichannel. Regarding the single-channel approach, the time-frame is divided into three phases: the sensing phase, the synchronization phase and the data transmission phase. During the sensing phase, all SUs perform spectrum sensing on the underlying channel. Since all SUs are assumed to be synchronized, the sensing results will only be referred to PUs. Then, only SUs whose sensing outcomes indicate an available channel proceed to the next phase (denoted as active SUs). In the synchronization phase, active SUs broadcast beacon messages for synchronization purposes. In the third phase, active SUs perform channel contention, using a backoff mechanism, and the wining SU can finally transmit the data during the last phase. All these stages occur in the same channel.

The multichannel CR MAC protocol proposed by the authors is somehow similar to the previous one (single-channel). However, in this case, there is one dedicated global common CC assumed to be always available, and several data channels owned by PUs with the possibility of being explored by SUs. Each SU is equipped with only one transceiver that can be tuned to the CC or vacant data channels. In this case there are four phases in each time-frame. In the sensing phase, all SUs perform spectrum sensing in all underlying channels. During the synchronization phase, all SUs tune to the common CC to exchange beacons to achieve synchronization. To participate in the data transmission phase, SUs must have at least one sensed free data channel. Then, SUs will compete for the common CC using a backoff mechanism, and the winning SU, using a two-way handshake (RTS/CTS) will negotiate the receiver about the data channels to use. In fact there will be no negotiation, because this CR MAC protocol assumes that the sender-receiver pair can use all common available data channels. In the end of the transmission, the receiver will confirm the data transmission using an ACK message.

The authors assume that SUs can perfectly synchronize with each other, *i.e.*, there is no synchronization error, which can be a rough system model assumption. Moreover, in the multichannel approach the authors make an assumption that is very expensive, computationally complex and somehow unreal: they admit that each SU is equipped with the same number of sensors as channels to sense. There is also another issue related with fairness: the proposed CR MAC protocol grants all the common available data channels to the contending winner pair. Maybe the authors should establish a maximum number of data channels to use, and then, the remaining available data channels would be negotiated using the same method, until there is no more available channels, or the end of the time-frame has been reached.

HC-MAC

The Hardware-Constrained MAC [64] protocol is quite different from the previous ones because it does not requires for a global synchronization, which means that there are synchronization only among neighboring SUs. However the main characteristic of HC-MAC protocol is that it tries to perform efficient spectrum sensing and spectrum access by considering several hardware constraints, such as, the aforementioned operational limitations of a single-transceiver, partial spectrum sensing and spectrum aggregation limits. Hardware constraints can be divided into two categories: sensing constraints, concerned with the tradeoff between the time taken for sensing and the resulting accuracy, and transmission constraints, related to the limitations posed by the Orthogonal Frequency Division Multiplexing (OFDM) technique that decides the bandwidth range, as well as the maximum allowed number of subcarriers.

One of the main contribution of HC-MAC is the sensing decision adopted by the SU. Limited by the hardware constraints of a single-transceiver, each SU must use an appropriate stopping rule for successive channel sensing. By choosing a greater number of channels, the available bandwidth increases, leading to a higher data rate. However, the cost of sensing, especially if the channel is found to be occupied must be also considered. So the goal is to find an optimal stopping rule such that the reward is maximized. In OFDM radios, this decision is constrained by the maximum spectrum holes that can be merged, and also by the limited width of the aggregate band. Since there are a maximum number of channels that a SU can sense before taking the stopping decision, it is said that the stoping problem has a finite horizon. This problem can be optimally solved by using the method of backward induction but, since this solution has an exponential complexity, the authors proposed a modified truncated version named *k-stage look-ahead*. Hence, it decides whether to stop or to continue sensing the channel, comparing the reward function value with throughput constraints.

The HC-MAC operation mode is divided into three phases, however in this case, the contention phase happens before the sensing phase. The last phase is reserved for data transmission. In the contention phase, C-RTS and C-CTS packets are sent over the common CC to gain access to the channel. The winning sender-receiver pair conducts sensing in each channel and exchange S-RTS and S-CTS on the common CC if that channel is available for both slides, *i.e.*, one set of S-RTS/S-CTS packet for each channel. The end of the sensing round is decided by the previous described stopping rule, and any common free data channel can be used. At the end of the transmission T-RTS and T-CTS messages are exchanged on the common CC, releasing the channel for other SUs. The major drawbacks of HC-MAC are: the high number of control messages that may saturate the common CC earlier, when compared to *classical* RTS-CTS based CR MAC protocols; the fact that remaining unused channels are not used by other SUs until the current transmission is ended, leading to inefficient spectrum utilization; and the Multi Channel Hidden Terminal Problem caused by the absence of a global synchronization together with a single-radio architecture.

3.3.2.2 Multi-transceiver CR MAC Protocols

Multi-transceiver CR MAC protocols, can also be divided into two categories depending on how the CC was implemented. [70] is an example of a multi-transceiver CR MAC protocol that adopts a sequenced assigned CC. On the other hand there are CR MAC protocols with a non-sequenced assigned CC that can assume a dedicated common CC (e.g. [61, 63, 99, 100]) or a dynamically configurable CC (e.g. [83, 85]). A common problem with MAC protocols that use dynamically configured CCs is the interruption of their MAC operations. These interruptions can occur quite frequently since the PU network can reclaim the CC at any time. Moreover, CRahNs do not have central support to gather the topology information and hence rely on local coordination for the same. The existence of a dedicated common CC makes it much easier to gather topology information compared to the case of CRahNs with dynamically configurable CCs. However, the majority of the CR MAC protocols with dedicated common CCs roughly assume that the CC is always available.

SYN-MAC

Kondareddy and Agrawal [70] proposed in 2008 a Synchronized MAC protocol for multi-hop CRNs, the SYN-MAC, which avoids the exploitation of a dedicated CC. Each SU is equipped with two radios, one dedicated to control signal exchange and the other for data transmission. In SYN-MAC the time is divided into time slots and each slot represents a data channel. Then, the control signal exchange occurs in the channels represented by the slots, while data transfer can occur in any channel that is found idle for both sender and receiver. It means that the control signaling is similar to slow frequency hopping. During the network initialization state, at the beginning of each time slot, SUs broadcast a beacon in all available channels to exchange information about the channels set and to synchronize their radios. The channel advertisement technique adopted by the SYN-MAC avoids the multichannel hidden terminal problem and keep track of primary user's presence. Interested SUs respond with their own list of available channels, and further communications will be made in one of those selected channels. The SYN-MAC operation mode is the following: a) when a SU wants to transfer data, it first chooses one of the channels that it shares with the receiver; b) then it waits for the time slot representing the selected channel, and starts a negotiation process using a backoff mechanism similar to IEEE 802.11 DCF; c) if it wins the contention then it starts the data exchange in that channel.

SYN-MAC protocol has the advantage of not using a dedicated common CC, and due to the dedicated listening is also able to avoid the multichannel hidden terminal problem. However, this protocol does not offer fast protection of PUs since their detection is notified to neighbors only in specific slots.

Opportunistic-MAC

Su and Zhang [74, 99] proposed an Opportunistic cognitive MAC protocol that permits the co-existence of a CRN with a PRN. SUs are equipped with two transceivers, one for a dedicated common CC and the other that can be dynamically tuned to any chosen available channel. The time structure of the Opportunistic MAC is slotted for data transmission over the licensed channels, while the common CC operation is partly slotted, followed by a random access negotiation phase. Then, the Opportunistic MAC protocol has an hybrid channel access scheme. The authors assume that the common CC is located in a licensed spectrum for SUs, and so it is always available. The two phases of the common CC are the reporting phase and the negotiation phase.

The reporting phase is divided into mini-slots, with many slots as channels to sense. At the beginning of each time slot, the SU senses only one channel: if the channel is sensed idle, the SU sends a beacon over the common CC during the mini-slot related to that channel; otherwise no beacons will be sent if no PU is detected. This means that beacons are only used to inform the neighbors about the PU activity, and not for synchronization purposes. To ensure that all the channels are sensed, each SU independently chooses a channel with equal probability. During the negotiation phase, SUs negotiate the channel for data communication via contention-based schemes and two-way handshakes (RTS/CTS). However, since the sensed channel is randomly chosen, the neighboring nodes do not have a priori knowledge of this choice and do not silence their own transmissions to improve the sensing accuracy. This fact increases the probability of false alarm and decreases the opportunistic spectrum utilization.

Opportunistic-MAC protocol adopts two different sensing policies: the Random Sensing Policy (RSP) and the Negotiating Sensing Policy (NSP). Using the RSP, at the beginning of each time slot, SUs randomly sense one of the licensed channel. If the channel is sensed idle, the SU transmits a beacon in the corresponding mini-slot, as previously mentioned. With this approach, and because all the channels have equal probability of being sensed, higher number of SUs lead to higher probability of sensing all channels. In the NSP the idea is to let SUs know which channels have been sensed by their neighbors. During the negotiation phase, SUs include the information about the channels sensed in the RTS/CTS messages. Then, in the next time slot, SUs randomly pick a new channel from the set of channels for which the SU has not received any beacon during the reporting phase, excluding those which have been reported as sensed. This sensing policy guarantees that the number of sensed channels increases with the time.

SCA-MAC

Hsu et al. [100] proposed a Statical Channel Allocation (SCA) MAC protocol for CRNs. The proposed protocol allows SUs to do opportunistic access to any continuous part of the spectrum, licensed or unlicensed. Like other CR MAC protocols, SUs collects information from the spectrum and generate statistics about the spectrum usage: based on this statistics, the authors can successfully increase the probability of transmission and reduce the probability of interference to PUs. SCA-MAC operation principle consists of three phases.

During the first phase, used for environment sensing and learning, SUs sense the spectrum continuously. In the second phase, the sender and the receiver will determine which channel should be used to communicate, by exchanging, in the common CC, Control RTS and Control CTS (CCTS) messages containing the information about the available channels. These messages are exchanged via contention-based mechanism, e.g. exponential backoff. The main contribution of SCA-MAC is the set of three parameters and rules considered to perform data channel selection: a) optimum transmission range, which specifies the proper spectrum usage that a node would search for transmission opportunities, i.e., SUs can change the range dynamically depending on the spectrum opportunity; b) maximum channel aggregation, which means the SUs can aggregate channels to transmit, depending on the available bandwidth; c) closest possible opening, which means that if more than one channel is necessary, then SUs will choose the channel that requires the most shortest waiting time. Once the pair successfully agree about the set of data channels to use, SUs can start to transmit after a period of contention to avoid collisions with other SUs that could have selected the same or overlapped data channels.

SCA-MAC's protocol description is not clear about the common CC synchronization. Since it has two transceivers, one dedicated to the common CC, it is not clear why the authors have implemented a contention period before sending data messages. SCA-MAC presents another drawback: the computational complexity for determining the successful transmission probabilities quickly increase with the number of licensed channels.

COMAC

In 2009, Salameh et al. [63] proposed a contention based CR and asynchronous MAC protocol that tries to satisfy QoS constraints by limiting the number of used channels per SU. The COgnititive MAC (COMAC) assumes that SUs are equipped with two or more transceivers, in which one of them is tuned in an always available CC, that can be dedicated to the CRN or in the unlicensed ISM band. The COMAC exploits the underlay spectrum access, i.e. allows SUs to use the licensed band while limiting the generated interference to PUs. Each SU determines the maximum transmission power over various channels such that the PU receiver outage probability is kept below a predefined value. For that, a stochastic model is proposed to aggregate interference within the PRN and for the primary to secondary interference.

Regarding the COMAC data channel negotiation, each SU maintains a list of its locally available channels (denoted by LAC), which is the set of channels that are not being used by any of its neighbor. This list is continuously updated through the exchange of control packets. When two SUs want to communicate, the pair exchanges their LACs over the CC, and the receiver selects the channels to support the communication according to the spectrum state information: the maximum allowable transmission power for channel and the requested data rate. According to these parameters, when the receiver gets the sender's RTS packet it will compare the sender LAC with its own, and select the common available channels. Then, the common available channels whose received Signal to Interference plus Noise Ratio (SINR) is below a fixed threshold are removed from the list, and the remaining channels are sorted in descending order of their data rate. Finally, selected channels are chosen from the sorted list until the request data rate is satisfied or the list is exhausted. If the data rate condition is satisfied the receiver transmits to the sender a CTS message including the list of the selected channels. Otherwise, the receiver will not answer, and the sender will reschedule a new RTS packet transmission.

While COMAC permits parallel transmissions to take place in the same neighborhood, the multichannel hidden terminal problem is not completely solved due to possible collisions on the common CC within neighbors SUs transmissions. Furthermore, as previously mentioned, the interference temperature implementation typically results in poor performance, leading the authorities (FCC) to lay down this concept.

MMAC-CR

Timmers et al. [61] presented in 2010 an energy-efficient distributed multichannel MAC protocol for CR networks (MMAC-CR). By allowing the SUs to enter a dozen state when no communication is taking place, the MMAC-CR protocol achieves energy-efficient communication. Furthermore, the MMAC-CR also presents a two-stage sensing algorithm: i) a low-power inaccurate scan and ii) a high-power accurate scan. By using the low-power scan for periodically scheduled scans and only using the accurate scan when the low-power scan detects a change, the sensing algorithm also achieves energy-efficient operation. MMAC-CR uses a dedicated CCC free of PUs to perform network synchronization and common information exchange.

In MMAC-CR each SU maintains two data structures: one is called the Spectral Image of PUs (SIP) list, and the other is denominated as SUs Channel Load (SCL) vector. The SIP vector contains the SUs' local view on the spectrum, and each channel has three different entries, depending on the results of the sensing: a) no PUs active in the channel, the entry is marked with 0; b) if there are PUs active in the channel then the entry is marked with 1; c) if the spectral image of the channel is uncertain, then it is marked with a 2. When a SU joins the CRN it performs a fast scan for every channel and the outcome of these scans is stored in the SIP vector. The SIP vector will be constantly updated with the scan results. On the other hand, the SCL list is used for choosing the communication channel. It contains the expected load of the communication on each channel. When a SU wishes to transmit, it picks the spectral opportunity with the lowest SCL entry value.

Regarding the operation mode, the MMAC-CR time structure is divided into fixed-lengths beacon intervals with two distinguish phases: a) the Ad-hoc Traffic Indication Message (ATIM) period and b) the DATA phase. During the ATIM phase, SUs participate in the following:

- **Synchronization:** Time synchronization is achieved using the IEEE 802.11 Timer Synchronization Function (TSF) [101]. After the start of the ATIM window, SUs compete to transmit their beacon frame, which is carrying their local time, on the CC. Following the TSF, a node only updates its time if the time carried in a received beacon frame is faster than its own local time. When a node receives a beacon frame, it drops its own beacon frame.

- **Fast Sensing:** The SUs randomly select one of the channels to perform a fast scan. This scan is used to update the SIP value of the scanned channel. It will either confirm the present SIP value or not. If it is confirmed, then no further action is taken. Otherwise, the SIP value is updated to 2, which stands for uncertainty.

- **Sharing Sensing Results:** In this phase, the channel is divided into many minislots as is the number of sensed channels. Then, after using a synchronization packet (Scan Result Packet (SRP)), each SU transmits a busy signal in the corresponding minislot for every channel where the SIP is not 0. If a minislot is sensed busy, then the corresponding channel is excluded for communication. Every node that either received or transmitted the SRP frame now knows the spectral opportunities. This procedure becomes an efficient physical OR implementation, since it avoids all MAC overheads as compared with the case where every node has to transmit its own SIP in a packet.

- **DATA Channel Negotiation:** Data channel negotiation is indicated with two-way handshakes. SUs that have buffered packets transmits ATIM frames on the CC during the ATIM window. In the ATIM frame, a SU will insert the channel with the lowest SCL for data transmission, together with its queue status. Each SU that overhears the ATIM frame accordingly updates its SCL vector. If the receiving SU agrees with the selected channel, then it replies with an ATIM-ACK frame. After the ATIM window, the SUs that have exchanged ATIM frames stay awake until they have completed data exchange. The SUs that neither transmitted nor received ATIM frames enter a doze state until the next beacon interval.

The DATA window is not only used for data exchange, but also to perform fine sensing. SUs that have set he SIP to uncertain will perform a fine scan on the corresponding channel, which is now free from SUs' communications. This can be done in parallel with a communication in another channel, because MMAC-CR was developed for multi-transceiver SUs. Then, the SIP value of this channel is updated with the result of the fine scan. The data transmission follows the normal IEEE 802.11 DCF procedure with RTS/CTS exchange. Additionally, SUs that ended the data transmission before the end of the DATA window are allowed to ender a doze state to save energy.

Besides the energy consumption awareness, the other contribution of this proposal is its evaluation using realistic power models of the SDR: the benefits of MMAC-CR are presented in a more realistic setting compared to other proposals. However, MMAC-CR reliability strongly depends on the quality of the CC, mainly during the procedure of sharing the sensing results.

CREAM-MAC

In 2011, Zhang and Su [65] proposed the Cognitive Radio-EnAbled Multi-Channel MAC (CREAM-MAC), a CR PHY-MAC framework that integrates the cooperative sequential spectrum sensing at PHY layer and packet scheduling at MAC sublayer over the wireless DSA networks. The CREAM-MAC protocol employs a CCC as the rendezvous channel where SUs can exchange the control packets for data channel communication: the authors state that the CCC can be either dedicated (located in the ISM band) or dynamically configurable. However, since the authors do not explain how the CCC is dynamically selected, we decided to classify it as a dedicated CCC. Under the CREAM-MAC protocol, each SU is equipped with a single half-duplex transceiver and multiple sensors. These sensors are only used to perform spectrum sensing, increasing the cost and implementation complexity of this protocol.

It is also important to notice that, this recent protocol, continues to consider the interference temperature limit by allowing SUs to transmit even with the presence of PUs, keeping the level of interference to a minimum tolerable. For that, each SU transmission can not exceed the maximum tolerable interference period, denoted by T_d^{max} and computed based on the assumption that the activity of PUs is modelled as an ON/OFF model, with birth and death rates following exponential distributions. Regarding the data channel negotiation, the

CREAM-MAC protocol uses two pairs of control packets, exchanged in the dedicated CCC: the already and often referred RTS/CTS and the Channel-State-Transmitter (CTS)/Channel-State-Receiver (CSR). The first pair is exchanged, after a period of contention, for control channel reservation and to solve the hidden terminal problem. The second pair aims to synchronize the vacant channel information between sender and receiver, by exchanging their lists of available channels. Once the negotiation is complete, sender and receiver can exploit all the common available free data channels.

The authors admit the use of the OFDM access technique, allowing SUs to aggregate discontinuous channels, hence increasing the capacity of the communication channel. To prevent the multichannel hidden terminal problem, CREAM-MAC imposed a rule stating that SUs which just finished the data transmission can only select the same channel group which they just released within a waiting interval of duration T_d^{max}. Although CREAM-MAC is equipped with multiple sensing sensors, the authors decided to implement a new distributed spectrum sensing scheme, based on the already studied energy detector sensing scheme (Subsection 2.2.). This is called *two-threshold based sequential sensing policy*, where the idea is to decrease the probability of false alarm while confining the missed detection probability to a predefined value[5]. The basic idea of the proposed sensing policy is to collect PUs' signal samples sequentially in multiple uncorrelated sensing rounds to enhance the decision process. Two thresholds β_1 and β_2, with $\beta_1 < \beta_2$, are used: when the energy detector output is larger than β_2, CREAM-MAC claims the presence of a PU; when the output is lower than β_1, CREAM-MAC assumes that PUs are inactive; otherwise, if the detector output is between β_1 and β_2 then the SUs need to take one more sensing round to collect more PU's signal samples.

The authors presented the CREAM-MAC with an extensive theoretical analysis and simulation about its performance under two distinct scenarios: with and without a saturated network, which has a strong influence in the performance of the dedicated CCC. In the saturated network scenario, the CREAM-MAC protocol throughput increases as the number os sensors also increases. However, there is a point when the number of free channels is so high that the number of control packets will also be too high for the capacity of the CCC, resulting in a saturated dedicated CCC. This is one of the main drawbacks of this proposal, shared by all the CR MAC protocols with a dedicated CCC. The other disadvantage is the existence of several sensors, increasing the cost and complexity of this proposal.

Regarding the dynamically configurable CC, it can be available either locally to a neighborhood or globally to all SUs in the CRN. Usually, clusters based CR MAC protocols, as is the case of [85], select a local CC because it is easier to reconfigure a local CC than a global one in case PUs reclaim the current CC. Furthermore, the effects of CC saturation or attack are less severe in the case of a local CC, because only a part of the network is affected by these situations. On the other hand, it is difficult to achieve global time synchronization and a good network coordination in a CRahN with local CCs. Moreover, a sensing period without transmissions from SUs can not be ensured without global time synchronization.

[5]Based on the traditional single-threshold energy-detection method, decreasing the missed detection probability is equivalent to increase the false alarm probability, and consequently decreasing the spectrum access opportunities [102].

CogMesh

The next two CR MAC protocols adopt dynamically configurable CCs. Chen et al. [85] presented CogMesh protocol, a cluster based MAC scheme for CRNs. The CogMesh can be seen as a multichannel multi-access network, in which the available channels of a user changes very often during the life time of the node, making it different from the traditional mesh [103] or *ad hoc* networks. The topology management of the CogMesh is affected by two main factors: i) a CC may not be always available for the whole networks, and ii) the topology of the network changes over time according to the presence of PUs and SUs.

In the CogMesh, the spectrum access is controlled by the leader of the group, called clusterhead. The coordination channels of the cluster is called as the master channel, equivalent of a common CC. The master channel is divided into MAC superframes composed by a random access period for control message exchange and a guaranteed access period for data transmission, making it an hybrid MAC scheme. The guaranteed access period is divided into a Beacon Period (BP), a Neighborhood Broadcasting Period (NBP) divided into mini-slots, and a Data Period (DP). During the BP, a beacon containing the time synchronization, control information and resource allocation information of the cluster is sent by the clusterhead to begin the MAC superframe. During the NBP each SU broadcasts its one hop neighbor list (each neighbor's identity and its corresponding channel list) in the corresponding mini-slot. Finally, during the DP SUs use TDMA to transmit their data over the data channels.

Each group uses its own private spreading code for the data channels. The intra-cluster and inter-cluster communications are carried out during the private random access period and a public random access period, respectively. For inter-cluster communication, the clusterhead selects gateway nodes which can listen to adjacent clusters. The gateway nodes of two clusters can be either overlapping (shared gateway) or non-overlapping (non-shared gateway). The shared gateway node contacts the clusterhead during the private random access period, whereas the non-shared gateways contact during the public random access period. An external SU may contact the clusterhead during the public random access period. SUs use slotted ALOHA in the random access period. The MAC superframe also includes a Quiet Period (QP) during which the nodes perform cooperative sensing. The size of the cluster is limited to a predefined value for efficient operation.

Regarding the dynamically configurable master channels, there are several situations that it needs to be changed: the current master channel is too crowd because several neighbor clusters share the same channel; the channel quality deteriorates due to the increasing interference; there are many 1-hop neighbors on other channels that can be merged into this cluster; and PU occupation. In these situations, the clusterhead is responsible for reconfigure a new dedicated local CC. For that, it informs its members about the operation and, after receiving the confirmation from all members, the clusterhead starts the beacon in the new master channel from the next superframe. The neighbor clusters of this cluster are informed after the shifting.

The CogMesh addresses issues like node discovery, spectrum access, inter-cluster and intra-cluster communications. However, the nature of the PU spectrum has a direct impact on the CogMesh network. The clusters in the CogMesh use private spreading codes to overcome the multichannel hidden terminal problem. However, it is not clear how the spreading codes are selected. This problem may be more expressive when the number of clusters keeps varying. Moreover, there is a moderated/heavy network reconfiguration traffic overhead to maintain the clusters in the secondary network.

DOSS

Ma et al. [83] proposed in 2005, the Dynamic Open Spectrum Sharing (DOSS) MAC protocol, a peculiar MAC scheme that requires three transceivers: one each for monitoring the dynamically configurable CC, data channels and the busy tone channels. The DOSS protocol applies the concept of using busy tones to tackle the multichannel hidden terminal problem: it exclusively allocates a narrow band for the busy tones and, for each used data channel a busy tone will be emitted in the frequency band associated with that channel. The DOSS protocol consists of five steps:

- **PU detection:** since SUs can only use the licensed spectrum when there are no PUs, SUs will have to perform spectrum sensing to identify spectrum opportunities.
- **Set-up of three frequency band/channels:** First, the bands used for data communication are determined, which could be a set of non-contiguous frequencies. Then, an out-of-band CC is chosen for control signaling. However, to alleviate the CC saturation, three techniques are proposed: limitation of the control traffic going through the CC; robust setup of the bandwidth ratio of the CC and data band, such that the CC do not become the bottleneck of the network; slow migration of the CC based on the traffic load over the current CC.
- **Spectrum mapping:** DOSS establishes a one-to-one mapping between the narrow band (low bit rate) busy tones and the wide band (high bit rate) data channels. By observing narrow band busy tones, a node knows all the data receiving activities in its neighborhood. The busy tone band has a frequency range $[f_l, f_u]$ and the data band is contained in $[F_l, F_u]$, as shown in Figure 8. Then, the receiver only needs to convert the spectrum over which it is receiving a busy tone and send the busy tone in order to inform other neighbors not to send.
- **Spectrum negotiation:** In the spectrum negotiation phase the sender sends a REQuest (REQ) packet with the available spectrum bands to the receiver SU. The receiver replies with a REQ ACKnowledge (REQ_ACK) packet containing the information of a mutually acceptable spectrum band. According to the spectrum mapping, it also issues the corresponding busy tone, telling its neighbors not to transmit on chosen data spectrum band. Upon receiving the REQ_ACK the receiver tunes its data transceiver into the negotiated band, issuing the corresponding busy tone.
- **Data transfer:** If a packet is correctly received, the receiver replies with a DATA_ACK packet and turns off the busy tone. By receiving the DATA_ACK packet, the sender knows the transmission was successful and also turns off the associated busy tone. Otherwise, in the case of a timeout, the transmitter will repeat the transmission.

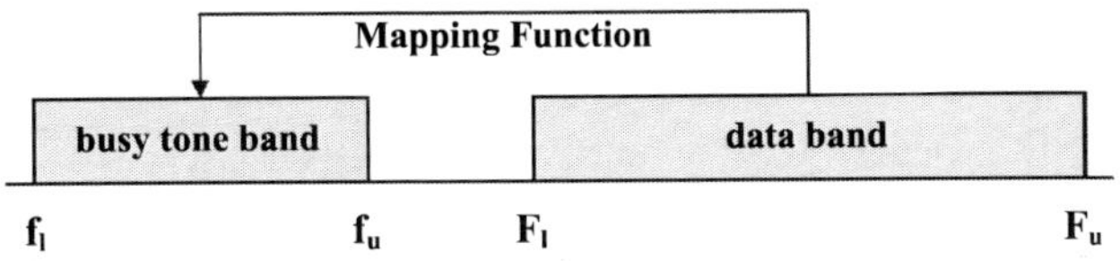

Figure 8. Spectrum Mapping in the DOSS MAC.

The DOSS scheme uses the concept of receiver initiated busy tones to mitigate the hidden and exposed multichannel terminal problems. However, several difficulties can arise when the busy tones come under the influence of PUs. For instance, a data channel might be available but its corresponding busy tone may become temporarily occupied by a PU. Besides, the device cost and complexity is increased due to the use of three transceivers.

3.4. CR MAC Standards

Although CR is still an emerging technology, a few standardization initiatives have already been approved to regulate the use of this technology. The first CR standards were designed to exploit the analogue television white spaces originated by the appearance of modern digital TV systems, as is the case of IEEE 802.22, 802.11af and ECMA-392. Others proposals try to explore unused satellite frequency bands, which include IEEE 802.11y and IEEE 802.11h protocols, or even to facilitate the coexistence of different wireless technologies, e.g. IEEE 802.16h.

3.4.1. IEEE 802.22 - Wireless Regional Area Networks

IEEE 802.22 or Wireless Regional Area Networks (WRAN) [79] was the first centralized CR PHY-MAC layer standard developed to exploit vacant TV spectrum bands (UHF and VHF bands). IEEE 802.22 standard divides the CRN into cells, each one composed by a BS, covering an area of radius spanning from 30 km to 100 km, and several WRAN end users, here denoted as Customer Premise Equipments (CPEs). The WRAN is designed to provide throughput of 1.5 Mb/s per CPE in the downstream, using Time Division Multiplexing (TDM), and 384 kb/s in the upstream using TDMA. OFDM modulation is used to overcome possibly excessive delay [104]. In addition, the standard provides PU protection including spectrum sensing and geolocation database for PU-SU coexistence, and also supports self-coexistence between WRANs via the Coexistence Beacon Protocol (CBP).

The WRAN standard specifies a time slotted operation, where each time-frame is denoted as superframe. The superframe is composed of multiple MAC frames preceded by the respective preamble. A preamble in the beginning of each superframe is used to synchronize the CPEs with the BS, followed by a Superframe Control Header (SCH) that is used to inform the SUs about the current available channels and supported bandwidths.

Each MAC frame comprises a DownStream (DS) frame and an UpStream (US) frame. The DS subframe contains a single packet burst from a given CPE, while the US subframe has multiple packets bursts, each one transmitted by different CPEs. The remaining fields have the following utility: in the DS subframe, the preamble deals with synchronization and channel estimation, the Frame Control Header (FCH) contains the size of the DS- and US-MAP fields, and the DS- and US-MAPs give the scheduling information for user bursts. In the US subframe, the Urgent Coexistence Situation (UCS) notification informs about recently detected PUs, while the other fields are used to derive the distance from the BS, and the individual bandwidth requests.

Regarding the protection to PUs and spectrum awareness, the IEEE 802.22 standard offers two methods to cope with these important tasks. In the first method, the knowledge of SUs location combined with a database of licensed transmitters can be used to determine which channels are locally available. This method is usually referred as geolocation/database. The other method, a Two-Stage Sensing (TSS) mechanism, consists of observing the spectrum to identify which channels are occupied by licensed transmissions. TSS's first stage, denominated Fast Sensing, is done at the rate of 1 $ms/channel$, and the sensing results are used to decide if a subsequent fine sensing stage is needed. This means that the sensing is completed quickly and with low accuracy. The second sensing stage, denoted as Fine Sensing, is performed on-demand at the rate of 25 $ms/channel$, which allows SUs to meet the strict QoS requirements by decreasing the rate of false alarm and misdetection. By using a distributed sensing mechanism, the BS delegates groups of CPEs to scan different channels, where the sensing information will be sent back to the BS.

Just like in the others CRNs, in the case of a PU appearance SUs will have to interrupt their communication and restart it in another channel. In order to minimize the effects of this interruption, the standard protocol offers a mechanism denoted as the Incumbent Detection Recovery Protocol (IDRP), that enables the CRN to restore normal operation with minimal performance degradation. In IRDP, backup channels are used to restore the communication in case a channel needs to be released after PU appearance. These channels, shared by the CPEs and BS, are kept in a list and used whenever a CPE looks for a BS during the recovery process.

IEEE 802.22 standard also allows the coexistence with other IEEE 802.22 CRNs through the Coexistence Beacon Protocol (CBP). The communication between cells is undertaken by the CBP beacons, which carry information about the cells and the DS/UP bandwidth allocations for the CPEs. These packets are exchanged using contention-based schemes during the periods labeled as self-coexistence windows.

3.4.2. ECMA-392

In December of 2009 the European Computer Manufacturers Association (ECMA) presented the first CR standard for personal/portable devices to exploit the television white spaces: the ECMA-392 [105]. In fact, this standard was started by the Cognitive Networking Alliance (CogNeA), and later on a draft specification was submitted to ECMA Technical Committee number 48 (related with High Rate Wireless Communications) [106]. The standard specifies PHY and MAC layers with the following characteristics: flexible network formation, support for real-time multimedia traffic and adaptation to different regulatory requirements.

3.4.3. IEEE 802.11af

IEEE is also developing a standard to exploit the television white spaces in a more domestic scenario (personal and portable devices). The IEEE 802.11af[6], or White-fi [107], aims to adapt the current IEEE 802.11 [108] to make use of the TV white spaces, just like the IEEE 802.22, however with a smaller range (20-30 m). The upcoming IEEE 802.11af standard has the following goal: the time to vacate the channel after PUs detection should be about $2s$ with 90% of probability of detection and 10% of probability of false alarm, at a SNR level as low as -20dB with a geolocation accuracy of about 50 m. It is also expected that IEEE 802.11af provides much higher speed and wider coverage than current IEEE 802.11, due to better propagation characteristics of the VHF/UHF bands [9].

3.4.4. IEEE 802.11h

IEEE 802.11h-2003, or just IEEE 802.11h, refers to the amendment added to the IEEE 802.11 standard [108] for Spectrum and Transmit Power Management Extensions[7]. The original purpose of IEEE 802.11h was to extend Wireless Local Area Network (WLAN) operation in Europe to the 5 GHz band, where WLAN devices (unlicensed users) need Dynamic Frequency Selection (DFS) and Transmit Power Control (TPC) to coexist with the band's primary users-radar and satellite systems (licensed users) [109]. According to European regulations, when a WLAN device in the 5 GHz band detects a radar signal on its current operational frequency channel, it must switch to another channel and when a WLAN device detects a satellite signal, it must limit its transmit power to 3 dB below the regulatory maximum [110, 111]. To meet these requirements, IEEE 802.11h defines DFS and TPC mechanisms on top of IEEE 802.11a PHY and MAC layers.

An IEEE 802.11h WLAN uses DFS for dynamically switching the operational frequency channel. Besides fulfilling the regulatory requirements, the change in the operational frequency channel can also occur when the condition of its current channel becomes poor due to the interference from neighboring devices. In an infrastructured scenario, the BS determines when to switch and which channel to switch to. For that, the BS monitors the status of the current channel, and it might also request that other stations measure and report the status of the current channel, or also those at other frequencies. There are three types of measurements: a) the basic measurement that determines whether a non-802.11 OFDM signal, and unidentified signal, or a radar signal is using the measured channel; b) the Clear-Channel Assessment (CCA), that measures the fraction duration of the channel's busy period during the total measurement interval; and c) the Received Power Indication (RPI), responsible for measuring the histogram of the quantized measures of the received energy power levels as seen at the antenna connector during the measurement interval. Based on these measurements, the BS decides if a new operational channel will be used.

The TPC mechanism was implemented to fulfill the following requirements: a) to improve the near-far problem[8]; b) to minimize interference to and from other cells/clusters;

[6]Project Authorization Request (PAR) was approved in the end of 2009.

[7]Actually integrated in the IEEE 802.11-2007 standard.

[8]The near-far problem occurs when all terminals transmit at the same power level, and the strongest received signals, usually those emitted by closer terminals, will be captured by the receiver. Then, it is impossible to detect the weak signals transmitted by distant terminals. This problem is characteristic of Code Division

and c) to improve the system performance on fading channels, by compensating for fading dips, i.e. samples with lower signal strength. The standard also specifies two TPC-related functions: i) the BS advertises the regulatory and local maximum-power level for the current frequency channel, by using TPC-Beacon/TPC-Probe beacons; ii) a reporting mechanism (TPC-Request/TPC-Report) to inform the BS about the power used to transmit the frame to the BS (TPC-Report), and also with the ratio of the received signal (TPC-Request) strength to the minimum desired by the station.

By using DSF and TPC mechanisms, it is possible to improve the performance of WLAN devices operating in the 5 GHz band in the Europe and simultaneously reduce the energy consumed by each terminal. By estimating the link quality between itself and the receiver, a terminal is able to determine the proper transmission power level, reducing the excessive energy wasted during its transmissions.

3.4.5. IEEE 802.11y

In 2005, FCC opened 3.65-3.7 GHz band, originally reserved for fixed satellite service networks, for public use. However, a regulatory policy was needed. Then, later in 2007, IEEE started to develop an amendment to the IEEE 802.11-2007 [108] that enables high powered Wi-Fi terminals to operate on a co-primary basis in the 3.65-3.7 GHz band in the United States, except when near a grandfathered satellite earth station. This amendment was approved for publication in 2008, and it is known as IEEE 802.11y [113] or "The US 3650 MHz rules".

This standard allows for registered station to operate at much high power than traditional Wi-Fi terminals (up to 20 W). Hence, by combining the increased power transmission with several new key elements and enhancements made to the IEEE 802.11-2007 MAC timing, it will be possible to develop IEEE 802.11 terminals operating with a distance of 5 km or more. As for the new key elements, IEEE 802.11y introduces two categories of stations:

- **Enabling Stations:** Enabling stations are high-powered fixed stations with authority to control when and how a dependent station can operate. An enabling station communicates an initial enabling signal to its dependent stations over the air, always including its location information. The GPS coordinates and altitude information of each enabling station is also registered in a public database to avoid interference with other enabling stations.

- **Dependent Stations:** Dependent stations are devices in the network that are not registered, but instead receive authorization to transmit from a registered enabling station. If a dependent stations does not receive the periodic enabling beacons, the dependent station should suspend the transmissions until it is re-enabled. A dependent station may be fixed or mobile.

Regarding the new enhancements, IEEE 802.11y adds three new concepts to 802.11-2007:

- **Dynamic Station Enablement (DSE):** Dynamic Station Enablement is the process by which an enabling station grants permission and dictates operational procedures

Multiple Access (CDMA) networks because all the transmitters share the same frequencies and transmission times[112].

to dependent stations. Beyond addressing the regulatory requirements for the 3.65 GHz frequency band, DSE offers the promise of other channel management and coordination benefits: for example, since the enabling station is not required to serve as the BS for each of its dependent stations, DSE can reduce the likelihood of a dependent station contributing to radio interference by allowing the dependent station to complete the enablement process via a geographically closer BS.

- **Contention Based Protocol Incorporating Regulatory Class Information:** Enhancements have been made to the carrier sensing and energy detection mechanisms of IEEE 802.11 in order to meet the FCC's definition of a contention based protocol: IEEE 802.11y devices can sense both IEEE 802.11 and non-802.11 devices and identify available spectrum as small as 5 MHz. Also, IEEE 802.11y BSs beacons identify the country and the regulatory domain for their physical location. By incorporating both channel use and regulatory class information, IEEE 802.11y devices can identify available channels and adjust operating parameters to the laws of the country in which the access point resides.

- **Extended Channel Switch Announcement (ECSA):** ECSA is used to coordinate a move from one channel to another with less contention or to change channel bandwidth. Specifically, an enabling station can identify the channel with the least aggregate interference to all of the stations that are connected to it on a completely dynamic basis. This capability ensures the best SNR and lowest power levels possible. ECSA is also being applied to IEEE 802.11n.

3.4.6. IEEE 802.16h - *The Cognitive WiMAX*

IEEE 802.16 [114], also denoted as Worldwide Interoperability for Microwave Access (WiMAX), is one of the two most principal candidates of 4G wireless communication technology, which enables mobile wireless devices to receive high-speed Internet service[9]. Although WiMAX is not the most successful 4G standard nowadays, IEEE is still improving it with new features: one of them, the amendment 2, introduces mechanisms for license-exempt operation, i.e. defines a set of cognitive radio capabilities for WiMAX networks, and is denoted by IEEE 802.16h [116].

In an real scenario a WiMAX network may coexist with licensed users (denoted in [116] as specific spectrum users) and other unlicensed users (denoted as non specific spectrum users) sharing the same frequency band. In such case, the interference that the WiMAX system may cause to each of the users can be different. IEEE 802.16h defines three possible levels of interference: a) acceptable interference, admissible in both licensed and unlicensed users, which does not cause degradation in the receiver performance for a given choice of modulation and/or coding; b) harmful interference, not admissible in licensed links, refers to strong interference that decreases the link performance; and c) destructive interference, when the receiver is not capable of decoding the receiver signal for any available modulation at the transmitter.

IEEE 802.16h proposes a PHY-MAC cross-layer framework characterized by two different profiles: the first one provides uncoordinated coexistence mechanisms

[9]The 3GPP Long Term Evolution (LTE) [115] is the other 4G technology frontrunner.

(WirelessMAN-UCP), without requiring much interaction among the different systems operating below 11 GHz in license-exempt bands, while the second profile offers coordinated coexistence mechanisms (WirelessMAN-CX), which addresses the required coordination of neighboring systems in order to reduce the interference among them [116].

In the uncoordinated profile, IEEE 802.16h allows distributed architectures for the radio resource management within the network formed by one IEEE 802.16 BS and its associated subordinated terminals. Each BS has a Distributed Radio Resource Management entity to execute the spectrum sharing policies and to build up a database for sharing information related to actual and intended future usage of radio spectrum. This database can be recovered from a master entity with the required information, or from different devices, e.g. using the GPS or informed by the operator.

On the other hand, with the coordinated coexistence mechanism, multiple secondary networks coexisting in the same region can collaborate in order to coordinate their transmissions and build a neighbor relationship. The amendment considers three basic mechanisms for achieving coexistence: a) MAC Frame Synchronization, for separating transmissions and enabling operation in synchronized zones; b) Dynamic Channel Selection (DCS) and Adaptive Channel Selection (ACS) for finding a less interfered or less used frequency; and c) Separation of the remaining interference in the time domain, by using a Coexistence Frame, coordinated scheduling, and a fairness approach, thus allowing the usage of a frequency channel by more than one system. In the coordinated profile a Coexistence Control Channel is used for inter-network coordination.

3.4.7. IEEE P1900

The IEEE 1900 projects form a new standard series within the IEEE with the objective of develop standards in the area of DSA, CR, interference management, coordination of wireless systems and advanced spectrum management. The IEEE 1900, or Dynamic Spectrum Access Networks (DySPAN) standards committee, was established in March 2005 jointly by the IEEE Communications Society and IEEE Electromagnetic Compatibility Society, and in 2007 it was placed under the Standard Coordinator Committee 41 (SCC 41).

The IEEE DySPAN standard committee is now divided into two already approved standards and five current Working Groups (WG) (IEEE 1900.x), each one responsible for addressing different CR and spectrum management issues. The IEEE 1900.1 "Standard Definitions and Concepts for Dynamic Spectrum Access: Technology Relating to Emerging Wireless Networks, System Functionality, and Spectrum Management" [117] was approved by the IEEE Standards Board in September 2008 and the IEEE 1900.2 "Recommended Practice for the Analysis of In-Band and Adjacent-Band Interference and Coexistence Between Radio Systems" [117] was completed in March 2008. Regarding the Working Groups, let us focus mainly in IEEE 1900.4 WG, which handles the "Architecture and Enablers for Optimized Radio and Spectrum Resource Usage" and IEEE 1900.6 WG, that deals with "Spectrum Sensing Interfaces and Data Structures for Dynamic Spectrum Access and other Advanced Radio Communication Systems".

As the title implies, the IEEE 1900.4 WG has aimed at standardizing architectures and functions for distributed decision making in order to optimize radio resource usage. IEEE 1900.4 is intended to apply to a heterogenous wireless environment, which may include multiple operators, multiple Radio Access Technologies (RATs), multiple Radio Access Networks (RANs), and multiple terminals. As the main subject of optimization, like other DSA standards, advanced spectrum management capabilities are considered in IEEE 1900.4: as an example, such a capability is that the assignment of spectrum to RANs can be dynamically changed, where "spectrum assignment" may be characterized as a carrier frequency, a signal bandwidth, among others. Terminals to which IEEE 1900.4 is applicable are reconfigurable and, for backward compatibility, IEEE 1900.4 does not rule out legacy terminals coexisting in the network. This way, a smooth progression to adoption of the standard is ensured.

IEEE 1900.4 categorizes the entities and interfaces of the management system. The two key management entities are the Network Reconfiguration Manager (NRM), responsible for deriving the radio resource usage constraints and evaluating them in order to meet a global objective, such as, global network power minimization or load balancing, and the Terminal Reconfiguration Manager (TRM) that manages the terminal within the framework defined by the NRM. As for the interfaces, the most important one is the interface between the NRM and TRM, which is a logical communication channel denoted as Radio Enable (RE). This RE may be mapped into: a) one or several legacy RANs already being used for data transmission (in-band solution) or b) one or several RANs or RATs dedicated to the RE (out-of-band solution) [118].

Regarding the IEEE 1900.6, the scope of this standard is to define the interfaces and structures enabling the information exchange between spectrum sensors and their clients, in radio communication systems [119]. By defining these entities, the standard facilitates interoperability between independently developed devices, allowing separate evolution of spectrum sensors and other system functions. The term clients includes the Cognitive Engines (CEs), Data Archive (DA) and Sensors (Ss). The CE is defined as the portion of the Cognitive Radio System (CRS) containing the policy-based control mechanism and the cognitive control mechanism, which must have knowledge about the current state and a set of attainable states, and also the cost associated with each state transition (cost of the reconfigurable radio platform). The DA is defined as a logical entity where sensing information obtained from spectrum sensors or other information sources, and regulatory and policy information are stored. The sensors can receive information from one or more sensors and forward it to a CE or the DA.

As for the interfaces, IEEE 1900.6 considers three different interfaces between spectrum sensors and their clients. The CE/DA-S interface is used for exchanging sensing information between a CE or DA and a sensor: the CE/DA-S interface is used in scenarios where a given CE or a DA obtains sensing information from one or several sensors, or a given sensor provides sensing information to one or several CEs or a DA. The S-S interface is used for exchanging sensing information between sensors; this is needed in cases where one sensor may not be able to obtain all required information, or in scenarios where sensors A and B exchange sensing information for collaborative or cooperative sensing. The CE-CE/DA interface is used for exchanging sensing information between CEs or between a CE and a DA: the CE-CE/DA interface is used in scenarios where two CEs exchange sensing information for collaborative or cooperative sensing. The CE-CE/DA interface is also used in scenarios where a CE obtains sensing information as well as policies and regulatory information from a DA.

Conclusion

This chapter provided an extensive and comprehensive literature review of the actual trends and common challenges in CRNs. Section 1. started by introducing the CR concept and its enabling technology. We have also identified the main actors of a CRN, the PUs and the SUs, and broadly divided the CRNs according to their architecture in infrastructured and infrastructureless (*ah hoc*). For each CRN architecture we have enumerated its characteristics, advantages and drawbacks, concluding that infrastructured CRNs can be more reliable to handle the SU's coordination, while infrastructureless CRNs can be more scalable and more flexible in terms of deployment, but require higher complexity in terms of synchronization. Finally, we have seen that in order to adapt to dynamic spectrum environment, the CR devices must adopt new spectrum-aware/access operations, forming the cognitive cycle which can be divided into four management functions: spectrum sensing, spectrum decision, spectrum sharing and spectrum mobility. Besides underlining the importance of each management function, we also compared the traditional PHY-MAC architecture with the CR PHY-MAC architecture, by analyzing the four CR management functions disposed in the classical OSI model perspective.

Section 2. described the main issues related with the spectrum sensing problem. We started by stressing the importance of performing an accurate spectrum sensing and several spectrum awareness methods were enumerated, including the geolocation/database, the use of periodic beacons and local radio spectrum monitoring. The last one was described in detail. Several challenges regarding the spectrum sensing operation including the choice of the sensing duration and frequency, how to avoid the hidden PU problem, or even the detection of spread spectrum PUs were also discussed. Moreover, three sensing techniques were briefly described. The most used one, the energy detector, is characterized by sensing the presence of licensed users based on the energy of the received signals. The major challenge of this technique is the computation of the interference threshold used to classify the spectrum state (idle/occupied). Other spectrum sensing techniques also described were the matched-filtering and the cyclostationarity-based sensing.

It is broadly assumed that the spectrum sensing task can only be performed by the PHY layer and adopting the methods aforementioned in this chapter. However, we believe that in some particular scenarios, e.g. in a stable radio environmental scenario, i.e. constant shadowing, fading and multi-path conditions, the MAC sublayer should be able to evaluate the PHY layer behavior by observing the probability of error during each transmission. By this way, the MAC sublayer can assist some tasks traditionally performed in the PHY layer, namely in situations when there exists high uncertainty about the presence/absence of PUs.

Section 3. overviews PHY-MAC cross-layered protocols for CRNs. The difference between classical MACs and CR MACs were enumerated and the most challenging implementation issues were discussed, including the configuration of the control channel, how to handle with the multichannel hidden terminal problem, or even the different QoS requirements for each PUs and SUs. Then, based on several intrinsic features such as the architecture, type of control channel, channel access method, among others, a CR MAC classification tree is presented. At least one CR MAC protocol is described for each classification, and its advantages and drawbacks are identified. Twenty three of the most important CR MAC protocols presented in the literature were described. Finally, several CR

MAC standards were also discussed. We have seen that most of them were proposed for centralized architectures, namely to explore the vacant frequencies left by the analog TV system, and also to cope with the coexistence of different wireless systems such as IEEE 802.11 (Wi-Fi) and IEEE 802.16 (WiMAX).

Although several PHY-MAC schemes have been proposed for distributed CRNs, requiring reliable mechanisms for exchanging control information, most of them assume that a CC is always available and free of PUs. However, this type of CC implementation represents a single point of failure, hence decreasing substantially the protocol's performance. Other CR MAC protocols proposed the use of a local CC to decrease the probability of CC's interference, but these solutions present low scalability and require tight synchronization. CR MAC protocols without CC proved to be computationally complex and do not guarantee full network access. Hence, we believe that is still possible to develop new CR MAC protocols adopting optimization techniques combined with channel fusion and considering other sources of information, such as channel's occupancy statistics and quality of service, which can outperform the existing ones.

Acknowledgments

This work was partially supported by COST Action IC0902 "Cognitive Radio and Networking for Cooperative Coexistence of Heterogeneous Wireless Networks", funded by the European Science Foundation, and by the Portuguese Science and Technology Foundation under the grant SFRH/BD/68367/2010 and the projects PTDC/EEA-TEL/115981/2009, PTDC/EEA-TEL/099074/2008, PEst-OE/EEI/LA0008/2011 and PEst-OE/EEI/UI0066/2011.

References

[1] Friedrich K. Jondral. Software-Defined Radio - Basics and Evolution to Cognitive Radio. *EURASIP Journal on Wireless Communication and Networking*, 3:275–283, 2005.

[2] J. Mitola III. *Cognitive radio-An integrated agent architecture for software-defined radio.* Ph.d. dissertation, Roy. Inst. Technol., Stockholm, Sweden, May 2000.

[3] Q. Zhao and B. M. Sadler. A Survey of Dynamic Spectrum Access. *Signal Processing Magazine, IEEE*, 24(3):79–89, May 2007.

[4] S. Haykin. Cognitive radio: brain-empowered wireless communications. *Selected Areas in Communications, IEEE Journal on*, 23(2):201–220, February 2005.

[5] III Mitola, J. Cognitive radio for flexible mobile multimedia communications. In *Mobile Multimedia Communications, 1999. IEEE International Workshop on*, pages 3–10, 1999.

[6] C. Cormio and K. R. Chowdhury. A survey on MAC protocols for cognitive radio networks. *Ad Hoc Networks*, 7(7):1315–1329, September 2009.

[7] Y.-C. Liang, K.-C. Chen, G. Y. Li, and P. Mahonen. Cognitive Radio Networking and Communications: An Overview. *Vehicular Technology, IEEE Transactions on*, 60(7):3386–3407, September 2011.

[8] I. F. Akyildiz, W.-Y. Lee, and K. R. Chowdhury. CRAHNs: Cognitive radio ad hoc networks. *Ad Hoc Networks*, 7(5):810–836, July 2009.

[9] Kang G. Shin, H. Kim, A. W. Min, and A. Kumar. Cognitive Radios for Dynamic Spectrum Access: From Concept to Reality. *IEEE Wireless Communications*, pages 64–74, December 2010.

[10] T. Yucek and H. Arslan. A Survey of Spectrum Sensing Algorithms for Cognitive Radio Applications. *IEEE Communications Surveys & Tutorials*, 11(1):116–130, 2009.

[11] Marko Höyhtyä, Atso Hekkala, Marcos Katz, and Aarne Mämmelä. Spectrum Awareness: Techniques and Challenges for Active Spectrum Sensing. In Frank H. P. Fitzek and Marcos D. Katz, editors, *Cognitive Wireless Networks*, pages 353–372. Springer Netherlands, 2007.

[12] S.J. Shellhammer, A.K. Sadek, and Wenyi Zhang. Technical challenges for cognitive radio in the TV white space spectrum. In *Information Theory and Applications Workshop, 2009*, pages 323–333, February 2009.

[13] M.J. Marcus. Unlicensed cognitive sharing of TV spectrum: the controversy at the Federal Communications Commission. *Communications Magazine, IEEE*, 43(5):24–25, May 2005.

[14] Y. Zhao, L. Morales, J. Gaeddert, K.K. Bae, J-S Um, and J.H. Reed. Applying Radio Environment Maps to Cognitive Wireless Regional Area Networks. In *New Frontiers in Dynamic Spectrum Access Networks, 2007. 2nd IEEE International Symposium on*, pages 115–118, April 2007.

[15] C.J.L. Arachchige, S. Venkatesan, and N. Mittal. An Asynchronous Neighbor Discovery Algorithm for Cognitive Radio Networks. In *New Frontiers in Dynamic Spectrum Access Networks, 2008. 3rd IEEE Symposium on*, pages 1–5, October 2008.

[16] FCC. Facilitating Opportunities for Flexible, Efficient and Reliable Spectrum Use Employing Cognitive Radio Technologies. ET Docket 03-322, December 2003.

[17] N.S. Shankar, C. Cordeiro, and K. Challapali. Spectrum agile radios: utilization and sensing architectures. In *New Frontiers in Dynamic Spectrum Access Networks, 2005. First IEEE International Symposium on*, pages 160–169, November 2005.

[18] Y. Yuan, P. Bahl, R. Chandra, P.A. Chou, J.I. Ferrell, T. Moscibroda, S. Narlanka, and Y. Wu. KNOWS: Cognitive Radio Networks Over White Spaces. In *New Frontiers in Dynamic Spectrum Access Networks, 2007. 2nd IEEE International Symposium on*, pages 416–427, April 2007.

[19] G. Ganesan and Y. Li. Cooperative spectrum sensing in cognitive radio networks. In *New Frontiers in Dynamic Spectrum Access Networks, 2005. First IEEE International Symposium on*, pages 137–143, November 2005.

[20] D. Cabric, S. M. Mishra, and R. Brodersen. Implementation issues in spectrum sensing for cognitive radios. In *IEEE Conference Record of the Thirty-Eighth Asilomar Conference on Signals, Systems and Computers*, pages 772–776, November 2004.

[21] J. G. Proakis. *Digital Communications*. McGraw-Hill, 4th edition, 2001.

[22] Y. Zhao, S. Mao, J. O Neel, and J.H. Reed. Performance Evaluation of Cognitive Radios: Metrics, Utility Functionsand Methodology. *Proceedings of the IEEE*, 97(4):642–659, April 2009.

[23] Ian F. Akyildiz, Won-Yeol Lee, Mehmet C. Vuran, and Shantidev Mohanty. NeXt generation/dynamic spectrum access/cognitive radio wireless networks: A survey. *Computer Networks*, 50(13):2127–2159, 2006.

[24] M. Subhedar and G. Birajdar. Spectrum Sensing Techniques in Cognitive Radio Networks: A Survey. *International Journal of Next-Generation Networks*, 3(2):37–51, 2011.

[25] B. Wild and K. Ramchandran. Detecting Primary Receivers for Cognitive Radio Applications. In *New Frontiers in Dynamic Spectrum Access Networks, 2005. First IEEE International Symposium on*, pages 124–130, November 2005.

[26] FCC. Establishment of an Interference Temperature Metric to Quantify and Manage Interference and to Expand Available Unlicensed Operation in Certain Fixed, Mobile and Satellite Frequency Bands. ET Docket 03-237, May 2007.

[27] J.J. Lehtomaki, M. Juntti, H. Saarnisaari, and S. Koivu. Threshold setting strategies for a quantized total power radiometer. *Signal Processing Letters, IEEE*, 12(11):796–799, November 2005.

[28] M. Luis, A. Furtado, R. Oliveira, R. Dinis, and L. Bernardo. Energy Sensing Parameterization Criteria for Cognitive Radios. In *Wireless Communication Systems (ISWCS), The Ninth International Symposium on*, July 2012.

[29] P. Wang, L. Xiao, S. Zhou, and J. Wang. Optimization of Detection Time for Channel Efficiency in Cognitive Radio Systems. In *Wireless Communications and Networking Conference, 2007. IEEE*, pages 111–115, March 2007.

[30] A. Tsertou and D.I. Laurenson. Revisiting the Hidden Terminal Problem in a CSMA/CA Wireless Network. *Mobile Computing, IEEE Transactions on*, 7(7):817–831, July 2008.

[31] S.D. Jones, E. Jung, X. Liu, N. Merheb, and I-Jeng Wang. Characterization of Spectrum Activities in the U.S. Public Safety Band for Opportunistic Spectrum Access. In *New Frontiers in Dynamic Spectrum Access Networks, 2007. 2nd IEEE International Symposium on*, pages 137–146, April 2007.

[32] Won-Yeol Lee and Ian F Akyildiz. Optimal Spectrum Sensing Framework for Cognitive Radio Networks. *IEEE Transactions on Wireless Communications*, 7(10):3845–3857, 2008.

[33] M. Gandetto and C. Regazzoni. Spectrum sensing: A distributed approach for cognitive terminals. *Selected Areas in Communications, IEEE Journal on*, 25(3):546–557, April 2007.

[34] E. Visotsky, S. Kuffner, and R. Peterson. On collaborative detection of TV transmissions in support of dynamic spectrum sharing. In *New Frontiers in Dynamic Spectrum Access Networks, 2005. First IEEE International Symposium on*, pages 338–345, November 2005.

[35] E. Peh and Y.-C. Liang. Optimization for Cooperative Sensing in Cognitive Radio Networks. In *IEEE Wireless Communications and Networking Conference, 2007.*, pages 27–32, March 2007.

[36] Z. Chair and P.K. Varshney. Optimal Data Fusion in Multiple Sensor Detection Systems. *Aerospace and Electronic Systems, IEEE Transactions on*, 22(1):98–101, January 1986.

[37] S.M. Mishra, A. Sahai, and R.W. Brodersen. Cooperative Sensing among Cognitive Radios. In *Communications, 2006. ICC IEEE International Conference on*, volume 4, pages 1658–1663, June 2006.

[38] I. F. Akyildiz, B. F. Lo, and R. Balakrishnan. Cooperative Spectrum Sensing in Cognitive Radio Networks: A Survey. *Physical Communications*, 4(1):40–62, November 2010.

[39] F. F. Digham, M.-S. Alouini, and M. K. Simon. On the Energy Detection of Unknown Signals Over Fading Channels. *Communications, IEEE Transactions on*, 55(1):21–24, January 2007.

[40] H. Tang. Some physical layer issues of wide-band cognitive radio systems. In *New Frontiers in Dynamic Spectrum Access Networks, 2005. First IEEE International Symposium on*, pages 151–159, November 2005.

[41] A. Sahai, N. Hoven, and R. Tandra. Some Fundamental Limits on Cognitive Radio. In *Communication, Control, and Computing. 48th Allerton Conference on*, October 2004.

[42] M. P. Olivieri, G. Barnett, A. Lackpour, A. Davis, and P. Ngo. A scalable dynamic spectrum allocation system with interference mitigation for teams of spectrally agile software defined radios. In *New Frontiers in Dynamic Spectrum Access Networks, 2005. First IEEE International Symposium on*, pages 170–179, November 2005.

[43] F. Weidling, D. Datla, V. Petty, P. Krishnan, and G. J. Minden. A framework for R.F. spectrum measurements and analysis. In *New Frontiers in Dynamic Spectrum Access Networks, 2005. First IEEE International Symposium on*, pages 573–576, November 2005.

[44] R. Tandra and A. Sahai. Fundamental limits on detection in low SNR under noise uncertainty. In *Wireless Networks, Communications and Mobile Computing, 2005 International Conference on*, volume 1, pages 464–469, June 2005.

[45] M. K. Tsatsanis and G. B. Giannakis. Transmitter induced cyclostationarity for blind channel equalization. *Signal Processing, IEEE Transactions on*, 45(7):1785–1794, July 1997.

[46] M. Ghozzi, F. Marx, M. Dohler, and J. Palicot. Cyclostatilonarilty-Based Test for Detection of Vacant Frequency Bands. In *Cognitive Radio Oriented Wireless Networks and Communications, 2006. 1st International Conference on*, pages 1–5, June 2006.

[47] J. Lunden, V. Koivunen, A. Huttunen, and H. V. Poor. Spectrum Sensing in Cognitive Radios Based on Multiple Cyclic Frequencies. In *Cognitive Radio Oriented Wireless Networks and Communications, 2007. 2nd International Conference on*, pages 37–43, August 2007.

[48] W. A. Gardner. Exploitation of spectral redundancy in cyclostationary signals. *Signal Processing Magazine, IEEE*, 8(2):14–36, April 1991.

[49] K. Maeda, A. Benjebbour, T. Asai, T. Furuno, and T. Ohya. Recognition Among OFDM-Based Systems Utilizing Cyclostationarity-Inducing Transmission. In *New Frontiers in Dynamic Spectrum Access Networks, 2007. 2nd IEEE International Symposium on*, pages 516–523, April 2007.

[50] P .D. Sutton, K. E. Nolan, and L. E. Doyle. Cyclostationary Signatures for Rendezvous in OFDM-Based Dynamic Spectrum Access Networks. In *New Frontiers in Dynamic Spectrum Access Networks, 2007. 2nd IEEE International Symposium on*, pages 220–231, April 2007.

[51] Y. Hur, J. Park, W. Woo, J.S. Lee, K. Lim, C.-H. Lee, H.S. Kim, and J. Laskar. A Cognitive Radio (CR) System Employing A Dual-Stage Spectrum Sensing Technique : A Multi-Resolution Spectrum Sensing (MRSS) and A Temporal Signature Detection (TSD) Technique. In *Global Telecommunications Conference, 2006. IEEE*, pages 1–5, November 2006.

[52] A. S. Tanenbaum. *Computer Networks*. Prentice Hall, 4th edition, 2003.

[53] H. Zhai, J. Wang, X. Chen, and Y. Fang. Medium access control in mobile ad hoc networks: challenges and solutions. *Wireless Communications and Mobile Computing*, 6(2):151–170, March 2006.

[54] A. C. V. Gummalla and J. O. Limb. Wireless Medium Access Control Protocols. *Communications Surveys & Tutorials, IEEE*, 3(2):2–15, 2nd quarter 2000.

[55] A. De Domenico, E. C. Strinati, and M.-G. Di Benedetto. A survey on mac strategies for cognitive radio networks. *To Be Published in Communications Surveys & Tutorials, IEEE*, 2012.

[56] A. K.-L. Yau, P. Komisarczuk, and P. D. Teal. On Multi-Channel MAC Protocols in Cognitive Radio Networks. In *Telecommunication Networks and Applications Conference, 2008. Australasian*, pages 300–305, December 2008.

[57] J. Mo, H.-S. W. So, and J. Walrand. Comparison of Multichannel MAC Protocols. *Mobile Computing, IEEE Transactions on*, 7(1):50–65, January 2008.

[58] Jungmin So and Nitin Vaidya. Multi-channel MAC for ad hoc networks: handling multi-channel hidden terminals using a single transceiver. In *Proc. of the 5th ACM international symposium on Mobile ad hoc Networking and Computing, ACM MobiHoc*, MobiHoc '04, pages 222–233, 2004.

[59] H. A. Bany Salameh and M. Krunz. Channel Access Protocols for Multihop Opportunistic Networks: Challenges and Recent developments. *Network, IEEE*, 23(4):14–19, July-August 2009.

[60] H. Wang, H. Qin, and L. Zhu. A Survey on MAC Protocols for Opportunistic Spectrum Access in Cognitive Radio Networks. In *Computer Science and Software Engineering, 2008 International Conference on*, volume 1, pages 214–218, December 2008.

[61] M. Timmers, S. Pollin, A. Dejonghe, L. V. der Perre, and F. Catthoor. A Distributed Multichannel MAC Protocol for Multihop Cognitive Radio Networks. *Vehicular Technology, IEEE Transactions on*, 59(1):446–459, January 2010.

[62] K. Ghaboosi, M. Latva-aho, and Y. Xiao. A Distributed Multi-channel Cognitive MAC Protocol for IEEE 802.11s Wireless Mesh Networks. In *Cognitive Radio Oriented Wireless Networks and Communications, 2008. 3rd International Conference on*, pages 1–8, May 2008.

[63] H. B. Salameh, M. Krunz, and O. Younis. MAC Protocol for Opportunistic Cognitive Radio Networks with Soft Guarantees. *Mobile Computing, IEEE Transactions on*, 8(10):1339–1352, October 2009.

[64] J. Jia, Q. Zhang, and X. Shen. HC-MAC: A Hardware-Constrained Cognitive MAC for Efficient Spectrum Management. *Selected Areas in Communications, IEEE Journal on*, 26(1):106–117, January 2008.

[65] X. Zhang and H. Su. CREAM-MAC: Cognitive Radio-EnAbled Multi-Channel MAC Protocol Over Dynamic Spectrum Access Networks. *Selected Topics in Signal Processing, IEEE Journal of*, 5(1):110–123, 2011.

[66] H.B. Salameh, M. Krunz, and O. Younis. Distance- and Traffic-Aware Channel Assignment in Cognitive Radio Networks. In *Sensor, Mesh and Ad Hoc Communications and Networks, 2008. 5th Annual IEEE Communications Society Conference on*, pages 10–18, June 2008.

[67] L Le and E. Hossain. OSA-MAC: A MAC Protocol for Opportunistic Spectrum Access in Cognitive Radio Networks. In *Wireless Communications and Networking Conference, 2008. IEEE*, pages 1426–1430, April 2008.

[68] D. Cabric, S.M. Mishra, D. Willkomm, R. Brodersen, and A. Wolisz. A Cognitive radio approach for usage of virtual unlicensed spectrum. In *Proc. of 14th IST Mobile Wireless Communications Summit*, June 2005.

[69] Q. Zhao, L. Tong, A. Swami, and Y. Chen. Decentralized cognitive MAC for opportunistic spectrum access in ad hoc networks: A POMDP framework. *Selected Areas in Communications, IEEE Journal on*, 25(3):589–600, April 2007.

[70] Y. R. Kondareddy and P. Agrawal. Synchronized MAC Protocol For Multi-hop Cognitive Radio Networks. In *Communications, 2008. IEEE International Conference on*, pages 3198–3202, May 2008.

[71] W. Hu, D. Willkomm, M. Abusubaih, J. Gross, G. Vlantis, M. Gerla, and A. Wolisz. COGNITIVE RADIOS FOR DYNAMIC SPECTRUM ACCESS - Dynamic Frequency Hopping Communities for Efficient IEEE 802.22 Operation. *Communications Magazine, IEEE*, 45(5):80–87, May 2007.

[72] N. C. Theis, R. W. Thomas, and L. A. DaSilva. Rendezvous for cognitive radios. *Mobile Computing, IEEE Transactions on*, 10(2):216–227, February 2011.

[73] Q. Chen, Y.-C. Liang, M. Motani, and W.-C. L. Wong. A Two-Level MAC Protocol Strategy for Opportunistic Spectrum Access in Cognitive Radio Networks. *Vehicular Technology, IEEE Transactions on*, 60(5):2164–2180, 2011.

[74] H. Su and X. Zhang. Cross-Layer Based Opportunistic MAC Protocols for QoS Provisionings Over Cognitive Radio Wireless Networks. *Selected Areas in Communications, IEEE Journal on*, 26(1):118–129, January 2008.

[75] P. Pawelczak, S. Pollin, Hoi-Sheung So, A. Bahai, R. Prasad, and R. Hekmat. Quality of service assessment of opportunistic spectrum access: a medium access control approach. *Wireless Communications, IEEE*, 15(5):20–29, October 2008.

[76] I.A. Getting. Perspective/navigation-The Global Positioning System. *Spectrum, IEEE*, 30(12):36–38, 43–47, December 1993.

[77] P. Enge and P. Misra. Special Issue on Global Positioning System. *Proceedings of the IEEE*, 87(1):3–15, January 1999.

[78] M. Thoppian, S. Venkatesan, R. Prakash, and R. Chandrasekaran. MAC-layer scheduling in cognitive radio based multi-hop wireless networks. In *World of Wireless, Mobile and Multimedia Networks, 2006. International Symposium on a*, pages 191–202, June 2006.

[79] IEEE 802.22. Working Group on Wireless Regional Area Networks. http://ieee802.org/22/, 2011.

[80] Q. Zhao and A. Swami. A Decision-Theoretic Framework for Opportunistic Spectrum Access. *Wireless Communications, IEEE*, 14(4):14–20, August 2007.

[81] D. Willkomm, M. Bohge, D. Hollos, J. Gross, and A. Wolisz. Double Hopping: A new approach for Dynamic Frequency Hopping in Cognitive Radio networks. In *Personal, Indoor and Mobile Radio Communications, 2008. IEEE 19th International Symposium on*, pages 1–6, September 2008.

[82] Q. Zhang and Y.-Q. Zhang. Cross-Layer Design for QoS Support in Multihop Wireless Networks. *Proceedings of the IEEE*, 96(1):64–76, January 2008.

[83] Liangping Ma, Xiaofeng Han, and Chien-chung Shen. Dynamic Open Spectrum Sharing MAC Protocol for Wireless Ad Hoc Networks. In *New Frontiers in Dynamic Spectrum Access Networks, 2005. First IEEE International Symposium on*, pages 203–213, November 2005.

[84] J. Zhao, H. Zheng, and G.-H. Yang. Distributed Coordination in Dynamic Spectrum Allocation Networks. In *New Frontiers in Dynamic Spectrum Access Networks, 2005. First IEEE International Symposium on*, pages 259–268, November 2005.

[85] T. Chen, H. Zhang, G. M. Maggio, and I. Chlamtac. CogMesh: A Cluster-Based Cognitive Radio Network. In *New Frontiers in Dynamic Spectrum Access Networks, 2007. 2nd IEEE International Symposium on*, pages 168–178, April 2007.

[86] S.C. Jha, M.M. Rashid, V.K. Bhargava, and C. Despins. Medium access control in distributed cognitive radio networks. *Wireless Communications, IEEE*, 18(4):41–51, August 2011.

[87] T.V. Krishna and A. Das. A survey on MAC protocols in OSA networks. *Computer Networks*, 53:1377–1394, June 2009.

[88] P. Pawelczak, S. Pollin, H.-S.W. So, A. Bahai, R.V. Prasad, and R. Hekmat. Performance Analysis of Multichannel Medium Access Control Algorithms for Opportunistic Spectrum Access. *Vehicular Technology, IEEE Transactions on*, 58(6):3014–3031, July 2009.

[89] Shao-Yu Lien, Chih-Cheng Tseng, and Kwang-Cheng Chen. Carrier Sensing Based Multiple Access Protocols for Cognitive Radio Networks. In *Communications, 2008. IEEE International Conference on*, pages 3208–3214, May 2008.

[90] C. Zou and C. Chigan. A Game Theoretic DSA-Driven MAC Framework for Cognitive Radio Networks. In *Communications, 2008. IEEE International Conference on*, pages 4165–4169, May 2008.

[91] M. M. Buddhikot, P. Kolodzy, S. Miller, K. Ryan, and J. Evans. DIMSUMNet: New Directions in Wireless Networking Using Coordinated. In *World of Wireless Mobile and Multimedia Networks, 2005. Sixth IEEE International Symposium on a*, pages 1–8, June 2005.

[92] S. C. Jha, U. Phuyal, M. M. Rashid, and V. K. Bhargava. Design of OMC-MAC: An Opportunistic Multi-Channel MAC with QoS Provisioning for Distributed Cognitive Radio Networks. *Wireless Communications, IEEE Transactions on*, 10(10):3414–3425, 2011.

[93] F. Hou, L. X. Cai, X. S. Shen, and J. Huang. Asynchronous Multichannel MAC Design With Difference-Set-Based Hopping Sequences. *Vehicular Technology, IEEE Transactions on*, 60(4):1728–1739, 2011.

[94] R. Bellman. A Markovian Decision Process. *Journal of Mathematics and Mechanics*, 6(5):679–684, 1957.

[95] Liangping Ma, Chien-chung Shen, and Bo Ryu. Single-Radio Adaptive Channel Algorithm for Spectrum Agile Wireless Ad Hoc Networks. In *New Frontiers in Dynamic Spectrum Access Networks, 2007. 2nd IEEE International Symposium on*, pages 547–558, April 2007.

[96] C. Cordeiro and K. Challapali. C-MAC: A Cognitive MAC Protocol for Multi- Channel Wireless Networks. In *New Frontiers in Dynamic Spectrum Access Networks, 2007. 2nd IEEE International Symposium on*, pages 147–157, April 2007.

[97] B. Hamdaoui and K.G. Shin. OS-MAC: An Efficient MAC Protocol for Spectrum-Agile Wireless Networks. *Mobile Computing, IEEE Transactions on*, 7(8):915–930, August 2008.

[98] L. T. Tan and L. B. Le. Distributed MAC Protocol for Cognitive Radio Networks: Design, Analysis and Optimization. *Vehicular Technology, IEEE Transactions on*, 60(8):3990–4003, 2011.

[99] H. Su and X. Zhang. Opportunistic MAC Protocols for Cognitive Radio Based Wireless Networks. In *IEEE 41st Annual Conference on Information Sciences and Systems*, volume 1, pages 363–368, March 2007.

[100] A. C.-C. Hsu, D. S. L. Wei, and C.-C. J. Kuo. A Cognitive MAC Protocol Using Statistical Channel Allocation for Wireless Ad-Hoc Networks. In *Wireless Communications and Networking Conference, 2007. IEEE*, pages 105–110, 2007.

[101] D. Zhou and T.-H. Lai. A compatible and scalable clock synchronization protocol in IEEE 802.11 ad hoc networks. In *Parallel Processing, 2005. International Conference on*, pages 295–302, June 2005.

[102] A. Ghasemi and E. S. Sousa. Collaborative spectrum sensing for opportunistic access in fading environments. In *New Frontiers in Dynamic Spectrum Access Networks, 2005. First IEEE International Symposium on*, pages 131–136, November 2005.

[103] R. Bruno, M. Conti, and E. Gregori. Mesh networks: commodity multihop ad hoc networks. *Communications Magazine, IEEE*, 43(3):123–131, March 2005.

[104] C. Stevenson, G. Chouinard, Z. Lei, W. Hu, S. Shellhammer, and W. Caldwell. IEEE 802.22: The first cognitive radio wireless regional area network standard. *Communications Magazine, IEEE*, 47(1):130–138, January 2009.

[105] ECMA. Standard ECMA-392: MAC and PHY for Operation in TV White Space. http://www.ecma-international.org/publications/standards/Ecma-392.htm, December 2009.

[106] M. S. Song, G. Ko, and S Hwang. Standardization of CogNeA on TV white spaces. In *Communications and Information Technology, 2009. 9th International Symposium on*, pages 820–823, September 2009.

[107] R. Kennedy and P. Ecclesine. Tutorial IEEE 802.11af. https://mentor.ieee.org /802.11/dcn/10/11-10-0742-00-0000-p802-11af-tutorial.ppt, July 2010.

[108] IEEE Computer Society. IEEE Standard for Information technology – Telecommunications and information exchange between systems – Local and metropolitan area networks – Specific requirements Part 11: Wireless LAN Medium Access Control (MAC) and Physical Layer (PHY) Specifications, 2010.

[109] A. Konsgen, Z. Hossain, and C. Gorg. Transmit Power Control Algorithms in IEEE 802.11h Based Networks. In *Personal, Indoor and Mobile Radio Communications, 2005. IEEE 16th International Symposium on*, volume 3, pages 1441–1445, September 2005.

[110] European Telecommunications Standards Institute. Broadband Radio Access Networks (BRAN) – 5 GHz high performance RLAN, 2008.

[111] Wi-Fi Alliance. Spectrum Sharing in the 5 GHz Band DFS Best Practices, 2007.

[112] D. Torrieri. *Principles of Spread-Spectrum Communications Systems*. Springer, 1st edition, 2005.

[113] IEEE Computer Society. IEEE Standard for Information technology – Telecommunications and information exchange between systems – Local and metropolitan area networks – Specific requirements Part 11: Wireless LAN Medium Access Control (MAC) and Physical Layer (PHY) Specifications Amendment 3: 3650 - 3700 MHz Operation in USA, 2008.

[114] IEEE Computer Society and IEEE Microwave Theory and Techniques Society. IEEE Standard for Local and metropolitan area networks – Part 16: Air Interface for Broadband Wireless Access Systems, 2009.

[115] E. Dahlman, S. Parkvall, and J. Sköld. *4G LTE/LTE-Advanced for Mobile Broadband*. Academic Press - Elsevier, 1st edition, 2011.

[116] IEEE Computer Society and IEEE Microwave Theory and Techniques Society. IEEE Standard for Local and metropolitan area networks - Part 16: Air Interface for Broadband Wireless Access Systems – Amendment 2: Improved Coexistence Mechanisms for License-Exempt Operation, 2010.

[117] IEEE Communications Society. IEEE Standard Definitions and Concepts for Dynamic Spectrum Access: Terminology Relating to Emerging Wireless Networks, System Functionality, and Spectrum Management, 2008.

[118] S. Buljore, M. Muck, P. Martigne, P. Houze, H. Harada, K. Ishizu, O. Holland, A. Mihailovic, K. A. Tsagkaris, O. Sallent, G. Clemo, M. Sooriyabandara, V. Ivanov,

K. Nolte, and M. Stametalos. Introduction to IEEE P1900.4 Activities. *Transactions on Communications, IEICE*, E91-B(1):2–9, January 2008.

[119] K. Moessner, H. Harada, C. Sun, Y.D. Alemseged, H. N. Tran, D. Noguet, R. Sawai, and N. Sato. Spectrum sensing for cognitive radio systems: technical aspects and standardization activities of the IEEE P1900.6 working group. *Wireless Communications, IEEE*, 18(1):30–37, February 2011.

In: Communication Systems: New Research
Editor: Vyacheslav Tuzlukov
ISBN: 978-1-62618-654-5

Chapter 3

RADIO RESOURCE MANAGEMENT AND FEMTOCELLS EMPLOYMENT IN LTE NETWORKS

Modar Shbat and Vyacheslav Tuzlukov *
School of Electronic Engineering, College of IT Engineering
Kyungpook National University, Daegu, South Korea

ABSTRACT

This chapter is devoted to discussion of the physical radio resource block (PRB) scheduling under femtocell employment in LTE networks. A simple feedback scheme and collaboration approach based on initial vision for femtocell-to-femtocell interface are proposed to speed up the handover procedure between femtocells, to reduce the signaling overhead, and to balance the traffic load among femtocells. The collaboration approach is compatible with different femtocell access modes. Other important aspects such as the power control, radio channel condition reporting, path loss models, and throughput definition are also discussed.

Keywords: Radio resource management (RRM), long term evolution (LTE) network, femtocell employment, interfernce, femtocell coverage, handover

3.1. INTRODUCTION

New generation of wireless broadband networks and the global mobility of user equipment (UE) motivated a great development of the communication technologies. The recent services are supported by diversity of solutions to fulfill the exploding growth of the mobile internet, expanded bandwidth, high capabilities in mobility, and radio resource management (RRM). The total requirements for the latest mobile generations and future network architectures are to increase the valuable spectrum efficiency using more flexible and

---

* Corresponding Author address: 1370 Sankyuk-dong, Buk-gu, Daegu, 702-701, South Korea. Email: tuzlukov@ee.knu.ac.kr.

effective spectrum allocation and radio scheduling schemes with the purpose to optimize the quality of service (QoS), and maximize system capacity. To achieve these goals, the RRM block exploits the advanced medium access control mixed with physical functions such as the resource sharing, channel quality indicator (CQI) or information reporting, link adaptation through adaptive modulation and coding, and hybrid automatic retransmission request. The third generation partnership project (3GPP) considers the long term evolution (LTE) networks as the latest standard in the mobile network technology.

The idea of LTE small cells that are called femtocells is able to provide additional advantages both for the users and for the vision of network development, for example, the service convergence. Based on Shannon theorem [1], it is evident that reducing the distance between the transmitter and receiver allows us to increase a capacity of wireless network under low transmission power. This fact generates the main femtocell principle defined in the 3GPP standards. The femtocell proposes a short range, low cost, and low power femtocell access point used as a base station (BS) emulating the evolved node B (eNB) in the LTE network topology. Femtocell access points are designed to solve the indoor mobile network coverage problems for limited areas and to achieve high and more reliable data rates for these environments. Owing to the low femtocell transmission power, femtocell access points can be deployed in higher density in comparison with macrocells in order to manage a large part of the network traffic leading to higher macrocell total throughput.

Self organizing network (SON) is an integral part of LTE networks. Some of the SON features are discussed and included in the standards such as 3GPP Rel-8, and Rel-9 [2,3]. The main SON solutions can be classified into two categories, namely, the self configuration and self optimization. The self configuration enables the LTE network to automatically perform the installation procedure, BS or eNB interface dynamic configuration, framework for physical cell identity (PCI) selection, and automatic neighbor cell discovery. The self optimization performs the basic mobility load balancing, the load information exchange between eNBs for interference mitigation, and enables the LTE network to autotune the operational parameters using UE, eNB, and performance measurements. The SON problems are more important and complicated in the case of LTE network extension employing the femtocells.

Femtocells now are a basic part of development strategy for the LTE networks and they require a total integration with the main network and optimum operation. The key LTE femtocell development factor is the physical radio resource block (PRB) scheduling process. This factor plays a very important role when the large volume of data is required to define a big number of femtocells with limited PRBs available in the network taking into account a tolerance degree of interference. Actually, initial LTE femtocell implementations for any form of the PRB scheduling methods could not satisfy all the requirements to SON, QoS, fairness, optimum power allocation, vertical handover, femtocell coverage problems, load balance among femtocells, interference, femtocell-to-femtocell handover procedure, and so on.

PRBs are scheduled every 1 msec in the 3GPP LTE network. Different frequency bandwidths or aggregated bandwidths can be assigned to an individual UE based on the channel condition and availability. Owing to rapidly and instantaneously changing nature of radio channel quality, we should have a sufficiently fast scheduling algorithm to compensate the changeable channel conditions. Before assigning the modulation technique and coding rate based on the transmission channel conditions to UE from eNB, the PRBs are appropriately allocated based on the used RRM technique. Thus, the problem of scheduling and distribution of the PRBs in 3GPP LTE among UEs is a complicated process. Speeding up the scheduling process is an important point with the purpose to achieve the proposed standard of PRB scheduling time without any sacrifices in other important network requirements like the fairness and security, especially in the case when different wireless networks and services are converged [4].

The basic difference between the macrocell and femtocell is the respective backhauls. According to the current implementation, the femtocell backhaul is an interface with the mobile core network through the public internet network, while the macrocell backhaul is a dedicated line to the mobile core network. In practice, it takes less than 100 msec for a handover process between macrocells [5], but the required time to transmit a single message via the public internet network could be over 200 msec which leads to slowly femtocell-to-femtocell- handover process (high delay). Owing to the small femtocell size in comparison with the macrocell one, the number of handovers (handover frequency) will be increased, especially, for the fast moving UE. Thus, the UE has great problems to be connected with the fast passing femtocells. The femtocell-to femtocell handover delay is also a serious problem that should be considered.

This chapter is devoted to an overview of key aspects to design the PRB allocation procedure for LTE networks employing the femtocells. Additionally, this chapter investigates the most important enabling technologies with the purpose to support the RRM and interference cancelation such as CQI reports using a simple feedback concept. Femtocell-to-femtocell interface is proposed to carry out the handover process between femtocells and to handle the load balancing procedure, in addition to controlling and managing the interference. Other important issues, namely, the power control, femtocell coverage, and path loss models are also discussed.

3.2. Macro-Femtocell Topology and LTE Main Technologies

According to the 3GPP LTE standards, the radio access technology based on the orthogonal frequency division multiplexing (OFDM) is employed by the downlink with flexible bandwidth. The single carrier frequency division multiple access is used by the uplink. Frequency division duplex and time division duplex multiple access techniques are employed. OFDM is robust to multipath fading and offers the high spectral efficiency and bandwidth scalability. The single carrier frequency division multiple access has been selected for the uplink based on the power consumption problem for UE. Additionally, the single carrier frequency division multiple access eliminates the high peak-to-average power ratio disadvantage when the instantaneous transmitted power is varied within the limits of a single OFDM symbol. PRBs are distributed among the UEs in a time-frequency domain that means several UEs share the OFDM subcarriers simultaneously. The eNB schedules the available radio resources among the downlink/uplink flows based on frequency reuse policy and scheduling technique. The frequency reuse increases the coverage and capacity, especially for the cell edge user, and reduces the intercell interference. The selection of the frequency reuse factor is a tradeoff between the capacity and signal-to-interference ratio as shown in Fig.3.1.

Generally, there are two main options for macro-femtocell topology in LTE networks. These two structures are shown in Fig. 3.2. In the first option, the femtocell access point or the home eNB (HeNB) is directly interfacing with the mobility management entity and serving gateway in the LTE network evolved packet core via the interface S1 (eNB-Core Network). In the second option, the HeNB is interfacing using the HeNB gateway that has a direct interface with mobility management entity and serving gateway emulating the wideband code division multiple access/high-speed packet access femtocell architecture standard.

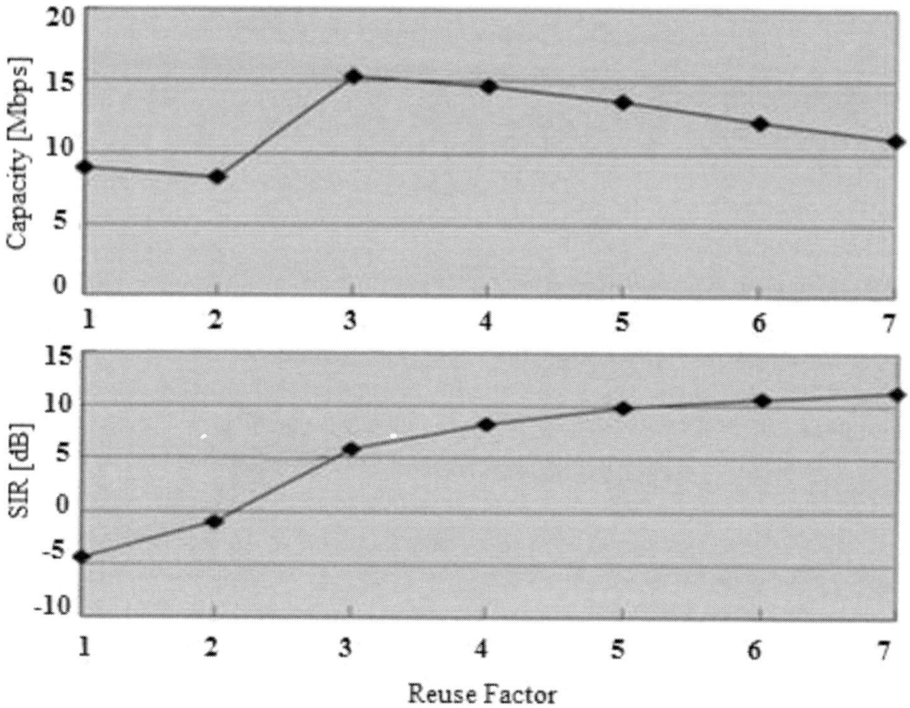

Figure 3.1. Cell edge SIR and channel capacity at different reuse factor.

It is well known that every eNB is defined by the sequence number called PCI. According to LTE standards, namely, 3GPP TS 36.211-840, there are for about 504 unique physical layer cell identities grouped into 168 unique physical layer cell identity groups and each group consists of three unique identities. The PCI of any eNB can be constructed based on primary and secondary synchronization identities as follows [6]:

$$PCI = 3ND_{ID}^{(1)} + ND_{ID}^{(2)}, \tag{3.1}$$

where $ND_{ID}^{(1)}$ covers the range 0~167 representing the physical layer cell identity group, and $ND_{ID}^{(2)}$ covers the range 0~2 representing the physical layer identity within the physical layer cell identity group. By the same way, a closed subscriber group (CSG) identity is used for femtocell identification in the LTE network. The CSG describes a limited set of users with connectivity access to a femtocell.

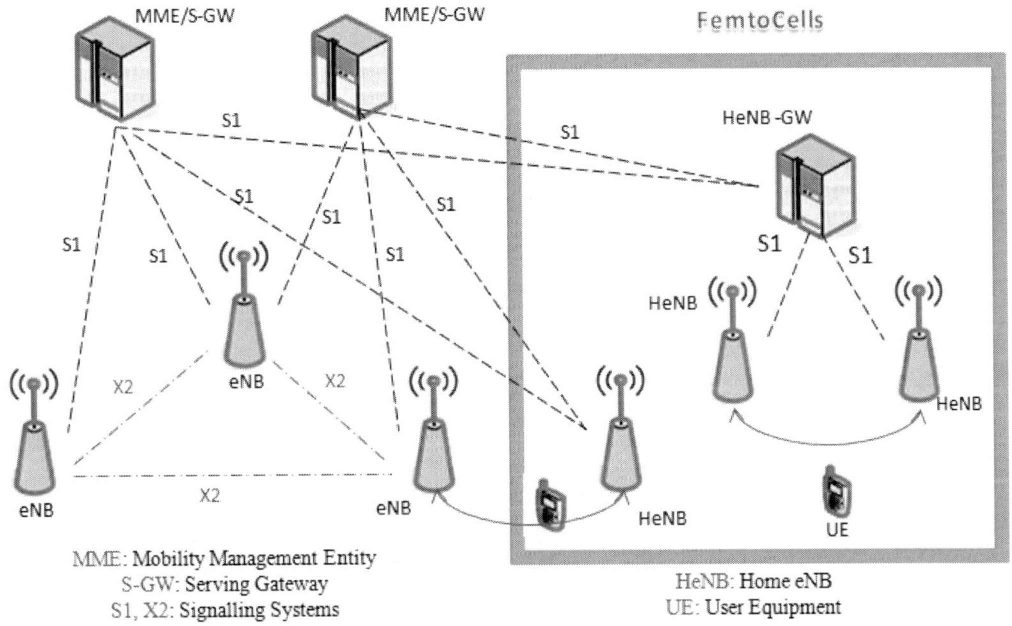

Figure 3.2. Basic macro-femtocell LTE network topology.

There are three major femtocell access modes, namely, the closed, open, and hybrid modes. In the closed mode, UEs included in the femtocell access control list are allowed to use the femtocell radio resources. The femtocell can also operate in open access mode, in which any UE possesses an access mode to femtocell. The hybrid access mode allows all the UEs to access the femtocell, but UEs included into the CSG are entitled to access and use the radio resources with priority. The UEs can obtain the CSG identities by distribution methods. For example, the CSG identities are provided for HeNBs from the operation and maintenance centre. In this case, HeNBs broadcast the CSG identities to the UEs. Each UE stores and keeps the required mode updating the information about PCI and CSG identities of the related eNB and HeNBs, respectively.

Two types of frame structure are proposed for LTE systems [7]. The first type is dedicated for the frequency division duplex mode and called the frame structure type 1. The second type is defined for the time division duplex mode and called the frame structure type 2 (see Fig. 3.3). The bandwidth of frame structure type 1 is divided into two parts to allow a transmission of the downlink and uplink data simultaneously. The frame structure type 2 is divided into two consecutive half frames (5 ms). Each half frame contains a special subframe that is not used for data transmission and is reserved for other purposes. In practical case, this subframe is used to send the downlink and uplink pilot symbols that are separated by a guard period.

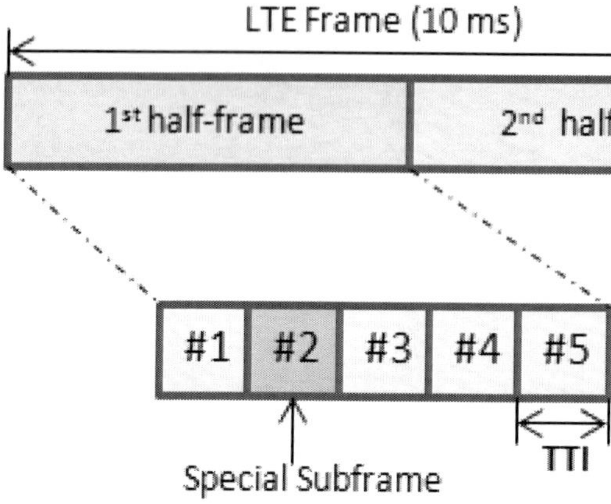

Figure 3.3. The LTE frame.

The OFDM subcarriers in the LTE network are spaced to be 15 KHz apart. Each subcarrier is modulated according to the adaptive modulation and coding that provides a flexibility to match one of the following modulation-coding schemes, namely, quadrature phase shift keying, 16- and 64-quadrature amplitude modulation based on the channel condition for each UE. The subcarriers are grouped into PRBs owing to the limited signaling channel resources when the scheduler needs to use the channel gain, i.e., the channel condition for each subcarrier and UE in order to optimize the PRB scheduling decisions. There are two main methods to create the PRBs, namely, the distributed method and localized method. In the distributed method, the PRB is defined by subcarriers distributed over the system bandwidth. In the case of localized method, the PRB is defined as contiguous subcarriers forming a certain subband in the system bandwidth. Figure 3.4 presents the PRBs creation methods.

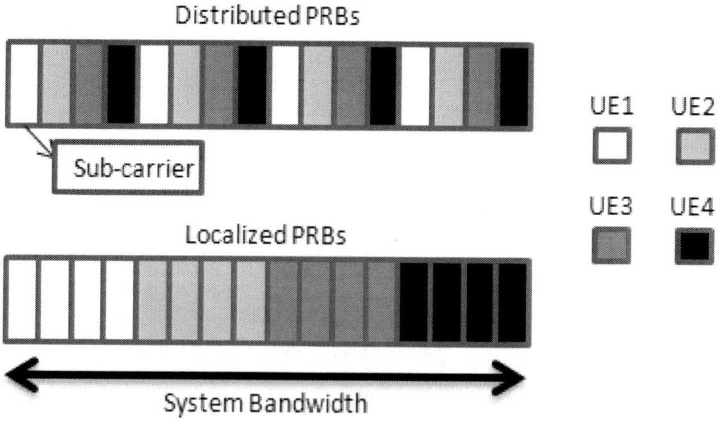

Figure 3.4. PRB creating methods.

Each PRB has time slot duration of 0.5 msec that corresponds to 6 ÷ 7 OFDM symbols and consists of 12 adjacent subcarriers. The smallest scheduled radio resource unit contains two consecutive PRBs and is called the scheduling block (SB). The SB is spanning a subframe time duration or transmission time interval of 1 msec at 180 KHz bandwidth. Figure 3.5 illustrates the scheduling block in the LTE downlink.

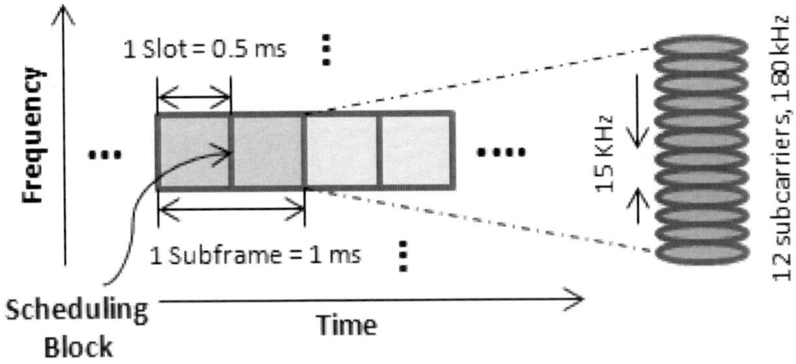

Figure 3.5. Scheduling block in the LTE downlink.

The PRB scheduling problem is to assign SBs to a subset of UEs in order to achieve some objectives such as maximization of the system throughput and the fairness or minimization of the packet drop rate. In order to simplify the scheduling process the schedulers operate using two levels or phases. The first phase is the time domain packet scheduling creating a list of UEs called the scheduling candidate set. UEs belonging to the scheduling candidate set can be assigned to PRBs or SBs in the current scheduling period. The previous phase is directly followed by the frequency domain packet scheduling (FDPS) that is the second phase defining the actual assignment of PRBs to the UEs in the created scheduling candidate set. In the majority of cases, the time domain packet scheduling uses the CQI factor to create the scheduling candidate set. Additionally, the use of other factors based on the scheduling technique is also possible.

3.3. RRM Considerations for LTE Femtocell Employment

3.3.1. General Statement

Femtocell HeNBs are implemented as an extension of the existing LTE network infrastructure. This condition means that there is no need to change the PRBs allocation in the network. It is expected that the femtocell frequency planning has self configuring ability. Thus, the femtocell automatically detects and negotiates the radio resources that can be used and should also resolve conflicts arising as a consequence of using the overlapping radio resources with the purpose to avoid interference problems. There are two approaches for femtocell employment in LTE networks to achieve an acceptable PRB scheduling performance with low interference levels [8]. The first approach is to set a chunk of the system spectrum aside for femtocells and use the remaining spectrum in the macrocell. This approach is called the dedicated channel deployment method. The second approach is to assign for femtocells the same spectrum as for the macrocells. This approach is called the co-channel deployment method. The advantage of the first approach is that the interference effects between the femtocells and macrocells can be eliminated and isolated, but this solution is not effective in terms of scarce spectrum cost and reduces the spectrum utilization efficiency. Obviously, the second approach suffers from the interference effects between the femtocells and macrocells. On the other hand, this approach is a cost effective solution and can improve the spectrum utilization efficiency in comparison with the dedicated channel approach.

3.3.2. RRM Design Issues

The motivations behind any PRB allocation strategies are based mainly on the tradeoff between the optimality of the scheduling decisions and the computational cost or complexity. The main RRM design factors taken into consideration the femtocells employment can be summarized in the following points:

1) Complexity and Scalability:

The PRB scheduler performs allocation decisions every transmission time interval (TTI) which is equal to 1 msec in order to be valid for the next TTI and then sends the allocation information to UEs via the physical downlink control channel. The downlink control information messages in the physical downlink control channel payload are used to inform the UEs about the allocated PRBs for their data transmission in the downlink and uplink directions. It is very important to ensure the scalability and low complexity in the PRB scheduling to limit the processing time and memory usage.

2) Spectral Efficiency:

Effective spectrum utilization is one of the main required goals to be achieved. Several performance indicators can be considered as maximization of the number of served UEs that have the best channel conditions within the limits of the given time interval.

3) Fairness among UEs:

Implementation of spectral efficient PRB allocation to maximize the total cell throughput may cause unfair sharing of radio resources between the UEs, especially, in the case of cell edge users or in more general case when there are UEs with bad channel conditions. Fairness is an important factor to guarantee a minimum performance for UEs with low CQI values.

4) QoS Level:

In the LTE systems, the QoS constraints vary based on application requirements and UEs related with CQIs. The QoS levels are mapped into some parameters such as the minimum guaranteed bit rate, the maximum delivery delay, and the packet loss rate.

5) Interference Management:

Under high density employment of femtocells and heavy traffic levels, the interference between neighboring femtocells and between the femtocells and related macrocells is a critical problem since the macrocells and femtocells share and utilize the same PRBs (see Fig. 3.6). As was mentioned before, in the co-channel approach case, the interference has a great effect and should be analyzed carefully under design of PRB scheduling technique. Under the use of femtocell, the transmit power control is the vital factor to reduce the effects of interference. Essential items of power control and interference elimination are presented as follows:

- The femtocell should use an appropriate transmit power. If the transmit power is too high, the femtocells can cause an interference with respect to the neighboring macrocells and femtocells. On the other hand, if the transmit power is low, it can be a reason to limit the coverage of the femtocells.

- The femtocell HeNB should have a small receiver gain. This allows us to prevent a passerby or neighboring UEs from camping onto the femtocell that increases the number of signaling messages in the core network if the UEs try to handover to the femtocell without any necessity.
- An upper limit of the transmit power of the UE must be configured. This idea is applied to ensure a handoff between the UE and macrocell instead of increase in the transmit power with the purpose to reach the femtocell HeNB.
- The RRM approach for dense femtocell deployment scenarios requires coordination between the neighboring femtocells in real time in order to reduce the interference effects and to perform radio resource negotiations.

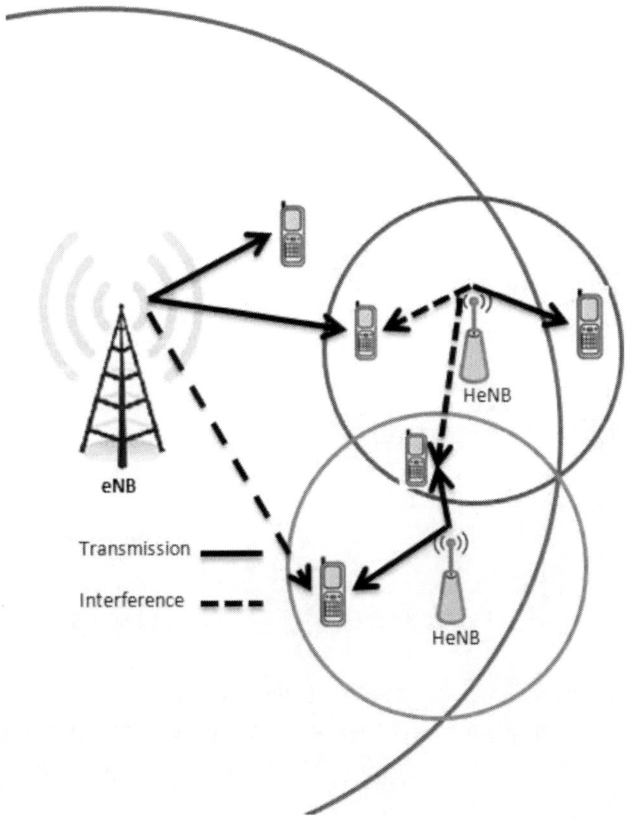

Figure 3.6. The macro-femtocell and femtocell-femtocell interference scenarios.

Finally, the most common interference reduction techniques used under femtocell employment in LTE networks are: the transmit power control, the randomize frequency use (for the low traffic or low dense femtocell scenarios), the resource allocation based on deduction when the traffic is predictable, and collaborative resource negotiation.

6) Femtocell Coverage:

The femtocell coverage involves the transmit power control and the coordination between femtocells. When the interference generated by the neighboring femtocells is identified, the suffering femtocell can negotiate the transmit power and the allocated PRBs to be used with these neighboring femtocells. The femtocell coverage optimization problem and other problems, for example, the coverage gaps and overlaps, can also be overcome using the transmit power control and collaborative PRB negotiation. Figure 3.7 shows the common femtocells coverage problems.

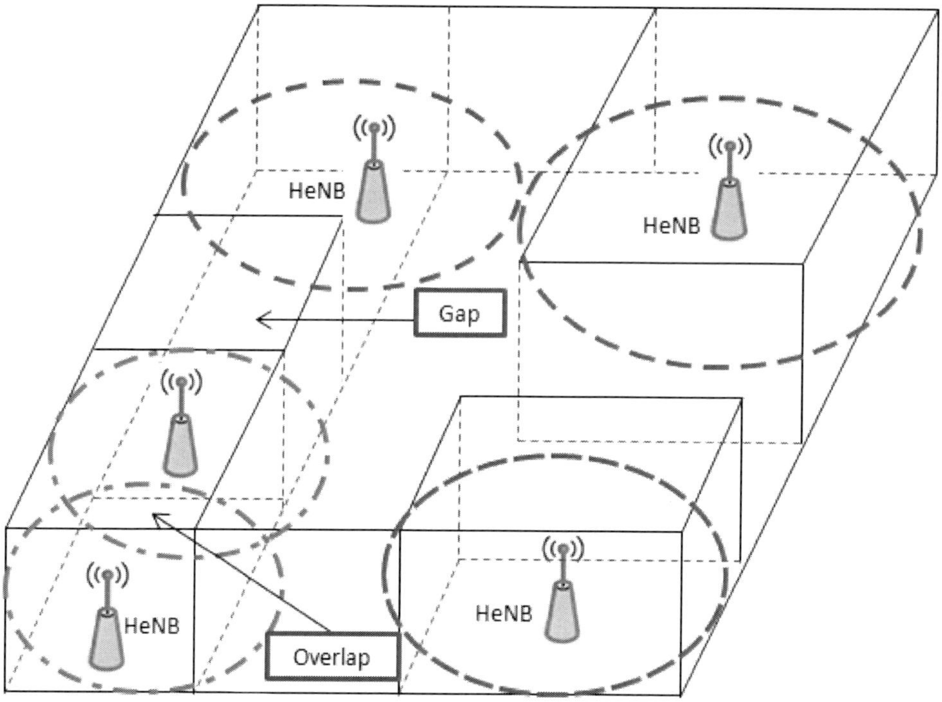

Figure 3.7. Femtocells coverage problems.

3.4. FDPS and CQI Reporting

Any PRB scheduler needs information about the radio channel condition in real time for all UEs and PRBs in order to achieve the high gain based on multi-user frequency diversity. This information is considered as the input for the adaptive frequency domain PRB allocation to UEs.

The FDPS is the frequency domain UEs multiplexing according to their instant radio conditions or their CQIs. The time domain scheduling characteristics are inherited and included in the FDPS. The main advantage of FDPS is an increase in the average system capacity and cell edge data rate [9].

The FDPS performance depends on the frequency domain scheduling resolution and the radio channel state report accuracy (CQI accuracy). The CQI report accuracy depends on two main factors assuming that a reliable uplink transmission channel is available. The first factor is the accuracy of PRB/CQI measurement and the second factor is the signaling resolution of the CQI reporting.

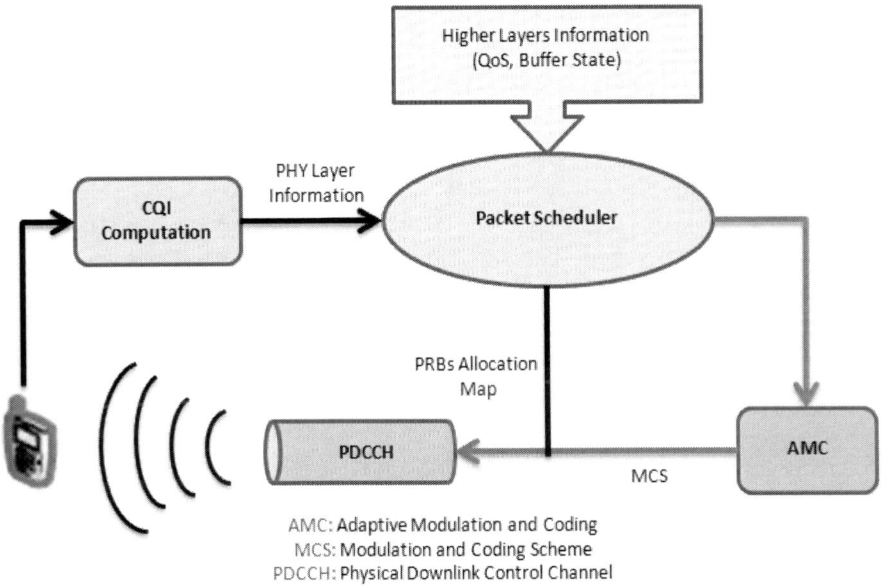

Figure 3.8. The FDPS model.

In the LTE network uplink, each UE transmits a sounding reference signal [10] to the BS (eNB or HeNB) considered as the PRB scheduling node. The sounding reference signal is used as a radio channel condition report or CQI. The PRB scheduler in the BS using the CQI reports for all UEs and PRBs performs the radio resources assignment process (PRBs to UEs) at each time slot (every 1 msec in LTE systems) according to the PRB scheduling approach. The FDPS model is presented in Fig. 3.8.

The scheduling decisions are made after computation of PRB metrics for each UE. This method allows us to reduce the scheduling complexity because each PRB is allocated to the UE with the highest metric independently of other PRBs. If we assume that the number of active UEs in the current TTI is equal to N_{User} and the number of available PRBs to be allocated to UEs is equal to N_{PRB}, the scheduler must calculate the number of metrics equal to $M = N_{PRB} N_{User}$ every TTI. The number of metrics M depends linearly on the number of PRBs and UEs and this linear function ensures the scheduling scalability requirements.

In general, each CQI is defined as a current measurement of the signal-to-interference-plus-noise ratio (SINR). The most important requirement of the CQI reporting method is to find an optimal tradeoff between the accurate radio channel quality estimation and signaling overhead. The scheduling complexity can also be reduced by separating the PRB allocation process and its power allocation process [11,12], since the computation cost of power allocation is high even if it is possible to increase the LTE network throughput.

Many frequency domain PRB scheduling techniques for uplink and downlink transmission in LTE networks have been proposed to be implemented including the well known techniques that are used and standardized for different types of wireless and mobile networks such as the round robin and proportional fair. It is very useful to describe briefly some PRB scheduling techniques taking into account the ability of adoption for femtocell employment in the LTE networks, for example, the aggressive subchannel allocation, the bandwidth assignment based on SNR (BABS), the amplitude craving greedy (ACG), and the simple dynamic subcarrier allocation (SDSA) techniques.

1) Aggressive Subchannel Allocation Algorithm:

The aggressive subchannel allocation algorithm is a dynamic PRB allocation approach considering conditionally all UE radio channel states or CQI reports with the purpose to maximize the throughput of the OFDM system [13]. Every UE sends COI reports of all PRBs to the BS using the uplink feedback channel. Each PRB is allocated to one UE only in order to avoid the intercell interference. The BS determines a UE to be scheduled according to fairness criteria [13]. After that the BS defines the suitable PRB allowing maximum data rate to the selected UE among the PRBs that are not allocated to other UEs. Two more assignment steps are performed by the scheduler in the BS presented in [13].

2) BABS Algorithm:

The BABS algorithm defines the minimum number of PRBs assigned to UE in order to achieve the required data rate for each UE [14]. Under considered approach, not all the PRBs are allocated to the UEs with the best channel conditions or high CQI values if the fairness between UEs is taken into account. In this algorithm, the BS defines the total data rate of each UE across all PRBs using the average feedback SINR before allocating a specific number of PRBs to the UEs. If the sum of allocated PRBs to all UEs exceeds the total number of PRBs, the PRBs allocated to the UEs at minimum level, i.e., the minimum number of PRBs, are cancelled one by one until the sum of all PRBs allocated to UEs is equal to the total number of PRBs. In the case when the sum of PRBs allocated to all UEs is smaller than the total number of PRBs, the reminder of PRBs is additionally assigned to the UE with minimum additional power requirements.

3) ACG Algorithm:

The ACG algorithm is a modified version of the rate craving greedy algorithm. The rate craving greedy algorithm scheduling approach starts with the estimation process of the UE transmission rate on each PRB and operates with the purpose to maximize the total transmission rate. ACG algorithm defines the number of required PRBs for each UE according to the BABS method [14]. The PRB is allocated to the UE with the maximum channel gain. This allocation process is performed sequentially from the first PRB to the last one. The ACG algorithm reduces the computation complexity. The total system throughput is not high. The ACG algorithm requires the ideal CQI reports that increase the bandwidth tradeoff to update the information about channel conditions.

4) SDSA Algorithm:

The SDSA algorithm proposes a new CQI scheme coupled with simple and dynamic PRB allocation scheme to reduce the bandwidth tradeoff associated with updating the information about radio channel conditions and the computation cost of dynamic subcarrier allocation [15]. The proposed CQI feedback contains only the average radio channel gain for the UEs and a single information bit per subcarrier. The average radio channel gain from the CQI feedback is used to set the number of subcarriers while a single information bit is used to assign the subcarriers to UEs (the CQI scheme contains 6 bits for average radio channel gain and 1 bit per subcarrier). The SDSA algorithm improves significantly the outage performance with very small throughput penalty.

For the femtocell employment in the LTE networks with the existed PRB scheduling approaches discussed above, additional modifications and mechanisms are needed. For example, in the case of dense femtocell scenarios, it is reasonable to assume that the list of neighboring HeNBs or femtocells is prepared and passed to each HeNB as a required parameter during the initial configuration stage by the central network management system of the LTE network. This list can be used for PRB allocation and interference management between femtocells. The list of neighboring HeNBs can also be obtained by dynamic discovering mechanism running among femtocells belonging to the same macrocell. Another example given in [16] is called the cognitive pilot channel (CPC) mechanism that can be used to identify possible interfering HeNBs and provide information about occupied frequencies or PRBs. The CPC is a concept of control channel dedicated to convey information associated with the frequencies used currently by HeNBs in a particular geographical area. The CPC can also provide useful information to enable the vertical handover or spectrum negotiation. Additionally, the femtocell can use the CPC to plan the allocation of PRBs and to avoid interference between other femtocells.

3.5. Power Control and Path Loss Models

3.5.1. Power Control

The power control is a dynamic process adjusting the transmission power for the radio channel or subcarrier with the purpose to compensate instantaneous variations of radio channel parameters. The power control allows us to save the power consumptions and, at the same time, to maintain a constant bit rate (power reduction). On the other hand, the power control can increase the bit rate by selecting the high modulation and coding scheme in the case of power boosting keeping the block error rate (BLER) under the threshold.

As was mentioned above, in the case of femtocells deployment, the power control is an important factor for coverage optimization and interference management. A self-organized downlink power control for interference management is presented in [17] for the case when HeNBs work in co-channel operation mode with the macrocell system. This solution assumes that there is no X2 interface between the eNBs and HeNBs under 3GPP standards release 11. Under this condition, the HeNB should make the decisions autonomously without any feedback in the macrocell network.

The HeNBs are modeled as a multiagent system and each HeNB is an independent agent that can learn through the reinforcement learning techniques the downlink power allocation strategy according to different interference situations. To recover the lack of the information in this power control model, the theory of partially observable Markov decision process [18] is considered. The partially observable Markov decision process theory is a suitable tool for decision making in the cases with some degree of uncertainty. The partially observable Markov decision process works on the basis of a set of beliefs that are built in HeNBs through spatial interpolation techniques based on the SINR measurements at their receivers. These beliefs are obtained by estimating the position of potential victim macrocell UEs and interpolating the SINR for these locations. The partially observable Markov decision process learning process is started based on the estimated information.

The power control concept presented in [17] seems to be complicated with high computation cost, and it is hard to predict its performance in the case of dense femtocell employment owing to the high interference level between femtocells and macrocell and between femtocells from the same macrocell or femtocells from different macrocells.

This chapter introduces a femtocell-to-femtocell interface allowing us to manage the interference between femtocells based on a simple feedback scheme. The proposed interface can also be used to perform the handover process between the femtocells and handle the load balancing.

3.5.2. Path Loss Models

The European wireless world initiative new radio (WINNER) project developed a new radio concept for the modern generation wireless communication systems used a channel bandwidth up to 100 MHz for one radio link and operation frequencies between 2 and 6 GHz. In this project, a new radio channel models are developed based on existing standards and channel measurements. The WINNER II path loss model is defined as the total path loss between two omnidirectional antennas (ideal antenna that radiates the power uniformly in all directions and is used as a reference antenna). The typical WINNER II path loss model can be presented for various scenarios in the following form [19]:

$$PL = A\lg d + B + C\lg\frac{f_c}{5} + X \quad [\text{dB}]\ , \tag{3.2}$$

where d is the distance between the transmitter and the receiver measured in meters, f_c is the system frequency measured in GHz, A is the fitting parameter including the path loss exponent, B is a fixed quantity defined based on empirical observations and determined by the free space path loss according to the reference distance, C is a parameter describing the path loss and it depends on the frequency, and X is a specific term of optional environment, i.e., wall attenuation in the non-line-of-sight scenario. This model can be applied for the frequencies between 2 and 6 GHz and for different antenna heights. The WINNER II path loss model is applicable in the case of outdoor femtocells employment scenario or in the case when there is a need to define the path losses between the eNB of the macrocell and the HeNB of the femtocell. Based on empirical observation of the variables *A, B, C* and *X* in (3.2), the practical measuring processes are described in [19].

In the case of indoor environment, the path losses and the femtocell coverage model are defined based on the LTE standards presented in [20] where the basic path loss is determined in the following form:

$$L_{total} = 20\lg f_c + N\lg d + L_f(u) - 28 \quad [\text{dB}]\ , \tag{3.3}$$

where f_c is the frequency in MHz, N is the power loss factor caused by distance d between the HeNB and the UE ($d > 1m$), L_f is the floor penetration factor in dB, and u is the number of floors between the HeNB and the UE for indoor environment ($u \geq 1$).

The maximum acceptable path losses ensuring an adequate high quality pilot channel signal are defined in the following form [20]:

$$L_{total} = L_{\max} - L_{FM} - L_W \ , \tag{3.4}$$

where L_{FM} are the shadow fade losses, L_W are the losses caused by the building outer wall, and $L_{\max}$ are the maximum acceptable path losses given by [20]

$$L_{\max} = 10 \lg \left[\left(\frac{P_{\max}}{N_{UE}} \right) \left(\frac{p_{CPICH}}{E_c/N_0} - 1 \right) \right] \ [\text{dB}] \ , \tag{3.5}$$

where $P_{\max}$ is the HeNB maximum transmitted power, N_{UE} is the UE receiver noise power, p_{CPICH} is the percentage of the femtocell power allocated to the pilot channel, and E_c/N_0 is the received power per chip of the pilot channel divided by the total noise power spectral density (the pilot channel signal quality). Table 3.1 presents the main simulation parameters of the femtocell coverage using the above-mentioned model for the path losses and femtocell coverage.

Table 3.1. Femtocell coverage simulation parameters

Parameter	Value
Carrier frequency	$f_c = 2$ **GHz**
E_c / N_0	-16 dB
L_f $(for\ 1\ floor)$	9 dB
L_f $(for\ 2\ floors)$	19 dB
L_W	10 dB
Number of floors (u)	2 floors
P_{CPICH}	0.1
UE noise figure	7 dB
L_{FM}	7.7 dB

Figure 3.9 presents the femtocell coverage range as a function of the maximum transmit power based on (3.3) – (3.5) using the introduced model for the femtocell coverage on the same floor and in the case of separation by one floor (indoor environment) [20]. The femtocell coverage range is presented in Fig 3.10 as a function of the pilot channel signal quality E_c/N_0 for a fixed maximum transmit power $P_{\max} = -5$ dBm using (3.3) – (3.5). As we can see from Fig. 3.10, the critical value of the E_c/N_0 is equal to -10 dB.

3.6. Femtocell-to-Femtocell Handover

3.6.1. General Statement

The handover (or handoff) approaches in wireless networks are divided on different classes based on the main factor used with the purpose to formulate the handover decision.

The main factor can be one of the following: the power level handover, the user population handover, the bandwidth handover, and the fuzzy logic handover [21]. In the case of LTE networks based on the macro-femtocell topology, there are three types of femtocell handover:

1) Inbound – the handover occurs from the macrocell to the femtocell;
2) Outbound – the handover occurs from the femtocell to the macrocell;
3) Femtocell – to – femtocell – the handover occurs between one femtocell and another femtocell.

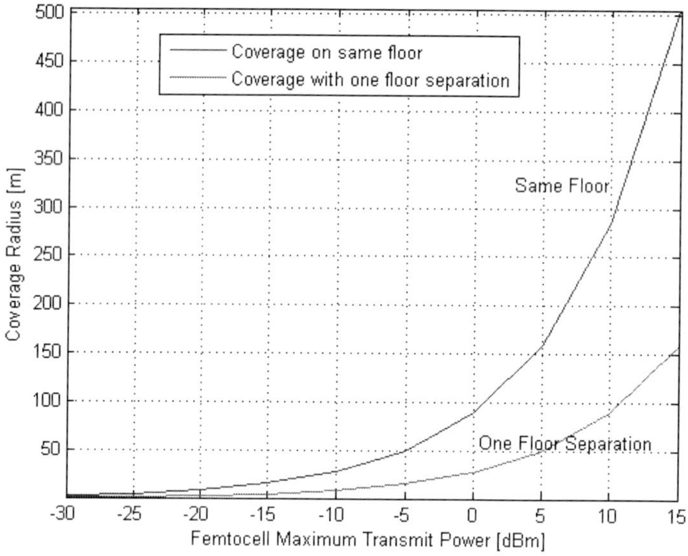

Figure 3.9. Femtocell coverage radius versus transmit power.

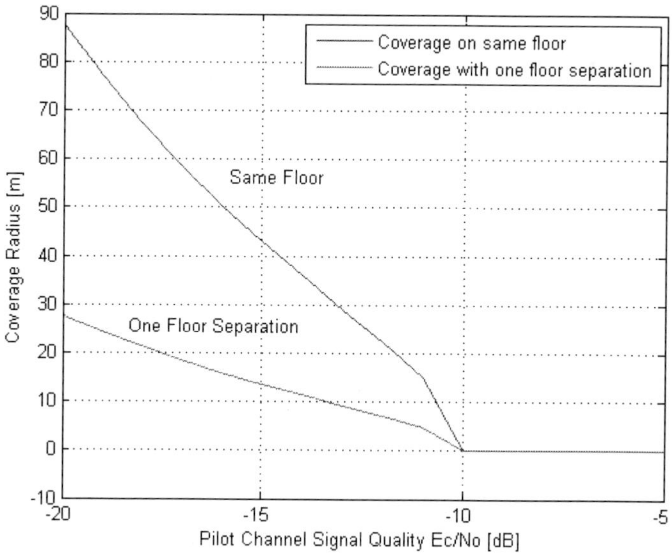

Figure 3.10. Femtocell coverage radius versus pilot channel signal quality.

In practice, it takes less than 100 ms for a handover process between macrocells [5], but the required time to transmit a single message via the public internet network could be over 200 ms that leads us to slowly femtocell-to-femtocell handover process (big delay). Owing to the femtocell small size in comparison with the macrocell one the number of handovers (handover frequency) will be increased, especially, for fast moving UE. Thus, the UE has great difficulties to be connected with the fast passing femtocells.

3.6.2. Standard Femtocell-to-Femtocell Handover Procedure

Any UE periodically scans and examines all available channels or PRBs in order to measure the signal power and reports about the measurements that are the feedback to its associated eNB or HeNB in a report message. A handover procedure is triggered by a positive handover decision that is happened every time when the measurement report or the feedback suggests that the best signal received by the UE is not sent by the current HeNB, but it is sent by another HeNB and the UE is not associated with this HeNB. When the current HeNB in the femtocell makes a positive handover decision, it sends a handover request message to the target HeNB that is the femtocell so that the UE is to be handed-over to via the mobility management entity. Then the target HeNB performs an admission control for the UE and responds with a positive handover response message. When the current HeNB receives the handover response, it sends the handover command to the UE. After that the UE disconnects or detaches from the associated femtocell and tries to handover to the new femtocell (target HeNB). The handover decision is completely independent of the speed of the mobile UE.

3.6.3. Non Standard Femtocell-to-Femtocell Handover Approaches

Recently, many suggested ideas about the femtocells handover have been considered as important contributions toward the possible solutions, for example, the proactively triggering handover procedures by predicting mobility of users [22], reducing the scanning time to identify associable femtocells by defining the records of recently visited cells [23], and reducing the unnecessary handovers by modifying the architecture and signal flow [24,25]. As was shown in [26], the UEs are classified into two modes based on the speed to insert the UE movement parameters in the handover decision function to apply the fast handover based on prefetch process. The first mode is called the swift mode when the UE speed is higher than a pre-defined threshold. Thus, the UE moves fast being a short time in the femtocell range. The second mode is called the free mode. That means the handover decision function allows a UE to handover to a femtocell only if it belongs to the free mode. The approximate UE speed is estimated based on measurement reports sent by UE. The femtocell-to-femtocell handover is carried out based on parts of the legacy handover process, which are segregated to prefetch high layer data to all HeNBs in the proximity of the UE.

The redundant handover (or unnecessary handover) problem is defined as the case when the handover procedure is initiated, but it is not completed before the next handover decision is made. If the handover is frequently repeated between two adjacent cells in short time intervals, this case is also considered as a redundant handover. This problem is usually occurs as a reason of short time radio channel variations caused by the fast fading or moving UE along edge (borders) of two neighboring cells. The redundant handover problem is more serious in the case of femtocells owing to the dense employment and the small femtocell coverage size that can significantly increase the amount of handover initiations since the UE can receive the signal both from eNB and from all HeNBs in the neighborhood.

The redundant handover elimination techniques are very important for femtocell employment in LTE networks [27]. The windowing technique (or signal averaging) [28] and the handover delay time [29] are modified to adapt their parameters based on the radio channel condition related to the UE position in the cell. Power control improvement allows us to reduce the number of redundant handovers in femtocells. These and other modifications allow the mentioned handover techniques to be adopted and used under femtocells employment and redundant handover elimination [25].

3.7. The Proposed Feedback Scheme and Handover Approach

3.7.1. The Feedback Scheme

In the macro-femtocell network topology, the HeNBs receive feedback information from the UE (see Fig. 3.11). This feedback allows us to define the effective SINR. The feedback corresponds to the estimated effective SINR defined by nonlinear averaging on the available PRBs or SBs in the subband of the femtocell as follows [30]:

$$SINR_{eff} = -\beta \ln\left(\frac{1}{K}\sum_{i=1}^{K} e^{-\frac{SINR_i}{\beta}}\right), \qquad (3.6)$$

where K is the total number of the averaged subcarriers, β is a parameter calibrated by means of link level simulation to fit the compression function to the block error rate defined in the case of additive white Gaussian noise (AWGN), and $SINR_i$ is the signal-to-interference-plus-noise ratio of the ith UE computed for each jth PRB and determined as:

$$SINR_i = \frac{P_{R,i,j}}{(FN_0 BW_{RB}) + I}, \qquad (3.7)$$

where F is the noise figure (the default value is 2.5), N_0 is the noise power spectral density (the default value is -174 dBm), BW_{RB} is the bandwidth of the resource block (180 kHz for SB), $P_{R,i,j}$ is the received power of the ith UE for the jth PRB depending on the propagation loss model, and I is the total power of interference caused by the HeNBs that share the same frequency resources. In the proposed scenario, it is possible to relate the number of feedback bits presenting the corresponding value of the $SINR_{eff}$ and forming the feedback indicator (FI) with the number of the femtocells included in the white list of any UE. We can assume that the UE sends a feedback to multiple femtocells related to the same eNB simultaneously (Fig. 3.12). This relation can be presented in the following form:

$$N_{femto} = 2^n \ , \qquad (3.8)$$

where n is the number of the feedback bits, and N_{femto} is the number of the femtocells that are belonged to the same eNB and are included in the UE white list. For example, if $n = 1$, the UE should sent the feedback to two HeNBs.

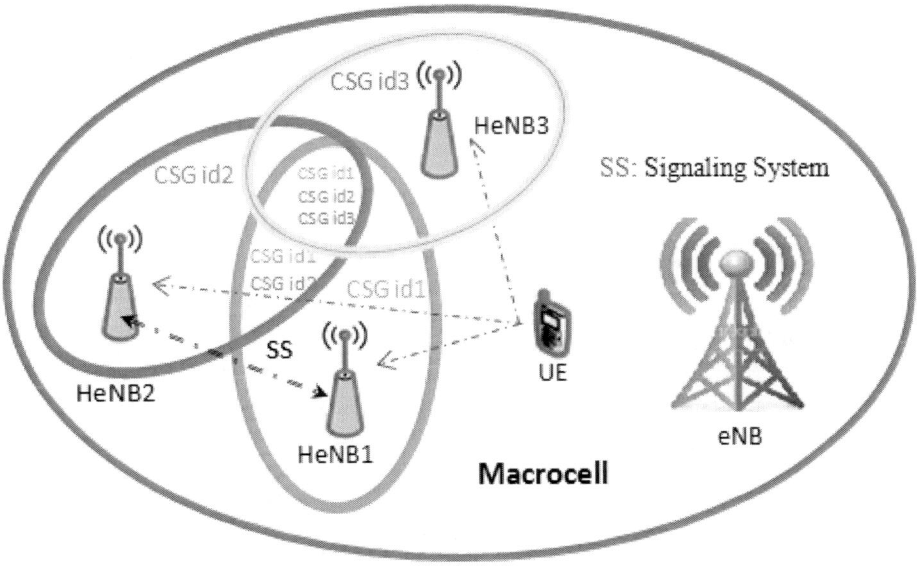

Figure 3.11. UE sends feedback to multiple HeNBs.

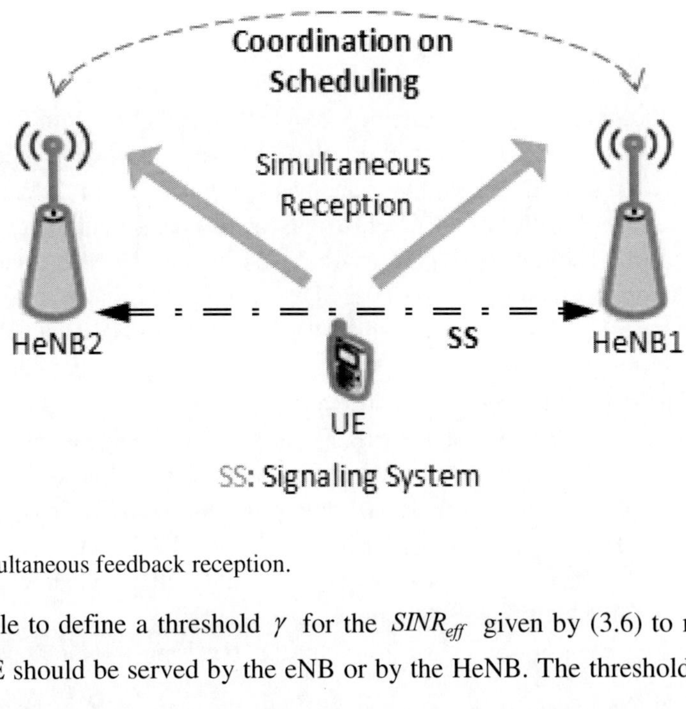

Figure 3.12. Simultaneous feedback reception.

It is possible to define a threshold γ for the $SINR_{eff}$ given by (3.6) to make a decision whether the UE should be served by the eNB or by the HeNB. The threshold γ can indicate that the UE should be served by the related eNB when $SINR_{eff} > \gamma$, otherwise the UE is served by one of the HeNBs from the white list.

3.7.2. The Proposed Femtocell-to-Femtocell Handover Approach

The proposed feedback scheme leads to automatic selection of the serving femtocell representing a specific CSG from the UE white list and allows the HeNB to update the priority list and reschedule the SBs assuming that all the HeNBs belonging to the UE white list receive the feedback simultaneously. These HeNBs exchange the signaling information via the signaling system representing a femtocell-to-femtocell interface (Fig. 3.12) to run the collaborative mode with the purpose to process the handover operation between the femtocells for a moving UE. The UE movement parameters do not play any role.

The femtocell-to-femtocell handover approach discussed in this chapter consists of two operation levels. The first level is applied locally to each HeNB of the related femtocell with the purpose to create a list of UEs that send feedback, which would be served within the limits of the upcoming TTI. Some of these UEs have been served by the same HeNB within the limits of the current TTI. In the UEs priority list of each HeNB, the UEs are ranked by ascending arrange of the FIs. The second level is applied between the femtocells in order to exchange the UEs priority lists created by each HeNB at the first level. Every HeNB sends this initial UEs priority list to other HeNBs in the same macrocell and receives back the initial priority lists from other HeNBs via the signaling system. After that, each HeNB creates the final version of UEs list based on completeness of comparison between the local UEs list and the received lists from other HeNB to drop and update the positions of UEs in the final UEs priority list. The changes take place only if UEs appear in more than one priority list or belong to more than one CSG [31].

Thus, if UE has the same position in the priority lists formed by two HeNBs, it is stored in the priority list of the HeNB in the case if this UE has been served by the same HeNB during the current TTI (no handover) and excluded from the other list. If the UE has two different ranks in the priority lists, it must be excluded from the list with the lower priority (lower FI). Thus, if this UE has been served by the HeNB possessing the highest FI, the UE should be served by the same HeNB and the handover is not needed. In the case if the UE has been served by the HeNB with the lower FI, it must be served by the other HeNB and the handover is performed. The handover process of the UE among HeNBs is carried out using the presented collaborative mode of the proposed UEs feedback.

For simplicity, we study the scenario when we have two HeNBs, namely, HeNB1 and HeNB2, and the number of active users $N = 6$ with the purpose to explain clearly the simple handover approach (Fig. 3.13). The UEs send feedback to different femtocells with different FI values. For the next TTI, when each HeNB has to schedule the available radio resources (PRBs), HeNB1 puts UE2 on the second position in the priority list according to its FI. It is known that UE2 has been served by this HeNB during the current TTI. Thus, we consider the HeNB1 as the current HeNB. On the other hand, HeNB2 puts UE2 on the first position in the priority list because UE2 has higher FI for this HeNB. The HeNB1 and HeNB2 exchange the priority lists and each HeNB compares the created list with the received one. HeNB1 excludes the UE2 from the priority list because it has the better position in the received list. In this case, the handover must be applied to the HeNB2 that is considered as the target HeNB. The handover process should be automatically completed without any needs to use the handover control messages if other factors, namely, the load balance, the interference management, and the PRB scheduling are not considered under the handover approach. If there is a need to use the handover control messages, for example, the handover request, handover response, etc, the interface between femtocells, e.g. signaling system, can handle these messages.

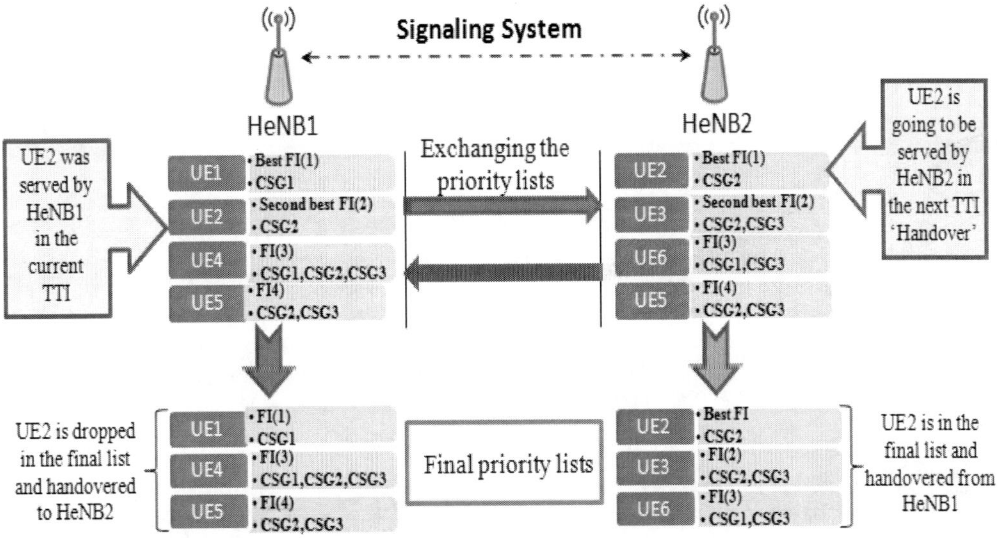

Figure 3.13. The handover approach: open access mode.

The case concerning the UE5 (see Fig. 3.13) is important for us because UE5 has the same rank in both priority lists. If UE5 has been served by HeNB1 during the current TTI, there is no need to perform a handover process and HeNB1 continues to serve UE5. In the case when UE5 has been served by HeNB2, there is a handover process from HeNB2 to HeNB1 for the next TTI. Thus, the last positive handover decision is made for load balancing between these two HeNBs. In the case of open access mode, if UE5 is a new user for both HeNBs, there is no handover process and HeNB1 decides to serve the UE5 to have the same number of UEs as HeNB2. This problem can be solved when more strict regulations concerning a feedback are applied as it is shown in (3.8) when the number of feedback bits n defines the maximum number of femtocells for that any UE can send feedback. Thus, UE should send different feedbacks to different femtocells. In this case, UE5 sends the feedback only to HeNB1 or HeNB2 and not both of them, since the FIs are the same. The scenario in Fig. 3.13 considers that all femtocells are operated in open access mode. In the case of closed access mode, the UEs registered by the CSG of the related HeNB are allowed to use the available PRBs only. If the femtocells are operated in the hybrid access mode, the conditions presented in Fig. 3.13 are changed and the new conditions are presented in Fig. 3.14.

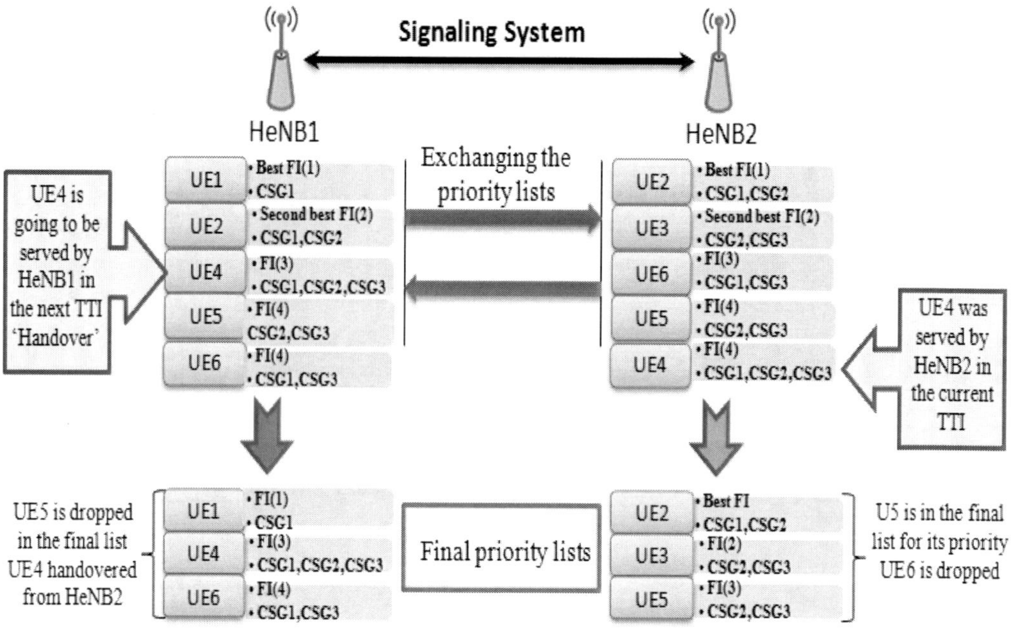

Figure 3.14. The handover approach: hybrid access mode.

The UE4 in Fig. 3.14 is a member of CSG1 and CSG2 related to HeNB1 and HeNB2, respectively, and UE4 is served by HeNB2 in the current TTI. During the current TTI, a positive handover decision is made. Thus, the UE4 is served by HeNB1 during the next TTI similarly to the UE4 case in Fig. 3.13. UE5 has the same FI value in both HeNBs. UE5 is a member of CSG2 only. By this reason, UE5 is excluded from the final priority list of HeNB1 and is included to the final priority list of HeNB2 according to the priority of serving the UEs that belong to the same CSG in the hybrid access mode. Owing to the fact that UE6 is a registered member of CSG1 and is not a member of CSG2, UE6 appears in the final list of HeNB1 even though when it has better FI for HeNB 2. A violation of the CGS membership priority can be done in the case when the load balance between the femtocells is taken into account.

Under the proposed feedback scheme and handover approach, the QoS criterion should consider the PRB scheduling process and load balance among the femtocells. The guaranteed bit rate (GBR) is considered as the only QoS criterion for all different services. Based on the GBR and FI of any UE, the required number of PRBs to be allocated to the UE is determined as [22]:

$$N_{PRB,i} = \frac{GBR_i}{M\,BW_{PRB}\,S_i}, \tag{3.9}$$

where $N_{PRB,i}$ is the required number of PRBs per TTI for ith UE, GBR_i is the GRB of ith UE, M is the number of OFDM symbols in the PRB, BW_{PRB} is the bandwidth of the PRB, and S_i is the spectral efficiency of the ith UE that can be presented in the following form [22]:

$$S_i = \log_2(1+\overline{SINR_i}) , \tag{3.10}$$

where $\overline{SINR_i}$ is the average SINR of ith UE over the whole frequency band of the femtocell. The basic admission control criterion can be presented as the sum of the number of PRBs per TTI required by new UE that requests admission and the number of PRBs per TTI required by active users in the femtocell. This sum should be less or equal to the total PRBs available in the femtocell. For instance, the admission control criterion can be presented in the following form:

$$\sum_{i=1}^{k} N_{PRB,i} + N_{PRB,new} \le N_{PRB,total} , \tag{3.11}$$

where $N_{PRB,i}$ is the number of PRBs required by the active ith UE, $N_{PRB,new}$ is the number of PRBs required by new UE, $N_{PRB,total}$ is the total number of PRBs available in the femtocell, and k is the number of active UEs in the femtocell.

It is important to note that the transmitted signal power defines the femtocell coverage area and has an impact on the interference, the handover process, and the UEs service off rate. In the downlink, the pilot signal power and the maximum transmit signal power used to limit the interference must be configured. In the proposed approach, the pilot signal power and the number of feedback bits for FI affect the performance of the handover procedure. The pilot signal power defines the femtocell range. Thus, for any moving UE with random initial position in the coverage area of multiple femtocells, the number of HeNBs in the UE white list varies from one position to another.

3.7.3. CQI and FI Mapping

According to the standards TS36.213-V8.2.0, the UE must perform the periodic and aperiodic feedbacks. A single CQI corresponds to index pointing a value in the CQI table. The CQI index is defined in terms of a channel coding rate and modulation scheme. Table 3.2 shows the corresponding parameters of each CQI index (4-bits CQI table).

In LTE networks, the adaptive modulation and coding have to ensure a block error ratio (BLER) less than 10^{-1}. Thus, the signal-to-noise ratio (SNR) and CQI mapping are required and can be obtained by plotting the BLER curves and choosing the SNR corresponding to BLER equal to 10^{-1}.

Fig. 3.15 shows the mentioned BLER curves as a function of SNR. The measured SINR and CQI mapping and the mapping model are presented in Fig. 3.16.

According to the LTE standards, there are three types of CQI report. As we can see from the Fig. 3.17, there are the CQI reports for specific UE and PRB [32]. The first CQI report type is the CQI report about the selected PRB. The second type is the CQI mean for the whole frequency band. The third type is the transport block CQI considered as a specification of transport block size, for which the UE would be able to receive data at BLER=10^{-1}.

Table 3.2. 4 bits CQI table

CQI Index	Mod	Coding rate X 1024	Efficiency
0	Out of Range (N/A)		
1	QPSK	78	0.1523
2	QPSK	120	0.2344
3	QPSK	193	0.3770
4	QPSK	308	0.6016
5	QPSK	449	0.8770
6	QPSK	602	1.1758
7	16QAM	378	1.4766
8	16QAM	490	1.9141
9	16QAM	616	2.4063
10	64QAM	466	2.7305
11	64QAM	567	3.3223
12	64QAM	666	3.9023
13	64QAM	772	4.5234
14	64QAM	873	5.1152
15	64QAM	948	5.5547

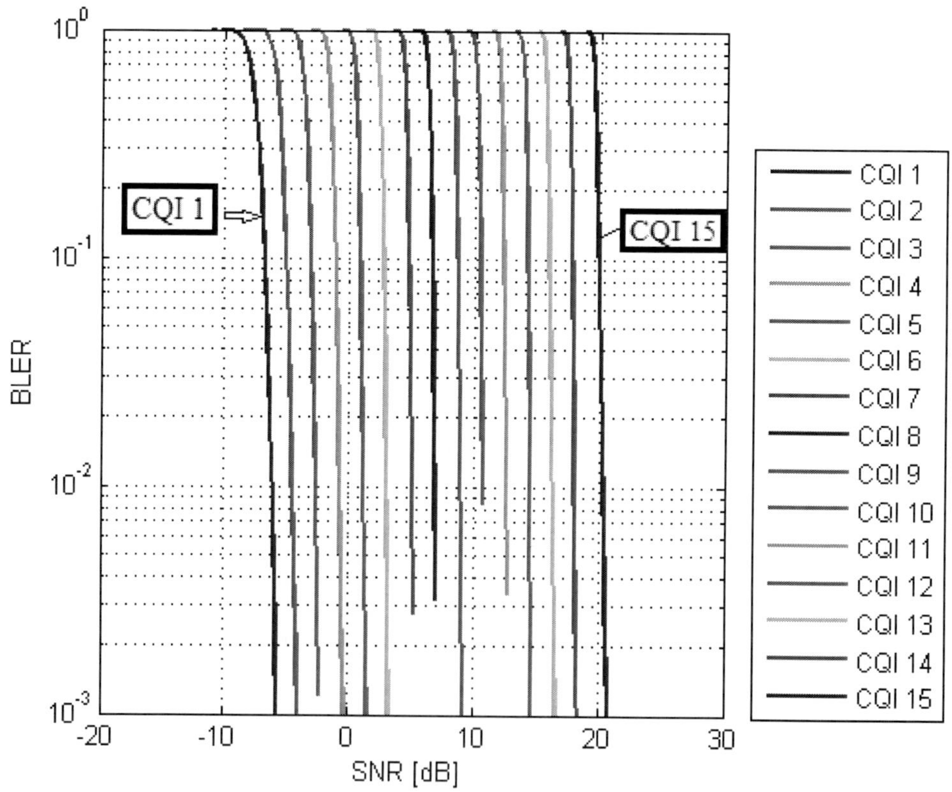

Figure 3.15. LTE BLER for CQI mapping.

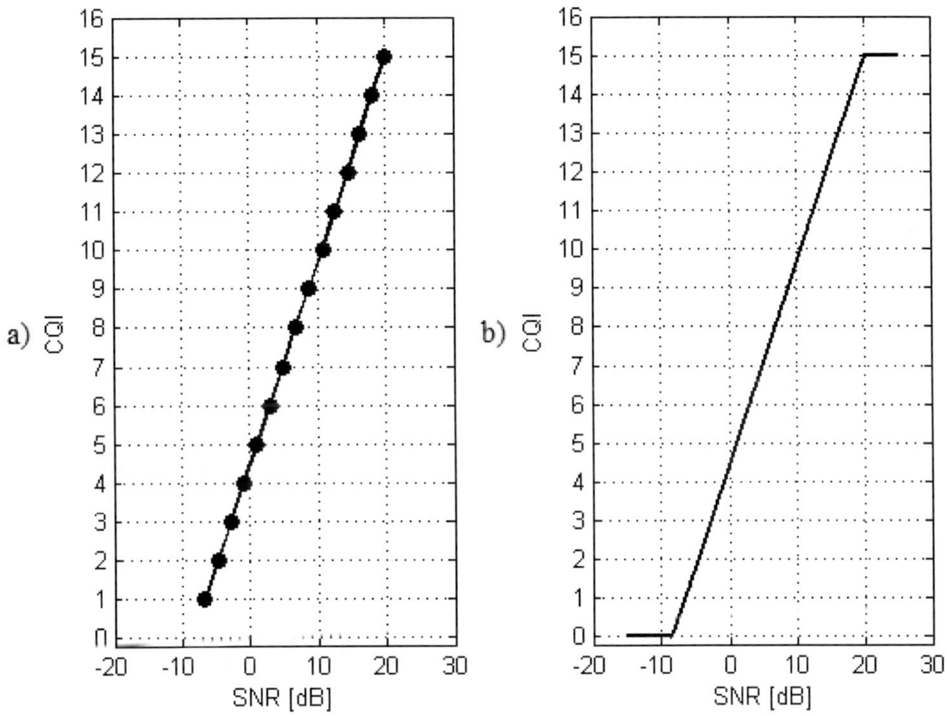

Figure 3.16. 4 bits SNR and CQI mapping: a) measured mapping; b) mapping model.

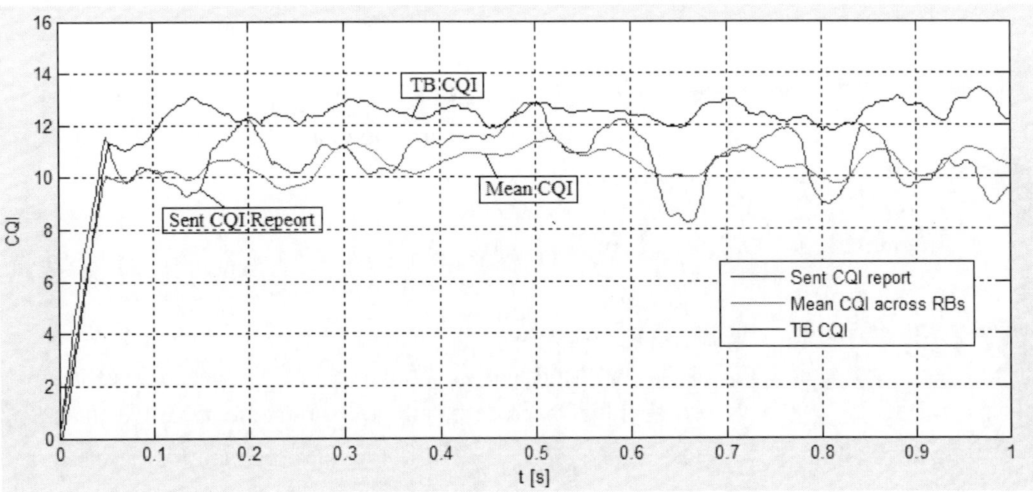

Figure 3.17. CQI report types.

Following the same way for the proposed feedback scheme and handover approach, it is possible to obtain a SINR and FI mapping model based on determination of the effective SINR ($SINR_{eff}$) for specific femtocell transmit power P_{femto} to design the SINR and FI mapping model. Figure 3.18 presents a simple example for the required SINR and FI mapping when the number of bits $n = 3$ and maximum $SINR_{eff} = 3$ dB.

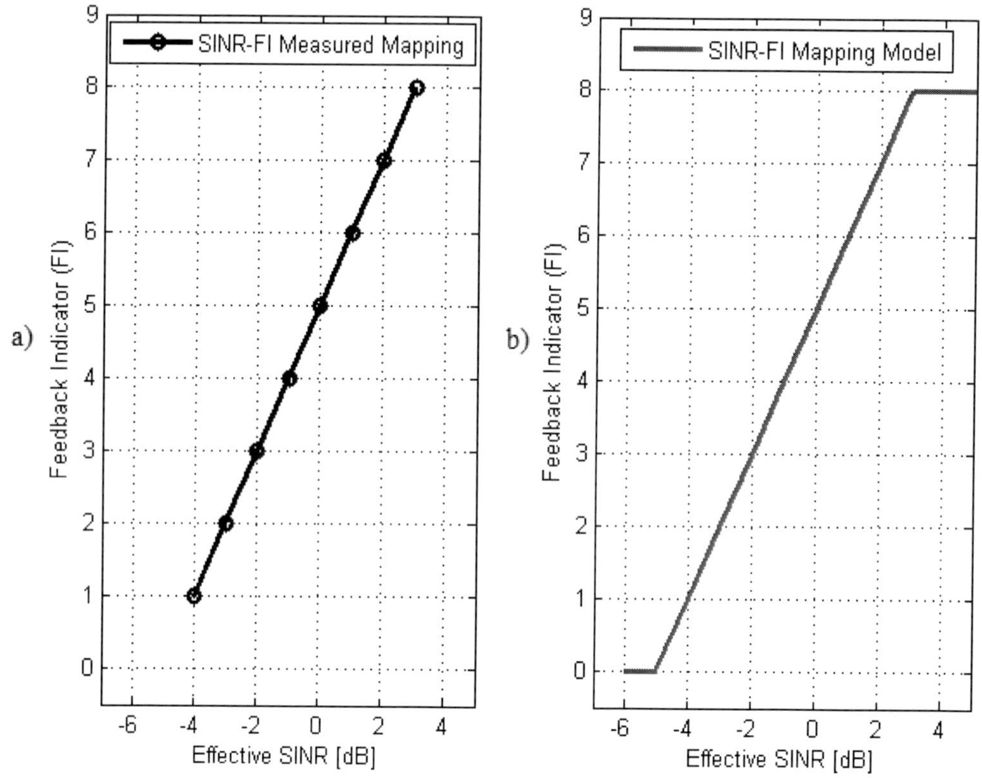

Figure 3.18. FI mapping: a) SINR and FI measured mapping; b) SINR and FI mapping model.

The femtocell transmit power P_{femto} has an average value equal to the power received from the closest macrocell at the target cell radius r corresponding to maximum power $P_{\max}$. The femtocell transmit power is defined in the following form [33]:

$$P_{femto} = \min\{P_{macro} + G(\theta) - L_{macro}(d) + L_{femto}(r),\ P_{\max}\}\ , \tag{3.12}$$

where P_{macro} is the transmit power of the sector where the femtocell is located, $G(\theta)$ is the antenna gain in the direction of the femtocell and θ is the angle of the received signal from the macrocell, $L_{macro}(d)$ is the average macrocell path loss at the femtocell distance d excluding any additional wall losses, and $L_{femto}(r)$ is the line of sight path loss at the target cell radius r excluding any wall losses and given by [33]

$$L_{femto}(r) = 38.5 + 20\lg r\,. \tag{3.13}$$

3.8. Throughput Definition

According to the basic Shannon theorem for the AWGN channel with average transmit power constraint, the capacity is defined in the following form [1]:

$$C = BW \log_2\left(1+\frac{P_t}{N_0 BW}\right), \tag{3.14}$$

where C is the capacity in bit per sec, BW is the bandwidth in Hz, P_t is the transmit power in watts, and N_0 is the noise power spectral density measured in watts. Increase in the transmit power leads to increase in the channel capacity. Increase in the capacity is diminishing by increase in the transmit power as shown in Fig. 3.19 when the noise power spectral density $N_0 = 1$ watt and the bandwidth $BW = 1$ Hz. Increase in the bandwidth means an increase in noise power both at the input and output of the receiver. There is no any sense to increase the bandwidth only in order to increase the capacity as shown in Fig. 3.20.

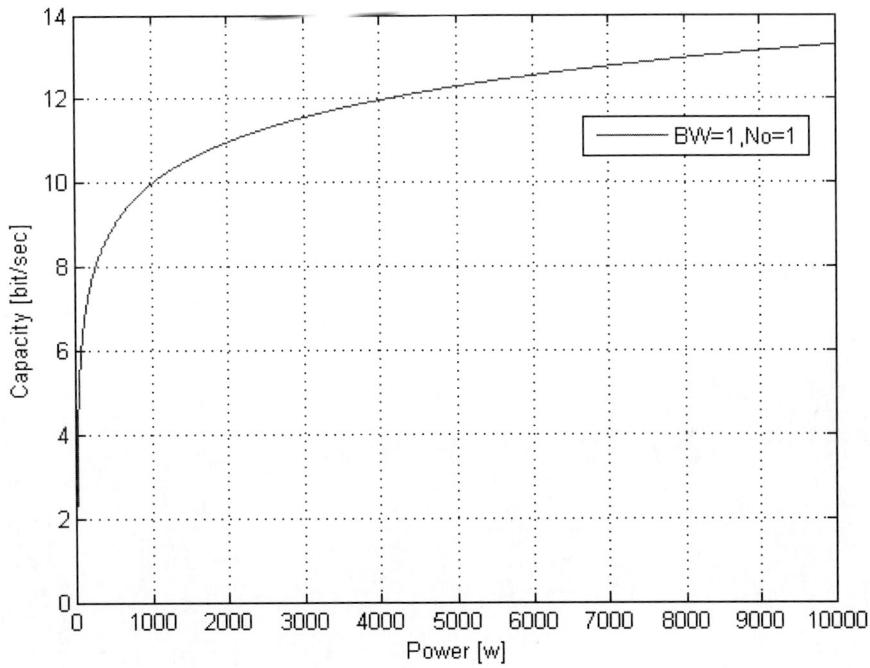

Figure 3.19. The channel capacity as a function of transmit power.

In practice, the variable in concern is the ratio between the energy per bit and the noise power spectral density E_b / N_0. Equation (3.14) can be written in another form to obtain the spectral efficiency assuming that $P_t = P_s = RE_b$, where R is the bit rate as follows:

$$\frac{R}{BW} < \log_2\left(1 + \frac{RE_b}{N_0 BW}\right). \tag{3.15}$$

Figure 3.21 shows the spectral efficiency as a function of E_b / N_0 . We can see that for constant bandwidth the bit rate increases with increasing in E_b / N_0.

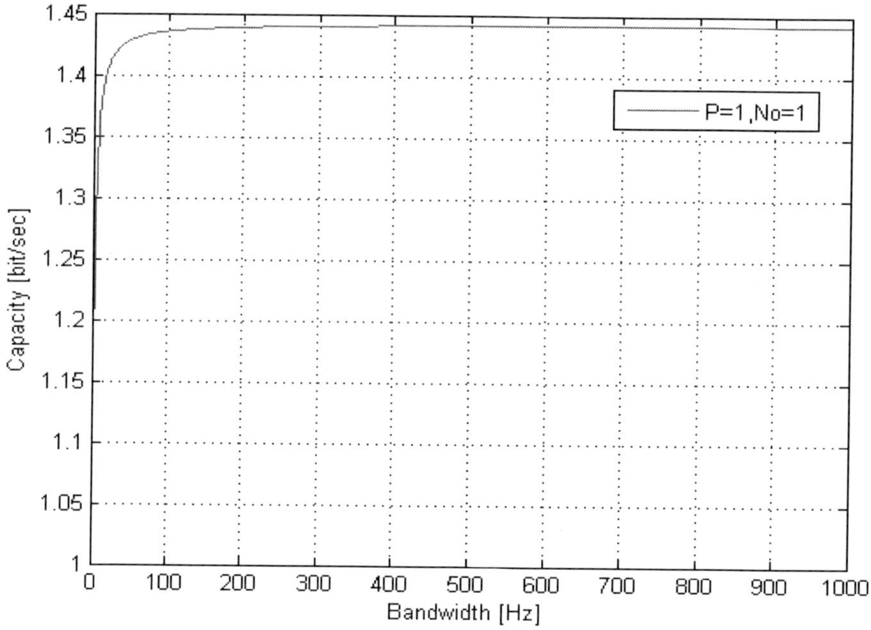

Figure 3.20. The channel capacity as a function of bandwidth.

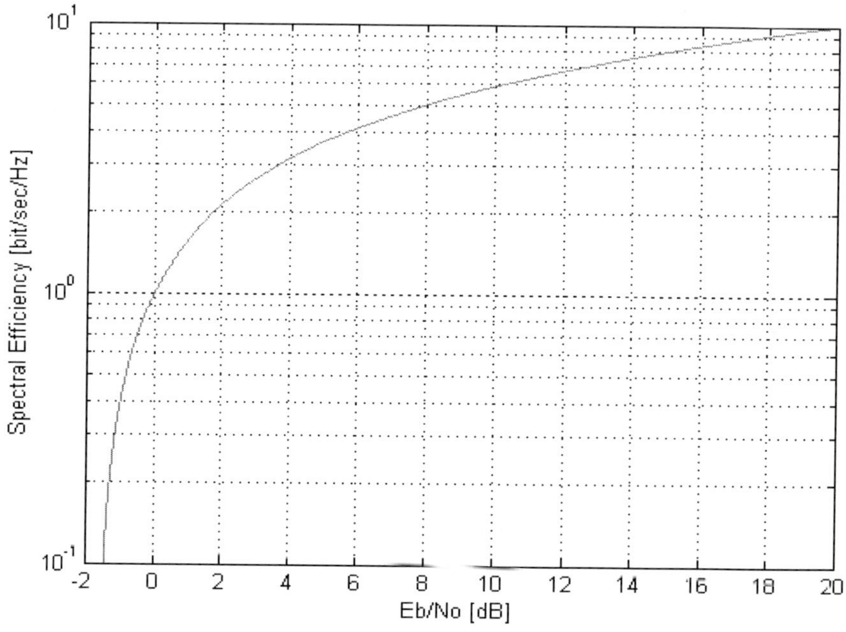

Figure 3.21. The spectral efficiency as a function of E_b / N_0 .

Many forms to define any communication system throughput are derived based on different factors and enabled technologies. In the case of using the frequency reuse, a significant improvement in the SINR can be achieved. Unfortunately, SINR improvement is accompanied by decreasing in the available spectrum per cell and, as a consequence, in the total system throughput. For this case, the system throughput can be estimated using the following form [1]:

$$THP_{FR} = \frac{BW}{FRF}\log_2(1 + SINR_{FR}) \;, \tag{3.16}$$

where FRF is the frequency reuse factor, $SINR_{FR}$ is the SINR in the case of frequency reuse. From the UE prospective, the expected throughput of the *i*th UE using the *j*th radio resource block depends on the expected SINR. Usually, the expected SINR is derived for the latest SINR measurement. Thus, the expected UE throughput can be presented in the following form [1]:

$$\hat{THP}_{i,j} = \Delta f \log_2\left(1 + S\hat{INR}_{i,j}\right), \tag{3.17}$$

where Δf is the resource block bandwidth, and $S\hat{INR}_{i,j}$ is the expected SINR of the *i*th UE using the *j*th resource block. Equation (3.17) is applicable in the case of macrocell UE as well as in the case of femtocell UE.

The total throughput of single or several macrocells under employment of femtocells based on the SINR is defined using the following form [1,34]:

$$THP = BW\log_2\left(1 + \frac{SINR}{10^{0.1\alpha}}\right). \tag{3.18}$$

where α is the operation factor (operation point). In general, the macrocell total throughput increases with increase in the number of femtocells and growth of SINR.

The practical capacity of macrocell *i*th UE on *m*th subcarrier for LTE advanced systems discussed in [35] can be given based on the estimated SINR by the following equation [35]:

$$C_{i,m} = \Delta f_{sc} \log_2\left(1 + \mu SINR_{i,m}\right), \tag{3.19}$$

where Δf_{sc} is the subcarrier spacing, $SINR_{i,m}$ is the estimated SINR of *i*th UE on an *m*th subcarrier, and μ is a constant determined by the desired bit error rate (BER). This constant can be given as

$$\mu = -1.5 / \ln(5 \times BER) \;. \tag{3.20}$$

The total throughput of any serving macrocell based on (3.19) and (3.20) is determined using the following form [36]:

$$THP_{macro} = \sum_{i}\sum_{m}\gamma_{i,m}C_{i,m} \ , \tag{3.21}$$

where $\gamma_{i,m}$ represents the subcarrier assignment for the macrocell UEs. For example, if $\gamma_{i,m} = 1$ it means the mth subcarrier is assigned to ith macrocell UE, otherwise, $\gamma_{i,m} = 0$. In the macrocell, each subcarrier is allocated to a single UE at every time slot. According to the orthogonal frequency division multiple access (OFDMA) characteristics, the following equality

$$\sum_{i=1}^{N_i}\gamma_{i,m} = 1 \ , \tag{3.22}$$

is always true, where N_i is the number of macro UEs in the macrocell.

3.9. Discussion

We assume that each user or UE gets access to the femtocell independently from other UEs. The total available bandwidth is 5 MHz. For simplicity, the frequency reuse factor is set to be $FRF = 1$. Thus, the available bandwidth or PRBs are shared equally by the macrocell and femtocells in the network (co-channel scenario). Table. 3.3 presents the main simulation parameters used to define the total macrocell throughput for different numbers of femtocells. The macrocell throughput is defined based on (3.18) while the estimation of the path losses from the femtocell site to the dominant macrocell boundary are defined using the following form [34]:

$$PL_{macro} = \begin{cases} 28 + 35\lg d + L_{walls} & [\text{dB}], \text{ for indoor} \\ 38.5 + 20\lg d & [\text{dB}], \text{ for outdoor} \end{cases} , \tag{3.23}$$

where is d the distance, and L_{walls} is the path losses caused by walls.

In general, the macrocell total throughput increases with increase in the number of femtocells and with the growth of SINR. Additionally, the throughput is related to the femtocell transmit power owing to the interference effects. Figure 3.22 presents the total macrocell throughput as a function of the number of femtocells in the case when SINR is defined for the macrocell radius equal to 800 m and femtocell radius equal to 30 m without considering the interference generated by other macrocells. The femtocell transmit power is 20 mW. As we can see from Fig. 3.22, the macrocell throughput increases with the number of femtocells under the condition when the interference is caused by the femtocells only.

Table 3.3. Parameters used to define the macrocell throughput

Parameter	Value
Maximum gain level ($G_{\max}$)	16 dB
Side-lobe gain level (G_s)	20 dB
Marco-cell transmit power	20 W
Femtocell transmit Power	20, 100 mW
FRF	1
BW	5 MHz
α	3 dB
d	10~1000 m
Number of interfering macrocells	10

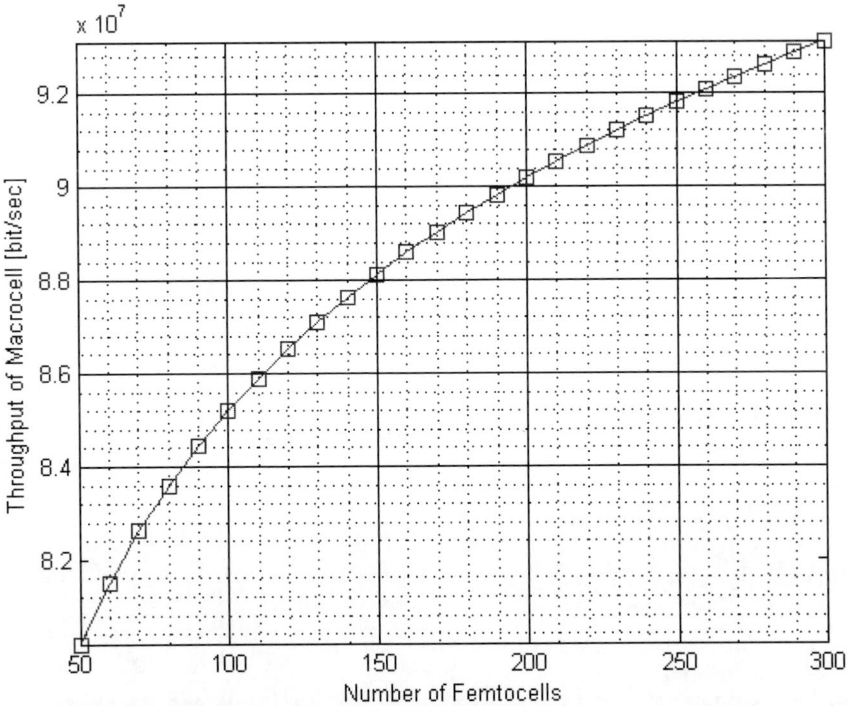

Figure 3.22. Macrocell throughput as a function of the number of femtocells: no interference from other macrocells.

In Fig. 3.23 the macrocell throughput is presented as a function of the number of femtocells taking into consideration the interference from femtocells and other macrocells. The SINR is defined when the femtocell transmit power is 100 mW, the number of interfering macrocells is equal to 10 and all of them are located from the subject macrocell on the distance equal to 500 m. In this case, Fig. 3.23 shows that the macrocell throughput decreases with increase in the number of femtocells owing to the effect of interference caused by other macrocells and femtocells since the transmit power of each femtocell is increased.

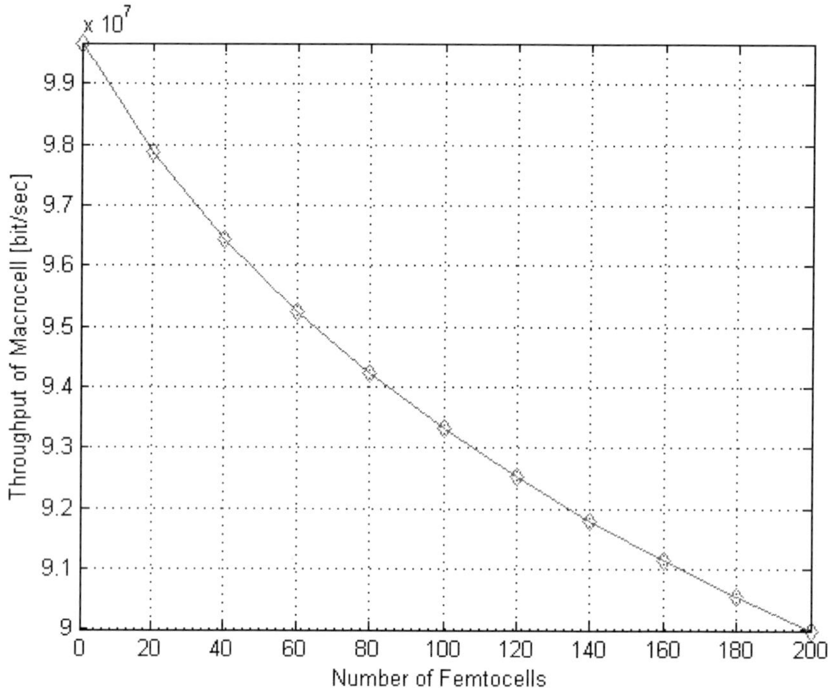

Figure 3.23. Macrocell throughput as a function of the number of femtocells: the interfernce caused by other macrocells is considered.

In the case of previous scenario concerning the femtocells employment according to co-channel spectrum planning and frequency reuse with $FRF = 1$, the FDPS scheme does not need excessive signaling overhead in the downlink and does not cause uncontrollable interference problems. The simulation results for FDPS in the 3GPP LTE network downlink is carried out under the conditions presented in Table. 3.4 assuming the typical urban model. The traffic model is a single 2 Mbit packet (UE downloads a fixed 2 Mbit packet). The PRB bandwidth is adjusted by changing the number of PRBs in 20 MHz total bandwidth.

Table 3.4. Simulation parameters of downlink throughput for 3GPP UTRAN-LTE cell

Parameter	Value
System bandwidth (BW)	20 MHz
Carrier frequency	2 GHz
Number of TTIs	100
Number of PRB (N_{PRB})	15~250
UDO	8, 10, 12
Modulation	QPSK
Traffic Model	Single 2 Mb packet
PRB bandwidth (BW_{PRB})	BW / N_{PRB}
Cell level UE distribution	Uniform
CQI reporting resolution	1 dB

The PRBs are scheduled based on the distributed proportional fair scheme. Figure 3.24 presents the cell throughput as a function of the number of PRBs within the limits of 20 MHz bandwidth for different user diversity order (UDO). The FDPS cell throughput is increased with the number of PRBs. FDPS performance is improved under increasing UDO.

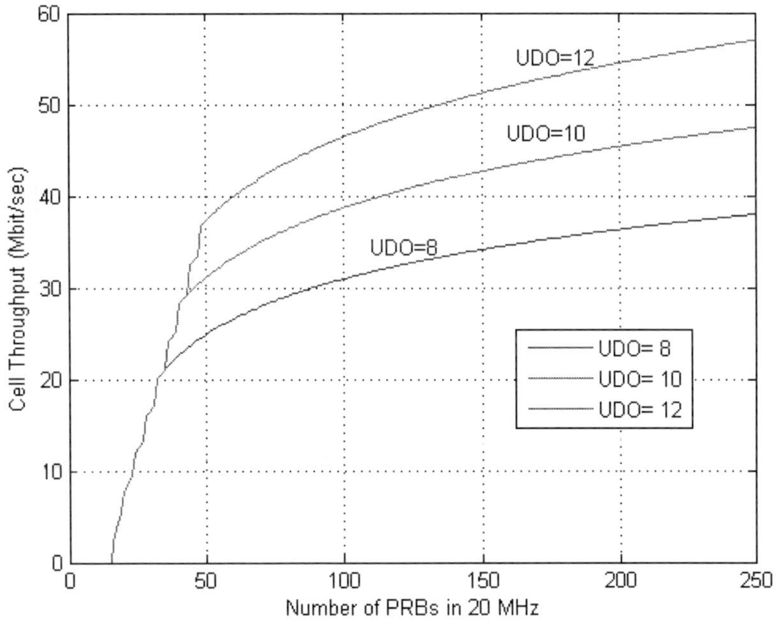

Figure 3.24. Cell throughput versus the number of PRBs in 20 MHz bandwidth under FDPS and distributed proportional fair scheduling scheme.

Table 3.5. Simulation parameters for cell throughput at different UDO and number of PRBs values

Parameter	Value
System bandwidth (BW)	$BW_{PRB} * N_{PRB}$
Carrier frequency	2 GHz
Number of TTIs	500
Number of PRB (N_{PRB})	12, 24, 48
UDO	2~20
Modulation	QPSK
Traffic Model	Single 2 Mb packet
PRB bandwidth (BW_{PRB})	375 kHz
Cell level UE distribution	Uniform
CQI reporting resolution	1 dB
Minimum distance	35 m
Maximum distance	500 m
Number of subcarriers corresponding to 375 KHz	25

In fact, increase in the number of PRBs within the limits of specific bandwidth leads directly to increase in the uplink CQI signaling and the downlink PRB allocation signaling overheads. As shown in Table. 3.5, the PRB bandwidth is limited to 375 KHz corresponding to 25 subcarriers. Another condition for FDPS performance factor with high number of UEs available for scheduling is discussed [9].

Figure 3.25 shows the average cell throughput as a function of UDO for the fixed number of PRBs equal to one of the following values: 12, 24, and 48. Figure 3.25 shows that the average cell throughput is increased with increase in the UDO with the fixed number of PRBs. Even at low values of UDO, the FDPS capacity gain is high, since each UE has optimization options in time and frequency domains. The best overall FDPS capacity performance is observed when the number of PRBs equals to 48. This result is consistent with the progress in LTE system [36].

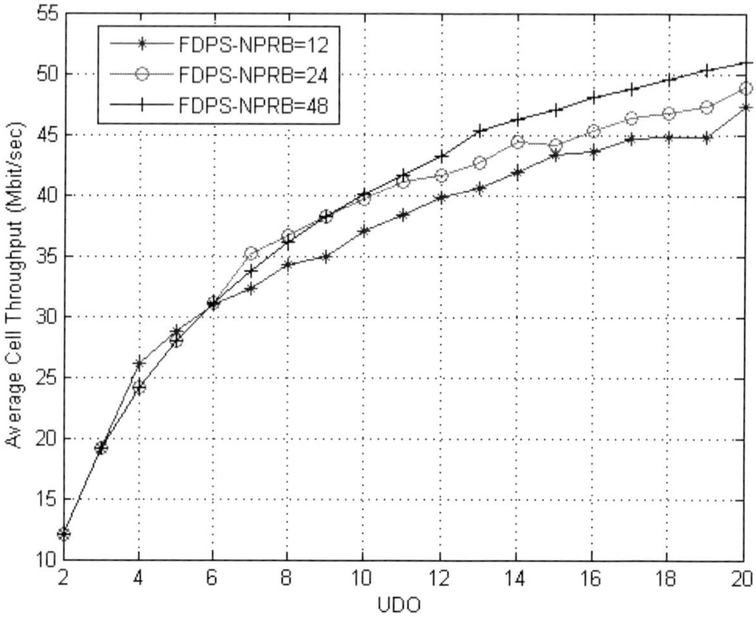

Figure 3.25. Average cell throughput versus UDO for the fixed number of PRBs.

As was discussed above, the femtocell coverage is determined according to the received power by UE from HeNB and the transmitted common pilot channel signal power. The common pilot channel signal is used by UE to estimate the channel quality, cell selection/reselection, and handover evaluation as both in the universal mobile telecommunication system and in the high speed downlink packet access networks. The power $P_{r_{i,n}}$ received by *i*th UE from *n*th HeNB can be presented in the following form [37,38]:

$$P_{r_{i,n}} = P_{t_{i,n}} - PL_{i,n} \geq \varepsilon \ , \tag{3.24}$$

where $P_{t_{i,n}}$ is the common pilot channel power transmitted from the HeNB of the nth femtocell to the ith UE, $PL_{i,n}$ is the path loss between the nth femtocell and ith UE, and ε is the minimum required power for UE to maintain a voice call, as it is recommended in [38,39].

The $P_{r_{i,n}}$ can be determined in term of the path loss coefficient between the HeNB and the UE as given by

$$P_{r_{i,n}} = G_{i,n} P_n \ , \tag{3.25}$$

where $G_{i,n}$ is the path loss coefficient between the nth HeNB and ith UE, and P_n is the HeNB transmitted power. The signal strength decreases inversely proportional to $d_{i,n}^4$, where $d_{i,n}$ is the distance between the UE and HeNB. In order to evaluate the introduced collaborative femtocell operation scheme, we employ the power assignment and PRB scheduling techniques based on the joint resource allocation algorithm (JRAA) [40].

The JRAA allocates, as many UEs as possible, to each available PRB in the system of arbitrary number of UEs and BSs (in our case HeNBs). The JRAA allows the BS and PRB assignment to be performed with the power control simultaneously. We consider that the configuration of the assumed number of UEs is already made by JRAA in terms of the association with the BSs or HeNBs. At this point, JRAA performs the PRBs allocation and the power control procedures for the UEs at each BSs or HeNBs. The employment of JRAA in the case of PRB allocation for femtocell UEs is restricted by interference. The channel or PRB reusing advantage is not possible without collaboration between femtocells. Assume that the number of available PRBs N_{PRB} is less than the number of active UEs N_{User} associated with specific number of HeNBs. In this case, i.e., $N_{User} > N_{PRB}$, some of the PRBs should be reused in order to serve all UEs. Owing to the interference restrictions, the PRBs reuse is possible if the proposed collaboration between femtocells by using the signaling system is applied. Otherwise, some UEs are not assigned to PRBs and, as a result, they are out of the service. The proposed collaboration and femtocell-to-femtocell handover approaches provide the required flexibility to assign and reuse the PRBs among UEs in the course of the PRB allocation process.

Table 3.6. Simulation parameters of JRAA

Parameter	Value
N_{User}	25
Number of HeNBs	4
N_{PRB} per UE	1
Desirable SINR	3 dB
PRBs Allocation	JRAA
Operation Model	Collaborative/ non collaborative
UE distribution	Random
Noise Power	0.1 W

The simulation results for the PRB scheduling under the use of JRAA method discussed in [40] are presented when the number of femtocells operated under open access mode is equal to 4, the number of UEs with random locations with respect to the HeNBs is equal to 25, and the desirable SINR with and without the collaborative mode based on the signaling system between the femtocells is equal to 3 dB (Fig. 3.26). Table. 3.6 presents the main simulation parameters used to obtain the results presented in Fig. 26.

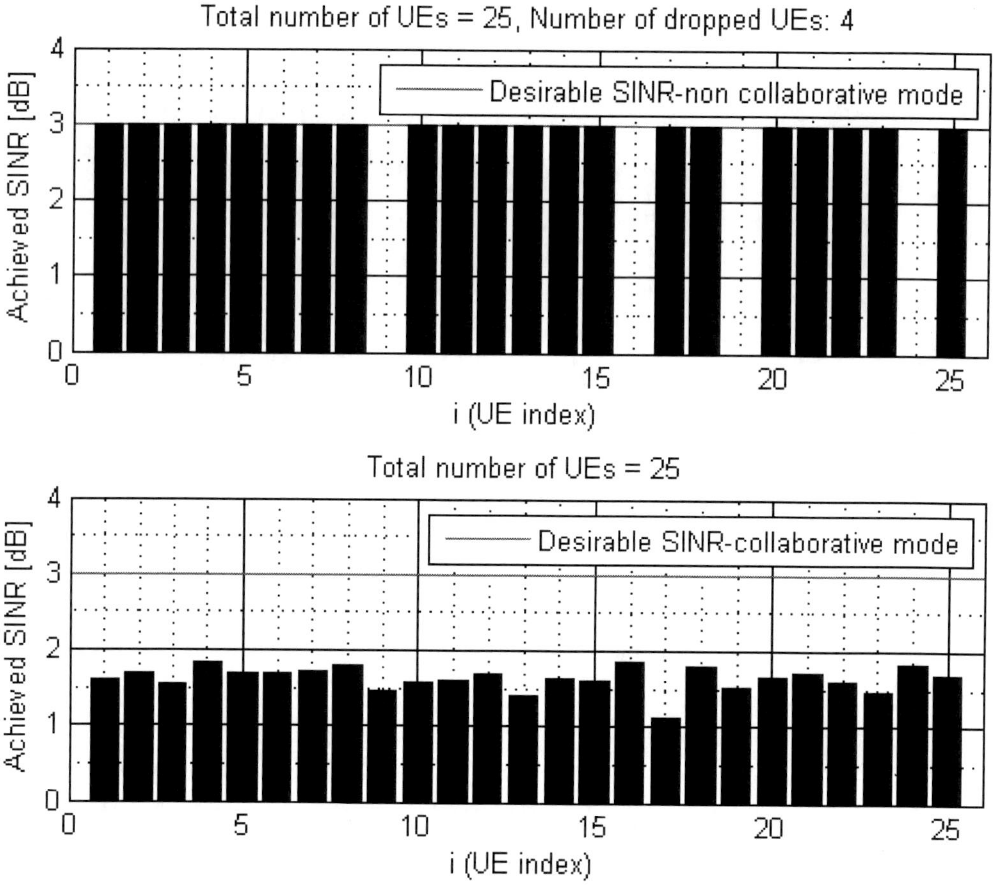

Figure 3.26. Comparison of power assignment and PRB scheduling among UEs between the non collaborative mode and collaborative mode.

When the collaborative mode is not applied between the femtocells (HeNBs), as we can see from Fig. 3.26, there are some service off UEs, because they are not assigned to PRBs owing to interference caused by applying PRB reuse strategy of JRAA in the case of femtocells. In the case of the collaborative mode, all the UEs are assigned to PRBs, i.e., all UEs are under service, and the PRB reuse is applied as a result of the collaboration between HeNBs in the femtocells. Thus, the proposed femtocell collaboration via the signaling system interface and handover scheme allows us to reduce the number of UEs excluded from the network service.

SUMMARY AND CONCLUSION

Recent investigations of data traffic show that the majority of mobile voice and data usage are concentrated in the indoor environments. Employment of femtocells in the LTE networks is required to enhance the transmission data rate for UEs, to extend and improve effectively the system coverage, to increase the whole network throughput, and to bring the network closer to the UEs. In the case of large scale and high density deployment of femtocells, many serious problems have to be taken into consideration, namely, the femtocell to macrocell, the macrocell to femtocell, and the femtocell to femtocell interference; the femtocell converge problems (gaps and overlaps); and the handover between the macrocell and femtocells as well as the handover between femtocells.

The development of SON techniques, especially, self configuration and self optimization, is an important problem under LTE femtocells deployment. The main RRM and PRB scheduling problems are discussed additionally to the power control, path loss models, CQI reporting, throughput definition methods, femtocell-to-femtocell handover problems.

The standard femtocell-to-femtocell handover procedure is applied using the femtocell backhaul that is an interface to the mobile core network through the public internet network. The required time to transmit a single message via the internet network, which can be over 200 msec, is high in comparison with the required time for handover process between macrocells. The small size of the femtocell leads to more frequent handovers for moving UE, in addition to the fact that handover control messages are sent via public internet. All these handover conditions make the handover procedure between femtocells slow and not efficient for fast moving UE. As a result, the modifications of femtocell-to-femtocell handover are needed.

The proposed femtocell-to-femtocell handover approach with the simple feedback scheme can help to accelerate the handover process, to manage the load between the femtocells, to overcome the drawback to use the slow public internet paths, to enhance the self optimization and self configuration capabilities, and to manage and avoid the interference in the case of sharing the PRBs among femtocells (co-channel scenario) or under applying the PRB reusing scheme. The assumed interface between femtocells in the absence of specific standardized features can be completed over the air (wireless link) or fast wired link between HeNBs (like X2 link). This interface makes the signaling between HeNBs fast, easy, and more efficient regardless of moving UE parameters.

The proposed simple feedback scheme and handover approach are flexible and can be integrated with different femtocell access modes, PRB scheduling, and power allocation techniques, for example, JRAA, cognitive radio resource management, and dynamic fractional frequency reuse radio resource management.

The JRAA is applied under the introduced conditions when the number of active femtocell UEs is more than the number of the available PRBs to be allocated for these UEs. Without the collaboration approach between the femtocells and owing to interference effects the probability that some UEs are out of the service or some UEs are not assigned to any PRBs is high. This problem could be solved applying the collaboration approach and handover scheme between femtocells.

The effectiveness to apply the discussed feedback scheme and handover approach with the radio resources scheduling techniques in the LTE networks for femtocell employment has to be confirmed after deep analysis in order to obtain a considerable reduction in the average scheduling delay and handover time.

REFERENCES

[1] Shannon, C., "A mathematical theory of communications," *Bell System Technical Journal*, 1948, Vol. 27, pt. 1, pp. 379–423.

[2] *Third generation partnership project (3GPP) LTE standards Rel-8, first LTE release, 2008.*

[3] *Third generation partnership project (3GPP) LTE standards Rel-9. WiMAX and LTE/UMTS interoperability, dual-cell HSDPA with MIMO, 2009.*

[4] Xie, B., Kumar, A., Zhao, D., Reddy, R., He, B., "On secure communications in intgrated heterogeneous wireless networks," *International Journal of Information Technology, Communications and Convergence (IJITCC)*, 2010, Vol. 1, No. 1, pp. 4–23.

[5] Dimou, K., Wang, M., Yang, Y., Kazmi, M., Larmo, A., Pettersson, J., Muller, W., Timner, Y., "Handover within 3GPP LTE: design principles and performance," *in Proceedings IEEE 70th Vehicular Technology Conference (VTC 2009-Fall)*, 20-23 September, 2009, Anchorage, Alaska, USA, pp. 1–5.

[6] 3GPP TS 36.211-840, *Evolved Universal Terrestrial Radio Access (E-UTRA): Physical Channels and Modulation.*

[7] 3GPP, Tech. Specif. *Group Radio Access Network; Physical Channel and Modulation (Release 8)*, 3GPP TS 36.211.

[8] Interference management in UMTS femtocells, femto forum white paper, *http://www.femtoforum.org/femto/ Index.php?id=69,* December, 2006.

[9] Pokhariyal, A., Kolding, T.E., Mogensen, P.E., "Performance of downlink frequency domain packet scheduling for the UTRAN long term evolution," *in Proceedings 17th Annual IEEE International Symposium on Personal, Indoor and Mobile Radio Communications (PIMRC'06)*, 11-14 September, 2006, Helsinki, Finland, pp. 1-5.

[10] *Physical Layer Aspects for Evolved Universal Terrestrial Radio Access* (UTRA), 3GPP TR 25.814.

[11] Shen, Z., Andrews, J.G., Evans, L.E., "Optimal power allocation in multiuser OFDM systems," *in Proceedings IEEE Global Telecommun. Conference*, 1-5 December, 2003, San Francisco, CA, USA,Vol. 1, pp. 337–341.

[12] Baamrani, K.E., Ouahman, A.A., Jimenez, V.P., Armada, A.G., S. Allaki, S., "Subcarrier and power allocation for the downlink of multiuser OFDM transmission," *Wireless Personal Communications*, 2006, Vol. 39, No. 4, pp. 457–465.

[13] Ko, S.J., Heo, J., Chang, K.H., "Aggressive subchannel allocation algorithm for efficient dynamic channel allocation in multiuser OFDMA system," *in Proceedings 17th Annual IEEE International Symposium on Personal, Indoor and Mobile Radio Communications (PIMRC'06)*, 11-14 September, 2006, Helsinki, Finland, pp. 1-5.

[14] Kivanc, D., Li, G., Liu, H., "Computationally efficient bandwidth allocation and power control for OFDMA," *IEEE Transactions on Wireless Communications*, 2003, Vol. 2, No. 6, pp. 1150-1158.

[15] Woo, C., Oh, E., Hong, D., "Simple dynamic subcarrier allocation with CQI feedback reduction for OFDMA systems," *IEEE Transactions on Vehicular Technology*, 2008, Vol. 57, No. 5, pp. 3299-3303.

[16] E3 work package 5, "A white paper by the FP7 project end-to-end efficiency (E3) - Support for heterogeneous standards using CPC," *https://ict-e3.eu/project/white papers/ CPC white paper FINAL.pdf*, June, 2009.

[17] Serrano, A.G., Giupponi, L., "Managing femto to macro interference without X2 interface support through POMDP," *Mobile Networks and Applications*, 2012, Vol. 17, No. 5, pp. 633–647, DOI 10.1007/s11036-012-0394-y.

[18] Littman, M.L., Cassandra, A.R., Kaelbling, L.P., "Learning policies for partially observable environments: scaling up," *in Proceedings 12th International Conference on Machine Learning*, 9-12 July, 1995, Tahoe City, California, USA, pp. 362–370.

[19] Kyosti, P., Meinila, J., Hentila, L., Zhao, X., Jamsa, T., Schneider, C., Narandzic, M., Milojevic, M., Hong, A., Ylitalo, J., Holappa, V.M., Alatossava, M., Bultitude, R., Jong, Y., Rautiainen, T., "Part I: channel models, IST-4-027756, WINNER II D1.1.2 V1.1," *WINNER II Channel Models,* 2007.

[20] Propagation data and prediction methods for the planning of indoor radio communication systems and radio local area networks in the frequency range 900 MHz to 100 GHz. *Rec. ITU-R P.1238.*

[21] Dang, M.S., Prakash, A., Anuekar, D.K., Kapoor, M., Shorey, R., "Fuzzy logic based handoff in wireless networks," *in Proceedings Vehicular Technology Conference (VTC2000)*, 15-18 May, 2000, Tokyo, Japan, pp. 2375-2379.

[22] Ulvan, A., Bestak, R., Ulvan, M., "The study of handover procedure in LTE-based femtocell network," *in Proceedings Wireless and Mobile Networking Conference (WMNC), Third Joint IFIP*, 13-15 October, 2010, Budapest, Hungary, pp. 1–6.

[23] Lee, H.Y., Lin, Y.B., "A cache scheme for femtocell reselection," *IEEE Communications Letters,* Vol. 14, No. 1, pp. 27–29.

[24] Chowdhury, M., Ryu, W., Rhee, E., Jang, Y.M., "Handover between macrocell and femtocell for UMTS based networks," *in Proceedings 11th International Conference on Advanced Communication Technology (ICACT 2009)*, 15-18 February, 2009, Gangwon-do, South Korea, Vol. 1, pp. 237–241.

[25] Kim, J.S., Lee, T.J., "Handover in UMTS networks with hybrid access femtocells," *in Proceedings 12th International Conference on Advanced Communication Technology (ICACT2010)*, 7-10 February, 2010, Bongpyeong-myeon, South Korea,Vol. 1, pp. 904–908.

[26] Rath, A., Panwar, S., "Fast handover in cellular networks with femtocells," *Center for Advanced Technology in Telecommunications (CATT) and the Wireless Internet Center for Advanced Technology (WICAT)*, research aricle, 2012.

[27] Becvar, Z., Mach, P., "Adaptive techniques for elimination of redundant handovers in femtocells," *in Proceedings 10th International Conference on Networks (ICN 2011)*, 23-28 January, 2011, St. Maarten, The Netherlands Antilles, pp. 230-234.

[28] Zonoozi, M., Dassanayake, P., Faulkner, M., "Optimum hysteresis level, signal averaging time and handover delay," *in Proceedings 47th IEEE Vehicular Technology*

Conference (VTC), 4-7 May, 1997, Arizona, USA, pp. 310 - 313, doi: 10.1109/VETEC.1997.596370.

[29] Becvar, Z., Zelenka, J., "Implementation of handover delay timer into WiMAX," *in Proceedings 6th Conference on Telecommunications (ConfTele 2007)*, 9-11 May, 2007, Peniche, Portugal, pp. 107-114.

[30] Ikuno, J.C., Wrulich, M., Rupp, M., "System level simulation of LTE networks," *in Proceedings 2010 IEEE 71st Vehicular Technology Conference (VTC2010)*, 16-19 May, 2010, Taipei, Taiwan, pp. 1-5.

[31] Shbat, M.S., Tuzlukov, V., "Handover technique between femtocells in LTE network using collaborative approach," *in Proceedings 18th Asia-Pacific Conference on Communications (APCC- 2012)*, 15-17 October, 2012, Jeju Island, South Korea, pp. 61-66.

[32] Ikuno, J.C., Rupp, M., Taranetz, M., Wrulich, M., "LTE system level simulator," Institute of Communications and Radio-Frequency Engineering, Vienna University of Technology, Austria, Web: *http://www.nt.tuwien.ac.at/ltesimulator*.

[33] Claussen, H., Ho, T.W.L., Samuel, L.G., "An overview of the femtocell concept," *Bell Labs Technical Journal*, 2008, Vol. 13, No. 1, pp. 221–246.

[34] Carroll, J.O., Claussen, H., Doyle, L., "Partial GSM spectrum reuse for femtocells," *in Proceedings 20th International Symposium on Personal, Indoor, and Mobile Radio Communications*, 13-16 September, 2009, Tokyo, Japan, pp. 2111-2116.

[35] Bouras, C., Kokkinos, V., Kontodimas, K., Papazois, A., "A simulation framework for LTE-A systems with femtocell overlays," in *Proceedings 7th ACM Workshop on Performance Monitoring and Measurement of Heterogeneous Wireless and Wired Networks (PM2HW2N'12)*, 21-25 October, 2012, Paphos, Cyprus, pp. 85-90.

[36] 3GPP Technical Report 25.814, version 1.0.3. *Physical Layer Aspects for Evolved UTRA*, Feb. 2006.

[37] Mohjazi, L.S., Al-Qutayri, M.A., Barada, H.R., Poon, K.F., "Performance evaluation of heuristic techniques for coverage optimization in femtocells," *in Proceedings 18th IEEE International Conference on Electronics, Circuits and Systems (ICECS)*, 11-14 December, 2011, Beirut , Lebanon, pp. 587-590.

[38] Mohjazi, L.S., Al-Qutayri, M.A., Barada, H.R., Poon, K.F., Shubair, R.M., "Self-optimization of pilot power in enterprise femtocells using multi objective heuristic," *Journal of Computer Networks and Communications*, Vol. 2012, Article ID 303465, 14 pages, doi:10.1155/2012/303465.

[39] 3GPP Technical Specification 25.104 v.7.9.0. *Base Station (BS) Radio Transmission and Reception (FDD)*, 2008.

[40] Papavassiliou, S., Tassiulas, L., "Improving the capacity in wireless networks through integrated channel base station and power assignment," *IEEE Transaction on Vehicular Technology*, 1998, Vol. 47, No. 2, pp. 417-427.

In: Communication Systems: New Research
Editor: Vyacheslav Tuzlukov

ISBN: 978-1-62618-654-5

Chapter 4

INTRUSION DETECTION IN VEHICULAR AD-HOC NETWORKS: A PHYSICAL LAYER APPROACH

Y. Abu Haeyeh, I. Abualhaol, Y. Iraqi and S. Muhaidat*

Khalifa University of Science, Technology and Research,
Abu Dhabi, United Arab Emirates

ABSTRACT

Vehicular ad-hoc networks (VANETs) are a critical component of the intelligent transportation and traffic management systems (ITTSs) which involve the application of advanced information processing, communications, sensors, and control technologies in an integrated manner to improve the functionality and the safety of both public and private transportation systems, providing drivers with timely information on road and traffic conditions including collisions, incidents, congestion, surface condition, etc., and achieving smooth traffic flow on the roads. In order to provide reliable and secure communications, intrusion detection systems (IDSs) can serve as a second defense wall after prevention-based approaches, such as encryption. This Chapter addresses the problem of detecting illicit wireless transmissions from the physical layer perspective, assuming the presence of regular ongoing legitimate transmissions. We consider energy detection-based spectrum sensing, which is a widely used spectrum sensing technique due to its simplicity. In particular, we assume all monitoring nodes perform local energy detection by collecting M samples of the received signals and, then, compare against a local threshold λ. A decision is made to activate an intrusion alarm, when the accumulated energy is greater than λ.

Keywords: VANETs, ITTSs, Intrusion Detection, Energy Detection

---

*Sami Muhaidat is a Visiting Reader at University of Surrey, UK. Email: muhaidat@ieee.org.

INTRODUCTION

The means of using moving vehicles as nodes in a wireless network have attracted researchers since 1980s. The increased number of wireless devices that can be used in vehicles including personal digital assistances (PDAs), laptops and cell phones brought more attention to research on VANETs in terms of routing, broadcasting, Quality of Service (QoS) and most importantly is security and privacy (Zeadally, Hunt, Chen, Irwin, & Hassan, 2010). In Vehicular Ad-Hoc Networks (VANETs) smart vehicles can benefit from the various applications of VANETs in Intelligent Transportation Systems (ITSs) which have significant impacts on our daily life starting from enhancing essential features like safety including collision avoidance systems, road conditions warning, and curve speed alarms, to allowing for transport efficiency and entertainment applications like web browsing and audio-video streaming that have been discussed briefly in the literature (Ezell, 2010).

There are two main components of vehicular networks, vehicles equipped with VANET technology and fixed roadside units (RSUs) that act as stationary nodes, receive information from passing vehicles and broadcast common messages. When vehicles share broadcast messages, this is called Inter-vehicle communication, while vehicles interacting with roadside unit correspond to vehicle-to-roadside communication. The main factors that would influence the configuration of these communication links is the reliability, privacy and security. Security is challenging in this unstructured network due to several reasons including: the lack of central coordination in several applications, the random connectivity between the vehicles, and the hidden terminals problem. (Hartenstein & Laberteaux, 2008)

Prevention based methods such as authentication services and access control cannot prevent all possible attacks. Thus intrusion detection is needed to act as a second defense wall to enhance the security of Vehicular networks, since any malicious intruder with access to the open medium of VANET can threaten the information security and as a consequence affects the passenger's safety. There is a need for a secure and reliable system to ensure that messages with life critical information will not be modified or discarded or even inserted by any attacker. In order to ensure secure dissemination of traffic flow messages and safety warnings from the physical layer perspective, a dynamic Intrusion detection system with multiple thresholds algorithm is presented and discussed in this chapter.

SYSTEM MODEL

In this chapter we address the problem of detecting malicious roadside attacker that transmit forged warning messages to the vehicles passing its transmission range to force them react in a certain manner while the Road Side Unit (RSU) is transmitting the real legitimate messages to the vehicles within its transmission range as shown in Fig.1. Hence, we assume the presence of one legitimate transmission and one Intrusion in each vehicle (node) monitoring range, we consider local detection as each vehicle (node) performs energy detection individually by collecting M number of samples which correspond to the time bandwidth product TW to produce its decision based on the collected energy, if it is above a known threshold it transmits an intrusion alarm to the RSU which takes its role in warning the rest of the vehicles in the area.

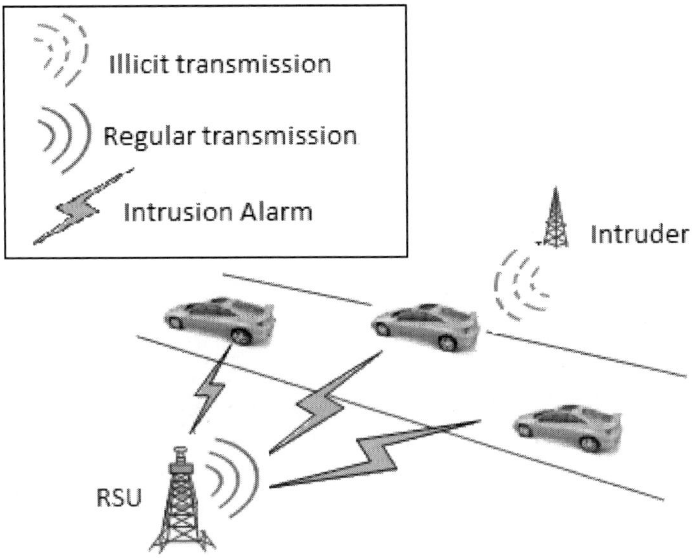

Figure 1. System Model.

The monitoring vehicle (node) detection performance depends on two factors. The first one is the probability of false alarm P_{fa}, when it detects an intrusion while the received signal is just an interfering noise or a regular legitimate signal. The second factor is the probability of miss detection P_{md} whenever the intruder is present while the node declares no intrusion. In the proposed detector design we target minimizing the joint probability P_e of P_{md} and P_{fa} by using a new detection scheme with multiple detection thresholds based on the knowledge of both legitimate and Intruder's transmission powers.

The Principle of Energy Detection

The most popular non-coherent spectrum sensing technique is the energy detection. It decides the presence of a signal based on comparing the estimated received energy with a predefined threshold derived from the statistics information available for the noise. Energy detector is a good choice when your target is an unknown deterministic signal (URKOWITZ, 1967) under Additive White Gaussian Noise (AWGN) channel due to its simplicity without the need for any prior information about the signal, even though it is not the most efficient detection technique since its performance degrades considerably under multipath fading channels especially at low Signal to Noise Ratio (SNR).

The energy detector function is to differentiate between the following hypotheses:

$$y_j = \begin{cases} n_j & H_0 \\ I_j + n_j & H_1 \end{cases} \quad j = 1, \ldots, M \tag{1}$$

Where M is the number of samples collected for energy measurement derived from the product of the sensing time with the filter bandwidth (TW). While hypothesis H_0 means no intrusion and H_1 corresponds to the presence of intrusion signal, n_j is the Additive White Gaussian Noise (AWGN), and I_j is the unknown Intrusion signal. When the signal has an unknown form, we treat the signal samples as random processes. Thus, the samples of the transmitted signal follow an independent and identically distributed (i.i.d.) random process with zero mean and variance $I?_I{}^2$, i.e. $I_j \sim \mathcal{N}\left(0\,, I?_I{}^2\right)$ and $n_j \sim \mathcal{N}\left(0\,, I?_n{}^2\right)$. When M samples are collected for energy measurements, the received energy Y will be:

$$Y = \sum_{j=1}^{M} y_j{}^2 = \begin{cases} \sum_{j=1}^{M} n_j{}^2 & H_0 \\ \sum_{j=1}^{M} \left[I_j + n_j\right]^2 & H_1 \end{cases} \tag{2}$$

As can be seen in the energy detector block diagram (Fig. 2), the signal is filtered with Band Pass Filter (BPF) to normalize the noise variance and digitized by the Analog to Digital Converter (ADC), then squared and integrated over the observation time interval T to produce an estimate value for the received signal energy to be compared with a known threshold value to decide the presence or absence of intruders.

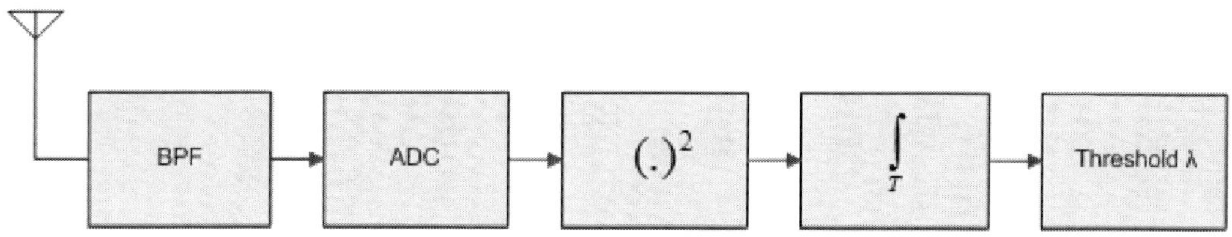

Figure 2. Energy detector Block diagram.

The decision statistics Y has a central Chi-square distribution with M degrees of freedom in hypothesis H_0 and non-central Chi-square distribution with M degrees of freedom and non-centrality parameter $(2I?)$ in hypothesis H_1, where $I?$ corresponds to Signal to Noise Ratio (SNR) (Digham, Alouini, & Simon, 2003),

$$Y \sim \begin{cases} \chi_M^2 & H_0 \\ \chi_M^2\,(2\gamma) & H_1 \end{cases} \tag{3}$$

For large number of samples, we can apply the central limit theorem to approximate the test statistics as Gaussian random variable, so we can evaluate the probability of detection P_d, and the probability of false alarm P_{fa} based on its mean and variance as

$$P_d = P\{Y > \lambda |\, H_1\} = Q\left(\frac{\lambda - M\left(\sigma_I^2 + \sigma_n{}^2\right)}{\sqrt{2M}\left(\sigma_I^2 + \sigma_n{}^2\right)}\right) \tag{4}$$

$$P_{fa} = P\{Y > \lambda |\, H_0\} = Q\left(\frac{\lambda - M\,\sigma_n{}^2}{\sqrt{2M}\sigma_n{}^2}\right) \tag{5}$$

where $Q(x) = \int_x^{\infty} \frac{1}{\sqrt{2\pi}} \exp\left(-\frac{u^2}{2}\right) du.$

In the Energy Detection literature there are two different ways for setting the detection threshold, by targeting a fixed false alarm rate P_{fa} which is called Constant False Alarm Rate (CFAR) or setting a target P_d which is called a Constant Detection Rate (CDR). Since the probability of detection P_d needs more knowledge about the Intrusion signal, it is more practical to set the threshold based on fixed P_{fa} , hence the detection threshold λ can be evaluated as

$$\lambda = \sigma_n{}^2 \left(\sqrt{2M}Q^{-1}(P_{fa}) + M\right) \quad (6)$$

It is important to know the sensing time available on hand by computing the critical number of samples M needed for a desired probability of detection P_d and a tolerable false alarm rate P_{fa} for a certain SNR value $I?$, which can be derived from the above formulas for P_d and P_{fa} where ($I? = \frac{I?_I^2}{I?_n{}^2}$) as following;

$$Q^{-1}(P_d) = \frac{\lambda - M\,(\sigma_I^2 + \sigma_n{}^2)}{\sqrt{2M}(\sigma_I^2 + \sigma_n{}^2)} = \frac{\sigma_n{}^2\,(\sqrt{2M}Q^{-1}(P_{fa}) + M) - M\,(\sigma_I^2 + \sigma_n{}^2)}{\sqrt{2M}\,(\sigma_I^2 + \sigma_n{}^2)}$$

$$Q^{-1}(P_d) = \frac{\sigma_n{}^2}{(\sigma_I^2 + \sigma_n{}^2)} Q^{-1}(P_{fa}) - \frac{M\sigma_I^2}{\sqrt{2M}\,(\sigma_I^2 + \sigma_n{}^2)} = \frac{1}{(\gamma+1)} Q^{-1}(P_{fa}) - \frac{\sqrt{M}\ \gamma}{\sqrt{2}\,(\gamma+1)}$$

$$\left(Q^{-1}(P_d) - \frac{1}{(\gamma+1)} Q^{-1}(P_{fa})\right)\left(\frac{\gamma+1}{\gamma}\right) = -\sqrt{\frac{M}{2}}$$

$$M = 2\left[\gamma^{-1}\left(Q^{-1}(P_d) - Q^{-1}(P_{fa})\right) + Q^{-1}(P_d)\right]^2 \quad (7)$$

A Novel Intrusion Detection Scheme for VANETs

When studying intrusion detection in VANETs, we need to consider the regular communications ongoing in the network, hence, assuming the presence of legitimate signal, the detection problem will become as the following:

$$H_0 : y_j \begin{cases} n_j & with\ 1-p \\ S_j + n_j & with\ p \end{cases} \text{No Intrusion} \quad (8)$$

$$H_1 : y_j \begin{cases} I_j + n_j & with\ 1-p \\ I_j + S_j + n_j & with\ p \end{cases} \text{With Intrusion} \quad (9)$$

Where p denotes the probability of legitimate transmission presence, which means that $(1-p)$ refers to the case of legitimate signal absence. Then the energy of the received signal at the energy detector is the test statistics, denoted as Y if M samples are collected in both cases:

$$H_0 : Y = \sum_{j=1}^{M} y_j^{\,2} = \begin{cases} \sum_{j=1}^{M} n_j^{\,2} & (1-p) \\ \sum_{j=1}^{M} [S_j + n_j]^2 & p \end{cases} \tag{10}$$

$$H_1 : Y = \sum_{j=1}^{M} y_j^{\,2} = \begin{cases} \sum_{j=1}^{M} [I_j + n_j]^2 & (1-p) \\ \sum_{j=1}^{M} [I_j + S_j + n_j]^2 & p \end{cases} \tag{11}$$

It is difficult to detect malicious intruders in the presence of regular legitimate transmissions especially if both are using the same transmission system, thus we propose a multi-thresholds detection strategy that improved the detection performance and enabled the detector to distinguish between the different received signals indirectly based on each signal's SNR.

In our system, we assume the legitimate signal is present in the network with probability p, while the Intrusion signal is present in the network with probability α. To compute the probability of detection in the presence of legitimate signal, we need to evaluate the expectation and variance of the received signal in both hypotheses.

The test statistics Y with Gaussian approximation will have the following mean and variance assuming the legitimate signal absence:

$$Y \sim \begin{cases} \mathcal{N}\,(\mathrm{M}\,\sigma_n^{\,2}, 2M\,\sigma_n^{\,4}) & H_0 \\ \mathcal{N}\,(M(\sigma_I^2 + \sigma_n^2), 2M(\sigma_I^2 + \sigma_n^{\,2})^2) & H_1 \end{cases} \tag{12}$$

On the other hand, when we take into account the legitimate transmission presence, the decision statistics will become:

$$YY \sim \begin{cases} \mathcal{N}(M(\sigma_S^{\,2} + \sigma_n^2)\,, 2M(\sigma_S^{\,2} + \sigma_n^2)^2) & H_0 \\ \mathcal{N}\,(\mathrm{M}(\sigma_I^{\,2} + \sigma_S^2 + \sigma_n^2), 2\mathrm{M}(\sigma_I^{\,2} + \sigma_S^2 + \sigma_n^2)^2) & H_1 \end{cases} \tag{13}$$

Applying Gaussian approximation, the probability of false alarm can be found from (13) in H_0 considering the presence of a legitimate signal as:

$$P_{fa} = P\{Y > \lambda \,|\, H_0\} = Q\left(\frac{\lambda - M\,\sigma_n^{\,2}}{\sqrt{2M}\sigma_n^{\,2}}\right)(1-p) + Q\left(\frac{\lambda - M\,((\sigma_S^2 + \sigma_n^{\,2})}{\sqrt{2M}((\sigma_S^2 + \sigma_n^{\,2})}\right)p \tag{14}$$

While the probability of detection based on the expectation and variance of Y in H_1 can be evaluated as:

$$P_d = P\{Y > \lambda \,|\, H_1\} = Q\left(\frac{\lambda - M\,(\sigma_I^2 + \sigma_n^{\,2})}{\sqrt{2M}(\sigma_I^2 + \sigma_n^{\,2})}\right)(1-p) + Q\left(\frac{\lambda - M\,((\sigma_I^2 + \sigma_S^2 + \sigma_n^{\,2})}{\sqrt{2M}((\sigma_I^2 + \sigma_S^2 + \sigma_n^{\,2})}\right)p \tag{15}$$

Each detection threshold λ relates to a pair of (P_d, P_{fa}) that represent the receiver operation characteristics (ROC) of the energy detector. When a legitimate signal exists in the network, we have to be aware of its transmission power to differentiate it from malicious attacks, thus deriving a one fixed threshold value based on the probability of false alarm in (14) will not reduce the false alarms caused by the legitimate presence, since a fixed value will not reflect the cases for different hypotheses tests: no transmission, just legitimate or just Intrusion, and the existence of both signals in the network at the same time. Hence, a new multi value threshold algorithm is proposed and explained in the next section.

MULTI-THRESHOLDS ALGORITHM FOR INTRUSION DETECTION

In energy detection, the detector cannot distinguish between two signals given that it performs blind detection based on the signal's energy level without any information about its structure, modulation type, or any signal features. This will cause a high false alarm rate by detecting the legitimate signal instead of the intruder. We know that the key enabling for raising the detection probability and reducing the false alarm rate is setting an optimal detection threshold that controls the tradeoff between the probability of miss detection (P_{md}) and the probability of false alarm (P_{fa}). Motivated by this requirement, the need for a multi-thresholds algorithm is essential.

Based on the transmission power of each signal, there are three possible cases: first case when the legitimate power is higher than the intruder's power $I?_S{}^2 > I?_I{}^2$ which represents the worst case with high false alarm rate based on the traditional detection threshold algorithm. When $I?_S{}^2 > I?_I{}^2$, the high legitimate power affect the detector ability to detect the lower power intruders, and will consider any signal with energy higher than the threshold as Intruder, hence we propose a new three-threshold algorithm to overcome the false alarms issue and improve the system performance.

As given in Fig. 3, we have three different thresholds λ_1, λ_2, and λ_3

$$\lambda_1 = \sigma_n{}^2 \left(\sqrt{2M} Q^{-1}(P_{fa}) + M\right) \tag{16}$$

$$\lambda_2 = (\sigma_S^2 + \sigma_n{}^2)\left(\sqrt{2M} Q^{-1}(P_{fa}) + M\right) \tag{17}$$

$$\lambda_3 = (\sigma_I{}^2 + \sigma_n^2)\left(\sqrt{2M} Q^{-1}(P_{fa}) + M\right) \tag{18}$$

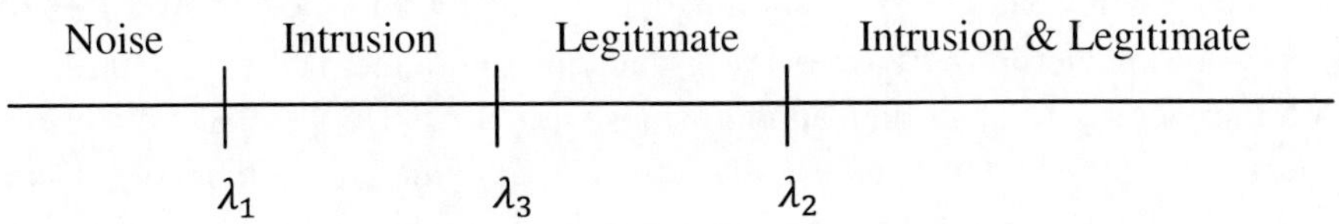

Figure 3. The proposed three-thresholds scheme when $\sigma_S{}^2 > \sigma_I{}^2$.

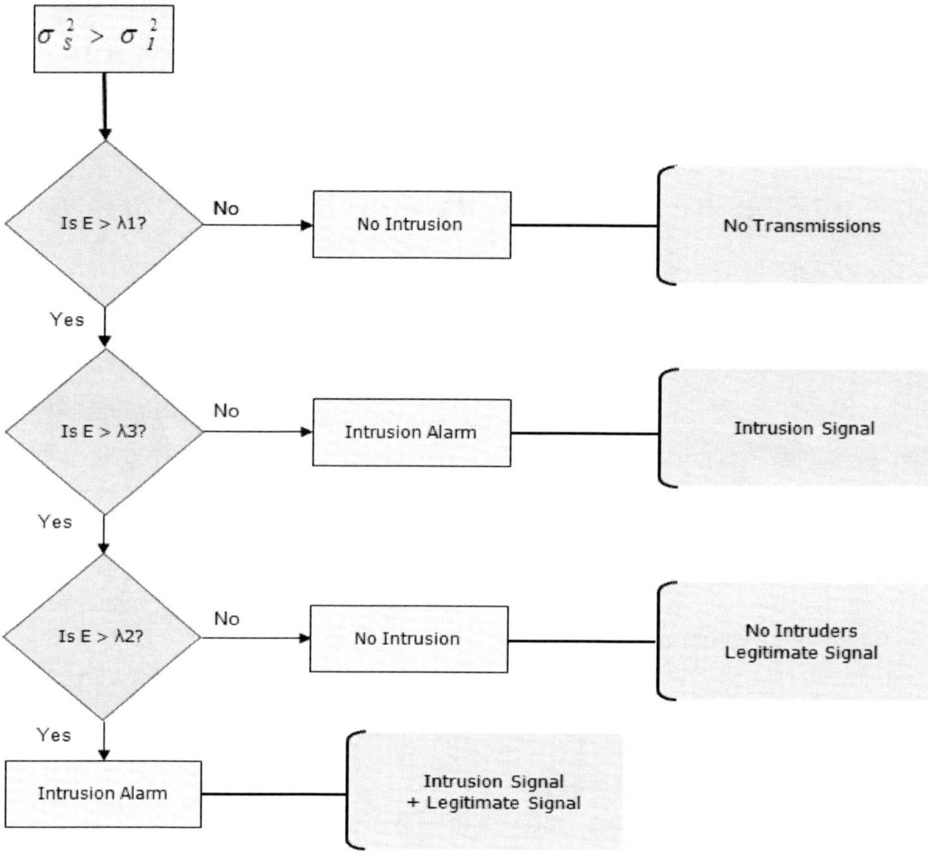

Figure 4. Flow chart of the proposed three-thresholds algorithm when $\sigma_S{}^2 > \sigma_I{}^2$.

The flow chart of the proposed three-thresholds algorithm when $\sigma_S{}^2 > \sigma_I{}^2$ is given in Fig. 4. In this algorithm, the detector compares the received energy with λ_1, if it is less, then for sure there are no intruders since there are no transmissions ongoing in the network. If the received energy is above λ_1 then there is a transmission either legitimate or intrusion or both. Then received energy is compared with λ_3 that depends on the Intruder's power (as shown in Equation 18) whose power in this case is lower than the legitimate, if it is less, it declares the presence of Intruder by transmitting an intrusion alarm to the RSU that will broadcast a warning message regarding the detected intruder. On the other hand if received energy is above λ_3, then the algorithm compares the energy with λ_2 to determine if the received signal is just a legitimate signal when the energy is lower than λ_2, then an intrusion alarm that announces detecting an intruder in the network is sent.

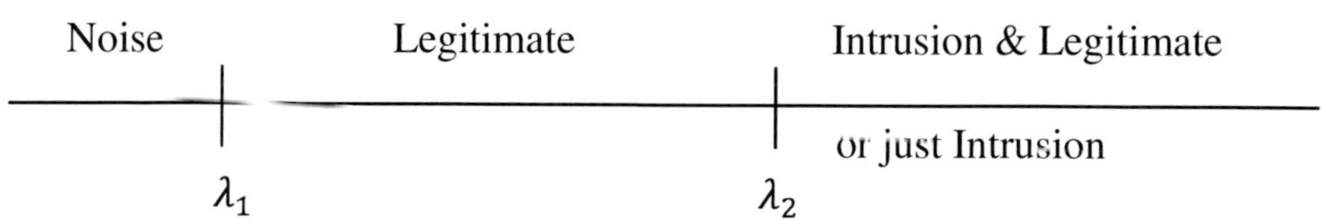

Figure 5. The proposed two-thresholds scheme when $\sigma_S{}^2 < \sigma_I{}^2$.

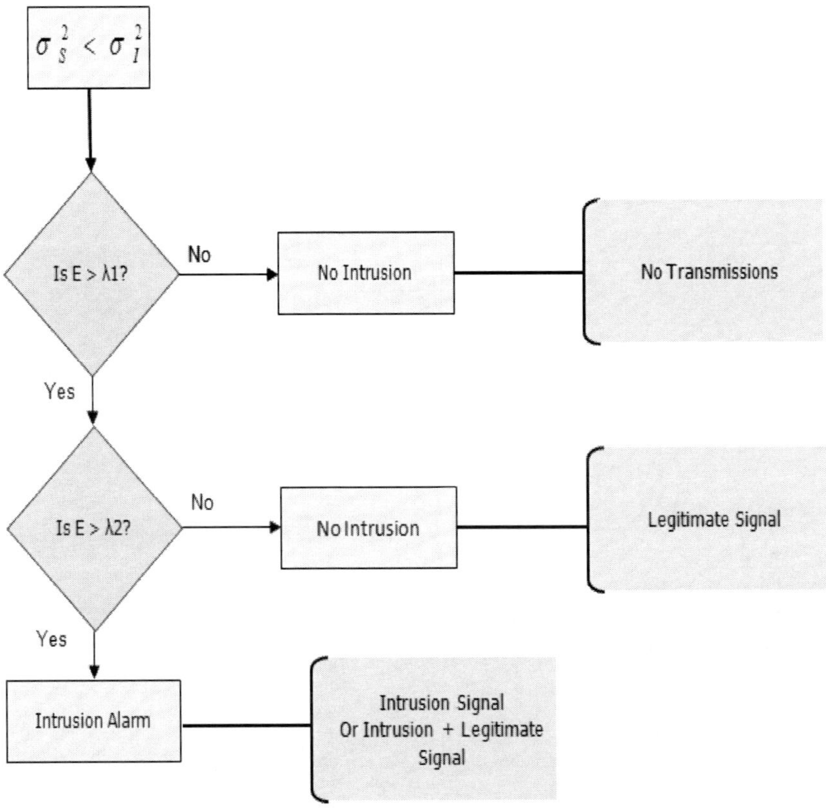

Figure 6. Flow chart of the proposed two-thresholds algorithm when $\sigma_S^2 < \sigma_I^2$.

In the second case the legitimate signal power is less than the intruder power ($\sigma_S^2 < \sigma_I^2$). In this case, two thresholds are enough for efficient detection as given in Fig.5, since it is easier to detect the higher power intrusion signal by comparing the received energy with λ_1 to check if there is a need to continue the detection process or stop it based on the presence of any transmissions, because there is no need to waste more sensing time while the medium is vacant. When there is a transmission the algorithm compares the energy with λ_2 to announce the detection of an Intrusion if it is above λ_2. If the energy is less than λ_2 , this means it is just a legitimate transmission and no attacks are predicted as given in Fig. 6.

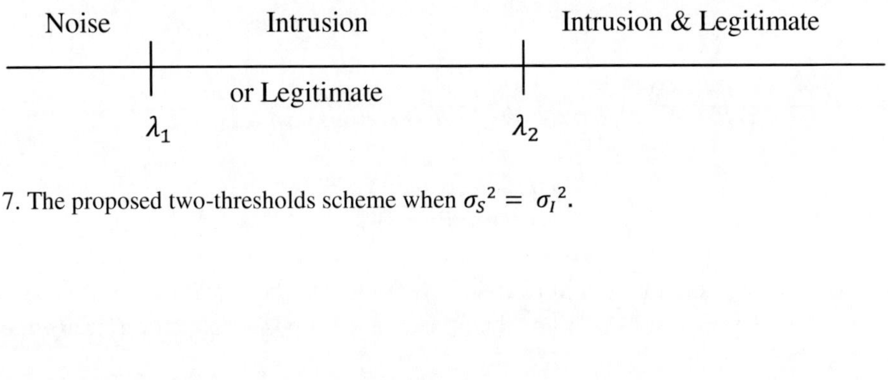

Figure 7. The proposed two-thresholds scheme when $\sigma_S^2 = \sigma_I^2$.

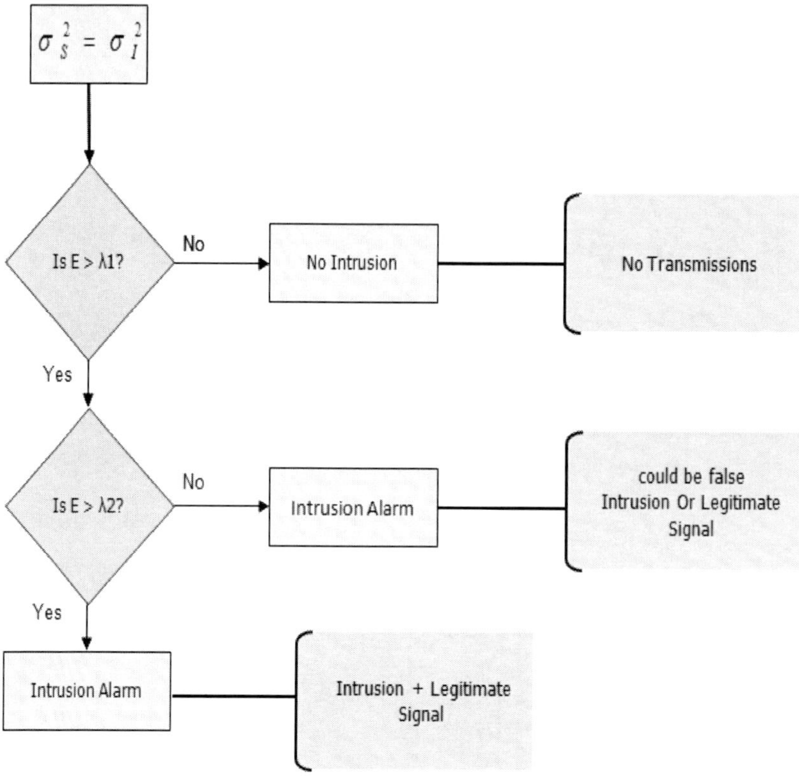

Figure 8. Flow chart of the proposed two-thresholds algorithm when ${\sigma_S}^2 = {\sigma_I}^2$.

Finally, in the third case the legitimate power equals the Intruders ${\sigma_S}^2 = {\sigma_I}^2$, which presents the case where λ_2 is equal to λ_3 , which basically means a higher false alarm rate because of the detector's inability to differentiate the Intruder from legitimate signals, similar to the previous cases, the algorithm starts comparing the received signal energy with λ_1 to ensure the existence of a transmission, while comparing with λ_2 will not affect its decision since it builds its decision based on the transmission power of each signal and they are equal (see Fig.8), having Intrusion and legitimate transmissions that reach the monitoring node (vehicle) with exactly the same power, if this happens then the detector would detect the legitimate as intruder falsely which lead to higher probability of false alarm.

The detection and false alarm probabilities P_d and P_{fa} with multiple thresholds can be given as:

$$P_d = Q\left(\frac{\lambda_1 - M\,(\sigma_I^2 + {\sigma_n}^2)}{\sqrt{2M}(\sigma_I^2 + {\sigma_n}^2)}\right)(1-p) + Q\left(\frac{\lambda_2 - M\,(\sigma_S^2 + \sigma_I^2 + {\sigma_n}^2)}{\sqrt{2M}(\sigma_S^2 + \sigma_I^2 + {\sigma_n}^2)}\right)p \tag{19}$$

$$P_{fa} = Q\left(\frac{\lambda_1 - M\,{\sigma_n}^2}{\sqrt{2M}{\sigma_n}^2}\right)(1-p) + Q\left(\frac{\lambda_2 - M\,((\sigma_S^2 + {\sigma_n}^2))}{\sqrt{2M}((\sigma_S^2 + {\sigma_n}^2))}\right)p \tag{20}$$

The effects of the new multiple thresholds algorithm on the system performance will be shown in the simulation results. The evaluation of the proposed algorithm performance in terms of the probability of false alarm P_{fa}, and the probability of miss detection P_{md} that formulate the probability of error P_e as

$$Pe = P(fa, H_0) + P(md, H_1) = P_{fa|H_0}P(H_0) + P_{md|H_1}P(H_1) \quad (21)$$

$$where \quad P_{md} = 1 - P_d \quad , \quad P(H_0) = 1 - \alpha \quad , \quad P(H_1) = \alpha$$

Results and Discussions

1. Simulation parameters

Table 1. Simulation parameters

Number of samples M	5000
Probability of false alarm P_{fa}	0.01
Probability of legitimate presence p	0.4
Probability of Intrusion presence α	0.5
Noise variance $\sigma_n{}^2$	1
Transmission system	BPSK

2. The conventional energy detector

To illustrate the behavior of the conventional energy detector, we plot the probability of detection versus the probability of false alarm with different Intrusion's SNR without considering the presence of legitimate transmissions as shown in Fig. 6. It is obvious that

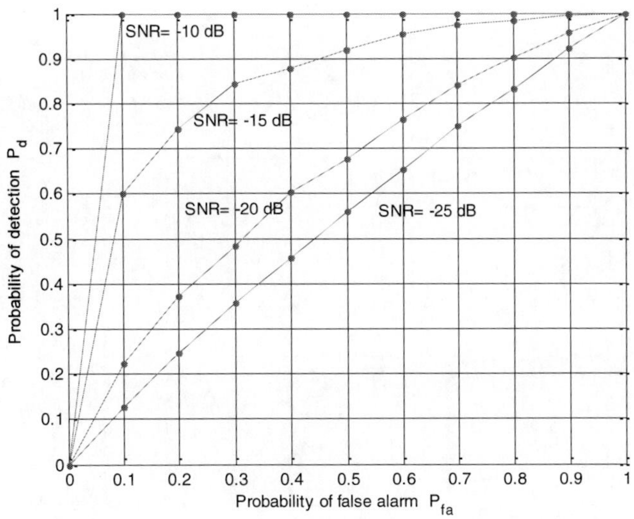

Figure 9. Conventional detector performance for different values of Intrusion's SNR.

increasing the probability of false alarm leads to higher probability of detection that includes both the right and the false detections. Fig. 9 implicitly highlights the fact that reducing the detection threshold will improve the detection probability for given false alarm probability.

To check at which SNR the detector performance reaches the maximum in terms of probability of detection and minimum false alarms, we plot P_d versus SNR by varying P_{fa} as seen in Fig. 10. The detector shows its optimum performance when SNR is greater than -10 dB when P_{fa} is 0.01 as stated in Table 1.

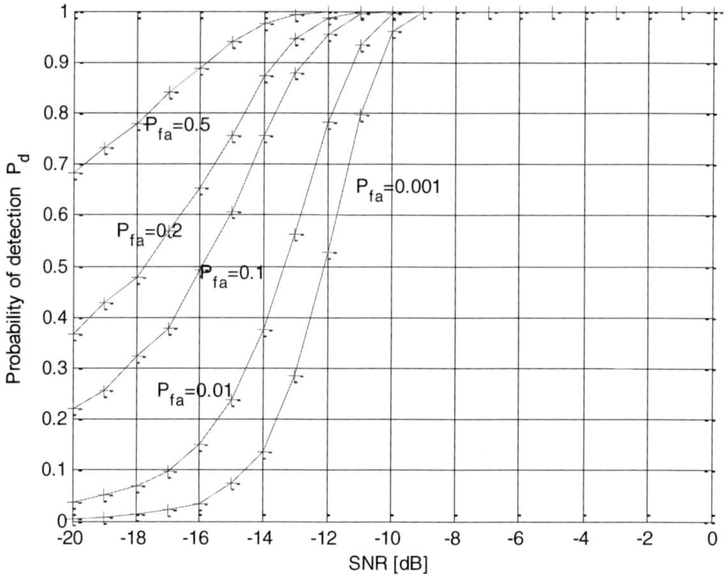

Figure 10. Conventional energy detector P_d vs. SNR.

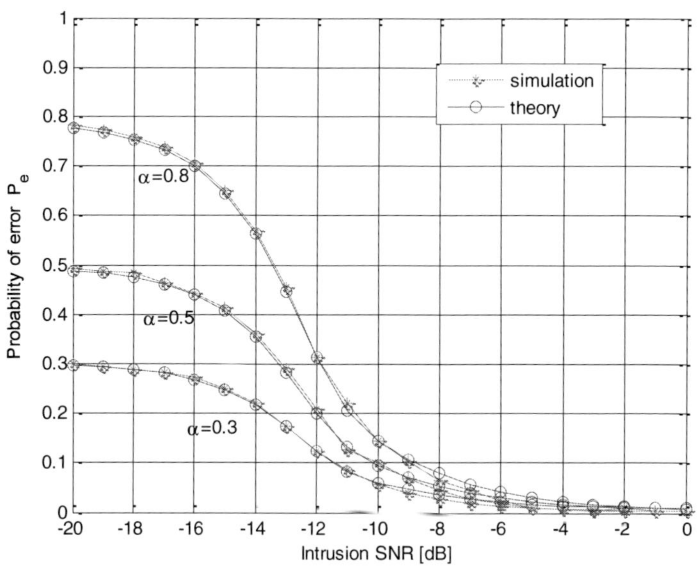

Figure 11. The proposed scheme P_e vs. intrusion SNR for different values of α.

3. The proposed intrusion detection algorithm

To test the proposed algorithm, the analytical and simulation probability of error for different values of α (the intruder's probability of presence) is given in Fig. 11 where the analytical curve coincides with the simulation. It is obvious from the Fig. 11 that for higher α, when the intruder is present more in the network, the probability of error increases due to the increase in the miss detection probability. Depending on the power of the intruder and the legitimate signal, we can decide which algorithm to use (i.e. two-thresholds or three-thresholds) based on its performance in different scenarios.

4. Performance Comparison

To evaluate the performance of the proposed intrusion detection system based on the proposed multi-threshold technique in the presence of regular legitimate transmissions, we compare its functionality based on three metrics; P_{md}, P_{fa}, and P_e respectively in the three cases (power of Intrusion and legitimate transmissions).

When $\sigma_S{}^2 > \sigma_I{}^2$, Fig.12 shows comparison between P_{md} based on traditional value threshold, two-thresholds, and three-thresholds. It is clear from Fig. 12 that the three-threshold algorithm shows higher P_{md}, while it presents lower false alarm rate in Fig. 13 which reduces the overall error probability in this case as shown in Fig. 14, hence it has the superior performance over the other threshold selection techniques in reducing the false alarm probability. It is also obvious that the performance of one-threshold is similar to two-threshold technique.

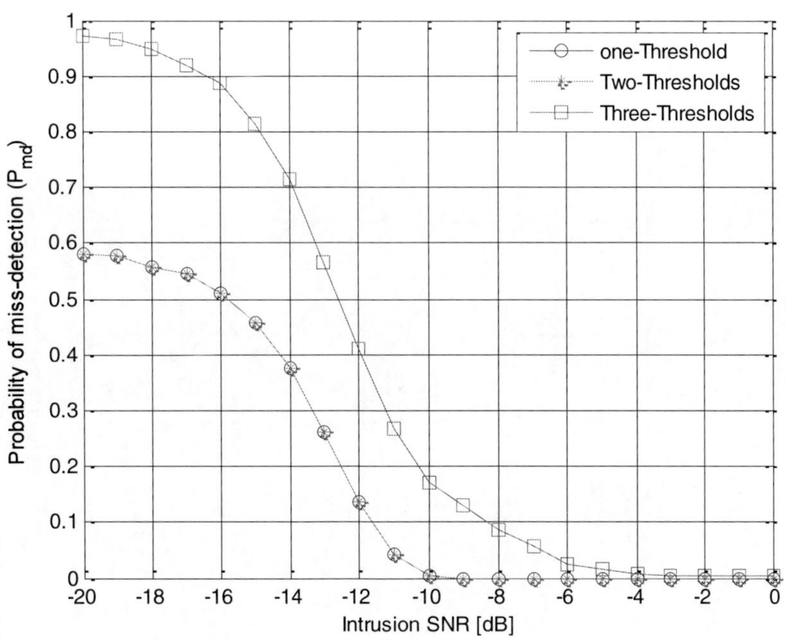

Figure 12. P_{md} vs. Intruder's SNR for different thresholds when $\sigma_S{}^2 > \sigma_I{}^2$.

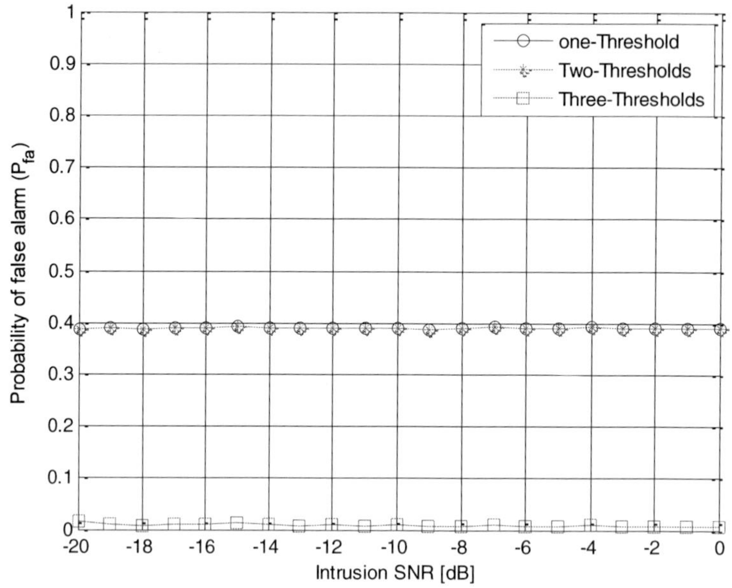

Figure 13. P_{fa} vs. Intruder's SNR for different thresholds when $\sigma_S{}^2 > \sigma_I{}^2$.

Both one-threshold and two-thresholds have high false alarm rate, but produce more reliable detection, which is more important in security, to have a reliable detection system instead of decreasing the false alarms. Hence the optimum detection performance is obtained by using just two-threshold scheme.

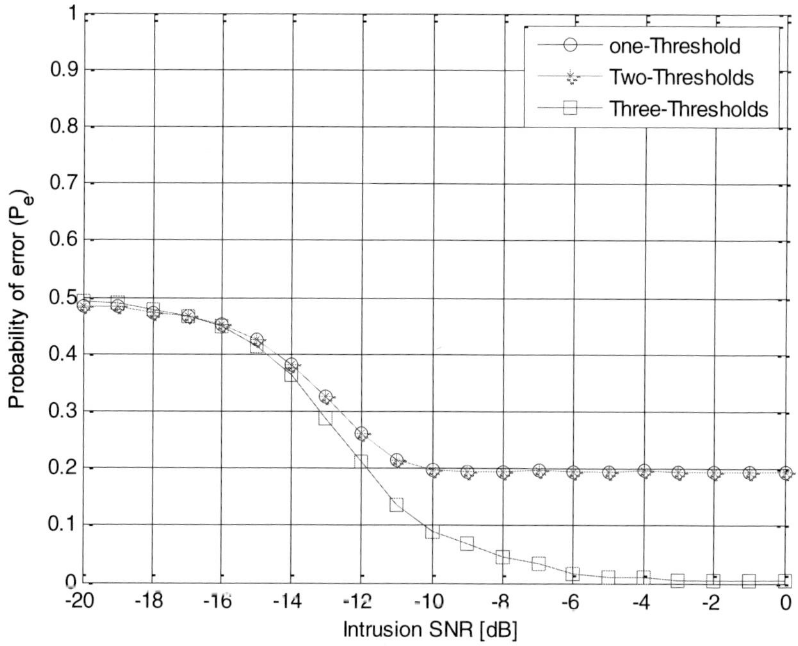

Figure 14. P_e vs. Intruder's SNR for different thresholds when $\sigma_S{}^2 > \sigma_I{}^2$.

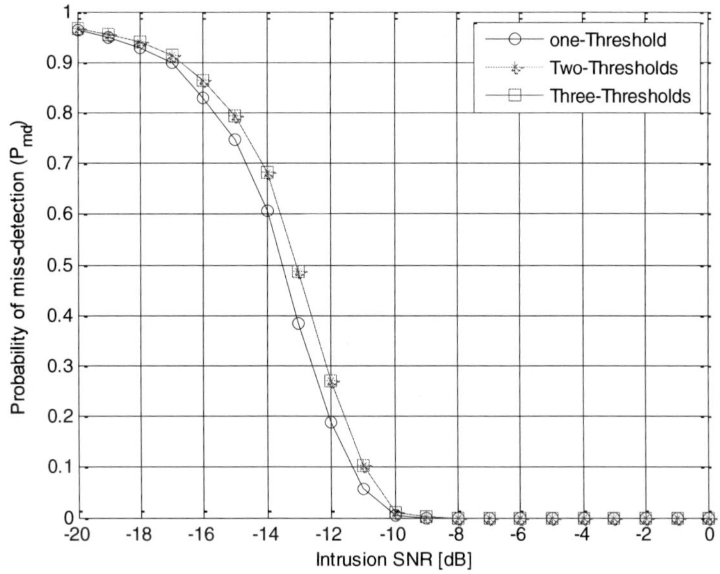

Figure 15. P_{md} vs. Intruder's SNR for different thresholds when $\sigma_S{}^2 < \sigma_I{}^2$.

When $\sigma_S{}^2 = \sigma_I{}^2$, it differs from the previous case where Figs 15-17 show that the two-thresholds algorithm and the three-thresholds have similar performance in terms of miss detection and false alarms which yield the same error probability, but their performance has superiority over the one-threshold scheme with minimum false alarm rate as given in Fig. 15. In this case, no need for three-threshold algorithm given that two-thresholds maintain similar improved performance.

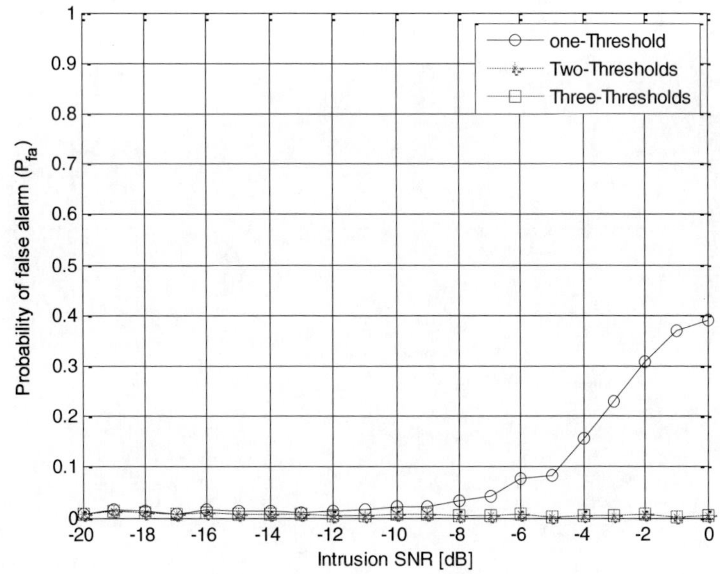

Figure 16. P_{fa} vs. Intruder's SNR for different thresholds when $\sigma_S{}^2 < \sigma_I{}^2$.

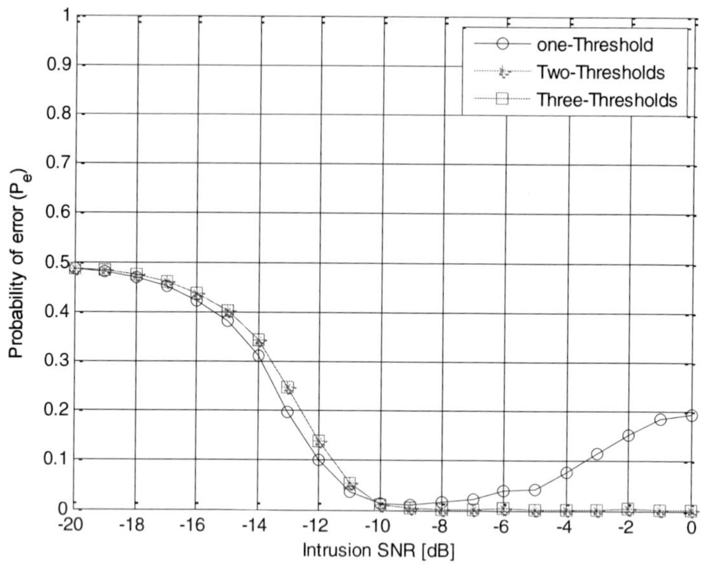

Figure 17. P_e vs. Intruder's SNR for different thresholds when $\sigma_S{}^2 < \sigma_I{}^2$.

When $\sigma_S{}^2 < \sigma_I{}^2$, they all have the same performance in terms of false alarm rate and probability of miss detection, hence the probability of error is minimized to 0.2 and does not reach zero because of the false alarm existence due to the detector's incapability to distinguish between the legitimate signal and the intrusion signal based on their equal average power (see Fig. 18-20).

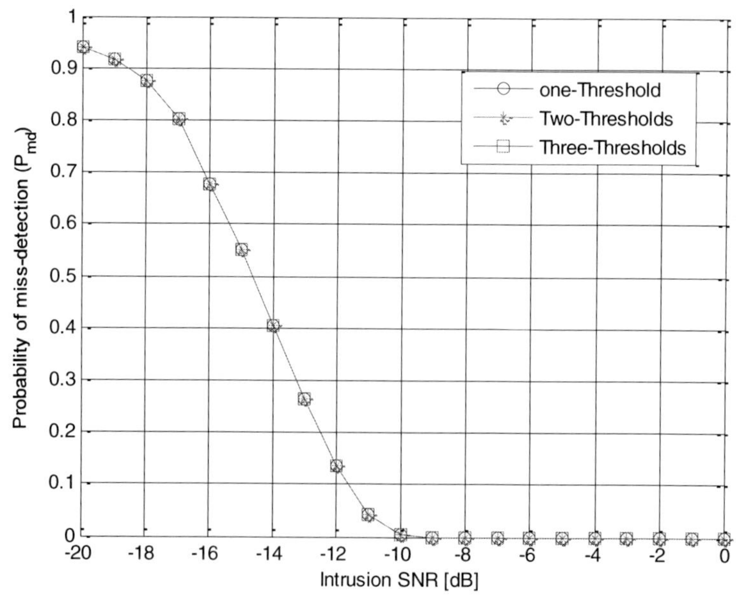

Figure 18. P_{md} vs. Intruder's SNR for different thresholds when $\sigma_S{}^2 = \sigma_I{}^2$.

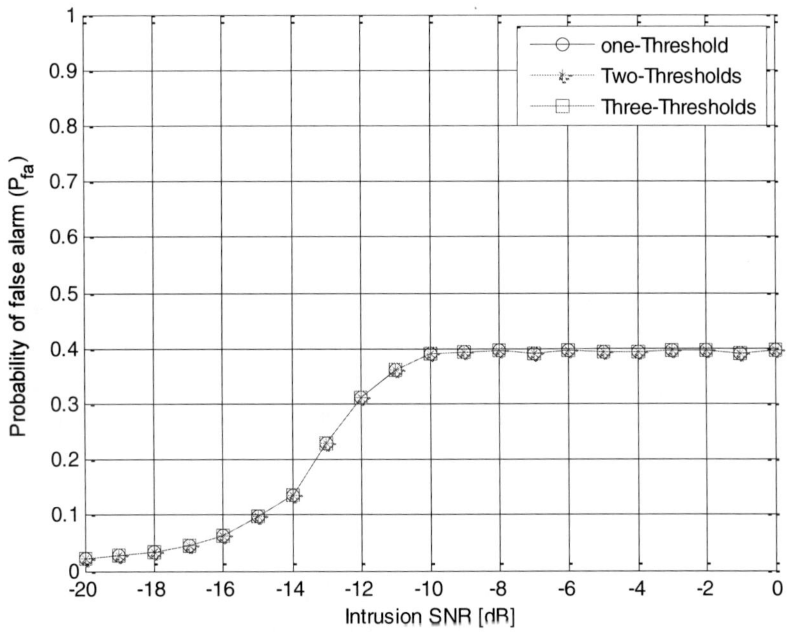

Figure 19. P_{fa} vs. Intruder's SNR for different thresholds when $\sigma_S{}^2 = \sigma_I{}^2$.

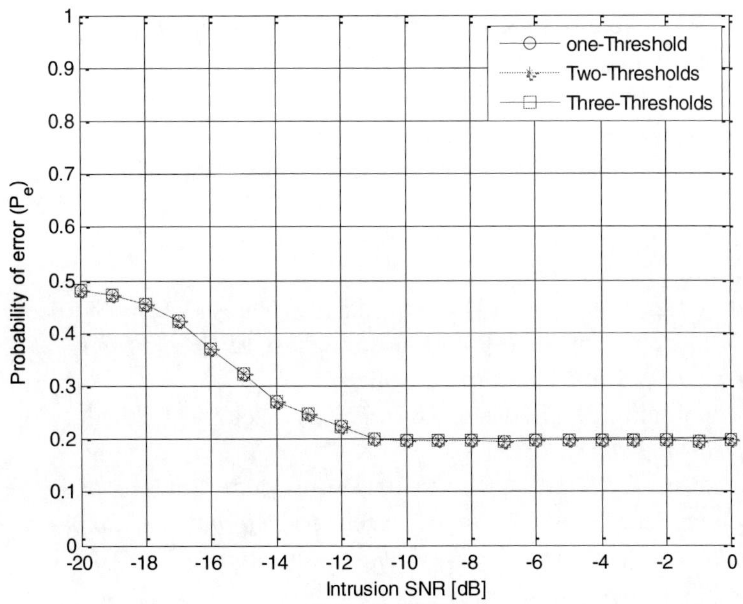

Figure 20. P_e vs. Intruder's SNR for different thresholds when $\sigma_S{}^2 = \sigma_I{}^2$.

CONCLUSION

In this chapter, a novel intrusion detection scheme based on multiple thresholds for vehicular networks is proposed assuming the presence of regular legitimate transmissions ongoing in the network. The proposed Multi-threshold scheme enhances the reliability of the intrusion detection in VANET network and improves its performance in terms of the probability of error which involves considering the probability of false alarm and the probability of miss detection. The new proposed Multi-threshold scheme performance is varying based on the transmission power of both legitimate and intrusion signals and the estimated noise variance. The trade-off in the proposed schemes includes providing better detection efficiency by reducing the amount of false alarms (this could distract the drivers while the network is secure and no attackers exist) and maintaining reliable detection (i.e. decreasing the misdetection probability) to ensure maximum security in network.

REFERENCES

Axell, E., Leus, G., Larsson, E. G., & Poor, H. V. (2012). Spectrum Sensing for Cognitive Radio. *IEEE Signal Processing Magazine*, 101-116.

Digham, F. F., Alouini, M. S., & Simon, M. K. (2003, May). On the Enrgy Detection of Unknown Signals Over Fading Channels. *Proc. IEEE*, 3575-3579.

Digham, F. F., Alouini, M.-S., & Simon, M. K. (2007). On the Energy Detection of Unknown Signals Over Fading Channels. *IEEE Transactions on Communications*, 21-24.

Ezell, S. J. (2010, January). Explaining International IT Application Leadership: Intelligent Transportation Systems. *The information Technology and Innovation Foundation.*

Hartenstein, H., & Laberteaux, K. P. (2008). A Tutorial Survey on Vehicular Ad Hoc Networks. *Topics in Ad Hoc and Sensor Networks*, 164-171.

Lee, J.-W., Kim, J.-H., Oh, H.-J., & Hwang, S.-H. (2008). Energy Detector using Adaptive-Fixed Thresholds in Cognitive Radio Systems. *APCC 2008. 14th Asia-Pacific Conference on Communications* (pp. 1-4). IEEE Conference Publications.

Ma, J., Zhao, G., & Li, Y. (2008). Soft Combination and Detection for Cooperative Spectrum Sensing in Cognitive Radio Networks. *IEEE Transactions on Wireless Communications*, 4502-4507.

Nair, P., Vinod, A., & Krishna, A. (2010). An Adaptive Threshold Based Energy Detector for Spectrum Sensing in Cognitive Radios at Low SNR. *IEEE International Conference on Communication Systems (ICCS)*, (pp. 574-578).

Raya, M., Papadimitratos, P., & Hubaux, J.-P. (2006). Securing Vehicular Communications. *Wireless Communications, IEEE*, 8-15.

Reisi, N., Ahmadian, M., & Salari, S. (2010). Performance Analysis of Energy Detection-based Spectrum Sensing over Fading Channels. *6th International Conference on Wireless Communications Networking and Mobile Computing (WiCOM)* (pp. 1-4). IEEE Conference Publications.

Shi, Q., & Comaniciu, C. (2010). Efficient cooperative detection for wireless sentinel networks. *44th Annual Conference on Information Sciences and Systems (CISS)*, (pp. 1-6).

Urkowitz, H. (1967). Energy Detection of Unknown Deterministic Signals. *Proceedings of the IEEE*, *55*, 523-531.

Xie, S., Shen, L., & Liu, J. (2009). Optimal threshold of energy detection for spectrum sensing in cognitive radio. *International Conference on Wireless Communications & Signal Processing (WCSP 2009)*, (pp. 1-5).

Xuping, Z., Haigen, H., & Guoxin, Z. (2010). Optimal threshold and weighted cooperative data combining rule in Cognitive Radio network. *12th IEEE International Conference on ommunication Technology (ICCT)*, (pp. 1464-1467).

Yucek, T., & Arslan, H. (2009). A Survey of Spectrum Sensing Algorithms for Cognitive Radio Applications. *Communications Surveys & Tutorials, IEEE, 11*, 116-130.

Zeadally, S., Hunt, R., Chen, Y.-S., Irwin, A., & Hassan, A. (2010). Vehicular ad hoc networks (VANETs): status, results, and challenges. *Telecommunications Systems*.

In: Communication Systems: New Research
Editor: Vyacheslav Tuzlukov
ISBN: 978-1-62618-654-5

Chapter 5

SINGLE CARRIER FREQUENCY DOMAIN EQUALIZATION FOR BROADBAND COMMUNICATIONS – A REVIEW

Homa Eghbali[1], Sami Muhaidat[2,3], Mehrdad Dianati[3] and Rahim Tafazolli[3]
[1]School of Engineering Science, Simon Fraser University, Burnaby, Canada
[2] Department of Electrical and Computer Engr., Khalifa University, Abu Dhabi, UAE
[3]Centre for Communications System Research (CCSR), University of Surrey, Guildford, United Kingdom

Abstract

A key challenge for high-speed broadband applications is the dispersive nature of frequency-selective fading channels, which causes the so-called intersymbol interference (ISI), leading to an inevitable performance degradation. More specifically, in the time domain equalization, when the channel time spread is larger than one symbol period, interference among consecutive transmitted symbols happen, which leads to ISI and consequently a distortion in the received signal. Single-carrier (SC) frequency domain equalization (FDE) was proposed as an alternative promising approach to OFDM for ISI mitigation. SC-FDE avoids the drawbacks associated with OFDM, including a large peak-to-average power ratio (PAPR), intolerance to amplifier nonlinearities, and high sensitivity to carrier frequency offset (CFO), while enjoying a comparable complexity and performance. Unlike time-domain equalization whose complexity grows exponentially with channel memory and spectral efficiency, SC-FDE enjoys low complexity due to its efficient FFT/IFFT operation-based transceiver structure, which is very similar to that of OFDM. In this chapter, we review the recent research on SC-FDE (conventional as well as distributed) and its applications. First, we review the basics of SC transmissions coupled with FDE, and discuss its advantages over OFDM. Then, we summarize recent studies on SC-FDE integration with space-time block coding (STBC), also known as STBC-SC-FDE. We next summarize recent studies on the performance of SC-FDE for cooperative STBC systems, also known as distributed (D)-STBC-SC-FDE, as an extension of conventional STBC to cooperative systems. Finally, we will conclude the chapter by providing *novel results* on performance of linear equalizers in a two-way relaying cooperative system with SC-FDE scheme.

Keywords: Single Carrier Frequency domain equalization, space-time coding, cooperative communications

1. Introduction

The growing demand for high data rate services for wireless multimedia and internet services has lead to intensive research efforts on high speed data transmission. A key challenge for high-speed broadband applications is the dispersive nature of frequency-selective fading channels, which causes the so-called ISI leading to inevitable performance degradation. An efficient approach to mitigate ISI is the use of OFDM which converts the ISI channel with AWGN into parallel ISI-free subchannels by implementing IFFT at the transmitter and FFT at the receiver side [1]. It has been shown that OFDM is an attractive equalization scheme for digital audio/video broadcasting (DAB/DVB) [2], and it has successfully been applied to high-speed modems over digital subscriber lines (DSL) [3]. Recently, it has also been proposed for broadband television systems and mobile wireless local area networks such as IEEE802.11a and HIPERLAN/2 (HL2) standards [4].

The IFFT precoding at the transmitter side and insertion of cyclic prefix (CP) enable OFDM with very simple equalization of frequency-selective finite impulse response (FIR) channels. To avoid InterBlock Interfernce (IBI) between successive FFT processed blocks, the CP is discarded and the truncated blocks are FFT processed so that the frequency-selective channels are converted into parallel flat-fading independent subchannels. In this way, the linear channel convolution is converted into circular convolution and the receiver complexity both in equalization and the symbol decoding stages is reduced [5]. However, since each symbol is transmitted over a single flat subchannel, the multipath diversity is lost along with the fact there is no guarantee for symbol detectability when channel nulls occur in the subchannels. More specifically, due to the transmission of modulation symbols at those frequency bands with deep amplitude depression, the uncoded error-rate performance of OFDM transmission over frequency selective channels is relatively poor.

A SC system transmits a modulated single carrier at a high symbol rate and is basically the frequency domain (FD) analog of what is done by conventional linear time domain (TD) equalizers. Note that FDE is performed on a block of data at a time, where the operations on this block involve an efficient FFT operation and a simple channel inversion operation, which makes it computationally more efficient than the corresponding TD equalization, for channels with severe delay spreads. Therefore, the SC system with FDE (SC-FDE) has essentially the same performance and low complexity as OFDM, when combined with FFT processing and the use of cyclic prefixing [6]. Fig. 1 illustrates the system block diagram for the OFDM and SC-FDE wireless systems.

1.1. SC-FDE Versus OFDM

To reduce the complexity due to the time domain equalization whose complexity grows exponentially with the channel memory length, frequency domain equalization was proposed as a tool to mitigate ISI in both OFDM and SC-FDE wireless systems. SC-FDE is the less popular but conceptually similar alternative approach to the OFDM scheme for ISI suppression. According to the system block diagrams for OFDM and SC-FDE wireless systems

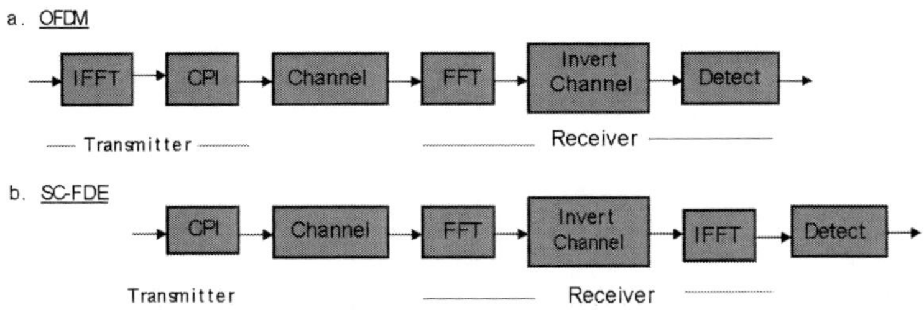

Figure 1. Block diagram of wireless systems: (a) OFDM, (b) SC-FDE.

depicted in Fig. 1, the only apparent difference between the two systems is that the IFFT block is moved to the end of the receiver chain in the SC-FDE system, rather than operating at the transmitter to create a multi-carrier waveform as in OFDM. More specifically, in OFDM wireless systems, Fourier transforms are employed for modulation and demodulation, whereas in the SC-FDE system, the FFT blocks are incorporated at the receiver side to convert time domain (TD) signal to the FD and back. Even though, complexity of SC-FDE system is similar to that of OFDM, SC-FDE system benefits from three main advantages over its competitor OFDM, as explained in the following.

Sensitivity to channel frequency nulls: In OFDM wireless systems, modulated symbols are confined within narrow frequency bands that may suffer from deep amplitude depressions and channel nulls. Since the decision makings are done in the FD, if the channel frequency response exhibits a deep amplitude null close to the subcarrier frequency, the associated information is lost. This means that the error-rate performance of uncoded OFDM is dominated by the subcarriers with the lowest signal-to-noise ratio (SNR), and the underlying diversity is lost. In SC-FDE system, however, the modulated symbols are transmitted over the whole available frequency band, which makes the spectral nulls having only a small impact on the overall system performance.

Sensitivity to CFO: Frequency synchronization is a critical task for OFDM systems. Note that any carrier frequency error between the OFDM system's transmitter side and the receiver side can destroy the mutual underlying orthogonality between the modulated OFDM subcarriers, which results in interference between adjacent subcarriers, also known as intercarrier interference (ICI).

High Peak-to-Average Power Ratio (PAPR): An OFDM signal is the sum of multiple sinusoids modulated by independent information symbols. As a result, the OFDM signal's envelope, for large FFT orders, is characterized by a wide dynamic range, which consequently increases the linearity requirements of the analog front-end, dramatically. In other words, the large variations in the instantaneous transmit power, i.e. large PAPR, has a negative impact on the transmitter power amplifier (PA) efficiency, which leads to increased transmit power consumption and increased PA cost. As a result, to decrease the average transmit power, the system data range is reduced. Note that, since having high PA efficiency is of great importance at the mobile terminals, the OFDM transmission scheme is more suitable for the downlink communication.

The rest of this chapter is organized as follows. We start in Section 2 by describing our model and assumptions and reviewing of the SISO SC MMSE-FDE. In Section 3, we review an Alamouti-like scheme for combining STBC and SC MMSE-FDE. D-STBC-SC-

FDE is discussed in Section 4, followed by SC-FDE for two-way relay (TWR) channels in Section 5. Section 6 reviews the SC-FDE for cognitive radio channels, and the chapter is concluded in Section VII.

Notation: In the following, $\overline{(.)}, (.)^{\mathrm{T}}, \mathrm{and} (.)^{\mathrm{H}}$ denote conjugate, transpose, and Hermitian transpose operations, respectively. $*$ denotes linear convolution, $\|.\|_F$ denotes the Frobenius norm, $\|\cdot\|$ denotes the Euclidean norm of a vector, $[.]_{k,l}$ denotes the $(k,l)^{th}$ entry of a matrix, $[.]_k$ denotes the kth entry of a vector, I_M denotes the identity matrix of size M, and $0_{M\times M}$ denotes all-zero matrix of size $M\times M$. $Q = Q_M^{-1} = Q_M^H$ represents the $M\times M$ IFFT matrix, whose (l,k) element is given by $Q(l,k) = 1/\sqrt{M}\exp(j2\pi lk/M)$ where $0 \le l,k \le M-1$. Bold upper-case letters denote matrices and bold lower-case letters denote vectors. Two functions $f(SNR)$ and $g(SNR)$ are exponentially equal, denoted by $f(SNR) \doteq g(SNR)$, when $\lim_{SNR\to\infty} \frac{\log f(SNR)}{\log SNR} \doteq \lim_{SNR\to\infty} \frac{\log g(SNR)}{\log SNR}$, where SNR denotes the transmission signal-to-noise ratio . The ordering operators $\dot{\leqslant}$, and $\dot{\geqslant}$ are defined accordingly.

2. SC-FDE for Single-input Single-output Systems

We consider single-carrier block transmission over an additive white Guassian noise (AWGN) frequency-selective channel, assuming a single receive antenna and single transmit antenna. The channel impulse response (CIR) over block j is modeled as a finite impulse response (FIR) filter with coefficients $h^j = \left[h^j[0],...,h^j[L]\right]^{\mathrm{T}}$, where L denotes the corresponding channel memory length. The random vector h^j is assumed to be independent zero-mean complex Gaussian with uniform power delay profile with variance $1/(L+1)$. Information symbols are parsed to streams of $M\times 1$ blocks x^j , and then multiplied by a zero-padding (ZP) matrix $\Psi = \left[I_M^{\mathrm{T}}, 0_{M\times L}^{\mathrm{T}}\right]^{\mathrm{T}}$ of size $N\times M$, where N is the frame length. As explained in [7], the use of zero-padding as the precoding method in a single-carrier transmission scenario ensures that the available multipath diversity is fully exploited. To further remove inter-block interference and make the channel matrix circulant, a CP with length L is added between adjacent information blocks. Due to the adopted precoding form, i.e. zero padding, we simply insert additional zeros at the start of the frame as CP.

Note that the received blocks j at the receiver terminal is given by

$$r^j = H^j d^j + n^j, \tag{1}$$

where $d^j = \Psi x^j$ denotes the zero-padded information vector, n^j is complex Gaussian with zero mean and variance of $N_0/2$ per dimension, and H^j is $N\times N$ circulant matrix with entries $[H^j]_{k,l} = h^j((k-l) \bmod N)$. First, the received signal (cf. (1)) is transformed to the FD by applying the Discrete Fourier Transform (DFT) [1] , i.e. multiplying by the Q matrix, as follows

$$Qr^j = QH^j d^j + Qn^j. \tag{2}$$

Exploiting the circulant structure of the channel matrices, we have

$$H^j = Q^{\mathrm{H}}\Lambda^j Q, \tag{3}$$

[1] In this Chapter, we assume that the DFT size N is a power of 2, thus, the terms FFT and DFT are interchangeable.

where Λ^j is a diagonal matrix whose (n,n) element is equal to the nth DFT coefficient of h^j. Using (3) and dropping the superscript j for brevity, we can write (2) in matrix form as follows,

$$Qr = \Lambda Qd + Qn. \tag{4}$$

we can multiply (4) by Λ^H , resulting in

$$r_{out} = |\Lambda_1|^2 Qd + n_{out}, \tag{5}$$

where n^j_{out} is a noise vector with each entry still Gaussian with zero-mean and variance of $|\Lambda|^2 N_0/2$ per dimension.The resulting data stream can be now detected by applying the MMSE equalizer, i.e., multiplying (5) by

$$\left(|\Lambda|^2 + \frac{1}{\text{SNR}} I_N\right)^{-1}, \tag{6}$$

where SNR represents the signal-to-noise ratio. Next, signals are multiplied by Q^H to be transferred back to the TD, and are further fed into a minimum Euclidean distance decoder, yielding $\hat{r}$, i.e., a estimated version of the transmitted symbols. Pay attention that in SC-FDE decision making is performed in the TD, whereas in OFDM, signals are detected in the FD.

3. SC-FDE for Space-Time Block Coded Systems

It has been shown in [8] that spatial diversity offers significant improvement in link reliability and spectral efficiency through the use of multiple antennas at the transmitter and/or receiver side. These gains are typically realized at the physical layer and require co-located antenna elements (i.e. physical antenna arrays) at the base station and/or the mobile terminal. Multiple-antenna techniques are very attractive for deployment in cellular applications at base stations and have already been included in the 3rd generation wireless standards. An effective and practical way to realize the benefits of MIMO systems is to employ Space-time block codes (STBCs) [9, 10]. While not demanding any channel state information (CSI) at the transmitter side, Alamouti's STBC scheme has been proved to achieve full spatial diversity at full transmission rate for 2 transmit antennas and any signal constellation. Most importantly, maximum likelihood decoding of STBC requires only simple linear processing, which is unique. Since STBCs have been proposed for frequency-flat fading channels, it is challenging to apply them over frequency-selective channels.

We consider STBC transmission, assuming a single receive antenna and two transmit antennas over a frequency-selective fading wireless channel, although extension to multiple receive and transmit antennas is straightforward. Specifically, we consider the single-carrier frequency domain equalization scheme proposed in [11]. The CIR over block j between the mth transmit antenna and the receive antenna is modeled as a FIR filter with coefficients $h^j_m = \left[h^j_m[0], ..., h^j_m[L_m]\right]^{\mathrm{T}}$, where $m = 1, 2$ and L_m denotes the corresponding channel memory length. The random vectors h^j_m are assumed to be independent zero-mean complex Gaussian with uniform power delay profile. The CIRs are assumed to be constant over two

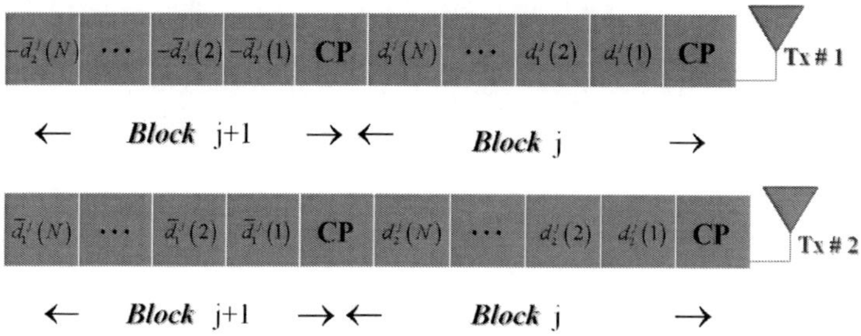

Figure 2. Frame structure of STBC-SC-FDE system.

consecutive blocks, i.e., j and $j+1$, and vary independently every two blocks. Information symbols are first parsed to two streams of $M \times 1$ blocks x_i^j , $i = 1,2,$. The transmitted blocks are generated by the encoding rule $d_1^{j+1} = -J\bar{d}_2^j$, $d_2^{j+1} = J\bar{d}_1^j$,, where we denote the zero-padded information vectors by $d_i^j = \Psi x_i^j$, $m = 1,2$ and $J = \mathrm{P}_N^M$ is a $N \times N$ a partial permutation matrix [12]. To further remove inter-block interference and make the channel matrix circulant, CP with length $L_{CP} = L_1 + L_2 + 1$ is added between adjacent information blocks. Due to the adopted precoding form, i.e., zero padding, we simply insert additional zeros at the start of the frame as CP. Fig. 2 depicts the block format for the STBC-SC-FDE system. With two transmit and one receive antenna and assuming that the channel coefficients remain constant over two consecutive blocks, i.e. $H_m^j = H_m^{j+1} = H_m$, received blocks j and $j+1$ and are given by

$$r_1^j = H_1 d_1^j + H_2 d_2^j + n_1^j \tag{7}$$

$$r_2{}^{j+1} = -H_1 J\bar{d}_2^j + H_2 J\bar{d}_1^j + n_2{}^{j+1} \tag{8}$$

where n_1^j and n_2^{j+1} are complex Gaussian with zero mean and variance of $N_0/2$ per dimension. In (7)-(8), H_1 , H_2 are $N \times N$ circulant matrices with entries $[H_m]_{k,l} = h_m((k-l) \bmod N)$ $m = 1,2$, respectively. After multiplying the received signals in (7) and (8) by the Q matrix, we obtain

$$Qr_1^j = QH_1 d_1^j + QH_2 d_2^j + Qn_1^j \tag{9}$$

$$QJ\bar{r}_2^{j+1} = -QH_1^{\mathrm{H}} d_2^j + QH_2^{\mathrm{H}} d_1^j + QJ\bar{n}_2^{j+1}. \tag{10}$$

Exploiting the circulant structure of the channel matrices, we have

$$H_m = Q^{\mathrm{H}} \Lambda_m Q \tag{11}$$

where $\Lambda_m, m = 1,2,$ is a diagonal matrix whose (n,n) element is equal to the nth DFT coefficient of h_m. Using (11) and dropping the superscript j for brevity, we can write (9) and (10) in matrix form as

$$\begin{bmatrix} Qr_1 \\ QJ\bar{r}_2 \end{bmatrix} = \begin{bmatrix} \Lambda_1 & \Lambda_2 \\ \bar{\Lambda}_2 & -\bar{\Lambda}_1 \end{bmatrix} \begin{bmatrix} Qd_1 \\ Qd_2 \end{bmatrix} + \begin{bmatrix} Qn_1 \\ QJn_2 \end{bmatrix} = \tilde{\Lambda} U + \tilde{N}. \tag{12}$$

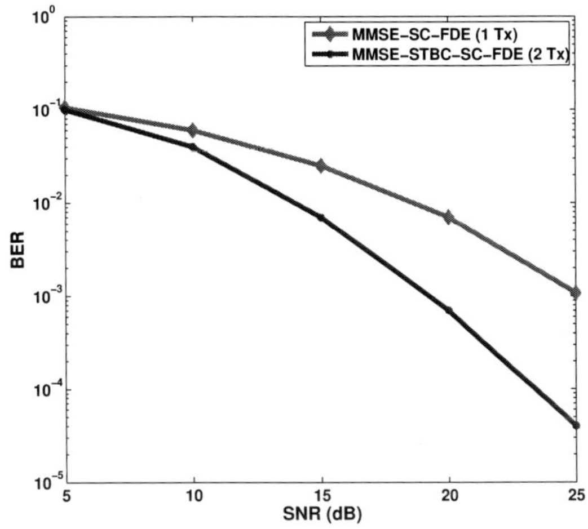

Figure 3. BER of MMSE-SC-FDE and MMSE-STBC-SC-FDE for EDGE TU channel with 8-PSK modulation and N = 64.

Since $\tilde{\mathbf{\Lambda}}$ is an orthogonal matrix of size $2N \times 2N$, we can multiply (6) by $\tilde{\Lambda}^H$, resulting in

$$r_{i,out} = \left(|\Lambda_1|^2 + |\Lambda_2|^2\right) Qd_i + n_{i,out} \tag{13}$$

where $n^j_{out,m}$ is a noise vector with each entry still Gaussian with zero-mean and variance of $\left(|\Lambda_1|^2 + |\Lambda_2|^2\right) N_0/2$ per dimension. The resulting data streams can be now detected by applying the MMSE equalizer, i.e., multiplying (13) by

$$\left(|\Lambda_1|^2 + |\Lambda_2|^2 + \frac{1}{\text{SNR}} I_N\right)^{-1}. \tag{14}$$

Results from (14) are further fed into a minimum Euclidean distance decoder, yielding $\hat{x}_1$ and $\hat{x}_2$, i.e., a decoded version of the transmitted symbols.

Fig. 3 illustrates the error performance of MMSE-SC-FDE versus MMSE-STBC-SC-FDE for the typical urban (TU) channel with a linearized GMSK transmit pulse shape, 8-PSK modulation, and a symbol duration of 3.69 s as in the proposed third generation TDMA cellular standard EDGE5 [11]. We assume $L_1 = L_2 = 3$, and $N = 64$. Fig. 3 proves significant performance improvement achieved in MMSE-STBC-SC-FDE as compared with the single transmit antenna system, especially at high SNR regime where effects of diversity are more pronounced.

4. SC-FDE for Cooperative Networks

Recently, it has been demonstrated that "cooperative diversity" provides an effective means of improving spectral and power efficiency of wireless networks as an alternative to multiple-antenna transmission schemes [13–22].

Although most of the current literature on SC-FDE is limited to point-to-point systems, there have been recent results reported on FDE in the context of cooperative scenarios [15, 23]. Distributed-STBC (D-STBC) SC-FDE has been discussed in [15], assuming the maximum likelihood sequence detection (MLSD) equalization. Although significant diversity gains were demonstrated, the complexity of MLSD equalizers increases with the channel memory, signal constellation size, and the number of transmit/receive antennas. This, in turn, places significant additional computational and power consumption loads on the receiver side. Therefore, low-complexity equalization schemes without sacrificing performance are particularly desirable, especially for cases where the receiver implementation is supposed to be simple and required to operate on a limited battery power. In this section, we study the MMSE-based receiver design for D-SC-FDE transmissions. Specifically, we are considering amplify-and-forward (AF) relay networks for cooperative scenarios in which either of $S \to R$ or $R \to D$ links are frequency selective Rayleigh fading.

A single-relay assisted cooperative communication scenario is considered. All terminals are equipped with single transmit and receive antennas. Any linear modulation technique such as QAM or PSK modulation can be used. We assume AF relaying and adopt the user cooperation protocol proposed by Nabar *et al.* [24]. Specifically, the source terminal communicates with the relay terminal during the first signaling interval. There is no transmission from source-to-destination within this period. In the second signaling interval, both the relay and source terminals communicate with the destination terminal.

The CIRs for $S \to R$, $S \to D$ and $R \to D$ links for the jth transmission block are given by $h_{SR}^{j} = \left[h_{SR}^{j}[0], \ldots, h_{SR}^{j}[L_{SR}]\right]^{\mathrm{T}}$, $h_{RD}^{j} = \left[h_{SD}^{j}[0], \ldots, h_{SD}^{j}[L_{SD}]\right]^{\mathrm{T}}$ and $h_{SD}^{j} = \left[h_{RD}^{j}[0], \ldots, h_{RD}^{j}[L_{RD}]\right]^{\mathrm{T}}$, respectively, where L_{SR}, L_{RD} and L_{SD} denote the corresponding channel memory lengths. The $S \to R$ link is assumed to be frequency selective Rayleigh fading, while the random vectors h_{RD}^{j} and h_{SD}^{j}, are assumed to be independent zero-mean complex Gaussian with power delay profile vectors denoted by $v_{RD} = [\sigma_{RD}^{2}(0), \ldots, \sigma_{RD}^{2}(L_{RD})]$ and $v_{SD} = [\sigma_{SD}^{2}(0), \ldots, \sigma_{SD}^{2}(L_{SD})]$ that are normalized such that $\sum_{l_{RD}=0}^{L_{RD}} \sigma_{RD}^{2}(l_{RD}) = 1$ and $\sum_{l_{SD}=0}^{L_{SD}} \sigma_{SD}^{2}(l_{SD}) = 1$. The CIRs are assumed to be constant over two consecutive blocks and vary independently every two blocks.

Information symbols are first parsed to two streams of $M \times 1$ blocks x_{i}^{j}, $i = 1, 2$ and then multiplied by a ZP matrix $\Psi = \left[I_{M}^{\mathrm{T}}, 0_{M \times \ell}^{\mathrm{T}}\right]^{\mathrm{T}}$ of size $N \times M$, where $\ell = \max(L_{SD}, L_{SR} + L_{RD} + 1)$ and N is the frame length. The use of zero-padding as the precoding method in a single-carrier transmission scenario ensures that the available multipath diversity is fully exploited. A CP with length ℓ is further added between adjacent information blocks. Due to the adopted precoding form, i.e., zero padding, we simply insert additional zeros at the start of the frame as CP. The transmitted blocks for $S \to R$ and $S \to D$ links are generated by the encoding rule $d_{1}^{j+1} = -J\bar{d}_{2}^{j}$, $d_{2}^{j+1} = J\bar{d}_{1}^{j}$ $j = 0, 2, 4, 6, \ldots$, where $d_{i}^{j} = \Psi x_{i}{}^{j}$, $i = 1, 2$, is the zero-padded information vector and $J = \mathrm{P}_{N}^{M}$ is a $N \times N$ partial permutation matrix.

The signal received at the relay terminal during the first signaling interval is

$$r_{R}^{j} = \sqrt{E_{SR}} H_{SR}^{j} d_{1}^{j} + n_{R}^{j}, \tag{15}$$

where n_{R}^{j} is the additive white Gaussian noise vector with each entry having zero-mean and variance of $N_{0}/2$ per dimension. The relay terminal normalizes each entry of the received signal $[r_R^j]_n$, $n = 1,2,...,N$, by a factor of $\mathrm{E}\left(|[r_R^j]_n|^2\right) = E_{SR} + N_0$ to ensure unit average energy and re-transmits the signal during the second time slot. After some mathematical manipulations, the received signal at the destination terminal in the second time slot is given by

$$r_1^j = \sqrt{\frac{E_{RD}E_{SR}}{E_{SR}+N_0}} H_{RD}^j H_{SR}^j d_1^j + \sqrt{E_{SD}} H_{SD}^j d_2^j + \tilde{n}_1^j. \tag{16}$$

In (16), each entry of the *effective* noise term $\tilde{n}_1^j$ (conditioned on h_{RD}) has zero mean and a variance of ρN_0 where ρ is defined by

$$\rho = 1 + \frac{E_{RD}}{E_{SR}+N_0} \sum_{m=0}^{L_3} |h_{RD}(m)|^2. \tag{17}$$

In (16), H_{SR}^j, H_{SD}^j and H_{RD}^j are $N \times N$ circulant matrices with entries $[H_i^j]_{k,l} = h_i^j((k-l) \bmod N)$, i denotes SR, SD, RD, respectively. Finally, E_{SR}, E_{SD} and E_{RD} are the average energies available at the respective terminal which take into account possibly different path loss and shadowing effects between the $S \to D$ and $R \to D$ links, respectively. The destination terminal normalizes the received signal by a factor of $\sqrt{\rho}$. This does not affect the SNR, but simplifies the ensuing presentation [24]. After normalization, we obtain

$$r_1^j = \sqrt{\gamma_1} H_{RD}^j H_{SR}^j d_1^j + \sqrt{\gamma_2} H_{SD}^j d_2^j + n_1^j \tag{18}$$

where n_1^j is complex Gaussian with zero mean and variance of $N_0/2$ per dimension and the scaling coefficients γ_1, γ_2 are defined as

$$\gamma_1 = \frac{(E_{SR}/N_0)E_{RD}}{1 + E_{SR}/N_0 + \sum_{m=0}^{L_3} |h_{RD}(m)|^2 E_{RD}/N_0} \tag{19}$$

$$\gamma_2 = \frac{(1 + E_{SR}/N_0)E_{SD}}{1 + E_{SR}/N_0 + \sum_{m=0}^{L_3} |h_{RD}(m)|^2 E_{RD}/N_0}. \tag{20}$$

Assuming that the channel coefficients remain constant over two consecutive blocks, i.e., $H_i^k = H_i^{k+1} = H_i$, i denotes SR, SD, RD, the received signals for blocks j and $j+1$ are given by

$$r_1^j = \sqrt{\gamma_1} H_{RD} H_{SR} d_1^j + \sqrt{\gamma_2} H_{SD} d_2^j + n_1^j, \tag{21}$$

$$r_2^{j+1} = -\sqrt{\gamma_1} H_{RD} H_{SR} J \bar{d}_2^j + \sqrt{\gamma_2} H_{SD} J \bar{d}_1^j + n_2^{j+1}. \tag{22}$$

Next, we transform the received signals to the FD as follows,

$$Qr_1^j = \sqrt{\gamma_1} Q H_{RD} H_{SR} d_1^j + \sqrt{\gamma_2} Q H_{SD} d_2^j + Q n_1^j, \tag{23}$$

$$QJ\bar{r}_2^{j+1} = -\sqrt{\gamma_1} Q H_{SR}^{\mathrm{H}} H_{RD}^{\mathrm{H}} d_2^j + \sqrt{\gamma_2} Q H_{SD}^{\mathrm{H}} d_1^j + QJ\bar{n}_2^{j+1}. \tag{24}$$

Exploiting the circulant structure of the channel matrices, we have

$$H_i^j = Q^{\mathrm{H}} \Lambda_i^j Q, \tag{25}$$

where Λ_i^j, i denotes SR, SD, RD, is a diagonal matrix whose (n,n) element is equal to the nth DFT coefficient of h_i^j. Using (25), we can write (23) and (24) in matrix form as

$$\begin{bmatrix} Qr_1^j \\ QJ\bar{r}_2^{j+1} \end{bmatrix} = \tilde{N} + \tilde{\Lambda} U = \begin{bmatrix} Qn_1^j \\ QJn_2^{j+1} \end{bmatrix} + \begin{bmatrix} \sqrt{\gamma_1}\Lambda_{RD}\Lambda_{SR} & \sqrt{\gamma_2}\Lambda_{SD} \\ \sqrt{\gamma_2}\bar{\Lambda}_{SD} & -\sqrt{\gamma_1}\bar{\Lambda}_{RD}\bar{\Lambda}_{SR} \end{bmatrix} \begin{bmatrix} u_1^j \\ u_2^{j+1} \end{bmatrix} \tag{26}$$

where $u_i^j = Qd_i^j$ for $i = 1,2$. Since $\tilde{\Lambda}$ is an orthogonal matrix of size $2N \times 2N$, we can multiply (26) by

$$K = \left(I_2 \otimes \left(\gamma_1 |\Lambda_{RD}|^2 |\Lambda_{SR}|^2 + \gamma_2 |\Lambda_{SD}|^2 \right)^{-1/2} \right) \tilde{\Lambda}^H. \tag{27}$$

The resulting output streams are now decoupled allowing us to write each output sequence as

$$r_{out,i}^j = \sqrt{\gamma_1 |\Lambda_{RD}|^2 |\Lambda_{SR}|^2 + \gamma_2 |\Lambda_{SD}|^2} Qd_i^j + n_{out,i}^j, \tag{28}$$

where $n_{out,i}^j$ is a noise vector with each entry still Gaussian with zero-mean and variance of $N_0/2$ per dimension. Dropping the subscript j for brevity, the decoded signal streams are further fed into a minimum Euclidean distance decoder yielding $\hat{x}_1$ and $\hat{x}_2$, i.e., a decoded version of the transmitted symbols.

4.1. Diversity gain Analysis

Here, we investigate the achievable diversity gain for the proposed receiver through the derivation of the pairwise error probability (PEP) expression.

Defining the transmitted codeword vector and the erroneously-decoded codeword vector as x and $\hat{x}$, respectively, and applying the standard Chernoff bound, we obtain

$$P(x, \hat{x} | h_{SR}, h_{RD}, h_{SD}) \leq \exp\left(-\frac{d^2(x, \hat{x})}{4N_0} \right), \tag{29}$$

where $d^2(x, \hat{x})$ denotes the Euclidean distance between x and $\hat{x}$. Following [13], the Euclidean distance in (29) can be expressed as

$$d^2(x, \hat{x}) = \gamma_1 \| H_{RD} H_{SR} e_1 \|^2 + \gamma_2 \| H_{SD} e_2 \|^2, \tag{30}$$

where $e_i = d_i - \hat{d}_i$, $i = 1,2$.

Having $\|H_i\|^2 = N \sum_{l_i=0}^{L_i} (|h_i(l_i)|)^2 \quad i = SR, SD$, where N is the frame length, (30) can be further simplified to

$$\begin{aligned} d^2(x, \hat{x}) \approx\ & \gamma_1 \textstyle\sum_{l_{RD=0}}^{L_{RD}} h_{RD}(l_{RD})^2 \| H_{SR} e_1 \|^2 + \gamma_2 \| H_{SD} e_2 \|^2 \\ \approx\ & \gamma_1 \textstyle\sum_{l_{SR=0}}^{L_{SR}} h_{SR}(l_{SR})^2 \| H_{RD} e_1 \|^2 + \gamma_2 \| H_{SD} e_2 \|^2 \end{aligned} \tag{31}$$

Introducing

$$\mathfrak{R}_i = \begin{bmatrix} [d]_0 & [d]_1 & \cdots & [d]_{N-1} \\ [d]_{N-1} & [d]_0 & \cdots & [d]_{N-2} \\ \vdots & \vdots & \cdots & \vdots \\ [d]_{N-L_i} & [d]_{N-L_i+1} & \cdots & [d]_{N-L_i-1} \end{bmatrix} \tag{32}$$

and $\mathfrak{R}_i = (\mathfrak{R}_i - \hat{\mathfrak{R}}_i)(\mathfrak{R}_i - \hat{\mathfrak{R}}_i)^H$, for $i = 1, 2$, (31) can be written as follows

$$d^2(x,\hat{x}) \approx \gamma_1 \sum_{l_{RD}=0}^{L_{RD}} h_{RD}(l_{RD})^2 h_{SR}^T \mathfrak{R}_1 (h_{SR}^T)^H + \gamma_2 h_{SD}^T \mathfrak{R}_2 (h_{SD}^T)^H. \tag{33}$$

Defining $\Theta_i = diag(v_i)$, $\mu_i = \Theta_i^{-1/2} h_i$, and $A_i = \Theta_i^{1/2} \mathfrak{R}_i \Theta_i^{1/2}$, $i = SR, SD$, we have

$$d^2(x,\hat{x}) \approx \gamma_1 \sum_{l_{RD}=0}^{L_{RD}} h_{RD}(l_{RD})^2 \mu_{SR}^T A_{SR} (\mu_{SR}^T)^H + \gamma_2 \mu_{SD}^T A_{SD} (\mu_{SD}^T)^H, \tag{34}$$

where A_i, $i = SR, SD$ are Hermitian and full rank [13]. Therefore, there exists a unitary matrix U_{A_i} such that $U_{\Lambda_i}^H A_i U_{A_i} = \Lambda_i$ where Δ_i is a diagonal matrix with elements equal to the eigenvalues of A_i. Therefore, (34) can be expressed as

$$d^2(x,\hat{x}) \approx \gamma_1 \sum_{l_{RD}=0}^{L_{RD}} h_{RD}(l_{RD})^2 \beta_{SR} \Delta_{SR} \beta_{SR}^H + \gamma_2 \beta_{SD} \Delta_{SD} \beta_{SD}^H, \tag{35}$$

where $\beta_i = \mu_i^T U_{A_i}$, $i = SR, SD$. Thus, (35) can be further simplified to

$$d^2(x,\hat{x}) \approx \gamma_1 \sum_{l_{RD}=0}^{L_{RD}} |h_{RD}(l_{RD})|^2 \sum_{l_{SR}=0}^{L_{SR}} \lambda_{SR}(l_{SR}) |\beta_{SR}(l_{SR})|^2 + \gamma_2 \sum_{l_{SD}=0}^{L_{SD}} \lambda_{SD}(l_{SD}) |\beta_{SD}(l_{SD})|^2, \tag{36}$$

where $\lambda_{SR}(l_{SR})(\lambda_{SD}(l_{SD}))$ denote the l_{SR}^{th} (l_{SD}^{th}) eigenvalue of codeword difference matrixes and β_{SR} (β_{SD}) are zero-mean complex Gaussian vectors with unit variance. Assuming perfect power control and sufficiently large $E_{SR}/N_0 > E_{SD}/N_0$ values, the scaling factors in (19) and (20) reduce to $\gamma_1 = \gamma_2 = E_{SD}$. Then, we can write (36) as

$$d^2(x,\hat{x}) \approx E_{SD} \sum_{l_{RD}=0}^{L_{RD}} |h_{RD}(l_{RD})|^2 \sum_{l_{SR}=0}^{L_{SR}} \lambda_{SR}(l_{SR}) |\beta_{SR}(l_{SR})|^2 + E_{SD} \sum_{l_{SD}=0}^{L_{SD}} \lambda_{SD}(l_{SD}) |\beta_{SD}(l_{SD})|^2. \tag{37}$$

In the following, we will derive the PEP expression under two different scenarios:

Case I: $L_{RD} > L_{SR}$

First, we define the random variables $Y_2 = \sum_{l_{SD}=0}^{L_{SD}} \lambda_{SD}(l_{SD}) |\beta_{SD}(l_{SD})|^2$ and $Y_1 = X_1 X_2$ with $X_1 = \sum_{lRD=0}^{L_{RD}} |h_{RD}(l_{RD})|^2$ and $X_2 = \sum_{l_{SR}=0}^{L_{SR}} \lambda_{SR}(l_{SR}) |\beta_{SR}(l_{SR})|^2$. Substituting (37) in (29) and averaging the resulting expression with respect to Y_1 and Y_2, we obtain

$$\begin{aligned} P(x,\hat{x}) &\le E_{Y_1}\left[\exp\left(-\tfrac{E_{SD}}{4N_0} Y_1\right)\right] E_{Y_2}\left[\exp\left(-\tfrac{E_{SD}}{4N_0} Y_2\right)\right] \\ &= \Phi_{Y_1}(\omega)\big|_{j\omega = -\frac{E_{SD}}{4N_0}} \times \Phi_{Y_2}(\omega)\big|_{j\omega = -\frac{E_{SD}}{4N_0}}, \end{aligned} \tag{38}$$

where $\Phi_{Y_1}(\omega)$ and $\Phi_{Y_2}(\omega)$ are the characteristic functions of Y_1 and Y_2 respectively. Since the entries of $|\beta_{SD}|$ are Rayleigh distributed, $\Phi_{Y_2}(\omega)$ is given as [25]

$$\Phi_{Y_2}(\omega)|_{j\omega=-\frac{E_{SD}}{4N_0}} = \prod_{l_{SD}=0}^{L_{SD}} \left(1+\frac{E_{SD}}{4N_0}\lambda_{SD}(l_{SD})\right)^{-1}. \tag{39}$$

$\Phi_{Y_1}(\omega)$ can be evaluated as [13]

$$\Phi_{Y1}(\omega) = \int_0^\infty f_{X_1}(x_1)\Phi_{X_2}(\omega x_1)dx_1, \tag{40}$$

where $f_{X_1}(x_1)$, $x_1 \geq 0$ is the probability density function (pdf) of X_1, and $\phi_{X_2}(\omega x_1)$ is the characteristic function of X_2. Having X_1 to be a chi squared random variable with $2(L_{RD}+1)$ degrees of freedom [13],

$$f_{X_1}(x_1) = \frac{(L_{RD}+1)^{L_{RD}+1}}{\Gamma(L_{RD}+1)} x_1^{L_{RD}} e^{-(L_{RD}+1)x_1} \tag{41}$$

and $\phi_{Y_1}(\omega x_1)$ can be evaluated as [13]

$$\Phi_{X_2}(\omega x_1)|_{j\omega=-\frac{E_{SD}}{4N_0}} = \prod_{l_{SR}=0}^{L_{SR}} (1-j\omega x_1\lambda_{SR}(l_{SR}))^{-1}. \tag{42}$$

Having (41) and (42), (40) can be evaluated as

$$\Phi_{Y_1}(\omega)|_{j\omega=-\frac{E_{SD}}{4N_0}} = \frac{(L_{RD}+1)^{L_{RD}+1}}{\Gamma(L_{RD}+1)}\left(\frac{E_{SD}}{4N_0}\right)^{-(L_{SR}+1)} \prod_{l_{SR}=0}^{L_{SR}} \frac{1}{\lambda_{SR}(l_{SR})}\times$$
$$\int_0^\infty \frac{x_1^{L_{RD}} e^{-x_1(L_{RD}+1)}}{\prod_{l_{SR}}^{L_{SR}}\left(\frac{1}{\lambda_{SR}(l_{SR})E_{SD}/4N_0}+x_1\right)} dx_1. \tag{43}$$

Assuming high SNR, i.e., $E_{SD}/4N_0 \gg 1$, (43) can be simplified as [26] [p.382, 3.351.3]

$$\Phi_{Y_1}(\omega)|_{j\omega=-\frac{E_{SD}}{4N_0}} = (L_{RD}+1)^{L_{SR}} \frac{\Gamma(L_{RD}-L_{SR}+1)}{\Gamma(L_{RD}+1)}\left(\frac{E_{SD}}{4N_0}\right)^{-(L_{SR}+1)} \prod_{l_{SR}=0}^{L_{SR}} \frac{1}{\lambda_{SR}(l_{SR})} \tag{44}$$

Substituting (39) and (44) in (38), we find the final PEP expression given as [13]

$$P(x,\hat{x}) \leqslant \left(\frac{E_{SD}}{4N_0}\right)^{-(L_{SR}+L_{SD}+1)} (L_{RD}+1)^{L_{SR}} \frac{\Gamma(L_{RD}-L_{SR}+1)}{\Gamma(L_{RD}+1)}\times$$
$$\prod_{l_{SR}=0}^{L_{SR}} \frac{1}{\lambda_{SR}(l_{SR})} \prod_{l_{SD}=0}^{L_{SD}} \frac{1}{\lambda_{SD}(l_{SD})}. \tag{45}$$

Case II: $L_{SR} > L_{RD}$

Noting that this case is similar to Case I with L_{SR} and L_{RD} interchanged, the PEP expression is given as [13]

$$P(\mathbf{x},\hat{\mathbf{x}}) \leqslant \left(\frac{E_{SD}}{4N_0}\right)^{-(L_{SR}+L_{SD}+1)} (L_{SR}+1)^{L_{RD}} \frac{\Gamma(L_{SR}-L_{RD}+1)}{\Gamma(L_{SR}+1)}\times$$
$$\prod_{l_{RD}=0}^{L_{RD}} \frac{1}{\lambda_{RD}(l_{RD})} \prod_{l_{SD}=0}^{L_{SD}} \frac{1}{\lambda_{SD}(l_{SD})}. \tag{46}$$

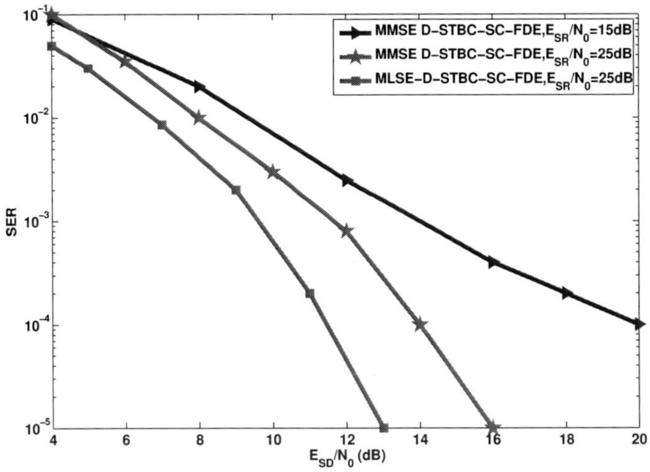

Figure 4. SER of MMSE/MLSE D-STBC-SC-FDE.

Case III: $L_{SR} = L_{RD}$

Following the same argument in Case 1, the PEP expression can be expressed as follows [13]

$$P(\mathbf{x},\hat{\mathbf{x}}) \leqslant \left(\frac{E_{SD}}{4N_0}\right)^{-(L_{RD}+L_{SD}+2)} \frac{(L_{RD}+1)^{L_{RD}+1}}{\Gamma(L_{RD}+1)}(-1)^{L_{RD}-1}\times$$
$$\prod_{l_{SD}=0}^{L_{SD}} \frac{1}{\lambda_{SD}(l_{SD})} \sum_{l_{SR}=1}^{L_{SR}+1} \frac{p_{l_1}}{\lambda_{SD}(l_{SR})^{L_{RD}+1}} e^{\beta} Ei(-\beta)\,. \quad (47)$$

Fig. 4 illustrates the symbol error rate (SER) performance of MLSE D-STBC-SC-FDE and MMSE D-STBC-SC-FDE considering a scenario with a nonfading $R \to D$ link which can be justified in a practical scenario by the existence of a strong line-of-sight path. $S \to D$ and $S \to R$ links are modeled as frequency-selective channels with memory lengths $L_{SR} = L_{SD} = 3$ and a uniform delay power profile. We further assume a perfect power control, i.e. the $S \to D$ and $R \to D$ links are balanced ($E_{SD}/N_0 = E_{RD}/N_0 \gg 1, E_{SR}/N_0 > E_{SD}/N_0$). It can be seen that with MLSE decoding, D-STBC-SC-FDE is able to provide a diversity order of $L_{SR} + L_{SD} + 2 = 8$, confirming our conclusions from the PEP analysis, cf. (46). Although MLSE is able to collect the maximum achievable diversity gain, it's complexity might be prohibitive for some cases. Therefore, we also consider MMSE in our simulations as an alternative suboptimum equalizer. Our simulation results indicate that the MLSE equalizer outperforms the MMSE equalizer by $\approx$ 2.7 dB at $SER = 10^{-4}$. This loss can be further reduced, if multiple antennas are employed at the destination [12]. It is also observed from Fig. 4 that the performance of MMSE D-STBC-SC-FDE is significantly degraded as E_{SR}/N_0 decreases. Note that this is due to the existance of the error floor. In the investigation of the maximum achievable diversity order, we have assumed that SNR in the $S \to R$ link is sufficiently large, i.e. $E_{SR}/N_0 > E_{SD}/N_0$. Considering the limiting case

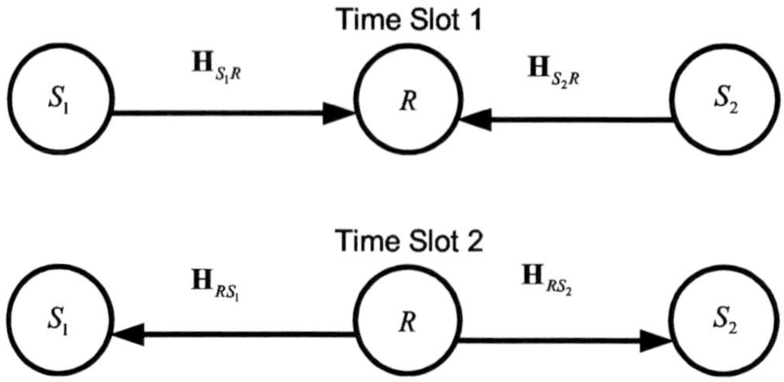

Figure 5. TWR-SC-FDE System Model.

$E_{SD}/N_0 \to \infty$, however, following(36), the PEP expression reduces to

$$P(\mathbf{x},\hat{\mathbf{x}}) \leqslant \left(\frac{E_{SR}}{N_0}\right)^{-(L_{SR}+L_{SD}+2)} \prod_{l_{SR}=0}^{L_{SR}} \frac{1}{\lambda_{SR}(l_{SR})} \prod_{l_{SD}=0}^{L_{SD}} \frac{1}{\lambda_{SD}(l_{SD})}. \tag{48}$$

It is observed that the performance becomes independent of E_{SD}/N_0, and is now governed by E_{SR}/N_0. Therefore, the performance is expected to deteriorate for low E_{SR}/N_0 resulting in error floors, which is an inherent disadvantage of the employed user cooperation protocol of [24]. It should be noted that such error floors can be avoided if Protocol I of [24] is used as the user cooperation protocol where the source-to-destination transmission is allowed within the first signaling interval. Moreover, relay-selection algorithms can be used to remove the error floor and to preserve the diversity order at an acceptable level.

5. SC-FDE for Two-Way Cooperative Relay Networks

In this section, we investigate cooperative SC-FDE for two-way relay networks assuming linear equalization, where two terminals exchange their information through a relay node in a bi-directional manner, as depicted in Fig. 5. In traditional half-duplex dual-hop cooperative networks, the source and destination nodes require four time-slots to complete both the downlink and uplink transmissions. In two-way relaying (TWR) protocols, however, only two time slots are needed to complete the downlink and uplink signal transmissions, which improves system's spectral efficiency [27–29]. In TWR schemes, two sources transmit their corresponding information signals to the intermediate relay terminal, simultaneously. In the second time slot, the relay terminal broadcasts an amplified version of the overlapped messages to the two sources. Since two sources know their own transmitted messages, the back-propagating self-interference can be readily subtracted before decoding. TWR mechanism has proven to not only increase the spectral efficiency, but also to improve the system's overall capacity [30–32]. There has been recent research efforts on the performance of two-way relay network employing OFDM over frequency selective channels [33]. However, performance of two-way relay networks for SC-FDE systems with linear equalizers has not received much attention in the literature. We consider a TWR channel, consisting of two source nodes S_1, and S_2 each equipped with a single antenna, and a relay node R. It is assumed that the TWR protocol uses two consecutive equal-duration time-slots for one round of information exchange between S_1 and S_2 via R , which linearly processes the received signal during the first time slot and then broadcasts the resulting signal to S_1 and S_2 during the second time slot.

The channel impulse responses (CIRs) for $S_i \to R$ and $R \to S_i$ links for the ith source terminal in the jth transmission block are given by $h_{S_iR}(j) = \left[h^j_{S_iR}[0], ..., h^j_{S_iR}[L_{S_iR}]\right]^{\mathrm{T}}$ and $h_{RS_i}(j) = \left[h^j_{RS_i}[0], ..., h^j_{RS_i}[L_{RS_i}]\right]^{\mathrm{T}}$, respectively, where L_{S_iR} and L_{RS_i} denote the corresponding channel memory lengths. The random vectors h_{S_iR} and h_{RS_i} are assumed to be independent zero-mean complex Gaussian with uniform power delay profile vectors.

The received signal at relay R during the first signaling phase, also known as the multiple access phase, can be expressed as below,

$$r_R(2n) = \sqrt{E_{S_1R}} H_{S_1R}(2n) x_1(2n) + \sqrt{E_{S_2R}} H_{S_2R}(2n) x_2(2n) + n_R(2n), \tag{49}$$

where $\sqrt{E_{S_iR}}$ denotes the average signal power transmitted by the source node S_i, H_{S_iR} is $N \times N$ circulant matrix with entries $[H_{S_iR}]_{k,l} = h_{S_iR}((k-l) \bmod N)$, and $n_R(2n)$ is the additive white Gaussian noise vector with each entry having zero-mean and variance of $\sigma^2_{n_R}$. Upon receiving the mixed signal from S_1 and S_2 , R processes it with AF relay operation, also known as linear analogue relaying, and then broadcasts the processed signal to S_1 and S_2 during the second signaling interval, with unit power. We can assume without loss of generality that the channel reciprocity holds for TWRC during uplink and downlink transmissions, i.e. the channel links $S_i \to R$ and $R \to S_i$ during the first and the second signaling intervals are symmetrical. Applying the Identity AF matrices, the received signals at S_1 can be written as,

$$\begin{aligned} y_{R_1}(2n+1) &= \sqrt{E_{S_1R}} H_{S_1R}(2n+1) H_{S_1R}(2n) x_1(2n) \\ &+ \sqrt{E_{S_2R}} H_{S_1R}(2n+1) H_{S_2R}(2n) x_2(2n) + H_{S_1R}(2n+1) n_R(2n) + n_{S_1}(2n+1), \end{aligned} \tag{50}$$

where $n_{S_1}(2n)$ is the additive white Gaussian noise vector with each entry having zero-mean and variance of $\sigma^2_{n_{S_1}}$. Without loss of generality, we will henceforth suppress the subscripts $2n$ and $2n+1$. Note that on the right hand side of (50), the first term is the self interference of S_1, while the second term contains the desired message from S_2. Assuming that both $H_{S_1R}H_{S_1R}$ and $H_{S_1R}H_{S_2R}$ are perfectly known at S_1 via training-based channel estimation prior to data transmission, S_1 can first subtract its self-interference from $y_{R_1}(2n+1)$ and then coherently demodulate $x_2(2n)$. The above practice is known as analogue network coding (ANC). Following similar analysis for the received signal at S_2, $y_{R_2}(2n+1)$, and further subtracting the back-propagating self-interference from $y_{R_1}(2n+1)$ and $y_{R_2}(2n+1)$, yields,

$$y_1 = \sqrt{E_{S_2R}} H_{S_1R} H_{S_2R} x_2 + H_{S_1R} n_R + n_{S_1}, \tag{51}$$

and similarly,

$$y_2 = \sqrt{E_{S_1R}} H_{S_2R} H_{S_1R} x_1 + H_{S_2R} n_R + n_{S_2}. \tag{52}$$

From this point on, we consider the processing performed at the first destination S_1 . Similar analysis is valid for S_2. We assume that the received signal is processed by an MMSE equalizer, represented by W , where its output is,

$$\tilde{x}_2 = W y_1 = \sqrt{E_{S_2R}} W H_{S_1R} H_{S_2R} x_2 + W H_{S_1R} n_R + W n_{S_1}. \tag{53}$$

5.1. MMSE SC-FDE in TWRC

In this section, we derive the optimum MMSE structure of the linear equalizer for cooperative SC-FDE receiver in a TWR network. We start by finding the MMSE linear equalizer for (52). Note that the error vector between the filtered received signal Wy_1 and the desired signal is

$$e = Wy_1 - x_2. \tag{54}$$

We first, examine the mean-squared error (MSE) of the considered system as follows,

$$\begin{aligned} &E\left[e^H e\right] = E\left[(Wy_1 - x_2)^H (Wy_1 - x_2)\right] = \\ &E\left[E_{S_2R} x_2^H H_{S_2R}^H H_{S_1R}^H W^H W H_{S_1R} H_{S_2R} x_2\right] - E\left[\sqrt{E_{S_2R}} x_2^H H_{S_2R}^H H_{S_1R}^H W^H x_2\right] \\ &+ E\left[n_R^H H_{S_1R}^H W^H W H_{S_1R} n_R\right] + E\left[n_{S_1}^H W^H W n_{S_1}\right] - \\ &E\left[\sqrt{E_{S_2R}} x_2^H W H_{S_1R} H_{S_2R} x_2\right] + E\left[x_2^H x_2\right] \end{aligned} \tag{55}$$

To simplify (55), we use the remarks given in the following lemma.

Lemma 1: Consider the quadratic structure of the form $Q = s^H A s$, which can be expanded as follows,

$$\begin{aligned} Q = &\bar{s}_1 A_{11} s_1 + \bar{s}_2 A_{21} s_1 + \cdots + \bar{s}_N A_{N1} s_1 + \\ &\bar{s}_1 A_{12} s_2 + \bar{s}_2 A_{22} s_2 + \cdots + \bar{s}_N A_{N2} s_2 + \\ &\bar{s}_1 A_{13} s_3 + \bar{s}_2 A_{23} s_3 + \cdots + \bar{s}_N A_{N3} s_3 + \\ &\vdots \qquad\qquad \vdots \qquad\qquad \vdots \\ &\bar{s}_1 A_{1N} s_N + \bar{s}_2 A_{2N} s_N + \cdots + \bar{s}_N A_{NN} s_N . \end{aligned} \tag{56}$$

With the assumption that the transmitted symbols as well as the noise samples are uncorrelated to each other within a block, the average power of one symbol is σ_s^2 , the average noise power is σ_n^2 , the matrix A is perfectly known, and therefore deterministic, i.e. we can express the expectation values as follows,

$$\begin{aligned} &E\left[\bar{x}_k A_{kl} x_l\right] = \begin{cases} 0 & k \neq l \\ \sigma_s^2 A_{kl} & k = l \end{cases} \\ &E\left[\bar{x}_k A_{kl} n_{R,l}\right] = E\left[\bar{x}_k A_{kl} n_{S_1,l}\right] = 0 \\ &E\left[\bar{n}_{R,k} A_{kl} n_{R,k}\right] = E\left[\bar{n}_{S_1,k} A_{kl} n_{S_1,k}\right] = \begin{cases} 0 & k \neq l \\ \sigma_n^2 A_{kl} & k = l \end{cases} . \end{aligned} \tag{57}$$

Using (57), we can simplify (55) as follows,

$$\begin{aligned} &E\left[e^H e\right] = E_{S_2R} \sigma_s^2 Trace\left\{H_{S_2R}^H H_{S_1R}^H W^H W H_{S_1R} H_{S_2R}\right\} - \\ &\sqrt{E_{S_2R}} \sigma_s^2 E\left\{H_{S_2R}^H H_{S_1R}^H W^H\right\} + \sigma_n^2 Trace\left\{H_{S_1R}^H W^H W H_{S_1R}\right\} + \\ &\sigma_n^2 Trace\left\{W^H W\right\} - \sqrt{E_{S_2R}} \sigma_s^2 Trace\left\{W H_{S_1R} H_{S_2R}\right\} + \sigma_s^2 N \end{aligned} \tag{58}$$

The expectation function can be minimized by taking partial differentiation with respect to W and setting this derivation to zero. For non-analytical complex function $f(X)$, containing X^H or $\bar{X}$, the following differentiation rule must be regarded,

$$\partial f(X) = \frac{\partial f(X)}{\partial X}\partial X + \frac{\partial f(X)}{\partial \bar{X}}\partial \bar{X}. \tag{59}$$

Furthermore, following rules for derivation must be applied,

$$\begin{aligned} \frac{\partial Trace\left(AX^H B\right)}{\partial \bar{X}} &= (BA)^T \\ \frac{\partial Trace\left(AXB\right)}{\partial X} &= (BA)^T. \end{aligned} \tag{60}$$

Therefore, the partial differentiation yields,

$$\begin{aligned} &\frac{\partial}{\partial W}\left\{E\left[e^H e\right]\right\} = \\ &= \begin{bmatrix} \left(E_{S_2R}\sigma_s^2 W H_{S_1R} H_{S_2R} H_{S_2R}^H H_{S_1R}^H + \sigma_n^2 W H_{S_1R} H_{S_1R}^H + \sigma_n^2 W - \sqrt{E_{S_2R}}\sigma_s^2 H_{S_1R} H_{S_2R}\right)^T \\ + \\ \left(E_{S_2R}\sigma_s^2 W H_{S_1R} H_{S_2R} H_{S_2R}^H H_{S_1R}^H + \sigma_n^2 W H_{S_1R} H_{S_1R}^H + \sigma_n^2 W - \sqrt{E_{S_2R}}\sigma_s^2 H_{S_1R} H_{S_2R}\right)^H \end{bmatrix} \end{aligned} \tag{61}$$

By setting (61) equal to zero, we obtain the optimal feed forward filter based on MMSE criterion, as given below

$$\begin{aligned} &W = \\ &\left\{\sqrt{E_{S_2R}}\sigma_s^2 \cdot \left(E_{S_2R}\sigma_s^2 H_{S_1R} H_{S_2R} H_{S_2R}^H H_{S_1R}^H + \sigma_n^2 H_{S_1R} H_{S_1R}^H + \sigma_n^2 I\right)^{-1} \cdot \left(H_{S_2R}^H H_{S_1R}^H\right)\right\} \\ &\cdot\left\{\left(E_{S_2R}\right)^{-\frac{1}{2}}\left[R_{x_2}^{-1} + H_{S_1R} H_{S_2R} R_\nu^{-1}\left(H_{S_1R} H_{S_2R}\right)^H\right]^{-1}\left(H_{S_1R} H_{S_2R}\right)^H R_\nu^{-1}\right\}, \end{aligned} \tag{62}$$

where, $\nu = H_{S_1R} n_R + n_{S_1}$, and $R_\upsilon = E\left[\upsilon\upsilon^H\right] = \sigma_n^2\left(H_{S_1R} H_{S_1R}^H + I\right)$.

Next, in order to characterize the effective noise covariance matrix, we define the diagonal matrix Λ_i which contains the N-point DFT of the channel matrix H_i, with the following eigen decomposition, $H_i = Q^H \Lambda Q$. In here,Q is the discrete Fourier transform, as written below,

$$Q(m,n) = \frac{1}{\sqrt{N}}\exp\left[-j\frac{2\pi}{N}(m-1)(n-1)\right], \quad m,n \in \{1,\ldots,N\} \tag{63}$$

Such that $Q^H Q = I$. Having said this, the diagonal elements of Λ_i are the channel's eigen values as explained below,

$$\lambda_n \triangleq \Lambda(n,n) = \sum_{i=0}^{N-1} h_i e^{-j\frac{2\pi i n}{N}}, \quad for\ n = 1,\ldots,N, \tag{64}$$

Therefore, the effective noise covariance matrix can be formulated as follows,

$$R_e = \frac{\frac{1}{E_{S_2R}} Q^H \left(\Lambda_{S_1R}\Lambda_{S_1R}^H + I\right) Q}{\frac{1}{E_{S_2R}\sigma_s^2} Q^H \left(\Lambda_{S_1R}\Lambda_{S_1R}^H + I\right) Q + \frac{1}{\sigma_n^2} Q^H \Lambda_{S_1R}\Lambda_{S_2R}\left(\Lambda_{S_1R}\Lambda_{S_2R}\right)^H Q} =$$
$$Q^H diag\left\{\frac{\frac{1}{E_{S_2R}} A}{\frac{1}{E_{S_2R}\sigma_s^2}(A+1) + \frac{1}{\sigma_n^2} B}\right\} Q = \tag{65}$$
$$Q^H diag\left\{\frac{\frac{1}{E_{S_2R}}\left(|\lambda_{S_1R,k}|^2 + 1\right)}{\frac{1}{E_{S_2R}\sigma_s^2}\left(|\lambda_{S_1R,k}|^2 + 1\right) + \frac{1}{\sigma_n^2}|\lambda_{S_1R,k}|^2 |\lambda_{S_2R,k}|^2}\right\} Q,$$

where $A = \sum_{k=0}^{N-1}\sum_{k'=0}^{N-1} h_{S_1R,k}\bar{h}_{S_1R,k'} e^{-j2\pi(k-k')/N}$, and $B = \sum_{l=0}^{N-1}\sum_{l'=0}^{N-1} h_{S_2R,l}\bar{h}_{S_2R,l'} e^{-j2\pi(l-l')/N}$.

Note that due to the underlying symmetry, the diagonal elements of $R(e)$ are identical. Following this analysis, one can characterize the optimum MMSE structure of the linear equalizer for cooperative SC-FDE receiver in a TWR network.

6. Conclusion

In this chapter, we have reviewed the recent advances in SC-FDE systems. First, we have introduced the so-called single-input-single-output (SISO)-SC-FDE with MMSE equalization and provided detail comparisons of SC-FDE with OFDM systems. We have then studied STBC-SC-FDE for conventional point-to-point systems, and D-STBC-SC-FDE for cooperative systems. We have further extended our discussion to TWR communications, and have presented recent studies on TWR SC-FDE.

References

[1] Z. Wang and G. B. Giannakis, "Linearly Precoded or Coded OFDM against Wireless Channel Fades?," *Proc. of the IEEE Workshop on Signal Processing Advances for Wireless Communications.*, 2001.

[2] Willie W. Lu, "Compact Multi-Dimensional Broadband Wireless - Convergence of Wireless Mobile and Access," *IEEE Commun. Mag.*, Nov. 2000, pp. 119–23

[3] S. Govardhanagiri, T. Karp, P. Heller, and T. Nguyen, "Performance analysis of multicarrier modulation systems using cosine modulated filter banks, " in *Proc. ICASSP '99*, vol. 3, 1999, pp. 1405–1408.

[4] B. Muquet, Z. Wang, G. B. Giannakis, M. de Courville, and P. Duhamel, "Cyclic prefix or zero padding for wireless multicarrier transmission?, " *IEEE Trans. Commun.*, vol. 50, pp. 2136–2148, Dec. 2002.

[5] D. P. Palomar, J. M. Cioffi, and M. A. Lagunas, "Joint Tx-Rx beamforming design for multicarrier MIMO channels: A unified framework for convex optimization," *IEEE Trans. Signal Process.*, vol. 51, no. 9, Sep. 2003.

[6] H. Sari, G. Karam, and I. Jeanclaude, "Transmission Techniques for Digital Terrestrial TV Broadcasting," *IEEE Commun. Mag.*, vol. 33, no. 2, Feb. 1995, pp. 100–109

[7] O. Edfors, M. Sandell, J.-J. van de Beek, S. K. Wilson, and P. O. Brjesson,"OFDM channel estimation by singular value decomposition" *IEEE Comm. Lett.*, vol. 46, no. 7, pp. 931-939, July 1998.

[8] J. G. Foschini, "Layered space-time architecture for wireless communication in a fading environment when using multi element antennas," *Bell Labs Tech. J.*, vol.2, pp.41-59, Autumn 1996.

[9] S. M. Alamouti, "Simple transmit diversity technique for wireless communications," *IEEE Journal on Select Areas in Communications*, vol. 16, pp. 1451-1458, 1998.

[10] V. Tarokh, H. J. Jafarkhani and A. R. Calderbank, "Space-time block codes from orthogonal designs," *IEEE Transactions on Information Theory*, vol.45, no.5, p.1456-1467, July 1999.

[11] N. Al-Dhahir, "Single Carrier Frequency Domain Equalization for Space Time Block Coded Transmissions over Frequency Selective Fading channels" *IEEE Comm. Lett.*, vol. 5, no. 7, July 2001.

[12] S. Zhou and G. B. Giannakis,"Space-Time Coding with Maximum Diversity Gains over Frequency-Selective Fading Channels," *IEEE Signal Processing Letter*s, vol. 8, no. 10, pp. 269 272, October 2001.

[13] H. Mheidat, M. Uysal, and N. Al-Dhahir, "Equalization Techniques for Distributed Space-Time Block Codes with Amplify-and-Forward Relaying", *IEEE Transactions on Signal Processing*, vol. 55, no. 5, part 1, p. 1839-1852, May 2007.

[14] A. Sendonaris, E. Erkip, and B. Aazhang, "User cooperation diversity. Part I. System description," *IEEE Trans. Commun.*, vol. 51, no. 11, pp. 1927-1938, Nov. 2003.

[15] –, "User cooperation diversity. Part II. Implementation aspects and performance analysis," *IEEE Trans. Commun.*, vol. 51, no. 11, pp. 1939-1948, Nov. 2003.

[16] J. N. Laneman and G. W.Wornell, "Distributed space-time-coded protocols for exploiting cooperative diversity in wireless networks," *IEEE Trans. Inf. Theory*, vol. 49, no. 10, pp. 2415-2425, Oct. 2003.

[17] M. Janani, A. Hedayat, T. E. Hunter, and A. Nosratinia, "Coded cooperation in wireless communications: space-time transmission and iterative decoding," *IEEE Trans. Signal Process.*, vol. 52, no. 2, pp. 362-371, Feb. 2004.

[18] H. Muhaidat, M. Uysal, and R. Adve, "Pilot-symbol-assisted detection scheme for distributed orthogonal space-time block coding," *IEEE Trans. Wireless Commun.*, vol. 8, no. 3, pp. 1057-1061, Mar. 2009.

[19] H. Mheidat, Murat Uysal and N. Al-Dhahir, "Comparative Analysis of Equalization Techniques for STBC with Application to EDGE", *IEEE VTC'04*-Spring, Milano, Italy, May 2004.

[20] Hakam Mheidat, Murat Uysal and N. Al-Dhahir, “Single-Carrier Frequency Domain Equalization for Broadband Cooperative Communications", *Wireless Communications and Networking Conference*, vol. 3, pp. 1578-1584, 2006.

[21] H. Eghbali, S. Muhaidat, and N. Al-Dhahir, “A Low Complexity two Stage MMSE-Based Receiver for Single-Carrier FrequencyDomain Equalization Transmissions over Frequency-Selective Channels," *IEEE GLOBECOM’09*, Honolulu, Hawaii, USA, November 2009.

[22] H. Eghbali, S. Muhaidat, and N. Al-Dhahir, “A Novel Receiver Design for Single-Carrier Frequency Domain Equalization in Broadband Wireless Networks with Amplify-and-Forward Relaying," *IEEE Trans.Wireless Commun.*, vol.10, no. 3, 721-727.

[23] D. Y. Seol, U. K. Kwon, G. H. Im, and E. S. Kim, “Relay-assisted SFBC single carrier transmission over uplink fast fading channels,” *in Proc. IEEE GLOBECOM*, Nov. 2007, pp. 3883-3887.

[24] R. U. Nabar, H. Bolcskei, and F.W. Kneubuhler, “Fading relay channels: Performance limits and space-time signal design,” *IEEE J. Sel. Areas Comm.*, vol. 22, no. 6, pp. 1099-1109, Aug. 2004.

[25] M. K. Simon and M. S. Alouini, Digital Communication Over Fading Channels: A Unified Approach to Performance Analysis . New York: Wiley-Intersci., 2000.

[26] I. S. Gradshteyn and I. M. Ryzhik, Table of Integrals, Series and Products. New York: Academic, 2000.

[27] J.Ping and S.H.Ting, “Rate Performance of AF Two-Way Relaying in Low SNR Region,” *IEEE Commun.Letters*, vol.13, pp.1-3, Apr.2009.

[28] K.J.Lee, K.W.Lee, H.Sung and I.Lee, “Sum-Rate Maximization for Two-Way MIMO Amplify-and-Forward Relaying Systems,” *IEEE 69th Vehic.Tech.Confer.*, 2009, pp.1-5, Apr.2009.

[29] T.Q.Duong, L.Hoang and V.N.Q.Bao, “On the Performance of Two-Way Amplify-and-Forward Relay Networks,” *IEICE Trans.Commun.*, vol.E92-B, pp.3957-3959, Dec.2009.

[30] J.Yang, P.Fan, T.Q.Duong and X.Lei, “Exact Performance of Two-Way AF Relaying in Nakagami-m Fading Environment,” *IEEE Trans.Wireless Commun.*, vol.10, no. 3, pp.1-7, Mar.2011.

[31] Y.Hu, K.H.Li and K.C.Teh, “Performance of Two-Way Amplify-and-Forward Relay Networks over Asymmetric Channels,” *IEEE MILCOM* 2009, pp.1-6, Oct.2009.

[32] L.Song, "Relay Selection for Two-Way Relaying With Amplify-and-Forward Protocols," *IEEE Trans.Vehic.Tech.*, vol.60, pp.1954-1959, May.2011.

[33] Y. Liu and M. Tao, "Optimal channel and relay assignment in OFDMbased multi-relay multi-pair two-way communication networks," *IEEE Trans. Commun.*, vol. 1, no. 2, April 2012

In: Communication Systems: New Research
Editor: Vyacheslav Tuzlukov
ISBN: 978-1-62618-654-5

Chapter 6

WIRELESS COMMUNICATIONS: GENERALIZED APPROACH TO SIGNAL PROCESSING

Vyacheslav Tuzlukov*
Department of Information Technologies and Communications,
School of Electronics Engineering, College of IT Engineering,
Kyungpook National University, Daegu, South Korea

ABSTRACT

The main idea discussed in this chapter is to use the generalized approach to signal processing (GASP) in noise in wireless communication systems. The GASP is based on a seemingly abstract idea: the intorduction of an additional noise source that does not carry any information about the signal and signal parameters in order to improve the qualitative performance of wireless communication systems. In the present chapter we would like to discuss the following aspects of GASP in wireless communication systems: a) the main theoretical statements and brief descripotion; b) array signal processing; c) multichannel adaptive signal processing; d) quadriphase DS-CDMA; e) optimal waveforms for MIMO wireless communication systems.

Keywords: Generalized approach to signal processing in noise; generalized detector; signal-to-noise ratio; detection performance; array signal processing; wireless communication systems; generalized likelihood ratio; Neyman-Pearson receiver.

ABBREVIATIONS

ACE adaptive coherence estimator;
AF – additional filter;
AGD – adaptive generalized detector;
AIC – Akaike information criterion;

* Corresponding Author address. Email: tuzlukov@ee.knu.ac.kr.

AMF – adaptive matched filter;
AR – autoregressive;
AWGN – additive white Gaussian noise;
BPSK – binary-phase-shifted-keying;
CFAR –constant false alarm rate;
DOA – direction of arrival;
DOF – degrees of freedom;
DS-CDMA – direct sequence code-division multiple-access;
DS-QPSK – direct sequence quadriphase shift keying;
FM – frequency modulation;
GASP – generalized approach to signal processing;
GD – generalized detector;
GLRT – generalized likelihood ratio;
GMVDR – generalized minimum variance distirtionless response;
IGA – improved Gaussian approximation;
MA – moving-average;
MAI – multiple-access interference;
MDL – minimum description lenght;
MF – matched filter;
MIMO – multipli-input multiple-output;
ML – maximum likelihood;
MVDR – minimum variance distirtionless response;
PAMF – parametric adaptive matched filter;
PF – preliminary filter;
PGD – parametric generalized detector;
ROC – receiveing operating characteristics;
QDS-CDMA – duadriphase direct-sequence code-division multiple-access;
RSNR – required signal-to-noise ratio;
SGA – standard Gaussian approximation;
SIGA – simplified improved Gaussian approximation;
SINR – signal-to-interference-plus-noise-ratio;
SIR – signal-to-interference ratio;
SNR – signal-to-noise ratio;
Sqrt-RC – square-root raised cosine;
WSCS – wide-sense cyclostationary;
WSS – wide-sense stationary;

6.1. Introduction

The additive and multiplicative noise exists forever in any wireless communication systems. Quality and integrity of any wireless communication systems are defined and limited by statistical characteristics of the noise and interference, which are caused by an electromagnetic field of environment.

The main characteristics of any wireless communication system are deteriorated as a result of effect of the additive and multiplicative noise. The effect of addition of noise and interference to the signal generates an appearance of false information in the case of the additive noise. For this reason, the parameters of the received signal, which is an additive mixture of the signal, noise, and interference, differ from the parameters of the transmitted signal. Stochastic distortions of parameters in the transmitted signal, attributable to unforeseen changes in instantaneous values of the signal phase and amplitude as a function of time, can be considered as the multiplicative noise. Under stimulus of the multiplicative noise, false information is a consequence of changed parameters of transmitted signals, for example, the parameters of transmitted signals are corrupted by the noise and interference. Thus, the impact of the additive noise and interference may be lowered by an increase in the signal-to-noise ratio (*SNR*). However, in the case of the multiplicative noise and interference, an increase in *SNR* does not produce any positive effects.

The main functional characteristics of any wireless communication systems are defined by an application area and are often specific for distinctive types of these systems. In the majority of cases, the main performance of any wireless communication systems are defined by some initial characterictics describing a quality of signal processing in the presence of noise: the precision of signal parameters measurement, the definition of resolution intervals of the signal parameters, and the probability of error.

The main idea discussed in this chapter is to use the generalized approach to signal processing (GASP) in noise in wireless communication systems. The GASP is based on a seemingly abstract idea: the intorduction of an additional noise source that does not carry any information about the signal and signal parameters in order to improve the qualitative performance of wireless communication systems. In other words, we compare statistical data defining the statistical parameters of the probability distribution densities (pdfs) of observed input stochastic samples from two independent frequency-time regions –a "yes" signal is possible in the first region and it is known a priori that a "no" signal is obtained in the second region. The proposed GASP allows us to formulate a decision-making rule based on determination of the *jointly sufficient statistics of the mean and variance* of the likelihood function (or functional). Classical and modern signal processing theories and approaches allow us to define *only the mean* of the likelihood function (or functional). Additional information about the statistical characteristics of the likelihood function (or functional) leads us to the better quality of signal detection and definition of signal parameters in compared with the optimal signal processing Algorithms of classical and modern theories and procedures.

Thus, for any wireless communication systems, we have to consider two problems – analysis and synthesis. The first problem (analysis) – the problem to study a stimulus of the additive and multiplicative noise on the main principles and performance using GASP – is an analysis of impact of the additive and multiplicative noise on the main characteristics of wireless communication systems, the receivers in which are constructed on the basis of GASP. This problem is very important in practice. This analysis allows us to define limitations on the use of wireless communication systems and to quantify the additive and multiplicative noise impact relative to other sources of interference present in these systems. If we are able to conclude that the presence of the additive and multiplicative noise is the main factor or one of the main factors limiting the performance of any wireless communication systems, then the second problem – the definition of structure and main parameters and characteristics of the generalized detector or receiver (GD or GR) under a dual stimulus of the additive and multiplicative noise – the problem of synthesis – arises.

Employment of GASP in wireless communication systems allows us to extend the well-known boundaries of the potential noise immunity set by the classical and modern signal processing theories and approaches. Implementation of wireless communication systems, the receivers of which are constructed on the basis of GASP, allows us to obtain the high-quality detection of signal and high accuracy of signal parameter definition with noise components present compared with that systems, the receivers of which are constructed on the basis of classical and modern theories and procedures. The optimal and asymptotic optimal signal processing algorithms of classical and modern theories and approaches, for signals with amplitude-frequency-phase structure characteristics that can be known and unknown a priori, are constituents of the signal processing algorithms that are designed on the basis of GASP.

In the present chapter, we would like to discus the following aspects of GASP for wireless communication systems:

- The main theoretical statements and brief description of GASP;
- Array signal processing and its features;
- Multichannel adaptive signal processing;
- Quadriphase direct-sequence code-division multiple access (DS-CDMA);
- Optimal waveforms for multiple-input multiple-output (MIMO) wireless communication systems.

As to array signal processing, we consider a problem of detecting a random spatially distributed signal source by an array of sensors based on GASP in noise. We derive some GD structures under several assumptions on the available statistics. The GD performance is evaluated and the effect of source angular spread is investigated. We notice the degrees of freedom of detection statistic distributions depend on both the signal angular spread and the number of data snapshots. At the high SNR and with small degrees of freedom, an increase of angular spread improves the detection performance. With large degrees of freedom the increase of angular spread reduces the detection performance. A comparison between GD and conventional beamformer is carried out by computer simulation. The results demostrate a superiority of GD implementation in wireless communication systems over the conventional beamformer detector structures as the angular spread becomes large.

Under multichannel adaptive signal processing based on GASP in wireles communication systems the parametric Rao test for multichannel adaptive signal detection by the adaptive GD constructed based on GASP in noise is derived by modeling the disturbed signal as a multichannel autoregressive process. The parametric Rao test takes a form identical to that of parametric GD for space-time adaptive processing in airborne surveillance radar systems and other similar applications. The equivalence offers new insights into the perfromance and GD implementation. Specifically, the Rao/GD is asymptotic, in the case of large samples, a parametric generalized likelihood ratio test GD (GLRT GD) due to an asymptotic equivalence between the Rao test and the GLRT GD. The asymptotic distribution of the Rao/GD test statistic is obtained in the closed form, which follows an exponential distribution under the null hypothesis – the received signal is absent – and, respectively, a noncentral Chi-squared distribution with two degrees of freedom under the alternative hypothesis – the received signal is present. The noncentrality parameter of the noncentral Chi-squared distribution is determined by the output signal-to-interference-plus-noise ratio (*SINR*) of a temporal whitening filter. Since the asymptotic distribution under the null hypothesis is independent of the unknown parameters, the Rao/GD asymptotically achieves the constant false alarm rate (CFAR) GD. Numerical results show that these results are superior in predicting the performance of the parametric adaptive matched filter detector even with moderate data support.

We consider the probability of bit-error P_{er} performance of asynchronous DS-CDMA wireless communication systems employing the GD. The effects of pulse shaping, quadriphase or direct sequence quadriphase shift keying (DS-QPSK) spreading, aperiodic spreading sequences are considered in DS-CDMA wireless communication systems based on GASP and compared with the coherent Neyman-Pearson receiver. An exact P_{er} expression and several approximations, namely, the expression using the characteristic function method, a simplified expression for the improved Gaussian approximation (IGA), and the simplified IGA (SIGA) are derived. Under conditions typically satisfied in practice and even with a small number of interferers, the standard Gaussian approximation (SGA) for the multiple-access interference component of the GD statistic and P_{er} performance is shown to be accurate. Moreover, the IGA is shown to reduce to the SGA for pulses with zero excess bandwidth. Second, the GD P_{er} performance of quadriphase DS-CDMA is shown to be superior to that of bi-phase DS-CDMA. Numerical examples by Monte Carlo simulation are presented to illustrate the GD P_{er} performance for square-root raised-cosine pulses and spreading factors of moderate to large values. Also, a superiority of GD implementation in DS-CDMA wireless communication systems over the Neyman-Pearson receiver is demonstrated.

The problem of waveform design for MIMO wireless communication systems employing the GD constructed in accordance with GASP is considered and discussed. We develop a procedure to design the optimal waveform that maximizes the *SINR* at the GD output. The optimal waveform requires knowledge of interference statistics. We also develop several suboptimal waveforms requiring knowledge of interference statistics. Thus, the transmit waveforms are adjusted based on clutter statistics. A model for received signals that incorporates the transmit waveforms is developed. The received signal detection problem is formulated for that model. Optimal and suboptimal signal processing algorithms are derived for designing the transmit waveforms under different assumptions regarding the statistical information available to GD. The performance of these algorithms is illustrated by computer simulation.

6.2. Main Theoretical Statements of GASP

6.2.1. Initial Premises

The optimal signal detection algorithm is based on the assumption that there is the frequency-time region Z of the noise, where a signal may be present. For example, there is an observed stochastic sample from this region, relative to which there is a need to make a decision a "yes" signal – the hypothesis $\mathcal{H}_1$ – or a "no" signal – the hypothesis $\mathcal{H}_0$ [1], i.e.

$$\begin{cases} \mathcal{H}_0 \Rightarrow \sum_{i=1}^{N} X_i a_i \le K_{\text{op}} , \\ \mathcal{H}_1 \Rightarrow \sum_{i=1}^{N} X_i a_i > K_{\text{op}} , \end{cases} \tag{6.1}$$

where X_i is the sample observed from the region Z; a_i is the expected signal from the region Z; and K_{op} is the threshold for decision-making.

We now proceed to modify the initial premises of the classical and modern signal detection theories. Suppose there are two independent frequency-time noise regions Z and Z^* belonging to the space A (see Fig. 6.1). The noise from these regions obeys the same pdf with equal statistical parameters. The same pdf and equality of the statistical parameters have been chosen for simplicity of this analysis. In general, the pdf and statistical parameters may not be the same. A "yes" signal is possible in the noise region Z as before. *It is known a priori that a "no" signal is obtained in the noise region* Z^*. Below we will cal the noise region Z^* the reference region, and, consequently, the observed sample from this region is called the reference sample. There is a need to make the decision a "yes" signal – the hypothesis $\mathcal{H}_1$ – or a "no" signal – the hypothesis $\mathcal{H}_0$ – in the observed input stochastic sample from the region Z, by comparing the statistical parameters of the pdf of this observed input stochastic sample with those of the observed input stochastic sample from the reference region Z^*.

The problem posed in the preceding sections of this chapter must be solved using the statistical decision-making theory. Thus, it is necessary to accumulate and compare statistical data defining the statistical parameters of the pdfs of the observed input stochastic samples from two independent frequency-time regions Z and Z^*. If the pdf statistical parameters for two samples are equal or agree with each other within the limits of a given accuracy, then the decision a "no" signal in the observed input stochastic sample from the region Z is made – the hypothesis $\mathcal{H}_0$. If the pdf statistical parameters of the observed input stochastic sample from the region Z differ from those of the reference sample from the region Z^* by a value that exceeds the prescribed error limit, then the decision a "yes" signal in the region Z is made – the hypothesis $\mathcal{H}_1$.

6.2.2. Likelihood Ratio

Now the problem is to obtain the jointly sufficient statistics to define the statistical parameters of pdfs. For this purpose let us avail ourselves of one of the well-known results [2]–[4]. It is known that a sufficient statistic is determined from the condition that the likelihood function has an extremum. In general, the condition of an extremum of the likelihood function, relative to the parameter to be determined with a prescribed accuracy, is defined in the following form:

$$\frac{\partial f_X(X_1,\ldots,X_N \mid \vartheta)}{\partial \vartheta} = 0 \ , \tag{6.2}$$

where N is the sample size determining the prescribed accuracy, and ϑ is the parameter to be determined. However, this equation is not used in practice. A simple mathematical procedure simplifies the representation of this equation. Since the logarithm is a monotonic function, the extrema of the functions $f_X(X_1,\ldots,X_N \mid \vartheta)$ and $\partial f_X(X_1,\ldots,X_N \mid \vartheta)$ are reached at the same values of the parameter ϑ. Therefore, the likelihood function equation is usually written in the following form:

$$\frac{\partial \ln f_X(X_1,\ldots,X_N \mid \vartheta)}{\partial \vartheta} = 0 \ , \tag{6.3}$$

As was shown in [4], it is easy to prove that the values $\sum_{i=1}^{N} X_i a_i$ and $\sum_{i=1}^{N} X_i^2$ are the jointly sufficient statistics of the likelihood function parameters for the observed input stochastic sample $X_1,\ldots,X_N$. The likelihood function for the reference sample $\eta_1,\ldots,\eta_N$ at the unit variance is determined in the following form:

$$f_\eta(\eta_1,\ldots,\eta_N) = \frac{1}{(2\pi)^{\frac{N}{2}}} \exp\left\{-\frac{1}{2}\sum_{i=1}^{N}\eta_i^2\right\} \ , \tag{6.4}$$

where $\sum_{i=1}^{N}\eta_i^2$ is the sufficient statistic of the likelihood function parameters for the reference sample $\eta_1,\ldots,\eta_N$. In the definition of the sufficient statistics using the input stochastic samples $X_1,\ldots,X_N$ and $\eta_1,\ldots,\eta_N$ the problem of their comparison arises.

Usually for this purpose, a diference is used (see Fig. 6.1). The resulting sufficient statistics are observed at the output of the difference device:

$$\ln f_X(X_1,\ldots,X_N) - \ln f_\eta(\eta_1,\ldots,\eta_N) = \frac{1}{2}\left\{\sum_{i=1}^{N} 2X_i a_i - \sum_{i=1}^{N} X_i^2 + \sum_{i=1}^{N}\eta_i^2 - \sum_{i=1}^{N} a_i^2\right\}. \tag{6.5}$$

It is customary to reference the last term on the right side of (6.5) to a threshold independent of the observed input stochastic sample, as in (6.1). Equation (6.5) obtained by definition of the resulting sufficient statistics is the logarithm of the likelihood function.

The signal detection algorithm based on two independent observed input stochastic samples, one of which is the reference sample with a priori information a "no" signal, follows from (6.5):

$$\ln f_X(X_1,\ldots,X_N) - \ln f_\eta(\eta_1,\ldots,\eta_N) = \ln\left\{\frac{f_X(X_1,\ldots,X_N)}{f_\eta(\eta_1,\ldots,\eta_N)}\right\}$$

$$= \frac{1}{2}\left\{\sum_{i=1}^{N} 2X_i a_i - \sum_{i=1}^{N} X_i^2 + \sum_{i=1}^{N} \eta_i^2 - \sum_{i=1}^{N} a_i^2\right\} = \ln C \tag{6.6}$$

or

$$\sum_{i=1}^{N} 2X_i a_i - \sum_{i=1}^{N} X_i^2 + \sum_{i=1}^{N} \eta_i^2 = \ln C + \sum_{i=1}^{N} a_i^2 = K_G \ , \tag{6.7}$$

where K_G is the threshold for making decision.

Proceeding from generally accepted concepts, it follows that the hypothesis $\mathcal{H}_1$ – a "yes" signal in the observed input stochastic sample $X_1,\ldots,X_N$ – is assumed if the following inequality is satisfied

$$\sum_{i=1}^{N} 2X_i a_i - \sum_{i=1}^{N} X_i^2 + \sum_{i=1}^{N} \eta_i^2 > K_G \ , \tag{6.8}$$

and the hypothesis $\mathcal{H}_0$ – a "no" signal in the observed input stochastic sample $X_1,\ldots,X_N$ – is assumed if the opposite inequality is satisfied. The first term on the left side of (6.8) with the factor 2 is the signal detection algorithm defining the Neyman-Pearson receiver or the matched filter. The more rigorous form of (6.8) is the following

$$\sum_{i=1}^{N} 2X_i a_i^* - \sum_{i=1}^{N} X_i^2 + \sum_{i=1}^{N} \eta_i^2 > K_G \ , \tag{6.9}$$

where a_i^* is the model signal.

Analysis of the signal detection algorithm in (6.8) and (6.9) shows that when considering the hypothesis $\mathcal{H}_1$ – a "yes" signal in the observed input stochastic sample $X_1,\ldots,X_N$

$$X_i = a_i + \xi_i \tag{6.10}$$

where ξ_i is the noise from the region Z, given $a_i = a_i^*$ and the left side of (6.8) and (6.9) has the following form:

$$2\sum_{i=1}^{N} [a_i + \xi_i] a_i^* - \sum_{i=1}^{N} [a_i + \xi_i]^2 + \sum_{i=1}^{N} \eta_i^2 = \sum_{i=1}^{N} a_i^2 + \sum_{i=1}^{N} \eta_i^2 - \sum_{i=1}^{N} \xi_i^2 \ , \tag{6.11}$$

where the term $\sum_{i=1}^{N} a_i^2 = E_a$ is the signal energy and the term $\sum_{i=1}^{N} \eta_i^2 - \sum_{i=1}^{N} \xi_i^2$ is the background noise.

When considering the hypothesis $\mathcal{H}_0$ – a "yes" signal in the observed input stochastic sample $X_1,\ldots,X_N$

$$X_i = \xi_i \tag{6.12}$$

and the condition $a_i = 0$ given $a_i = a_i^*$ and the left side of (6.8) and (6.9) has the following form:

$$\sum_{i=1}^{N} \eta_i^2 - \sum_{i=1}^{N} \xi_i^2 \,. \tag{6.13}$$

Subsequent analysis of the signal detection algorithm given by (6.8) and (6.9) will only be performed under the condition $a_i = a_i^*$. This statement is very important for further understanding of GASP in the presence of noise. How we do this becomes clear in the discussion of the experimental results published in [5] and discussed in detail in [6].

It must be emphasized that in the statistical sense

$$\sum_{i=1}^{N} \eta_i^2 - \sum_{i=1}^{N} \xi_i^2 \to 0 \quad \text{as} \quad N \to \infty \,, \tag{6.14}$$

since the processes ξ_i and η_i are independent and uncorrelated and have the same power spectral density of the additive Gaussian noise $0.5N_0$ owing to initial conditions.

In this way, it has been shown that the signal detection algorithm given by (6.1) based on the observation of the only input stochastic sample $X_1,\ldots,X_N$ and the signal detection algorithm given by (6.8) and (6.9) based on the observation of the two independent and uncorrelated samples $X_1,\ldots,X_N$ and $\eta_1,\ldots,\eta_N$ have the same approach and are defined by the likelihood function using the statistical theory of decision making. The difference is that the numerator and denominator of the likelihood function used for synthesis of the signal detection algorithm given by (6.1) involve the same observed input stochastic sample $X_1,\ldots,X_N$, but a "yes" signal is assumed in the numerator and a "no" signal is assumed in the denominator. The numerator of the likelihood function used for synthesis of the signal detection algorithm given by (6.8) and (6.9) involves the observed input stochastic sample $X_1,\ldots,X_N$ where a "yes" signal may be present, and the denominator involves the reference sample $\eta_1,\ldots,\eta_N$ which is known a priori to contain a "no" signal.

On this basis, it may be stated that only the sufficient statistic $\sum_{i=1}^{N} X_i a_i$ has been applied to define the mean of the likelihood function in the signal detection algorithm given by (6.1). In the signal detection algorithm given by (6.8) and (6.9) the jointly sufficient statistics

$2\sum_{i=1}^{N} X_i a_i$ and $\sum_{i=1}^{N} \eta_i^2 - \sum_{i=1}^{N} \xi_i^2$ are used to define the mean and variance of the likelihood function. This fact permits us to obtain more complete information in the decision-making process in comparison with the signal detection algorithm given by (6.1). The signal detection algorithm given by (6.8) and (6.9) is free from a number of conditions unique to the signal detection algorithm given by (6.1).

As the signal detection algorithm given by (6.1) is a component of the signal detection algorithm determined by (6.8) and (6.9), the last of which has been called *the generalized signal detection algorithm.*

6.2.3. Engineering Interpretation

The technical realization of independent sampling from the regions Z and Z^* obeying the same pdf with equal statistical parameters is not difficult. The solution of the problem of detecting the signal a_i with the additive Gaussian noise w_i is well-known [1]–[3]. The observed input stochastic process X_i is examined at the output of the linear section of the receiver which has an ideal impulse response and the bandwidth W. It is supposed that the noise at the input of the receiver front-end or input receiver linear system is the additive white Gaussian noise with the correlation function $0.5N_0\delta(t_{i+1} - t_i)$, where $\delta(x)$ is the Dirac delta function.

The signal a_i is assumed to be completely known and the signal energy is taken to be equal to 1. The spectral power density $0.5N_0$ is considered an a priori indeterminate parameter. The gain of the linear system of the receiver is equal to 1 for simplicity. Through analysis, the problem is reduced to testing the complex hypothesis with the decision function

$$\operatorname{Re}\left\{\sum_{i=1}^{N} \dot{X}_i \dot{a}_i^*\right\} > K(P_{\text{er}})\sqrt{\sum_{i=1}^{N} |\dot{X}_i|^2}\ , \tag{6.15}$$

where $\dot{a}_i^*$ is the impulse response of the filter matched with the signal or, in other words, the matched filter; $K(P_{\text{er}})$ is the threshold defined by the probability of error; and $\sum_{i=1}^{N} |\dot{X}_i|^2$ is the statistics defining the decision function. It turns out that the signal detector constructed in accordance with the above decision function renders the probability of error P_{er} stable given an unknown noise power and has the greatest probability of detection P_D for any *SNR*.

Let us interpret this problem. We use two input linear systems of the receiver instead of a single input linear system for our set of statistics. These input linear systems will be called the preliminary (PF) and additional (AF) filters. The impulse responses of the PF and AF must obey the same law for simplicity of consideration. In a general case the impulse response of the PF and AF are differed, of course. The centered or resonant frequencies of the PF and AF must be detuned with respect to each other by a value given from the well-known results published in [7] and [8]. For the purpose of providing the uncorrelated statistics at the PF and AF outputs the detuning value between the resonant or centered frequencies of the PF and AF must exceed the effective signal bandwidth *W*. As is well known from [7] and [8], if this value reaches $4 \div 5W$, the coefficient of correlation between the statistics at the PF and AF outputs does not exceed 0.05 and tends to approach zero. In practice, these statistics may be regarded as uncorrelated. The PF effective bandwidth is equal to that of the signal frequency spectrum and can be even greater, but this is undesirable since the noise power at the PF output is proportional to the effective bandwidth value *W*. The AF effective bandwidth may be smaller than the PF effective bandwidth. However, for simplicity of consideration, in this chapter the AF effective bandwidth is assumed to be the same as the PF effective bandwidth *W*.

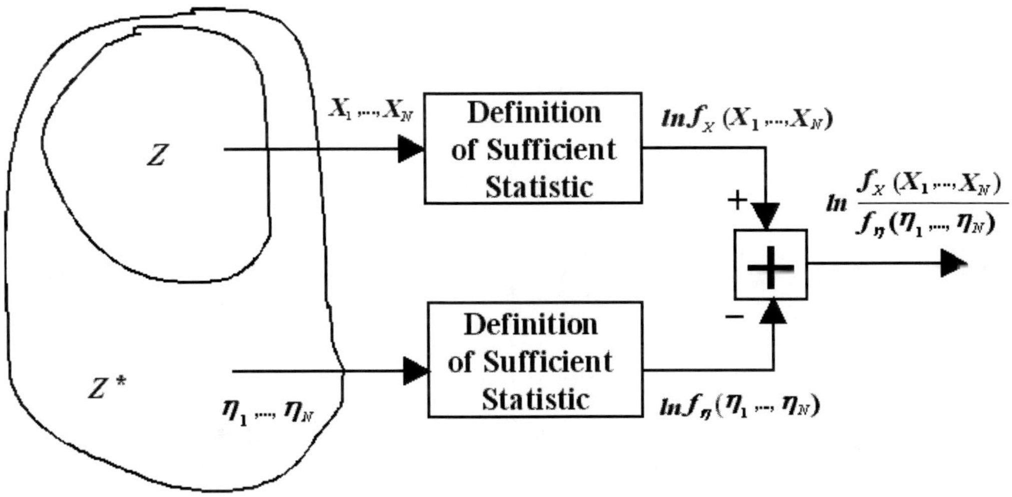

Figure 6.1. Definition of jointly sufficient statistics.

Thus, we can assume that the uncorrelated samples of the observed input stochastic process are formed at the PF and AF outputs. These samples obey the same pdf (this assumption is made for simplicity of analysis) with equal statistical parameters given that the same process is present at the PF and AF inputs, even if this process is the additive white Gaussian noise with the correlation function $0.5N_0\delta(t_{i+1} - t_i)$. The engineering interpretation of the signal detection algorithm given by (6.8) and (6.9) is the following:

- The AF can serve as the source of the reference sample $\eta_1,\ldots,\eta_N$ observed from the noise region Z^*. The AF centered or resonant frequency is detuned with respect to the carrier frequency of the signal by a value that can be determined on the basis results discussed in [7] and [8] depending on the specific practical conditions.
- The PF serves as the source of the sample $X_1,\ldots,X_N$ of the observed input stochastic process from the noise region *Z*. The PF bandwidth is matched with the effecttive signal bandwidth. The PF bandwidth value is equal to the AF bandwidth one for simplicity of analysis.
- The first term of the generalized signal detection algorithm given by (6.8) and (6.9) corresponds to the Neyman-Pearson receiver with the twice the gain.

- The second term of the generalized signal detection algorithm given by (6.8) and (6.9) corresponds to the energy detector coupled with PF.
- The third term of the generalized signal detection algorithm given by (6.8) and (6.9) corresponds to the energy detector coupled with AF.

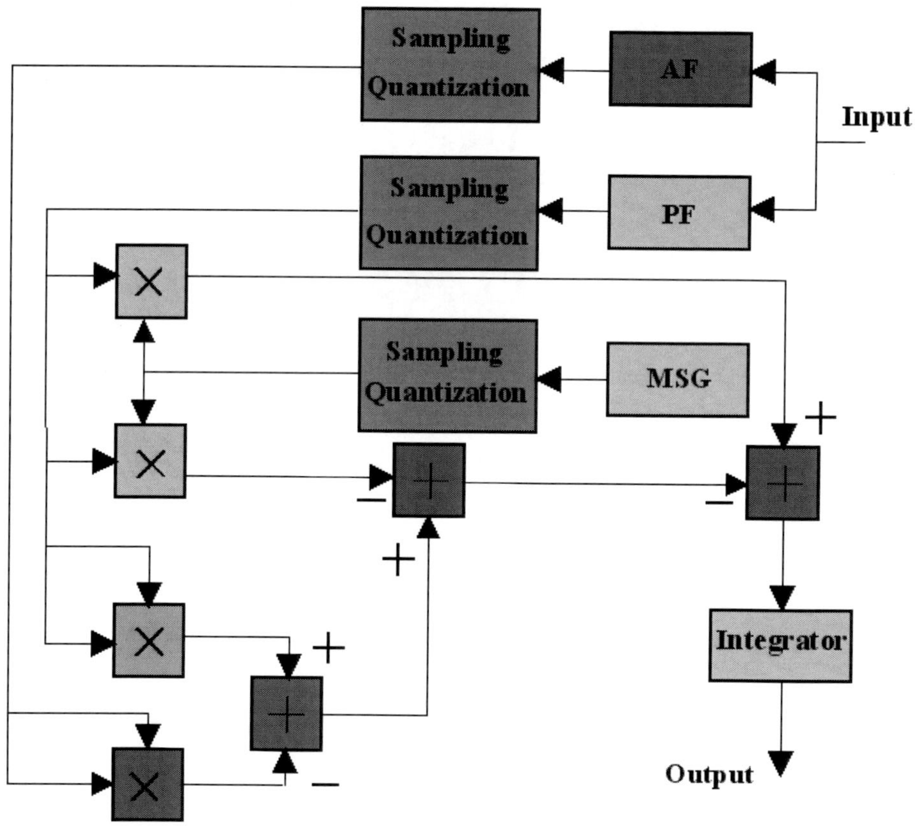

Figure 6.2. GD principal flowchart.

- The statistics of the energy detector coupled with PF is subtracted from the statistics of the energy detector coupled with AF. As a result, when a "no" signal in the input stochastic process we obtain the background noise only (6.14) that tends to approach zero in the statistical sense.
- The statistics of the energy detector coupled with PF is subtracted from the statistics of the Neyman-Pearson receiver. As a result, when a "yes" signal in the input stochastic process a complete compensation of the noise component $\sum_{i=1}^{N} a_i^* \xi_i$ that is a characteristic of the Neyman-Pearson receiver is achieved in the statistical sense if the condition $a_i = a_i^*$ is satisfied.

The detector presented in Fig. 6.2 is based on the engineering interpretation of GASP in noise stated above.

There is a need to make some comments regarding the noise forming at PF and AF outputs. If the Gaussian noise $w(t)$ comes in at the AF and PF inputs (the GD linear system front end), the noise forming at the AF and PF outputs is Gaussian, too, because AF and PF are the linear systems and, in a general case, take the following form:

$$\xi_{PF_i}(t) = \int_{-\infty}^{\infty} h_{PF}(\tau) w_i(t-\tau) d\tau \text{ and } \eta_{AF_i}(t) = \int_{-\infty}^{\infty} h_{AF}(\tau) w_i(t-\tau) d\tau \ . \tag{6.16}$$

If the additive white Gaussian noise (AWGN) with zero mean and two-sided power spectral density $0.5\mathcal{N}_0$ is coming in at the AF and PF inputs, then the noise forming at AF and PF outputs is Gaussian with zero mean and variance given by [9]

$$\sigma_n^2 = \frac{\mathcal{N}_0 \omega_0^2}{8\Delta_F}, \tag{6.16a}$$

where, in the case if AF (or PF) is the RLC oscillatory circuit, the AF (or PF) bandwidth Δ_F and resonance frequency ω_0 are defined in the following manner

$$\Delta_F = \pi\beta, \quad \omega_0 = \frac{1}{\sqrt{LC}}, \quad \beta = \frac{R}{2L} \ . \tag{6.16b}$$

The main GD functioning condition is an equality over the whole range of parameters between the model signal forming at the GD MSG output and the expected signal forming at the GD input liner system (the PF) output. How we can satisfy this condition in practice is discussed in detail in [5] and [6]? Detailed discussion about a choice of PF and AF and their impulse responses is given in [5], [6], [9]–[11] (see also http://www.sciencedir-ect.com/science/journal/10512004, click "Volume 8, 1998", "Volume 8, Issue 3", and "A new approach to signal detection theory").

It is of special interest to compare the above mentioned statements to the statements and analysis discussed in [12]. Some opponents of GASP in the presence of noise erroneously believe that this approach is the same as the one-input two-sample signal detection procedure. For this purpose, we briefly recall the main statements of the one-input two-sample signal detection procedure.

- Firstly, the signal sample is generated only at the time instants that correspond to the expected signal. In other words, the signal sample is generated only at the time instants when the expected signal may appear in time-frequency space.
- Secondly, the noise channel is formed within the limits of the time intervals, in which it is a priori known that the signal is absent.

The GASP in the presence of noise is based on the following statements:

- The first sample is generated independently of the time instants that correspond to the emergence of the signal in the time-frequency space.

- The reference sample – the second sample – is formed simultaneously with the first sample at the same time intervals as the first sample and exists without any limitations in time or readings, but it is known a priori that a "no" signal exists in the reference sample owing to the conditions of generation of the reference sample.
- The sample sizes of the first (signal) and second (reference) samples are *the same.*

These differences between the GASP in the presence of noise and the one-input two-sample signal detection procedure are very important. In addition we can see that the engineering interpretation of GASP in the presence of noise differs greatly from the one-input two-sample signal detection procedure discussed in detail in [12].

6.3. Array Signal Processing

6.3.1. Brief Review

In the majority of cases, a large class of modern array processing techniques has been designed for point sources, i.e. spatially discrete sources of electromagnetic energy. In many applications, the transmitter is best modeled as a distributed, rather than a point source. The distributed sources appear to have certain angular spread with a mean direction of arrival (DOA). The point source model is only an approximation of the practical situation when there is a large distance between the source and the receiver array.

The principal mechanism for making the source appear to be distributed in space is diffuse (irresolvable) and specular (resolvable) multipath caused by scattering of the propagation waves. For example, experimental results obtained in urban wireless communications reported significant angular scattering distributions due to local scattering and reflection from mobile stations [13]–[15] and base stations [16] and [17]. The characterization of the power azimuth spectrum shows that angular spreads as large as 25° have been observed. The amount of angular spread is highly dependent on the scattering around the mobile, the height of the base station, and the distance between the base station and the mobile station.

A secondary, but equally important, mechanism is the transmitter motion. If the source moves significantly during the observation interval or coherent integration time, it will appear to be distributed rather that discrete. Angular spread has a significant impact on any array processing algorithms [18], i.e. the *SNR* gain of the array reduces as the angular spread increases [19], causing possible performance degradation. In passive array signal processing area, the problems under study concern the extraction of information from measurements using an array of sensors.

Given the observations of the sensor outputs, the objective is to estimate the unknown parameters associated with the waveforms corrupted by noise. We start with a simple and computationally efficient detection scheme. If the "noise only" hypothesis is rejected, other algorithms are used to estimate the number of the sources and their unknown parameters, such as range and bearing. Prior work on distributed sources focuses primarily on source localization and DOA estimation [20]–[23]. Estimation of the number of distributed sources has also been studied in [20]. Subspace detectors have been studied in [24] and [25] for the cases where the signal lies in a deterministic subspace.

The case of detecting Gaussian signals with a low-rank covariance matrix and matched subspace detectors have been developed based on the generalized likelihood ratio test (GLRT). In both cases, the subspace, in which the signals lie, is assumed to be known. In other words, the dimension and rank of the signal subspace are assumed to be known a priori. This assumption does not hold in practical situations. The rank, orientation, and strength of the signal subspace vary along with the signal angular spread, DOA, and the energy distribution function. In this case, the unknown parameters may be estimated based on the maximum likelihood principle.

In this section, we discuss the GD constructed based on GASP in noise [5], [6], [9]–[11] for distributed sources and study the GD performance depending on the angular spread [26]. We first look at the case of known parameters. In this case, the GD can be approximated as a subspace beamformer. The detection performance depends on the detection statistic distribution and the sensor *SNR*. The degrees of freedom (DOF) of the detection statistics are determined by the angular spread and the number of data snapshots. If we fix the number of data snapshots, we show that the increase of angular spread reduces the mean of the detection statistics, which degrades the performance, and at the same time reduces its variance, which improves the performance. Within a certain range of DOF, the detection performance is improved with increasing the angular spread, up to a point, and then degrades slowly.

We consider the case when the various parameters, such as signal direction, angular spread, power, and noise power are unknown. We derive the GLRT GD and evaluate its performance by computer simulation. We show the GLRT GD has a significant performance advantage compared with conventional beamformers as the angular spread becomes large.

6.3.2. Problem Statement

6.3.2.1. Signal Model

Assume that we have an array with P sensors with an array response vector $\mathbf{a}(\phi)$, where ϕ denotes the azimuth. The array and all the sources are in the same plane. We assume a narrowband model for all signals and all the signals are defined within the limits of baseband. The signal received by the array from a single source is modeled as

$$\mathbf{X}_k = \mathbf{S}_k + \mathbf{n}_k \ , \qquad k = 1, \ldots, N \tag{6.17}$$

where N is the sample size and $\mathbf{X}_k = [x_{k_1}, \ldots, x_{k_P}]$ is the array output at sample reading k; $\mathbf{S}_k$ is the signal received at the array elements, assumed to be complex Gaussian with zero mean and covariance $E_S \mathbf{R}_S$, where E_S is the total signal energy and $\mathbf{R}_S$ is defined later in (6.19); $\mathbf{n}_k$ is the complex Gaussian noise with zero mean and covariance $\mathbf{R}_n$. The signal and noise are uncorrelated from sample to sample. The correlation matrix of observation vector $\mathbf{X}_k$ is given by

$$\mathbf{R}_1 = E\{\mathbf{X}_k, \mathbf{X}_k^H\} = E_S \mathbf{R}_S(\phi, \beta) + \mathbf{R}_n \ , \tag{6.18}$$

where $(\cdots)^H$ denotes the complex conjugate transpose and the signal covariance matrix is given by

$$\mathbf{R}_S(\phi,\beta) = \int_{-\pi}^{\pi} E(\theta;\phi,\beta)\mathbf{a}(\theta)\mathbf{a}^H(\theta)d\theta \ , \tag{6.19}$$

where ϕ is the source azimuth angle; β is the source angular spread with $0 \le \beta \le 2\pi$; $\mathbf{a}(\theta)$ is the array manifold at angle θ; $\mathbf{R}_S(\phi,\beta)$ is the normalized so that $Tr[\mathbf{R}_S(\phi,\beta)] = P$; $Tr[\cdots]$ denotes the trace of a matrix; $E(\theta;\phi,\beta)$ is the spatial energy distribution of source at the azimuth ϕ. More specifically, we may assume that $E(\theta;\phi,\beta) = E(\theta;-\phi,\beta)$ with $\int_{-\pi}^{\pi} E(\theta;\phi,\beta)d\theta = 1$. The shape of the energy distribution function depends on the angular spread parameter β. If $\beta = 0^\circ$, then $E(\theta;\phi,\beta)$ is a unit pulse. As β increases, the energy distribution function becomes wider.

For simplicity, we consider the uniformly distributed source model, i.e.

$$E(\theta;\phi,\beta) = \begin{cases} \dfrac{1}{\beta}, & \theta \in [\phi - 0.5\beta, \phi + 0.5\beta] \\ 0, & \text{otherwise}. \end{cases} \tag{6.20)}$$

One can also consider a Gaussian type distribution such that

$$E(\theta;\phi,\beta) = \begin{cases} \dfrac{1}{\sqrt{2\pi\beta^2}}\exp\left\{-\dfrac{(\theta-\varphi)^2}{2\beta^2}\right\}, & \theta \in [\phi - 0.5\beta, \phi + 0.5\beta] \\ 0, & \text{otherwise}. \end{cases} \tag{6.21}$$

6.3.2.2. Signal Subspace

By performing an eigen-decomposition of the matrix $\mathbf{R}_S(\phi,\beta)$, we get [27]

$$\mathbf{R}_S(\phi,\beta) = \mathbf{V}\,\mathbf{\Lambda}\,\mathbf{V}^H = (\mathbf{V}_r\ \mathbf{V}_n)\begin{pmatrix} \mathbf{\Lambda}_r & 0 \\ 0 & \mathbf{\Lambda}_n \end{pmatrix}\begin{pmatrix} \mathbf{V}_r^H \\ \mathbf{V}_n^H \end{pmatrix}, \tag{6.22}$$

where $\mathbf{V} = (\mathbf{V}_r\ \mathbf{V}_n)$, $\mathbf{\Lambda} = diag(\mathbf{\Lambda}_r, \mathbf{\Lambda}_n)$ and $\mathbf{\Lambda}_r = diag(\lambda_1,\ldots,\lambda_r)$. Here, $\mathbf{\Lambda}_r$ consists of the largest eigenvalues of $\mathbf{R}_S(\phi,\beta)$ in descending order and $\mathbf{V}_r = (\mathbf{v}_1,\ldots,\mathbf{v}_r)$ contains the corresponding orthonormal eigenvectors, $1 \le r \le P$. If $\mathbf{R}_S(\phi,\beta)$ is a low-rank matrix with rank *r*, then $\mathbf{R}_S(\phi,\beta) = \mathbf{V}_r\,\mathbf{\Lambda}_r\,\mathbf{V}_r^H$. More generally, we will assume that $\mathbf{R}_S(\phi,\beta)$ can be approximated by a rank r matrix. The number of dominant eigenvalues may be defined as the minimum number of eigenvalues whose sum exceeded τP, where $0 \le \tau \le 1$ is close to unity, for example $\tau = 0.95$. Thus,

$$\mathbf{R}_S(\phi,\beta) \approx \mathbf{V}_r \, \mathbf{\Lambda}_r \, \mathbf{V}_r^H \ . \tag{6.23}$$

The range space of $\mathbf{V}_r$ is called the signal subspace and its orthogonal complement, called the noise subspace, is spanned by $\mathbf{V}_n$.The effective rank of subspace is the number of dominant eigenvalues. The case where $\beta = 0^\circ$ corresponds to a point source where the rank $r = 1$, and the signal covariance becomes $\mathbf{R}_S(\phi,0) = \mathbf{a}(\phi)\mathbf{a}^H(\phi)$.

6.3.2.3. Detection Problem

The detection problem of a random Gaussian signal in the presence of noise and interference can be presented in the following form:

$$\mathbf{X}_k = \begin{cases} \mathbf{n}_k \to \mathcal{H}_0 \ , & k = 1,\ldots,N \\ \mathbf{S}_k + \mathbf{n}_k \to \mathcal{H}_1 \ , & k = 1,\ldots,N . \end{cases} \tag{6.24}$$

Covariance matrix of array output under the hypotheses $\mathcal{H}_0$ and $\mathcal{H}_1$ is $\mathbf{R}_0 = \mathbf{R}_n$ and $\mathbf{R}_1$ given by (6.17), respectively. Let us assume that the interference has characteristics similar to that of the noise and can be absorbed into the noise vector $\mathbf{n}_k$. We assume that the GD input noise is spatially white or $\mathbf{R}_n = \sigma_n^2\mathbf{I}$, where σ_n^2 is the noise variance and the noise power under the hypotheses $\mathcal{H}_0$ and $\mathcal{H}_1$ is not differed. If the noise is spatially colored or $\mathbf{R}_n = \sigma_n^2\mathbf{R}_{nn}$, where $\mathbf{R}_{nn}$ is a known positive definite Hermitian matrix, the detection is preceded by a prewhitening filter $\mathbf{R}_{nn}^{-0.5}$. We have the binary hypotheses with $\mathbf{X}_k$ being a whitened data vector. We assume that the noise vectors $\{\mathbf{n}_1,\ldots,\mathbf{n}_N\}$ are independent and identically distributed (i.i.d.). By grouping all the N snapshots of observation vectors into an observation matrix $\mathbf{X} = [\mathbf{x}_1,\ldots,\mathbf{x}_N]$, we have $\mathbf{X} = \mathbf{S} + \mathbf{N}$, where the signal matrix $\mathbf{S} = [\mathbf{s}_1,\ldots,\mathbf{s}_N]$ and the noise matrix $\mathbf{N} = [\mathbf{n}_1,\ldots,\mathbf{n}_N]$. The detection statistics depend on a set of parameters

$$p = \{E_S, \phi, \beta, \sigma_n^2\} \ . \tag{6.25}$$

If all parameters are known, this is a standard detection problem whose optimal solution is the GD likelihood ratio [5], [6]. If the parameters are unknown, we use the GLRT GD involving the replacement of unknown parameters by their maximum likelihood estimates under each hypothesis. Some GDs can be constructed depending on what parameters are known or unknown. We consider the GLRT GD in which the conventional beamformers are the counterparts with the same set of unknown parameters so that a fair comparison between the GLRT GD and conventional beamformer can be made. We consider the GD beamformer type 1 – all the signal parameters are known and this GD is used as a reference; the GLRT GD type 2 – the parameters ϕ, β are unknown; the GLRT GD type 3 – the parameters $E_S, \sigma_n^2, \phi, \beta$ are unknown; the GD beamformer type 4 – the parameter ϕ is unknown; and the GD beamformer type 5 – the parameters ϕ, E_S, σ_n^2 are unknown. We carry out a comparative analysis between detectors of types 1, 2, 3, 4, and 5 with the corresponding conventional beamformer. The conventional beamformer is designed assuming the point sources (zero angular spread, i.e. $\beta = 0^\circ$).

6.3.3. GD Detector

According to GASP in noise [5], [6], [9]–[11], the detection problem for the GD can be presented in the following form:

$$\mathbf{X}_k = \begin{cases} \boldsymbol{\eta}_k \to \mathcal{H}_0 , & k = 1,\ldots,N \\ \mathbf{S}_k + \boldsymbol{\xi}_k \to \mathcal{H}_1 , & k = 1,\ldots,N , \end{cases} \tag{6.26}$$

where the elements of the vectors $\boldsymbol{\xi}_k$ and $\boldsymbol{\eta}_k$ are given by (6.16). Assume, the received signals under the hypotheses $\mathcal{H}_0$ and $\mathcal{H}_1$ are complex Gaussian with zero mean and covariance $\mathbf{R}_0$ and $\mathbf{R}_1$, respectively. The probability density functions (pdfs) for the N observations are given by

$$\begin{cases} f(\mathbf{X};\mathcal{H}_0) = [\pi^P \mid \mathbf{R}_0 \mid]^{-N} \exp\{-N \times Tr[\mathbf{R}_0^{-1}\mathbf{Q}']\} , \\ f(\mathbf{X};\mathcal{H}_1) = [\pi^P \mid \mathbf{R}_1 \mid]^{-N} \times \exp\{-N \times Tr[\mathbf{R}_1^{-1}\mathbf{Q}]\} . \end{cases} \tag{6.27}$$

Here,

$$\begin{cases} \mathbf{Q} = \dfrac{1}{N}\sum_{k=1}^{N}[\mathbf{X}_k^{\mathcal{H}_1}][\mathbf{X}_k^{\mathcal{H}_1}]^H = \dfrac{1}{N}[\mathbf{X}^{\mathcal{H}_1}][\mathbf{X}^{\mathcal{H}_1}]^H , \\ \mathbf{Q}' = \dfrac{1}{N}\sum_{k=1}^{N}[\mathbf{X}_k^{\mathcal{H}_0}][\mathbf{X}_k^{\mathcal{H}_0}]^H = \dfrac{1}{N}[\mathbf{X}_k^{\mathcal{H}_0}][\mathbf{X}_k^{\mathcal{H}_0}]^H \end{cases} \tag{6.28}$$

are the sample covariance matrices under the hypotheses $\mathcal{H}_1$ and $\mathcal{H}_0$, respectively. We emphasize that we do not impose the constraint that $N \geq P$, i.e. detection can be carried out based on a single measurement $(N = 1)$ or multiple measurements $(N > 1)$. Based on the Neyman-Pearson theorem, when all the parameters are known, the GD test is the logarithm of the likelihood function ratio under the hypotheses $\mathcal{H}_1$ and $\mathcal{H}_0$, i.e.

$$\mathcal{L}(\mathbf{X}) = Tr\left\{2\mathbf{S}^H[\mathbf{R}_0^{-1} - \mathbf{R}_1^{-1}]\mathbf{X}^{\mathcal{H}_1} - [\mathbf{X}^{\mathcal{H}_1}][\mathbf{R}_0^{-1} - \mathbf{R}_1^{-1}][\mathbf{X}^{\mathcal{H}_1}]^H + [\mathbf{R}_0^{-1} - \mathbf{R}_1^{-1}]\mathbf{Q}' + N\ln\frac{|\mathbf{R}_0|}{|\mathbf{R}_1|}\right\} . \tag{6.29}$$

Next, we consider the case where some parameters are unknown. The unknown parameters under the hypotheses $\mathcal{H}_0$ and $\mathcal{H}_1$ are denoted by p_0 and p_1, respectively. In this case, the detection statistics is given by the GLRT [4]

$$l(\mathbf{X}, \hat{p}_0, \hat{p}_1) = \frac{\max_{p_1} f(\mathbf{X}, \mathcal{H}_1)}{\max_{p_0} f(\mathbf{X}, \mathcal{H}_0)} \tag{6.30}$$

or its logarithm

$$\mathcal{L}(\mathbf{X}, \hat{p}_0, \hat{p}_1) = Tr\Big\{2\mathbf{S}^H[\mathbf{R}_0^{-1}(\hat{p}_0) - \mathbf{R}_1^{-1}(\hat{p}_1)]\mathbf{X}^{\mathcal{H}_1} - [\mathbf{X}^{\mathcal{H}_1}][\mathbf{R}_0^{-1}(\hat{p}_0) - \mathbf{R}_1^{-1}(\hat{p}_1)][\mathbf{X}^{\mathcal{H}_1}]^H$$

$$+ [\mathbf{R}_0^{-1}(\hat{p}_0) - \mathbf{R}_1^{-1}(\hat{p}_1)]\mathbf{Q}' + N\ln\frac{|\mathbf{R}_0(\hat{p}_0)|}{|\mathbf{R}_1(\hat{p}_1)|}\Big\}, \tag{6.31}$$

where $\hat{p}_0$ and $\hat{p}_1$ are the parameters maximizing the likelihood function under the hypotheses $\mathcal{H}_0$ and $\mathcal{H}_1$, respectively.

6.3.3.1. Subspace GD Beamformer – Known Parameters

Ignoring the constant term $N\ln\{|\mathbf{R}_0|/|\mathbf{R}_1|\}$ in (6.29), we can write

$$\mathcal{L}(\mathbf{X}, p) = Tr\Big\{2\mathbf{S}^H\mathbf{W}\mathbf{W}^H\mathbf{X}^{\mathcal{H}_1} - [\mathbf{X}^{\mathcal{H}_1}]\mathbf{W}\mathbf{W}^H[\mathbf{X}^{\mathcal{H}_1}]^H + \mathbf{W}\mathbf{W}^H\mathbf{Q}'\Big\}, \tag{6.32}$$

where

$$\mathbf{W}\mathbf{W}^H = \mathbf{R}_0^{-1}(p) - \mathbf{R}_1^{-1}(p) \tag{6.33}$$

and p is defined in (6.25). The above decomposition of $\mathbf{W}\mathbf{W}^H$ is possible because $\mathbf{R}_0^{-1} - \mathbf{R}_1^{-1}$ is a non-negative definite matrix. In general, $\mathbf{W}$ is a $P \times r$ matrix, where $1 \le r \le P$. The rank of $\mathbf{W}$ depends on rank of the signal covariance matrix $\mathbf{R}_S$. Utilizing (6.20) and (6.23), we obtain

$$\mathbf{R}_1 \approx E_S\mathbf{V}_r\,\mathbf{\Lambda}_r\,\mathbf{V}_r^H + \mathbf{R}_0\,. \tag{6.34}$$

Employing the matrix inversion lemma, we obtain

$$\mathbf{R}_1^{-1} \approx \mathbf{R}_0^{-1} - \mathbf{R}_0^{-1}\mathbf{V}_r[E_S\,\mathbf{\Lambda}_r + \mathbf{V}_r^H\mathbf{R}_0^{-1}\mathbf{V}_r]^{-1}\mathbf{V}_r^H\mathbf{R}_0^{-1}\,. \tag{6.35}$$

Hence, we can rewrite (6.33) as follows:

$$\mathbf{W}\mathbf{W}^H \approx \mathbf{R}_0^{-1}\mathbf{V}_r[E_S\,\mathbf{\Lambda}_r + \mathbf{V}_r^H\mathbf{R}_0^{-1}\mathbf{V}_r]^{-1}\mathbf{V}_r^H\mathbf{R}_0^{-1} \tag{6.36}$$

and then, the subspace matrix is a $P \times r$ matrix

$$\mathbf{W} = \frac{\mathbf{R}_0^{-1}\mathbf{V}}{\sqrt{E_S\,\mathbf{\Lambda} + \mathbf{V}^H\mathbf{R}_0^{-1}\mathbf{V}}} = [\mathbf{w}_1, \ldots, \mathbf{w}_r] \ . \tag{6.37}$$

If the signal subspace is indeed low rank with rank r, the approximation sign is replaced by an equality sign. This leads to an interpretation of the GD as a bank of beamformers. We emphasize, this subspace beamformer is an approximation to the true GD. In our case, the noise covariance matrix

$$\mathbf{R}_n = \sigma_n^2\mathbf{I} \tag{6.38}$$

that leads us to a simpler form

$$\mathbf{W} \approx \frac{\mathbf{V}_r}{4\sigma_n^4\sqrt{\dfrac{\mathbf{V}_r^H\mathbf{V}_r}{4\sigma_n^4} + E_S\,\mathbf{\Lambda}}} \ . \tag{6.39}$$

We denote $\mathbf{Z}$ as an output matrix of beamformers

$$\mathbf{Z} = \mathbf{W}^H\mathbf{X} = [\mathbf{z}_1, \ldots, \mathbf{z}_N] \ , \tag{6.40}$$

where

$$\mathbf{z}_k = \mathbf{W}^H\mathbf{x}_k \ . \tag{6.41}$$

We will refer to this as a subspace beamformer to distinguish it from the conventional beamformer, where $\mathbf{w}$ is a $P \times 1$ vector. The GD output statistics is given by

$$\mathscr{L}(\mathbf{X}, p) = Tr\{2\mathbf{Z}\mathbf{S}^H - \mathbf{Y}\mathbf{Y}^H + \mathbf{Q}'\} = \sum_{k=1}^{N} \| 2\mathbf{z}_k\mathbf{s}_k^H - \mathbf{z}_k\mathbf{z}_k^H + [\mathbf{x}_k^{\mathcal{H}_0}][\mathbf{x}_k^{\mathcal{H}_0}]^H \|^2 \ . \tag{6.42}$$

6.3.3.2. GLRT GD – Unknown Parameters $p = \{\phi, \beta, E_S, \sigma_n^2\}$

The parameter $p = \{\phi, \beta\}$. In this case, the unknown parameter $p = \{\phi, \beta\}$ needs to be estimated. Note that the likelihood function of the GD background noise under the hypothesis $\mathcal{H}_0$ is independent of the signal $\mathbf{S}$ and its parameters p, and thus, it needs to be estimated only under the hypothesis $\mathcal{H}_1$. The detection statistics are constructed as follows by maximizing the argument p:

$$\mathscr{L}(\mathbf{X}, \hat{p}_1) = Tr\Big\{2\mathbf{S}^H[\mathbf{R}_0^{-1} - \mathbf{R}_1^{-1}(\hat{p}_1)][\mathbf{X}^{\mathcal{H}_1}] - [\mathbf{X}^{\mathcal{H}_1}][\mathbf{R}_0^{-1} - \mathbf{R}_1^{-1}(\hat{p}_1)][\mathbf{X}^{\mathcal{H}_1}]^H$$

$$+[\mathbf{R}_0^{-1} - \mathbf{R}_1^{-1}(\hat{p}_1)]\mathbf{Q}' + N \ln \frac{|\mathbf{R}_0|}{|\mathbf{R}_1(\hat{p}_1)|}\Big\} , \qquad (6.43)$$

where

$$\hat{p}_1 = \arg \max_{\phi \in (-0.5\pi, 0.5\pi), \beta \in (0, \beta_0)} \mathscr{L}(\mathbf{X}, p_1) . \qquad (6.44)$$

The estimation is done by a numerical maximization over the range $-\pi \le \phi \le \pi$ or $-0.5\pi \le \phi \le 0.5\pi$ (for a linear array), and $0 \le \beta \le \beta_0$, β_0 is the predetermined angular spread range. The maximization is carried out by the following steps.

- *Step1*: We divide a two-dimensional search range $\{\phi, \beta\}$ into small grids. The size of the grid $\{\delta\phi, \delta\beta\}$ is a fraction of an array beam width. We choose $\delta\beta = \alpha_1 BW$ and $\delta\phi = \alpha_2 BW$, where $0 < \alpha_1, \alpha_2 < 1$. We find that $\alpha_1, \alpha_2 = 0.5$ produce quite reliable results.
- *Step 2*: For every pair $\{\phi_i, \beta_i\}$, $\mathbf{R}_S(\phi_i, \beta_i)$ is constructed based on (6.21).
- *Step 3*: The GD output statistics $\mathscr{L}[\mathbf{X};(\phi_i, \beta_i)]$ is evaluated.
- *Step 4*: This process is repeated, and the largest value of $\mathscr{L}[\mathbf{X};(\phi_i, \beta_i)]$ is selected.

The parameter $p = \{\phi, \beta, E_S, \sigma_n^2\}$. In this case, all the parameters are unknown. Maximum likelihood (ML) procedure requires a nonlinear optimization. The GD searches the global maximum of likelihood function over the unknown parametric sets. The maximization can be carried out as follows:

$$l(\mathbf{X}, \hat{E}_S \hat{\sigma}_n^{-2}, \hat{\sigma}_S^2, \hat{\sigma}_n^2, \hat{p}_1) = \max_{p_1} l(\mathbf{X}, \hat{E}_S \hat{\sigma}_n^{-2}, \hat{\sigma}_S^2, \hat{\sigma}_n^2, \hat{p}_1)$$

$$= \max_{p_1} \max_{\rho \in (0, \rho_0)} l(\mathbf{X}, E_S \sigma_n^{-2}, \hat{\sigma}_S^2, \hat{\sigma}_n^2, p_1) . \qquad (6.45)$$

We first maximize the likelihood function over $(E_S \sigma_n^{-2}, \sigma_n^2)$, assuming that $p_1 = \{\phi, \beta\}$ is known. We maximize the result over $\{\phi, \beta\}$. The derivation of the maximum likelihood estimators of (E_S, σ_n^2) is provided at the fixed p_1. The GLRT GD likelihood function takes the following form:

$$l(\mathbf{X}, \hat{E}_S \hat{\sigma}_n^{-2}, \hat{\sigma}_S^2, \hat{\sigma}_n^2, p_1) = \frac{1}{|\hat{E}_S \hat{\sigma}_n^{-2} \mathbf{R}_S(p_1) + \mathbf{I}|^N} \times \left\{ \frac{Tr[\mathbf{Q}] Tr[(\hat{E}_S \hat{\sigma}_n^{-2} \mathbf{R}_S(p_1) + \mathbf{I})]}{\mathbf{Q}'} \right\}^{PN} . \qquad (6.46)$$

Taking the logarithm, we obtain the following GD output statistic

$$\mathscr{L}(\mathbf{X}, \hat{E}_S\hat{\sigma}_n^{-2}, \hat{\sigma}_S^2, \hat{\sigma}_n^2, p_1)$$

$$= PN \ln\left\{\frac{Tr[\mathbf{Q}]Tr[(\hat{E}_S\hat{\sigma}_n^{-2}\,\mathbf{R}_S(\hat{p}_1)+\mathbf{I})]}{\mathbf{Q}'}\right\} - N\ln|\hat{E}_S\hat{\sigma}_n^{-2}\mathbf{R}_S(\hat{p}_1)+\mathbf{I}|, \qquad (6.47)$$

where

$$\hat{p}_1 = \arg \max_{-0.5\pi<\phi<0.5\pi, 0\le\beta\le\beta_0} \mathscr{L}(\mathbf{X}, \hat{E}_S\hat{\sigma}_n^{-2}, \hat{\sigma}_S^2, \hat{\sigma}_n^2, p_1)\ . \qquad (6.48)$$

In summary, the maximization is carried out as follows.

- *Step 1*: Follow Steps 1 and 2 for the case when the parameter $p = \{\phi, \beta\}$.
- *Step 2*: For each pair of $\{\phi_i, \beta_i\}$, a one-dimensional search over ρ is carried out to select the maximal value of $\mathscr{L}(\mathbf{X}, \{\phi_i, \beta_i\} E_S\sigma_n^{-2})$. The search can be implemented through a binary search or an exhaustive search and p_0 is a predetermined search range.
- *Step 3*: This process is repeated for all the values of $\{\phi_i, \beta_i\}$ and the largest value of $\mathscr{L}(\mathbf{X}, \{\phi_i, \beta_i\} E_S\sigma_n^{-2})$ is selected.

6.3.3.3. GD Beamformer – Unknown Parameters $p = \{\phi, E_S, \sigma_n^2\}$

The parameter $p = \{\phi\}$. GD beamformer searches for the maximum energy by sweeping over all possible directions. This is the optimal solution (maximum likelihood) for point sources. If the source is distributed, only a fraction of the energy is captured by the beamformer. Degradation in performance will occur. For a linear array, the beamformer searches over $-0.5\pi \le \phi \le 0.5\pi$. The GD output statistic takes the following form

$$\mathscr{L}(\mathbf{X}) = \| 2\mathbf{S}^H\mathbf{w}\mathbf{w}^H(\hat{\phi})\mathbf{X}^{\mathscr{H}_1} - \mathbf{X}^{\mathscr{H}_1}\mathbf{w}\mathbf{w}^H[\mathbf{X}^{\mathscr{H}_1}]^H + \mathbf{w}\mathbf{w}^H\mathbf{Q}'\|^2\ , \qquad (6.49)$$

where

$$\hat{\phi} = \arg \max_{-0.5\pi<\phi<0.5\pi} \| \mathbf{w}^H(\hat{\phi})\mathbf{X}\|^2$$

and $\mathbf{a}(\phi)$ is the steering vector pointing to direction ϕ. The implementation of this GD beamformer is quite straightforward. The GD beamformer $\mathbf{w}$ is a normalized steering vector. The maximal value of GD beamformer output statistic is chosen by sweeping the GD beamformer from -0.5π to 0.5π.

The parameter $p = \{\phi, E_S, \sigma_n^2\}$. GD beamformer searches the maximum energy output over the range from -0.5π to 0.5π. The signal energy is divided by an estimate of the GD background noise variance that is calculated by projecting GD array output on the subspace orthogonal to the rank 1 steering vector. To exclude the signal power that "leaks" through the beams adjacent to the main beam, we estimate the GD background noise power from beams that are further away from the main beam. We can choose to exclude the β_0 / BW beams adjacent to the main beam, where β_0 is a predetermined angular spread range.

This seemingly *ad hoc* approach is the modified maximum likelihood estimate of the GD background noise variance. It is well known that the GD background noise variance estimate is given by [6]

$$\sigma_g^2 = \| \mathbf{P}_{\mathbf{A}}^{\perp} \mathbf{Q}' \|^2 (P - r_{\mathbf{A}})^{-1} , \tag{6.50}$$

where $\mathbf{P}_{\mathbf{A}}^{\perp}$ is the projection on the signal subspace $\mathbf{A}$ and $r_{\mathbf{A}}$ is the signal subspace rank. The accuracy of the GD background noise estimate will degrade when a smaller subspace is used due to a leakage of the signal power onto the noise subspace. We construct a sub-space $\mathbf{B} \subseteq \mathbf{A}$ so that the signal power is excluded from the estimation i.e. the subspace

$$\mathbf{B}(\phi, \beta_0) = \{\mathbf{a}(\phi - 0.5\beta_0), \ldots, \mathbf{a}(\phi + 0.5\beta_0)\} \tag{6.51}$$

is generated. The power that is projected onto the null space of $\mathbf{B}(\phi, \beta_0)$ is considered to be the GD background noise power, i.e.

$$\hat{\sigma}_g^2(\phi, \beta_0) = Tr\{\mathbf{P}_{\mathbf{B}(\phi,\beta_0)}^{\perp} \mathbf{Q}'\}(P - r_{\mathbf{B}})^{-1} , \tag{6.52}$$

$$\mathbf{P}_{\mathbf{B}(\phi,\beta_0)}^{\perp} = \mathbf{I} - \mathbf{P}_{\mathbf{B}(\phi,\beta_0)} , \tag{6.53}$$

and $r_{\mathbf{B}}$ is the rank of $\mathbf{B}$. The GD output statistic possessing CFAR properties is given by

$$\mathscr{L}(\mathbf{X}) = \max_{\phi \in (-0.5\pi, 0.5\pi)} \frac{Tr\{2\mathbf{S}^H \mathbf{P}_{\mathbf{a}(\phi)} \mathbf{X}^{\mathcal{H}_1} - \mathbf{X}^{\mathcal{H}_1} \mathbf{P}_{\mathbf{a}(\phi)} (\mathbf{X}^{\mathcal{H}_1})^H + \mathbf{P}_{\mathbf{a}(\phi)} \mathbf{Q}'\}}{\hat{\sigma}_g^2(\phi, \beta_0)} , \tag{6.54}$$

where $\mathbf{P}_{\mathbf{a}(\phi)}$ is the projection matrix on $\mathbf{a}(\phi)$. The implementation of this GD type involves three steps:

- *Step 1*: For every angle ϕ, we construct a projection matrix $\mathbf{P}_{\mathbf{a}(\phi)}$. The subspace $\mathbf{B}(\phi, \beta_0)$ and null projection matrix $\mathbf{P}_{\mathbf{B}(\phi,\beta_0)}$ are constructed accordingly.
- *Step 2*: The GD output is evaluated based on (6.50) for every angle ϕ.
- *Step 3*: The largest value of $\mathscr{L}(\mathbf{X}, \{\phi_i, \beta_i\} \rho)$ is selected.

6.3.4. Performance Analysis: Known Parameters

6.3.4.1. SNR Gain versus Angular Spread

The *SNR* gain is defined as the ratio of the array output to the input *SNR*. For point sources, the *SNR* gain depends on the number of sensors. For a linear array with *P* sensors *SNR* gain is equal to *P*. We derive *SNR* gain for the subspace beamformers. The subspace GD beamformer output takes the following form

$$\mathbf{z}_k = 2\mathbf{S}^H\mathbf{W}\mathbf{W}^H\mathbf{x}_k^{\mathcal{H}_1} - \mathbf{x}_k^{\mathcal{H}_1}\mathbf{W}\mathbf{W}^H(\mathbf{x}_k^{\mathcal{H}_1})^H + \boldsymbol{\eta}_k\mathbf{W}\mathbf{W}^H\boldsymbol{\eta}_k^H \ . \tag{6.55}$$

Let the output *SNR* can be presented as the ratio between the average signal and average background noise powers at the beamformer output

$$SNR_{\text{GD}}^{\text{out}} = \frac{Tr\{\mathbf{W}^H E[\mathbf{s}_k\mathbf{s}_k^H]\mathbf{W}\}}{Tr\{\mathbf{W}^H E[\mathbf{Q}']\mathbf{W}\}} = \frac{Tr\{E_S^2\mathbf{W}^H\mathbf{R}_S\mathbf{W}\}}{Tr\{\mathbf{W}^H\mathbf{R}_n\mathbf{R}_n^H\mathbf{W}\}} = \frac{E_S}{\sigma_n^2}\frac{Tr\{E_S\mathbf{W}^H\mathbf{R}_S\mathbf{W}\}}{Tr\{\mathbf{W}^H\mathbf{W}\}} \ . \tag{6.56}$$

The input *SNR* is defined as the ratio E_S/σ_n^2. The *SNR* gain is given by

$$\frac{SNR_{\text{GD}}^{\text{out}}}{SNR_{\text{GD}}^{\text{in}}} = \frac{Tr\{E_S\mathbf{W}^H\mathbf{R}_S\mathbf{W}\}}{Tr\{\mathbf{W}^H\mathbf{W}\}} \tag{6.57}$$

We are interested in how the *SNR* gain of the subspace GD beamformer varies with the signal angular spread. Note that

$$\mathbf{W}\mathbf{W}^H = 0.25\sigma_n^{-4}\mathbf{I} - [4\sigma_n^4\mathbf{I} + E_S\mathbf{R}_S(\beta)]^{-1}$$

$$\approx \mathbf{V}_r\{0.25\sigma_n^{-4}\mathbf{I} - [4\sigma_n^4\mathbf{I} + E_S\boldsymbol{\Lambda}_r]^{-1}\}\mathbf{V}_r^H = \mathbf{V}_r\mathbf{D}\mathbf{V}_r^H \ , \tag{6.58}$$

where

$$\mathbf{D} = 0.25\sigma_n^{-4}\mathbf{I} - [4\sigma_n^4\mathbf{I} + E_S\boldsymbol{\Lambda}_r]^{-1} \ . \tag{6.59}$$

Then, the *SNR* gain is defined as

$$\frac{SNR_{\text{GD}}^{\text{out}}}{SNR_{\text{GD}}^{\text{in}}} = \frac{Tr\{\mathbf{D}\boldsymbol{\Lambda}_r\}}{Tr\{\mathbf{D}\}} = \frac{\sum_{i=1}^{r}\lambda_i^2[1 + SNR\times\lambda_i]^{-1}}{\sum_{i=1}^{r}\lambda_i[1 + SNR\times\lambda_i]^{-1}} \ . \tag{6.60}$$

For simplicity, we assume that the dominant eigenvalues are approximately equal, i.e. $\lambda_i = P/r$, $i = 1,\ldots,r$. This approximation is certainly not very accurate, but it reveals some insights of the *SNR* gain behavior. Therefore, we obtain

$$\frac{SNR_{\text{GD}}^{\text{out}}}{SNR_{\text{GD}}^{\text{in}}} \approx \frac{P}{r}. \tag{6.61}$$

This result shows that the *SNR* gain for the subspace GD beamformer depends not only on the number of sensors *P* but on the source angular spread β through the subspace rank *r* as well. It is a monotone decreasing function of *r*. The following can be concluded as two special cases:

- In the point source case $\beta = 0^\circ$ or $\mathbf{R}_S = \mathbf{a}\mathbf{a}^H$, we get the *SNR* gain equal to *P*;
- In the case $\beta = 180^\circ$ or $\mathbf{R}_S = \mathbf{I}$, we get the unit *SNR* gain.

6.3.4.2. Receiver Operating Characteristics (ROC)

We derive an analytical expression of pdfs under two hypotheses $\mathcal{H}_0$ and $\mathcal{H}_1$ .By $\chi_M^2(z)$ and $f_M(z)$ we denote the central Chi-squared distribution with *M* degrees of freedom and the pdf of the central Chi-squared distribution with *M* degrees of freedom, respectively. Log-likelihood ratios at the hypotheses $\mathcal{H}_0$ and $\mathcal{H}_1$ can be represented by sum of quadratic forms in complex Gaussian random variables

$$\mathcal{L}(\mathbf{X},\mathcal{H}_0) = \mathbf{W}\mathbf{W}^H\mathbf{Q}' - \mathbf{X}_{\mathcal{H}_0}^H\mathbf{W}\mathbf{W}^H\mathbf{X}_{\mathcal{H}_0} = \sum_{k=1}^{N}\boldsymbol{\eta}_k^H\mathbf{R}_0^{-1}\boldsymbol{\eta}_k - \sum_{k=1}^{N}\boldsymbol{\xi}_k^H\mathbf{R}_0^{-1}\boldsymbol{\xi}_k$$

$$= \sum_{k=1}^{N}\Big[\sum_{j=1}^{r}\lambda_j\,|\mathbf{v}_j^H\boldsymbol{\eta}_k|^2 - \sum_{j=1}^{r}\lambda_j\,|\mathbf{v}_j^H\boldsymbol{\xi}_k|^2\Big]; \tag{6.62}$$

$$\mathcal{L}(\mathbf{X},\mathcal{H}_1) = 2\mathbf{S}^H\mathbf{R}_S\mathbf{X}_{\mathcal{H}_1} + \mathbf{W}\mathbf{W}^H\mathbf{Q}' - \mathbf{X}_{\mathcal{H}_1}^H\mathbf{W}\mathbf{W}^H\mathbf{X}_{\mathcal{H}_1}$$

$$= \sum_{k=1}^{N}\mathbf{s}_k^H(\mathbf{R}_0^{-1} - \mathbf{R}_1^{-1})\mathbf{s}_k + \sum_{k=1}^{N}\boldsymbol{\eta}_k^H\mathbf{R}_0^{-1}\boldsymbol{\eta}_k - \sum_{k=1}^{N}\boldsymbol{\xi}_k^H\mathbf{R}_0^{-1}\boldsymbol{\xi}_k$$

$$= \sum_{k=1}^{N}\left\{\sum_{j=1}^{r}\lambda_j\,|\mathbf{v}_j^H\mathbf{s}_k|^2 + \sum_{j=1}^{r}\lambda_j\,|\mathbf{v}_j^H\boldsymbol{\eta}_k|^2 - \sum_{j=1}^{r}\lambda_j\,|\mathbf{v}_j^H\boldsymbol{\xi}_k|^2\right\}, \tag{6.63}$$

where λ_j are the nonzero eigenvalues of the matrix $\mathbf{W}\mathbf{W}^H$, $\mathbf{v}_j$ are the associated eigenvectors. Thus, we can write

$$\mathscr{L}(\mathbf{X},\mathscr{H}_i)=\sum_{j=1}^{r}\gamma_{ji}\,|\,z_{jik}\,|^2\ ,\quad i=0,1 \tag{6.64}$$

where z_{jik} are independent zero mean and unit variance complex Gaussian random variables and $\gamma_{ji}, j=1,\ldots,r$ are the r real nonzero weights.

More precisely,

$$\gamma_{j0}=E_S\lambda_j(E_S\lambda_j+4\sigma_n^4)^{-1},\quad \gamma_{j1}=0.25E_S\lambda_j\sigma_n^{-4}\quad ,\ j=1,\ldots,r \tag{6.65}$$

where λ_j are the eigenvalues of $\mathbf{R}_S$. As extension of the results in [28] from real to complex numbers, the quadratic form $\mathscr{L}(\mathbf{X},\mathscr{H}_i)$ is distributed approximately as the scaled Chi-squared random variable, or

$$\mathscr{L}(\mathbf{X},\mathscr{H}_i)\sim\delta_i\chi^2_{M_i}\ , \tag{6.66}$$

where the scaling factor δ_i and the degrees of freedom M_i are given by

$$\begin{cases}\delta_i=\sum_{j=1}^{r}\gamma_{ji}^2\left\{\sum_{j=1}^{r}\gamma_{ji}\right\}^{-1},\\ M_i=2N\left\{\sum_{j=1}^{r}\gamma_{ji}\right\}^2\left\{\sum_{j=1}^{r}\gamma_{ij}^2\right\}^{-1}.\end{cases} \tag{6.67}$$

Plugging (6.65) into the last expression, we obtain

$$\delta_0=\frac{\sum_{j=1}^{r}\left[\frac{SNR^{in}\lambda_j}{SNR^{in}\lambda_j+4\sigma_n^2}\right]^2}{\sum_{j=1}^{r}\frac{SNR^{in}\lambda_j}{SNR^{in}\lambda_j+4\sigma_n^2}}\quad\text{and}\quad M_0=2N=2N\frac{\left\{\sum_{j=1}^{r}\frac{SNR^{in}\lambda_j}{SNR^{in}\lambda_j+4\sigma_n^2}\right\}^2}{\sum_{j=1}^{r}\left[\frac{SNR^{in}\lambda_j}{SNR^{in}\lambda_j+4\sigma_n^2}\right]^2}\ ; \tag{6.68}$$

$$\delta_1=\frac{SNR^{in}}{4\sigma_n^2P}\sum_{j=1}^{r}\lambda_j^2\quad\text{and}\quad M_1=\frac{2NP^2}{\sum_{j=1}^{r}\lambda_j^2}\ . \tag{6.69}$$

Utilizing the Jacobean transformation, we can show that the pdf of detection statistic can be presented as follows

$$\mathscr{L}(\mathbf{X},\mathcal{H}_i) = |\delta_i|^{-1} f_{M_i}(z_i\delta_i^{-1}) \ . \tag{6.70}$$

Note that the absolute sign can be ignored because δ_i is a positive number. This leads to an approximation of the probability of false alarm

$$P_{FA} \approx \int_{K_g}^{\infty} f_{M_0}(z)dz \tag{6.71}$$

and the probability of detection

$$P_D \approx \int_{\frac{\delta_0}{\delta_1}K_g}^{\infty} f_{M_1}(z)dz \ , \tag{6.72}$$

where the pdf is given by

$$f_{M_i}(u) = 0.5^{0.5M_i}[\Gamma(0.5M_i)]^{-1}u^{0.5M_i-1}\exp\{-0.5u\} \ . \tag{6.73}$$

Let

$$F_M(u) = \int_0^u f_M(z)dz \tag{6.74}$$

is the probability distribution function of Chi-squared random variable with M degrees of freedom and $F_M^{-1}(u)$ is the inverse probability distribution function. The probability of false alarm P_{AF} and the probability of detection P_D of quadratic form $\mathscr{L}(\mathbf{X},\mathcal{H}_i)$ can be written as

$$P_{AF} \approx 1 - F_{M_0}(u) \quad \text{and} \quad P_D \approx 1 - F_{M_1}(\delta_0 u/\delta_1) \ . \tag{6.75}$$

For a fixed level of false alarm rate, we have a closed-form expression of the probability of detection

$$P_D \approx 1 - F_{M_1}[\delta_0 F_{M_0}^{-1}(1-P_{FA})/\delta_1] \ . \tag{6.76}$$

Computer simulation shows that the analytical result is matched with the Monte Carlo simulation result quite well for different number of measurements, signal angular spreads, and false alarm rates. The asymptotic performance of the probability of detection P_D as N is sufficiently large can also be obtained in that the Chi-squared pdf with M degrees of freedom tends to become a Gaussian normal pdf with the mean M and variance $2M$. If M is sufficiently large

$$F_M(u) \to \Psi[u - M(2M)^{-0.5}]\ , \tag{6.77}$$

where

$$\Psi(u) = \int_{-\infty}^{u} (2\pi)^{-0.5} \exp\{-0.5u^2\} du\ . \tag{6.78}$$

We obtain

$$P_D \approx 1 - \Psi\{[\delta_0 \delta_1^{-1}[M_0 + \sqrt{2M_0}\Psi^{-1}(1 - P_{FA})] - M_1](2M_1)^{-0.5}\}\ . \tag{6.79}$$

6.3.4.3. Required SNR (RSNR)

The ROC describes a behavior of the probability of detection P_D as the *SNR* changes given the probability of false alarm P_{FA}. To compare different GD performance it is sometimes convenient to define a scalar performance measure, rather than use the entire ROC curve. We define the *RSNR* as the *SNR* needed to produce the probability of target detection at a given false alarm rate. Another quantity of interest is the output *SNR* required to achieve the same target probability of detection P_D for the fixed probability of false alarm P_{FA}.

From (6.68) and (6.69) we obtain

$$\frac{\delta_1}{\delta_0} = \frac{F_{M_1}^{-1}(1 - P_{FA})}{F_{M_0}^{-1}(1 - P_D)}\ . \tag{6.80}$$

Simplification leads us to

$$RSNR = \frac{P}{\sum_{j=1}^{r} \lambda_j^2} \times \frac{\left\{\sum_{j=1}^{r} \frac{SNR^{in}\lambda_j}{SNR^{in}\lambda_j + 4\sigma_n^2}\right\}^2}{\sum_{j=1}^{r}\left[\frac{SNR^{in}\lambda_j}{SNR^{in}\lambda_j + 4\sigma_n^2}\right]^2} \times \frac{F_{M_1}^{-1}(1 - P_{FA})}{F_{M_0}^{-1}(1 - P_D)}\ . \tag{6.81}$$

The general behavior of the quantity *RSNR* depends on eigenvalues, which has certain distributions [28]. Eq. (6.81) is quite complicated. To gain some insights of this quantity, we look at the case where all the principal eigenvalues of $\mathbf{R}_S$ are approximately equal, i.e. $\lambda_i = \lambda = P/r,\ i = 1,\ldots,r$. This approximation is certainly not accurate but serves for simplicity to illustrate the performance of *RSNR* as the degrees of freedom changes. A more accurate result may be obtained by a numerical approach. With this assumption, we have $M_0 \approx 2Nr, M_1 \approx 2Nr$. Let $v = 2Nr$ denotes the degrees of freedom, and after some transformations, we get

$$RSNR = \frac{v}{2P} \times \left[\frac{F_v^{-1}(1 - P_{FA})}{F_v^{-1}(1 - P_D)} - 1\right] \tag{6.82}$$

and the quantity *RSNR* is given as

$$RSNR = \frac{SNR^{out}}{SNR^{in}} \approx N \times \left[\frac{F_v^{-1}(1-P_{FA})}{F_v^{-1}(1-P_D)} - 1 \right] . \tag{6.83}$$

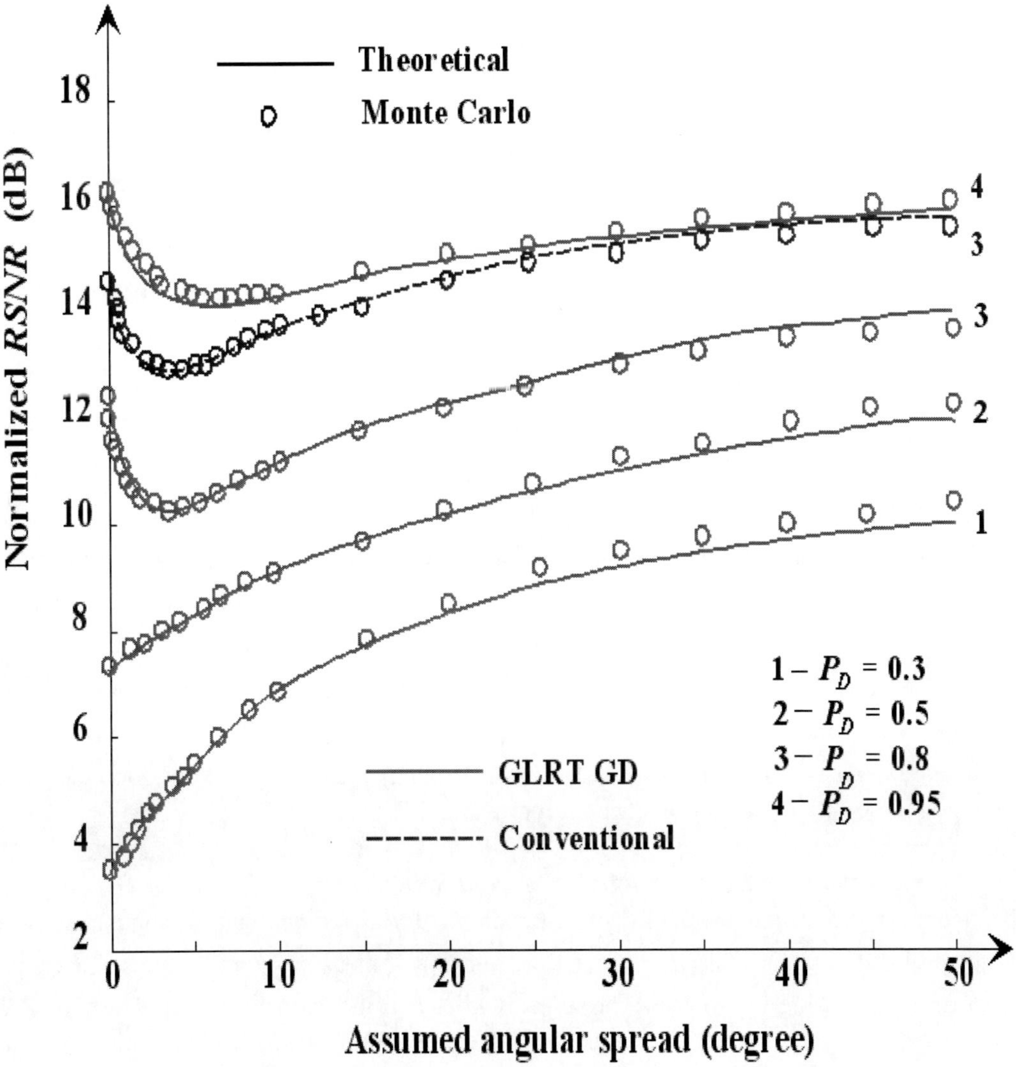

Figure 6.3. Normalized *RSNR* versus degrees of freedom $v \in [1,50]$ at various P_D ; $P_{FA} = 0.001$.

The quantity *RSNR* that is a function of (v, P_D, P_{FA}) can be easily evaluated by numerical method. For the fixed probability of false alarm P_{FA}, Fig.6.3 depicts the *RSNR* for different values of degrees of freedom v and the probability of detection P_D. To eliminate the effect of size of the linear array, we multiply the *RSNR* by the array size *P*. It is easy to see that the normalized *RSNR* (e.g., $P \times RSNR$) is a quantity that depends on the degrees of freedom v, the probability of detection P_D, and the probability of false alarm P_{FA}. This observation indicates that although the increase of the angular spread causes a reduction of the *SNR* gain, it changes the distribution of the detection statistics by increasing its degrees of freedom. Figure 6.3 also explains the different behavior of detection performance presented in Figs. 6.4 and 6.5. GD functioning in a meaningful detection range requires the large probability of target detection P_D. In this case, the *RSNR* performance improves as the degrees of freedom increase to a certain point, and then, it starts dropping. Large *N* or large angular spread gives rise to an asymptotic performance of the *RSNR*. As the degrees of freedom become large, either due to a large number of measurements or large angular spread, the detection performance degrades, but the degradation is not significant.

6.3.5. Numerical Results

We present computer simulations to illustrate the analytical performance results discussed above. We consider a uniform linear array with *P* sensors and half wavelength spacing. The array manifold is given by

$$\mathbf{a}(\phi) = \left[1, \exp\{j2\pi d\lambda^{-1}\sin\phi\}, \ldots, \exp\{j2\pi(P-1)d\lambda^{-1}\sin\phi\}\right]^T, \qquad (6.84)$$

where λ is the target return signal wavelength, $d = 0.5\lambda$ is the element spacing, and ϕ is the azimuth angle. We find that is convenient to normalize the angular spread by the array beam width [29].We choose $P = 20$. Thus, the beam width of this linear array is equal to 6.03°. We assume that the source is located at 20° relative to the broadside of the array. Throughout the simulation, the probability of false alarm $P_{FA} = 10^{-3}$ and the probability of detection $P_D = 0.8$. We are interested in the case if there are a few snapshots available for detection. In the simulation, we will study the cases where $N = 1$ and $N = 20$, respectively. More detailed discussion of detection by multiple measurements is in [3].

6.3.5.1. Parameters are Known – ROC, RSN and SNR Gain

Figure 6.4 depicts the probability of detection P_D for different values of angular spreads β, the probability of false alarm P_{FA}, and the number of measurements *N*. We see both the analytical results using the formulas presented earlier and the Monte Carlo trial results. We observe that there exists a very good match between two sets of results. We notice that the behavior of the probability of detection P_D curves is quite different when multiple measurements are used in comparison with that of the single measurement detection.

For a single measurement, $N = 1$ there is a crossover point of ROC curves for different angular spreads. We observe that as the angular spread increases from 0° to 30°, the detection performance improves up to a point, and then, it starts decreasing. If multiple measurements are used, the detection performance drops as the angular spread increases. This phenomenon is due to the change of the pdf shape, whose degrees of freedom are affected by both the angular spread and the number of measurements.

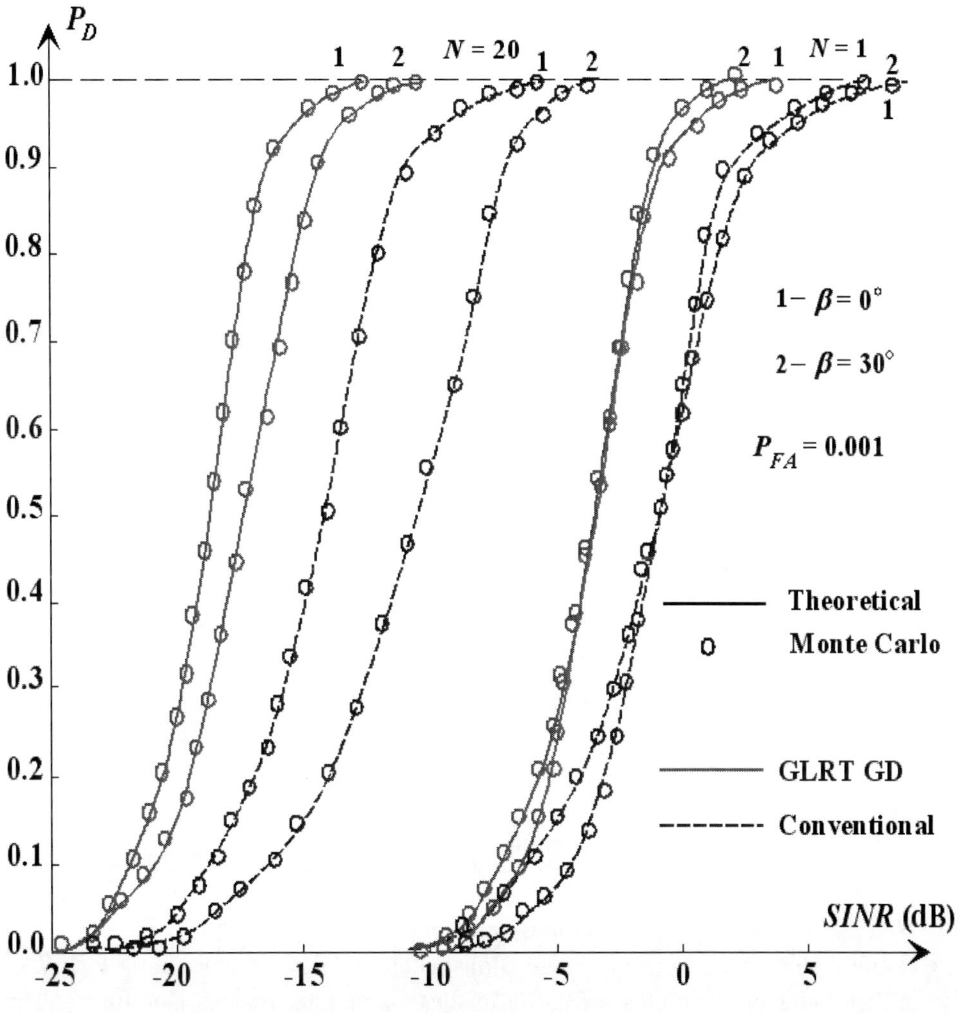

Figure 6.4. P_D versus SNR; all parameters are known; the number of sensors $P = 20$.

Alternative way to look this phenomenon is to study the *RSNR* versus the angular spread. Figure 6.5 shows the behavior of the *SNR* gain versus the angular spread The *SNR* gain plot decreases monotonically from 17 to 5 dB as the angular spread increases from $\beta = 0^{\circ}$ to $\beta = 360^{\circ}$. GD output *RSNR* shows a performance improvement caused by the changes of the GD output statistics pdf as the angular spread increases. As the angular spread and the effective rank of the GD subspace increases, the pdf of detection statistics becomes a Gaussian distribution. Once the pdf is sufficiently close to Gaussian one, no more improvement occurs. The *RSNR* plot, which is the difference of these two plots, is decreased up to a point and increased afterwards. Figure 6.4 demonstrates a superiority of GLRT GD employment in comparison with the conventional beamformer.

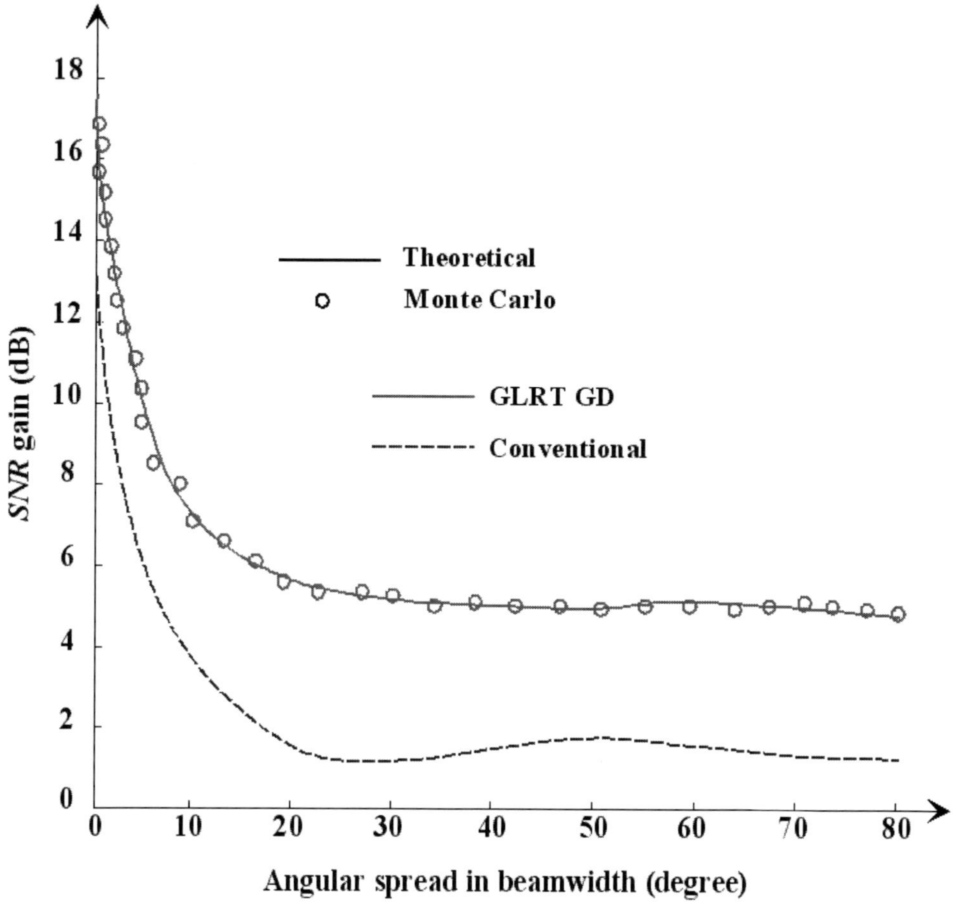

Figure 6.5. *SNR* gain versus normalized angular spread for an array with $P = 20$ sensors.

6.3.5.2. Detection with Angular Spread Mismatch

We examine the effects of angular spread mismatch on GD performance. In Fig. 6.6, the upper plot depicts the *SNR* gain loss factor defined as the ratio of the *SNR* gain for a subspace GD beamformer using an incorrect angular spread to that of a beamformer that uses the correct angular spread of 20°. This is due to the fact that the GD background noise power projected on the detector subspace is proportional to the effective rank of the subspace that increases with the assumed angular spread. The lower plot depicts *RSNR* loss factor as the assumed angular spread changes. GD with the correct angular spread has the best performance. The conventional beamformer detector assumed a zero angular spread and experienced a loss of approximately 7 dB in this case.

6.3.5.3. Detection Performance versus Angular Spread

At the fixed probability of false alarm $P_{FA} = 10^{-3}$ we study the probability of detection P_D GLRT GD detector searches over ranges of direction ϕ and angular spread β the maximal value of likelihood function. Searching is carried out over the range $-90^{\circ} \le \phi \le 90^{\circ}$ and $\beta \le \beta_0$, where β_0 is set to be 60°.

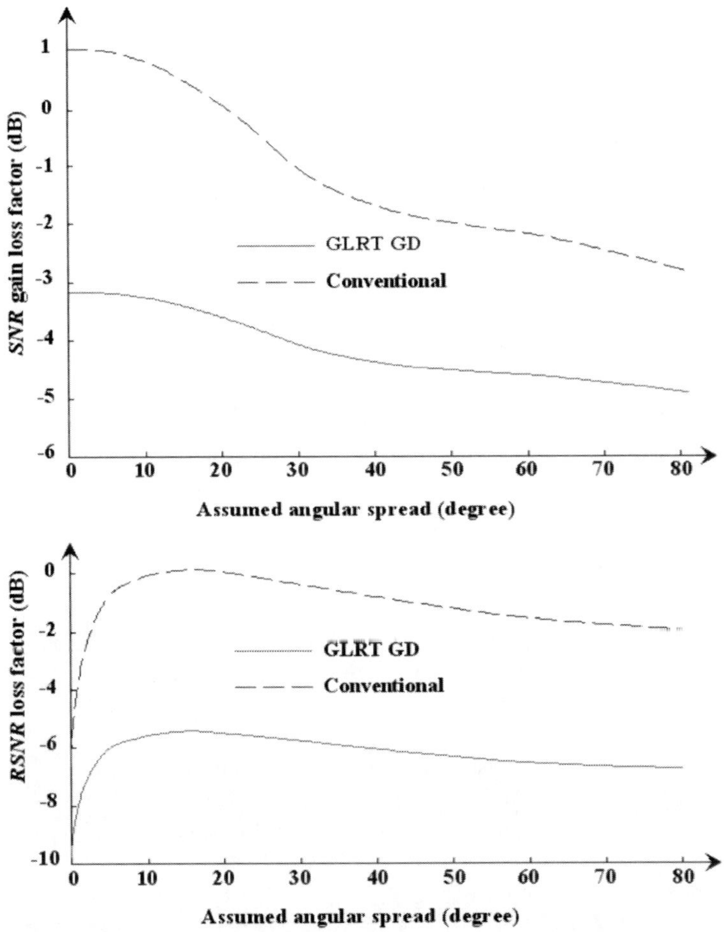

Figure 6.6. RSNR and SNR gain loss factor versus assumed angular spread when there is angular mismatch; the number of sensors $P = 20$; the signal angular spread $\beta = 20^\circ$.

Figures 6.7 and 6.8 depict the probability of detection P_D and *RSNR* versus angular spread for different detectors with single measurement, i.e. $N = 1$. The results show that when the angular spread is small there is a little difference between the GLRT GD and GD beamformer because the source is appeared to be a point-like source. The main beam of the GD beamformer is able to capture all that of the GD beamformer due to the excessive search which results in an increase of detection threshold. As the angular spread gets larger, the performance of both GLRT GD and GD beamformer degrades. The degradation of GD beamformer is much more significant than that of GLRT GD. In the case of large angular spread the GLRT GD demonstrates the better probability of detection P_D in comparison with the GD beamformer. We plot the results in the case of known parameters to see how much the detection performance degrades than occurs for other suboptimal detectors. We study the GD performance if multiple measurements, for example $N = 20$, are used. We compare the probability of detection P_D and *RSNR* when the parameters are unknown (see subsections 3.2 and 3.3) at $N = 20$ data snapshots. We see that the detection performance is appeared to be quite different than at $N = 1$. If $N = 20$, as the angular spread increases, the detection performance degrades monotonically. This observation suggests that large degrees of freedom caused either by multiple measurements or by a large angular spread, bring down the detection performance. The GD beamformer detection performance degrades faster in comparison with GLRT GD one.

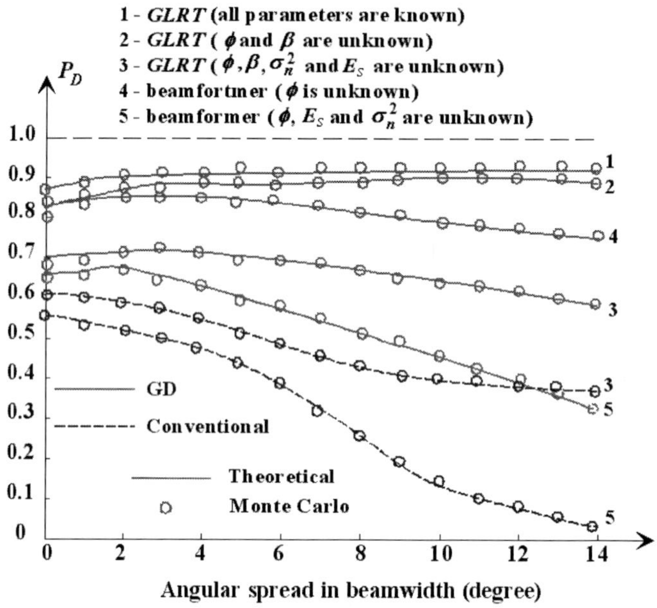

Figure 6.7. P_D versus normalized angular spread; the number of measurement is $N = 1$, $SNR = 2\text{dB}, P_{FA} = 0.001$.

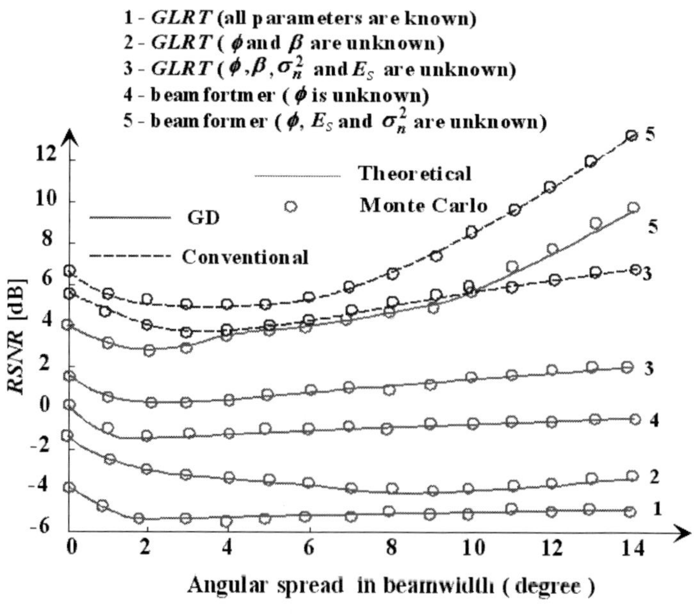

Figure 6.8. *RSNR* versus normalized angular spread; the number of measurement is $N = 1$, $P_D = 0.8, P_{FA} = 0.001$.

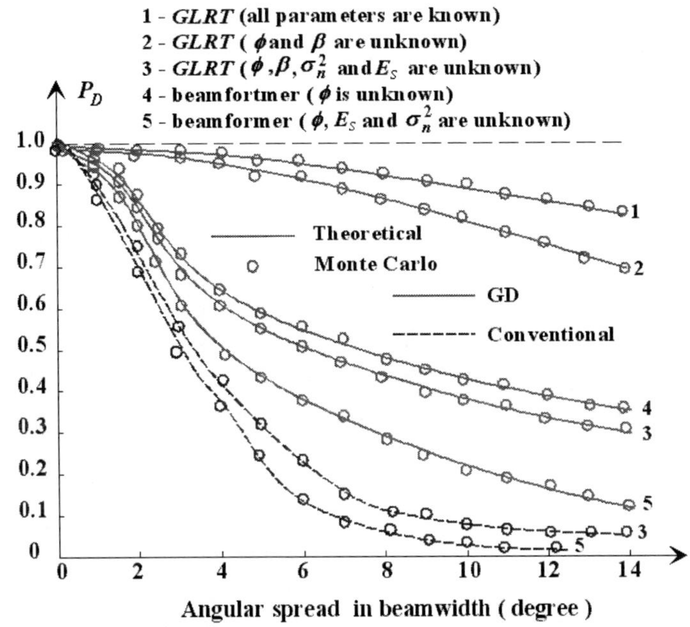

Figure 6.9. P_D versus normalized angular spread; the number of measurements is $N = 20$, . $N = 20, SNR = -8\text{dB}, P_{FA} = 0.001$.

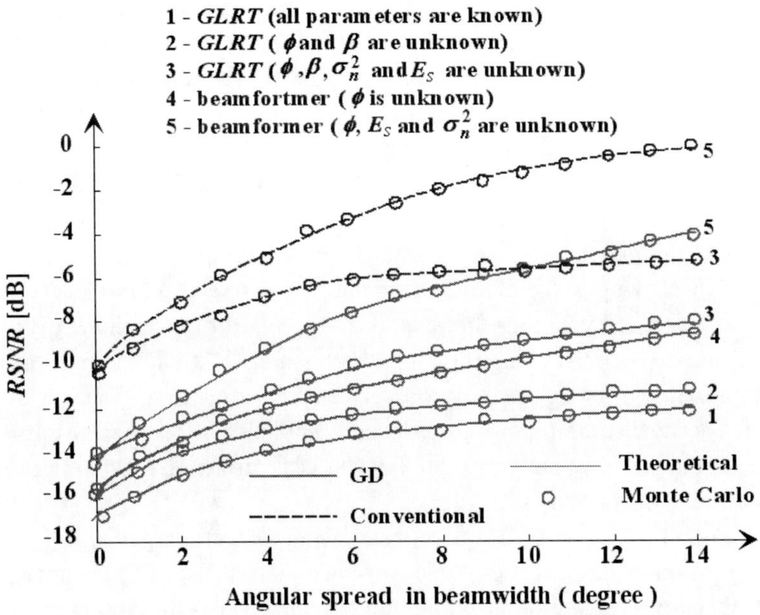

Figure 6.10. *RSNR* versus normalized angular spread; the number of measurements is $N = 20$, $P_D = 0.8, P_{FA} = 0.001$.

Simulation results in Figs. 6.9 and 6.10 also demonstrate the better GLRT GD and GD beamformer performance in comparison with conventional beamformers as the angular spread increases.

6.4. Multichannel Adaptive Signal Processing

6.4.1 Brief Review

Multichannel signal detection is encountered in a wide variety of applications. In wireless communications with radar systems, sensor arrays are often used to facilitate the so-called space-time adaptive processing, which offers enhanced target discrimination capability compared with space- or time- only processing [30] and [31]. In remote sensing systems, multispectral and hyperspectral sensors are used to collect spectral information across multiple spectral bands, which can be exploited for classification of different materials or detection of man-made objects on the ground [32] and [33]. Other examples of applications include wireless communications, relocation, sonars, audio and speech processing, and seismology [1], [34], and [35].

Multichannel signal detection based on space-time adaptive processing has been successfully used to mitigate the effects of clutter and/or interference in radar, remote sensing, and wireless communication systems [30]–[34]. Traditional detectors based on space-time adaptive processing, including the well-known RMB (Reed, Mallett, and Brennan) detectors [36], Kelly's generalized likelihood ratio test (GLRT) [37], the adaptive matched filter (AMF) detector [38]–[40], the adaptive coherence estimator (ACE) detector [41] and [42], and the adaptive generalized detector (AGD) [43]–[45] usually involve estimating and inverting a large-size space-time covariance matrix of the disturbance signal, i.e., the clutter, jamming, and noise, for each test cell using the target-free training data. This entails high complexity and large training requirement.While the first difficulty may create real-time implementation burdens, the second implies that such covariance matrix based techniques may not be used in heterogeneous (due to varying terrain, high platform altitude, Bistatic geometry, conformal array, among others) or dense-target environment that offers limited training data.

Addressing the above issues has become an important topic in recent multichannel signal detection research. One effective way to reduce the computational and training requirement is to utilize a suitable parametric model for the disturbance signal and exploit the model of signal detection. For example, the multichannel autoregressive (AR) models have been found to be very effective in representing the spatial and temporal correlation of the disturbance [46]–[49]. Parametric GD (PGD) based on a multichannel AR model is developed here. PGD detector has been shown to significantly outperform the aforementioned covariance matrix based detectors for small training size at reduced complexity. Specifically the PGD detector models the disturbance as a multichannel AR process driven by a temporally white but spatially colored multichannel noise.

While traditional space-time adaptive processing detectors perform joint space-time whitening using estimate of the space-time covariance matrix, the PGD detector adopts a two-step approach that involves temporal whitening via an inverse moving-average (MA) filter followed by spatial whitening. The parameters that need to be estimated are the AR coefficient matrices and the spatial covariance matrix of the driving multichannel noise, which are significantly fewer than what are involved in estimating the space-time covariance matrix. This is the essence behind the training and computational efficiency of the PGD detector. The PGD detector was obtained in a heuristic approach by modifying the AGD test statistic. Specifically, it replaces the joint space-time whitening incurred by the AGD detector with two separate whitening procedures in time and space. The test threshold and false alarm and detection probabilities were determined primarily by computer simulation, due to limited analysis available for the PGD detector.

In this section, we apply the parametric Rao test for multichannel signal detection based on GASP in noise [5], [6], [9]–[11] or in another words, the PGD [50]. The generic Rao test is known to offer a standard solution to a class of parameter testing problems. It is easier to derive and implement than the GLRT GD, and is also asymptotically, large-sample in the number of temporal observations and /or training signals, equivalent to the latter. Rao test was used to develop detectors for several other problems [51] and [52].

Our parametric Rao/GD differs from the generic one for multichannel signal detection in that we make explicit use of a multichannel AR model for the disturbance signal. We show that the parametric Rao/GD takes a form identical to that of the PGD detector. The only difference is that we use a maximum likelihood (ML) based estimator that involves using both test and training signals for parameter estimation, whereas the estimators in [47] use only training signals for parameter estimation. If the ML estimator is utilized, the parametric Rao/GD is asymptotically a parametric GLRT GD. Under the conditions stated in Section 4.2, the asymptotic distribution of the test statistic under both hypotheses is obtained in closed form, which can be used to set the test threshold and compute the corresponding detection and false alarm probabilities. Since the asymptotic distribution under the hypothesis $\mathcal{H}_0$ is independent of the unknown parameters, the parametric Rao/GD outperforms asymptotically MF detector. Presented numerical results demonstrate that our asymptotic results are accurate in predicting the performance of the Rao/GD even when the data size is modest.

6.4.2. System Model and Problem Statement

The problem considered involves detecting a known multichannel signal with unknown amplitude in the presence of spatially and temporally correlated disturbance

$$\begin{cases} \mathcal{H}_0: & \mathbf{X}_0(n) = \mathbf{D}(n), & n = 0,1,\ldots,N-1 \\ \mathcal{H}_1: & \mathbf{X}_0(n) = \alpha\mathbf{S}(n) + \mathbf{D}(n), & n = 0,1,\ldots,N-1 \end{cases} \tag{6.85}$$

where all the vectors are the $J \times 1$ vectors, J denotes the number of spatial channels, and N is the number of temporal observations. Henceforth, $\mathbf{X}_0(n)$ is called the test signal, $\mathbf{S}(n)$ is the signal to be detected with amplitude α, and $\mathbf{D}(n)$ is the disturbance signal that may be correlated in space and time. In addition to the test signal, it is assumed that a set of target - free training or secondary data vectors $\mathbf{X}_k(n), k = 1,2,\ldots,K$ and $n = 0,1,\ldots,N-1$, are available to assist signal detection.

Define the following $JN \times 1$ space-time vectors:

$$\begin{cases} \mathbf{S} = [\mathbf{S}^T(0), \mathbf{S}^T(1), \ldots, \mathbf{S}^T(N-1)]^T\ ; \\ \mathbf{D} = [\mathbf{D}^T(0), \mathbf{D}^T(1), \ldots, \mathbf{D}^T(N-1)]^T\ ; \\ \mathbf{X}_k = [\mathbf{X}_k^T(0), \mathbf{X}_k^T(1), \ldots, \mathbf{X}_k^T(N-1)]^T\ ; \quad k = 0, 1, \ldots, K\ , \end{cases} \tag{6.86}$$

where $(\cdots)^T$ means a transpose. Equation (6.85) can be written more compactly as

$$\begin{cases} \mathcal{H}_0: & \mathbf{X}_0 = \mathbf{D}, \\ \mathcal{H}_1: & \mathbf{X}_0 = \alpha\mathbf{S} + \mathbf{D}. \end{cases} \tag{6.87}$$

The composite hypothesis testing problem (6.85) or (6.87) is also a two-sided parameter testing problem that tests $\alpha = 0$ against $\alpha \neq 0$. The general assumptions are the following:

1. The signal vector **S** is deterministic and known to the detector.
2. The signal amplitude α is complex valued, deterministic, and unknown.
3. The secondary data $\{\mathbf{X}_k\}_{k=1}^K$ and the disturbance signal **D**, equivalently $\mathbf{X}_0$ under the hypothesis $\mathcal{H}_0$ are i.i.d. processes with the normal Gaussian distribution law $\mathcal{CN}(\mathbf{0}, \mathbf{R})$, where **R** is the unknown space-time covariance matrix.

In particular, the above signal detection problem occurs in an airborne space-time adaptive signal processing radar system with J array channels and a coherent processing interval (CPI) of N pulse repetition intervals (PRIs). The disturbance $\mathbf{D}(n)$ consists of ground clutter, jamming, and thermal noise, while $\mathbf{S}(n)$ is the target space-time steering vector. For a uniform equispaced linear array, the steering vector is [47]

$$\mathbf{S}(\omega_S, \omega_D) = \mathbf{S}_t(\omega_D) \otimes \mathbf{S}_S(\omega_S)\ , \tag{6.88}$$

where $\mathbf{S}_S(\omega_S)$ denotes the $J \times 1$ spatial steering vector:

$$\mathbf{S}_S(\omega_S) = \frac{1}{\sqrt{J}}\left[1, \exp\{j\omega_S\}, \ldots, \exp\{j\omega_S(J-1\}\right]^T \tag{6.89}$$

and $\mathbf{S}_t(\omega_D)$ denotes the $N \times 1$ temporal steering vector:

$$\mathbf{S}_t(\omega_D) = \frac{1}{\sqrt{N}}\left[1, \exp\{j\omega_D\}, \ldots, \exp\{j\omega_D(N-1\}\right]^T\ , \tag{6.90}$$

here ω_S and ω_D denote the normalized target spatial and Doppler frequencies, respectively.

While the previous assumptions 1) – 3) are standard, we further assume that

4. The disturbance signal $\mathbf{D}(n)$ can be modeled as a multichannel AR (P) process with known model order P but unknown AR coefficient matrices and spatial covariance.

Additional comments on this assumption are the following. We clarify that our goal here is not to justify whether AR models are appropriate or not for the space-time adaptive processing applications. As was shown in [47], the low-order multichannel AR models are very powerful and efficient in capturing the temporal and spatial correlation of the disturbance and, hence, can greatly help signal detection in airborne space-time adaptive si-gnal processing systems.

Our problem is how to exploit such a parametric model for the GD to solve the composite testing problem. The assumption that the model order P is known only used to simplify our presentation. In practice, the model order selection techniques, such as the Akaike information criterion (AIC) and the minimum description length (MDL) based techniques discussed in [53] are available for this task. Since such techniques may overestimate or underestimate the true model order, a relevant problem is how the proposed detector performs when overestimation or underestimation occurs (see [54]).

Finally, it is also possible to formulate the problem to include P as another parameter to be estimated. We do not follow such an approach in order to focus on the relations between the parametric Rao/GD test and the PAMF detector, which also assumes that a priori estimate of P, is available. Based on this assumption, the secondary data $\{\mathbf{X}_k\}_{k=1}^{K}$ are represented as

$$\mathbf{X}_k(n) = -\sum_{p=1}^{P} \mathbf{A}^H(p)\mathbf{X}_k(n-p) + \varepsilon_k(n) , \qquad k = 1,2,\ldots,K \tag{6.91}$$

where $\{\mathbf{A}^H(p)\}_{p=1}^{P}$ denote the $J \times J$ coefficient matrices and $\varepsilon_k(n)$ denote the driving multichannel spatial noise vectors, which are temporally white but spatially colored Gaussian noise

$$\varepsilon_k(n) \cong \mathcal{CN}(\mathbf{0},\mathbf{Q}) , \tag{6.92}$$

where $\mathbf{Q}$ denotes the $J \times J$ signal covariance matrix.

Meanwhile, the test signal $\mathbf{X}_0$ is given by

$$\mathbf{X}_0(n) - \alpha\mathbf{S}(n) = -\sum_{p=1}^{P} \mathbf{A}^H(p)\{\mathbf{X}_0(n-p) - \alpha\mathbf{S}(n-p)\} + \varepsilon_0(n) , \tag{6.93}$$

where $\alpha = 0$ under the hypothesis $\mathcal{H}_0$, $\alpha \neq 0$ under the hypothesis $\mathcal{H}_1$, and

$$\varepsilon_0(n) \cong \mathcal{CN}(\mathbf{0},\mathbf{Q}) . \tag{6.94}$$

Let $\widetilde{\mathbf{S}}(n)$ is a regression on $\mathbf{S}(n)$ and $\widetilde{\mathbf{X}}_0(n)$ is a regression on $\mathbf{X}_0(n)$ under the hypothesis $\mathcal{H}_1$:

$$\widetilde{\mathbf{S}}(n) = \mathbf{S}(n) + \sum_{p=1}^{P} \mathbf{A}^H(p)\mathbf{S}(n-p) \ ; \tag{6.95}$$

$$\widetilde{\mathbf{X}}_0(n) = \mathbf{X}_0(n) + \sum_{p=1}^{P} \mathbf{A}^H(p)\mathbf{X}_0(n-p) \ . \tag{6.96}$$

Then, the driving noise in (6.93) can be alternatively expressed as

$$\varepsilon_0(n) = \widetilde{\mathbf{X}}_0(n) - \alpha\widetilde{\mathbf{S}}(n) \ . \tag{6.97}$$

The problem of interest is to develop a decision rule for the above composite hypothesis testing problem using the test and training signals, as well as exploiting the multichannel parametric AR model.

6.4.3. A Priori Solutions

The number of solutions to the above problem has been developed. If the space-time covariance matrix $\mathbf{R}$ is known exactly, the matched filter (MF) statistics takes the following form [39]:

$$MF^{\text{out}} = \frac{|\mathbf{S}^H\mathbf{R}^{-1}\mathbf{X}_0|^2}{\mathbf{S}^H\mathbf{R}^{-1}\mathbf{S}} \underset{\mathcal{H}_0}{\overset{\mathcal{H}_1}{\gtrless}} K_{MF} \ , \tag{6.98}$$

where K_{MF} denotes the MF threshold. The MF detector is obtained by a GLRT approach [1], by which the ML estimate of the unknown amplitude α is the first estimated and then substituted back into the likelihood ratio to form a test statistics. It should be that the MF can not be implemented in real applications since $\mathbf{R}$ is unknown. However, it provides a baseline for performance comparison when considering any realizable detection scheme. In practice, the unknown $\mathbf{R}$ can be replaced by some estimate, such as the sample covariance matrix obtained from the secondary data:

$$\hat{\mathbf{R}} = \frac{1}{K}\sum_{k=1}^{K}\mathbf{X}_k\mathbf{X}_k^H \ . \tag{6.99}$$

Using $\hat{\mathbf{R}}$ in (6.98) leads us to the so-called AMF

$$AMF^{\text{out}} = \frac{|\mathbf{S}^H \hat{\mathbf{R}}^{-1} \mathbf{X}_0|^2}{\mathbf{S}^H \hat{\mathbf{R}}^{-1} \mathbf{S}} \underset{\mathcal{H}_0}{\overset{\mathcal{H}_1}{\gtrless}} K_{AMF} \ , \tag{6.100}$$

where K_{AMF} is the AMF threshold [38]–[40].

Alternatively, one can treat both α and $\mathbf{R}$ as unknowns and estimate them successively by ML. Such a GLRT approach was pursued by Kelly [37], which gives the following Kelly test:

$$KELLY^{\text{out}} = \frac{|\mathbf{S}^H \hat{\mathbf{R}}^{-1} \mathbf{X}_0|^2}{[\mathbf{S}^H \hat{\mathbf{R}}^{-1} \mathbf{S}][K + \mathbf{X}_0^H \hat{\mathbf{R}}^{-1} \mathbf{X}_0]} \underset{\mathcal{H}_0}{\overset{\mathcal{H}_1}{\gtrless}} K_{KELLY} \ , \tag{6.101}$$

where K_{KELLY} is the corresponding threshold.

The AMF and KELLY tests are both the CFAR detectors, what is desirable property in radar systems. However, they also entail a large training requirement. In particular, the sample covariance matrix $\hat{\mathbf{R}}$ has to be inverted, which imposes a constraint on the training size

$$K \geq JN \ , \tag{6.102}$$

to ensure a full-rank $\hat{\mathbf{R}}$.

The Reed-Brennan rule [38] suggests that at least $K \geq (2JN - 3)$ target-free secondary data vectors are needed to obtain the expected performance within 3 dB from the optimum MF. Such a training requirement may be difficult to meet, especially in non-homogeneous or dense-target environments. Besides excessive training, the computational complexity of these detectors is also high, since $\hat{\mathbf{R}}$ has to be computed and inverted for each CPI.

While the AMF and KELLY tests may be called the covariance matrix based techniques as they both involve computing and inverting $\hat{\mathbf{R}}$, the PAMF detector [47] utilize a multichannel AR (P) model that allows a spatial/temporal whitening to be implemented in a multichannel time-series fashion

$$PAMF^{\text{out}} = \frac{\left| \sum_{n=P}^{N-1} \hat{\tilde{\mathbf{S}}}_P^H(n) \hat{\mathbf{Q}}_P^{-1} \hat{\tilde{\mathbf{X}}}_{0,P}(n) \right|^2}{\sum_{n=P}^{N-1} \hat{\tilde{\mathbf{S}}}_P^H(n) \hat{\mathbf{Q}}_P^{-1} \hat{\tilde{\mathbf{S}}}_P(n)} \underset{\mathcal{H}_0}{\overset{\mathcal{H}_1}{\gtrless}} K_{PAMF} \ , \tag{6.103}$$

where $\hat{\mathbf{Q}}_P$ denotes an estimate of the spatial covariance matrix $\mathbf{Q}$, $\hat{\tilde{\mathbf{X}}}_{0,P}(n)$ and $\hat{\tilde{\mathbf{S}}}_P(n)$ are the temporally whitened test signal and steering vector, respectively. These are whitened using an inverse AR(P) filter, i.e. the multichannel MA filter, whose parameters, along with $\hat{\mathbf{Q}}_P$, are estimated from the secondary data.

In contrast to simultaneous spatio-temporal whitening used in the AMF and KELLY tests, the PAMF detector performs whitening in two distinct steps: temporal whitening followed by spatial whitening. The parametric approach offers savings in both training and computation, since the parameters to be estimated are significantly fewer, compared with the covariance matrix based approaches. Applying the above principles to the case of parametric GD [9] according to GASP in noise [5] and [6], we have

$$GD^{\text{out}} = \frac{\left|2\sum_{n=P}^{N-1}\hat{\tilde{\mathbf{S}}}_P^H(n)\hat{\mathbf{Q}}_P^{-1}\hat{\tilde{\mathbf{X}}}_{0,P}(n) - \sum_{n=P}^{N-1}\hat{\tilde{\mathbf{S}}}_P^H(n)\hat{\mathbf{Q}}_P^{-1}\hat{\tilde{\mathbf{S}}}_P(n) + \sum_{n=P}^{N-1}\mathbf{Q}_P^H\hat{\mathbf{Q}}_P^{-1}\mathbf{Q}_P\right|}{\sum_{n=P}^{N-1}\hat{\tilde{\mathbf{S}}}_P^H(n)\hat{\mathbf{Q}}_P^{-1}\hat{\tilde{\mathbf{S}}}_P(n)} \underset{\mathcal{H}_0}{\overset{\mathcal{H}_1}{\gtrless}} K_{GD} \ . \tag{6.104}$$

6.4.4. Parametric Rao/GD Test

6.4.4.1. Test Statistics

Parametric Rao/GD test with based on ML estimates of the nuisance parameters, i.e. parameters associated with the disturbance signal based on discussion in [31] is given by

$$Rao/GD^{\text{out}} = \frac{\left|4\sum_{n=P}^{N-1}\hat{\tilde{\mathbf{S}}}^H(n)\hat{\mathbf{Q}}^{-1}\hat{\tilde{\mathbf{X}}}_0(n) - \sum_{n=P}^{N-1}\hat{\tilde{\mathbf{S}}}^H(n)\hat{\mathbf{Q}}^{-1}\hat{\tilde{\mathbf{S}}}(n) + \sum_{n=P}^{N-1}\mathbf{Q}^H\hat{\mathbf{Q}}^{-1}\mathbf{Q}\right|}{\sum_{n=P}^{N-1}\hat{\tilde{\mathbf{S}}}^H(n)\hat{\mathbf{Q}}^{-1}\hat{\tilde{\mathbf{S}}}(n)} \underset{\mathcal{H}_0}{\overset{\mathcal{H}_1}{\gtrless}} K_{Rao/GD} \ , \tag{6.105}$$

where $K_{Rao/GD}$ is the test threshold that can be set by using the results discussed in *Subsection 6.4.4.2*; $\hat{\tilde{\mathbf{S}}}(n)$ and $\hat{\tilde{\mathbf{X}}}_0(n)$ denote, respectively the steering vector and test signal that have been whitened temporally; and the additional spatial whitening is provided by $\hat{\mathbf{Q}}^{-1}$ that is the inverse of the ML estimate of the spatial covariance matrix to be specified next.

Specifically, the temporally whitened steering vector and test signal in (6.105) are obtained as follows:

$$\hat{\tilde{\mathbf{S}}}(n) = \mathbf{S}(n) + \sum_{p=1}^{P}\hat{\mathbf{A}}^H(p)\mathbf{S}(n-p) \ ; \tag{6.106}$$

$$\hat{\tilde{\mathbf{X}}}_0(n) = \mathbf{X}_0(n) + \sum_{p=1}^{P}\hat{\mathbf{A}}^H(p)\mathbf{X}_0(n-p) \ , \tag{6.107}$$

where $\hat{\mathbf{A}}^H(p)$ is the ML estimate of the AR coefficient matrix $\mathbf{A}^H(p)$.

To present the ML estimates more compactly, let

$$\mathbf{A}^H = [\mathbf{A}^H(1), \mathbf{A}^H(2), \ldots, \mathbf{A}^H(P)] \in \mathcal{C}^{J \times JP} \tag{6.108}$$

which contains all the coefficient matrices involved in the P-th order AR model, and

$$\mathbf{Y}_k(n) = [\mathbf{X}_k^T(n-1), \mathbf{X}_k^T(n-2), \ldots, \mathbf{X}_k^T(n-P)]^T , \quad k = 0,1,\ldots,K , \tag{6.109}$$

which contains the regression subvectors formed from the test signal $\mathbf{X}_0$ or the k-th training signal $\mathbf{X}_k$.

We first compute the following correlation matrices:

$$\hat{\mathbf{R}}_{XX} = \sum_{n=P}^{N-1} \sum_{k=0}^{K} \mathbf{X}_k(n)\mathbf{X}_k^H(n) \ ; \tag{6.110}$$

$$\hat{\mathbf{R}}_{YY} = \sum_{n=P}^{N-1} \sum_{k=0}^{K} \mathbf{Y}_k(n)\mathbf{Y}_k^H(n) \ ; \tag{6.111}$$

$$\hat{\mathbf{R}}_{YX} - \sum_{n=P}^{N-1} \sum_{k=0}^{K} \mathbf{Y}_k(n)\mathbf{X}_k^H(n) \ . \tag{6.113}$$

Then, the ML estimates of the AR coefficients $\mathbf{A}^H$ and the spatial covariance matrix $\mathbf{Q}$ are given by

$$\hat{\mathbf{A}}^H = -\hat{\mathbf{R}}_{YX}^H \hat{\mathbf{R}}_{YY}^{-1} \ ; \tag{6.113}$$

$$\hat{\mathbf{Q}} = \frac{1}{(K+1)(N-P)} [\hat{\mathbf{R}}_{XX} - \hat{\mathbf{R}}_{YX}^H \hat{\mathbf{R}}_{YY}^{-1} \hat{\mathbf{R}}_{YX} \ . \tag{6.114}$$

The PGD also involves estimating the AR coefficients $\mathbf{A}^H$ and the spatial covariance matrix $\mathbf{Q}$. Several estimators were suggested including the Strand-Nuttall algorithm and the least squares (LS) estimators. The LS estimator was observed to yield the better performance.

Our ML estimator is similar to the LS estimator except that we use both the test and training signals to obtain parameter estimates, whereas the latter utilizes only the training signals for parameter estimation. The subscript "P" is therefore used for the parameter estimates in (6.105) to indicate the difference. Note that with the ML estimator, it is possible to derive the parameter estimates exclusively from the test signal, thus obviating the need for training. This could be advantageous where it is difficult to obtain training signals that are i.i.d. with respect to the disturbance in the test signal.

The detection performance of the parametric Rao/GD in the absence of training will be explored elsewhere. We would like to point out that approach is similar to Kelly's GLRT detector [37], which also employs both the test and training signals for parameter estimation. However, we stress that Kelly's GLRT detector does not exploit the multichannel parametric model as shown in (6.91) and (6.93). By comparing the parametric Rao/ GD (6.105) and PGD test statistic (6.103), we can quickly see that if both detectors use the ML estimator for parameter estimation, they are identical except for a scaling factor of 2.

Hence, under the conditions stated in Section 6.4.2, the PGD is a parametric Rao/GD. It should be noted that similar to other space-time adaptive signal processing detectors, the parametric Rao/GD test is adaptive in that the detector is the data dependent detector, for example, the correlator. This shall not be confused with recursive adaptive implementation. Although a recursive adaptive implementation of the parametric Rao/GD test is of interest in a real-time system, it is beyond the scope of the current section.

6.4.4.2. Asymptotic Analysis

The asymptotic distribution law of the Rao/GD test statistic can be presented in the following form [6]

$$Rao/GD^{\text{out}} \overset{\alpha}{\cong} \begin{cases} \mathcal{H}_0 \Rightarrow \chi_2^2\,; \\ \mathcal{H}_1 \Rightarrow \chi_2'^2(\lambda), \end{cases} \tag{6.115}$$

where χ_2^2 is the central Chi-squared distribution with 2 degrees of freedom and $\chi_2'^2(\lambda)$ is the noncentral Chi-squared distribution with 2 degrees of freedom and noncentrality parameter λ:

$$\lambda = 2|\alpha|^2 \sum_{n=P}^{N-1} \widetilde{\mathbf{S}}^H(n)\mathbf{Q}^{-1}\widetilde{\mathbf{S}}(n)\,, \tag{6.116}$$

where $\widetilde{\mathbf{S}}(n)$ is the temporally whitened steering vector given by (6.95). Note that the value of λ is related to the *SINR* at the output of the temporal whitening filter.

Recall that χ_2^2 is the random variable with the pdf given by

$$f_{\chi_2^2}(x) = \frac{1}{2}\exp\{-0.5x\}, \quad x \ge 0. \tag{6.117}$$

The pdf of $\chi_2'^2(\lambda)$ is given by [1]

$$f_{\chi_2'^2(\lambda)}(x) = \frac{1}{2}\exp\{-0.5(x+\lambda)\}I_0\left(\sqrt{\lambda x}\right), \quad x \ge 0 \tag{6.118}$$

Where $I_0(u)$ is the modified Bessel function of the first kind and zero order defined by

$$I_0(u) = \frac{1}{\pi}\int_0^{\pi} \exp\{u\cos\theta\}d\theta = \sum_{k=0}^{\infty} \frac{[0.25u^2]^k}{(k!)^2}\,. \tag{6.119}$$

The above distributions can be employed to set the Rao/GD test threshold for a given probability of false alarm P_{FA}, as well as to compute the probability of detection P_D. For a given threshold, the probability of false alarm P_{FA} is determined as

$$P_{FA} = \int_{K_{Rao/GD}}^{\infty} f_{\chi_2^2}(x)dx = \exp\{-0.5K_{Rao/GD}\} \tag{6.120}$$

that can easily be inverted to find the test threshold $K_{Rao/GD}$ for the given P_{FA} probability of false alarm.

In addition, the probability of detection is given by

$$P_D = \int_{K_{Rao/GD}}^{\infty} f_{\chi_2'^2}(x)dx = \int_{K_{Rao/GD}}^{\infty} \frac{1}{2}\exp\{-0.5(x+\lambda)\}I_0(\sqrt{\lambda x})dx \tag{6.121}$$

for the given test threshold $K_{Rao/GD}$.

The asymptotic distribution under the hypothesis $\mathcal{H}_0$ is independent of the unknown parameters. The probability of false alarm P_{FA} given by (6.120) depends only on the test threshold, which is a design parameter. It is evident that the Rao/GD test asymptotically achieves GD. The above analysis holds under assumptions 1) to 4) made in of Section 6.4.2 with one exception. In particular, since the ML parameter estimates are asymptotically Gaussian irrespective of the observed data distribution, the above analysis still holds if the Gaussian assumption in the assumption 3) of Section 6.4.2 is dropped. This also explains why it has been observed in several studies that the PGD obtains good performance even under the non-Gaussian observations.

6.4.5. Performance of the MF and AMF Detectors

Consider the MF detector given by (6.98) first. Let $\mathbf{R}^{-0.5}$ be the square root of the space-time covariance matrix **R**. Define

$$\widetilde{\mathbf{S}} = \mathbf{R}^{-0.5} \times \mathbf{S} \text{ and } \widetilde{\mathbf{X}}_0 = \mathbf{R}^{-0.5}\mathbf{X}_0 \tag{6.122}$$

those are the spatially and temporally whitened steering vector and test signal, respectively. Since the rank of $\widetilde{\mathbf{S}}\widetilde{\mathbf{S}}^H$ is one, we have the following eigen-decomposition:

$$\widetilde{\mathbf{S}}\widetilde{\mathbf{S}}^H = \mathbf{U}\mathbf{B}\mathbf{U}^H , \tag{6.123}$$

where

$$\mathbf{B} = diag\{\widetilde{\mathbf{S}}^H\widetilde{\mathbf{S}},0,\ldots,0\} \text{ and } \mathbf{U}\mathbf{U}^H = \mathbf{I} . \tag{6.124}$$

Let

$$\overline{\mathbf{X}}_0 = \mathbf{U}^H \widetilde{\mathbf{X}}_0 \text{ and } \overline{\mathbf{S}} = \mathbf{U}^H \widetilde{\mathbf{S}} , \tag{6.125}$$

which are rotated versions of $\widetilde{\mathbf{X}}_0$ and $\widetilde{\mathbf{S}}$, respectively. The test statistic can be defined as

$$MF^{\text{out}} = \frac{|\widetilde{\mathbf{S}}^H \widetilde{\mathbf{X}}_0|^2}{\widetilde{\mathbf{S}}^H \widetilde{\mathbf{S}}} = \frac{\widetilde{\mathbf{X}}_0^H \mathbf{UBU}^H \widetilde{\mathbf{X}}_0}{\widetilde{\mathbf{S}}^H \widetilde{\mathbf{S}}} = \overline{X}_{0,1}^* \overline{X}_{0,1} =| \overline{X}_{0,1}|^2 , \tag{6.126}$$

where $\overline{X}_{0,1}$ and $\overline{S}_1$ are the first elements of $\overline{\mathbf{X}}_0$ and $\overline{\mathbf{S}}$, respectively, and $(\cdots)^*$ means the complex conjugate.

It is clear from assumptions 1)–3) made in Section 6.4.2 that $\overline{X}_{0,1}$ is a complex Gaussian variable: $\overline{X}_{0,1} \sim \mathcal{CN}(\alpha\overline{S}_1, 1)$ with $\alpha = 0$ under the hypothesis $\mathcal{H}_0$ and $\alpha \neq 0$ under the hypothesis $\mathcal{H}_1$. Hence, (6.126) is subjected to the central Chi-squared distribution with 2 degrees of freedom under the hypothesis $\mathcal{H}_0$ and, respectively, the noncentral Chi-squared distribution with 2 degrees of freedom and the noncentrality parameter

$$\lambda_{MF} = 2|\alpha\overline{S}_1|^2 \tag{6.127}$$

under the hypothesis $\mathcal{H}_1$. It is noted that the distribution of the PAMF test statistic is similar to that of the parametric Rao/GD test statistic with the only difference of the noncentrality parameter under the hypothesis $\mathcal{H}_1$.

Hence, the probability of false alarm P_{FA} and the probability of detection P_D can be similarly computed as in (6.120) and (6.121). The performance of the AMF detector given by (6.100) was analyzed in [40] and it is summarized below. The density of a loss factor ρ defined in [41] is given by

$$f(\rho) = f_\beta(\rho; L-1, JN-1) , \tag{6.128}$$

where $L = K - JN + 1$ and the central Beta density function is

$$f_\beta(x; n, m) = \frac{(n+m-1)!}{(n-1)!(m-1)!} x^{n-1}(1-x)^{m-1} . \tag{6.129}$$

The probability of false alarm P_{FA} is given by

$$P_{FA}^{AMF} = \int_0^1 \frac{f(\rho) = f_\beta(\rho; L-1, JN-1)}{(1+\eta\rho)^L} d\rho ; \tag{6.130}$$

$$\eta = \frac{K_{KELLY}}{1 - K_{KELLY}} \tag{6.131}$$

is the test threshold of Kelly's GLRT (6.101). Meanwhile, the probability of detection P_D is given by

$$P_D^{AMF} = 1 - \int_0^1 \frac{1}{(1+\eta\rho)^L} \sum_{m=1}^{L} \binom{L}{m} (\eta\rho)^m G_m\left(\frac{\xi\rho}{1+\eta\rho}\right) f(\rho)\, d\rho \ , \tag{6.132}$$

where

$$\xi = \mathbf{S}^H \mathbf{R}^{-1} \mathbf{S} \tag{6.133}$$

and $G_m(\cdots)$ is the incomplete Gamma function given by

$$G_m(y) = \exp\{-y\} \sum_{k=0}^{m-1} \frac{y^k}{k!} \ . \tag{6.134}$$

The integrals can be computed by numerical integration.

6.4.6. Simulation Results

In the following, we present our numerical results of the parametric Rao/GD obtained by computer simulation and by the above asymptotic analysis. In addition, the performance of the PAMF (6.101) and AMF (6.98) detectors that can be computed analytically, is included for comparison. The disturbance signal is generated as a multichannel AR (2) process with randomly generated AR coefficients **A** and a spatial covariance matrix **Q**. In particular, **A** and **Q** are selected to ensure that **Q** is a valid covariance matrix and **A** is chosen to ensure that the resulting AR process is stable. Once **A** and **Q** are selected, they are fixed in all trials.

The signal vector **S** is generated as in (6.86) with randomly chosen normalized spatial and Doppler frequencies. The *SINR* is defined in the following form

$$SINR = |\alpha|^2 \, \mathbf{S}^H \mathbf{R}^{-1} \mathbf{S} \ , \tag{6.135}$$

where **R** is the $JN \times JN$ joint space-time covariance matrix of the disturbance **D**, which can be determined once **A** and **Q** are selected. To numerically set the threshold for the parametric Rao/GD, a total of 5×10^4 trials are run. Meanwhile, to determine the probability of detection P_D for a given threshold, a total of 10^4 trials are run for each *SINR*.

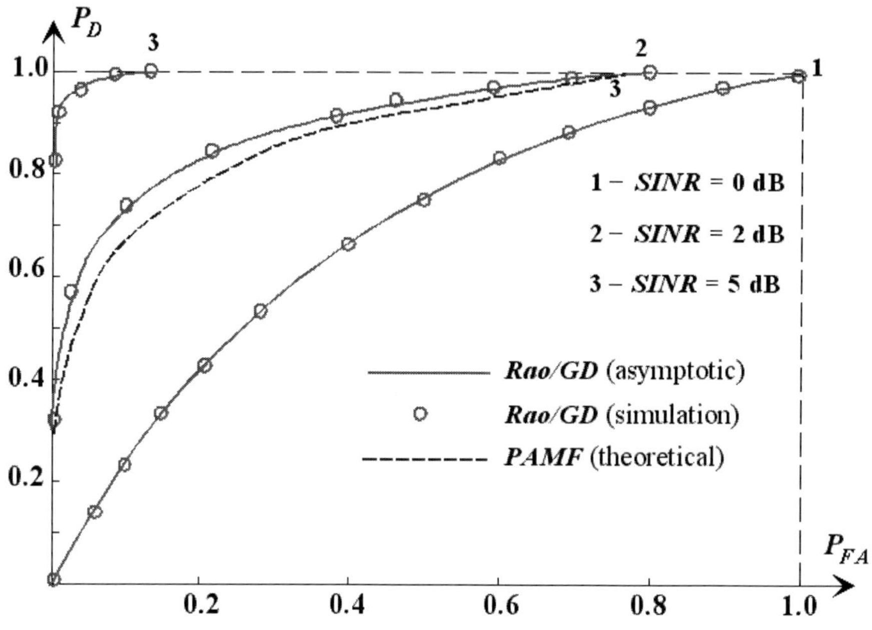

Figure 6.11. ROC curves.

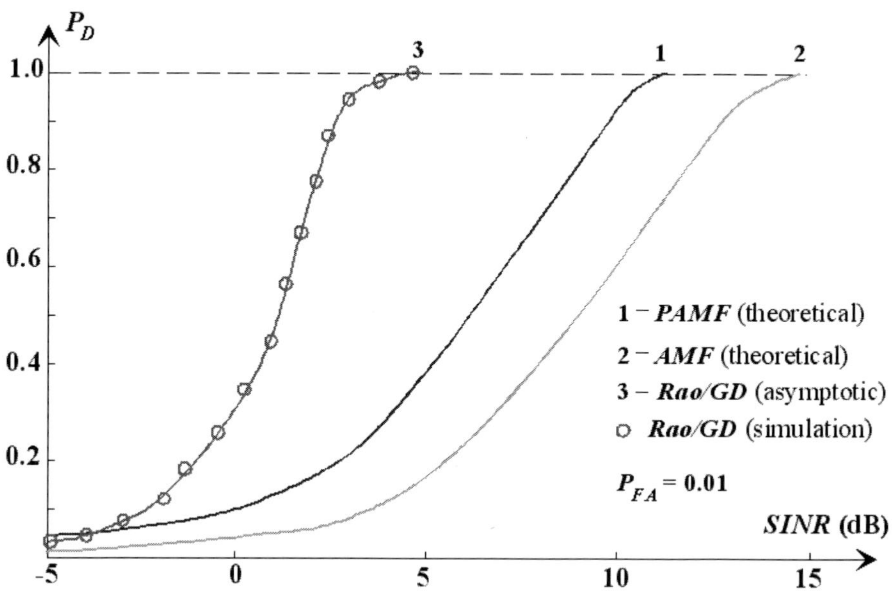

Figure 6.12. Probability of detection versus input *SINR* $J = 4,\ N = 32,\ K = 256$.

We examine the ROC of the parametric Rao/GD. The parameters used in the simulation are $J = 4, N = 32$,and $K = 256$. Figure 6.11 demonstrates the ROC curves for the parametric Rao/GD obtained by simulation and asymptotic analysis for *SINR* values of 0, 2, and 5 dB. It is seen that the simulation results match those obtained by asymptotic analysis. PAMF ROC is also presented for comparison. Superiority of the parametric Rao/ GD over the PAMF is evident. Figures 6.12–6.15 depict the probability of detection P_D versus *SINR* for the PAMF, AMF, and the parametric Rao/GD under various conditions. In particular, Figs. 6.12 and 6.14 correspond to the case with adequate training, for which the Reed-Brennan rule is satisfied, whereas Figs. 6.13 and 6.15 correspond to the case with a limited training, for which the AMF detector does not even exist, since the training size $K = 8$ is too small to meet the minimum training condition (6.100).

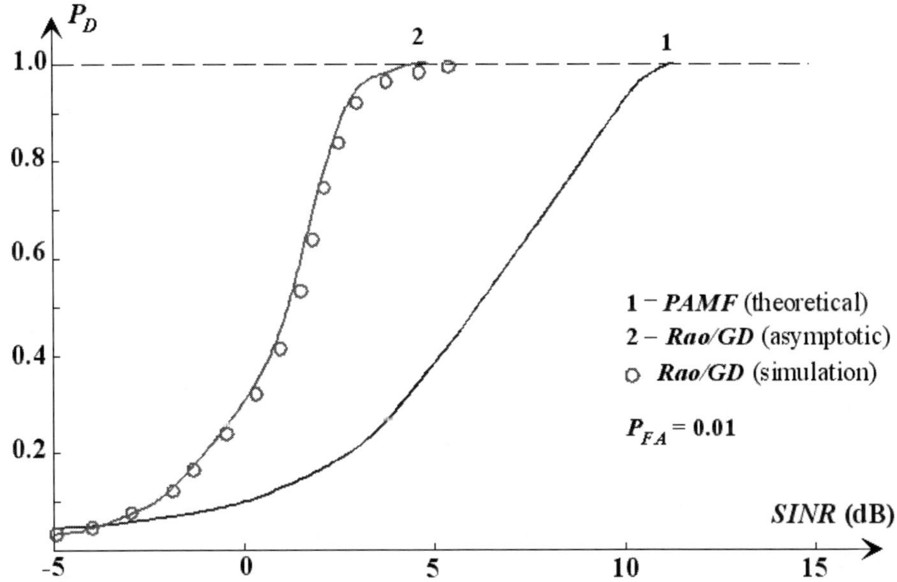

Figure 6.13. Probability of detection versus input *SINR* $J = 4,\ N = 32,\ K = 8$.

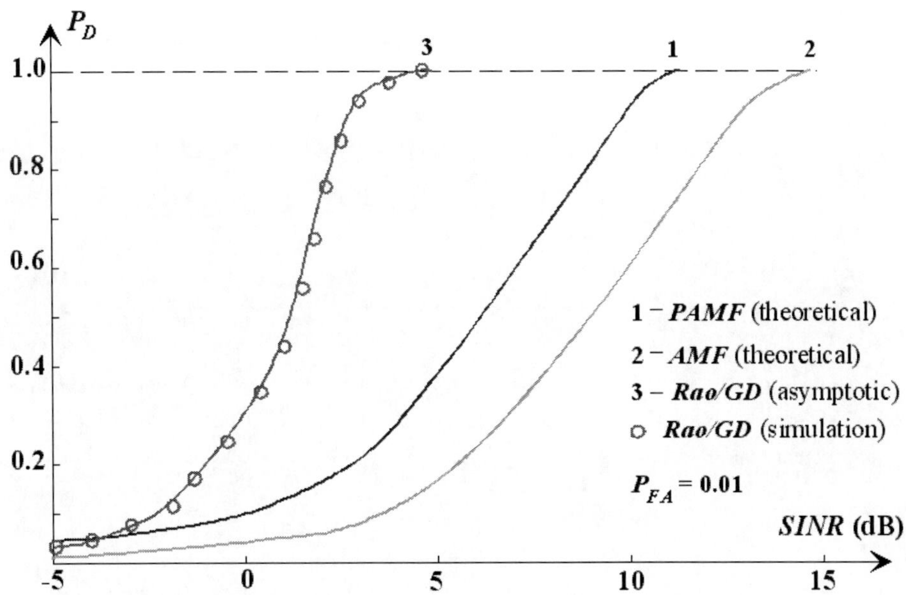

Figure 6.14. Probability of detection versus input *SINR* $J = 4,\ N = 16,\ K = 28$.

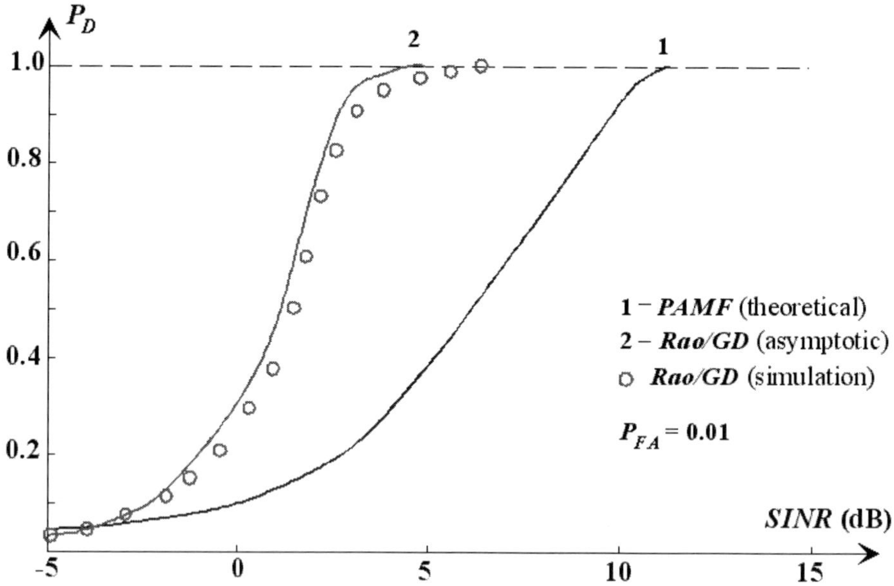

Figure 6.15. Probability of detection versus input *SINR* $J = 4$, $N = 16$, $K = 8$.

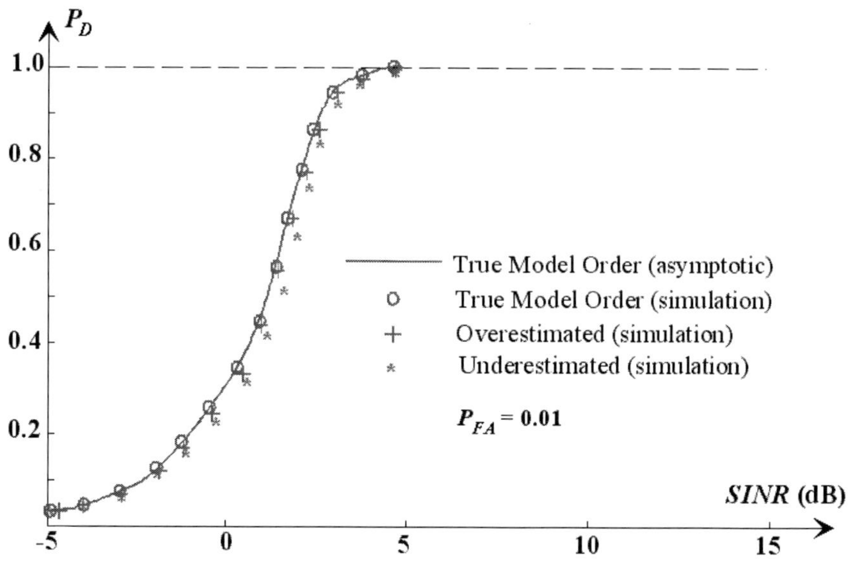

Figure 6.16. Probability of detection versus input *SINR* of parametric Rao/*GD* when model order of multichannel AR process used or computing test statistic is true $(P = 2)$, underestimated (assuming $P = 1$), and overestimated (assuming $P = 3$), along with $J = 4$, $N = 32$, $K = 256$.

An examination of these figures reveals the following:

- When the assumptions made in Section 6.4.2 are met, the asymptotic analysis provides a quite accurate prediction of the performance of the parametric Rao/GD.The gap between the asymptotic and simulated results is seen to widen as K

and/or N decreases. But even for the most challenging case with $K = 8$ and $N = 16$, the gap is about 0.5 dB, Fig.6.15;

- The parametric Rao/GD is very close to the theoretical GD. The gap between two detectors closes with increasing K and/or N;
- The parametric Rao/GD outperforms the PAMF and AMF detectors.

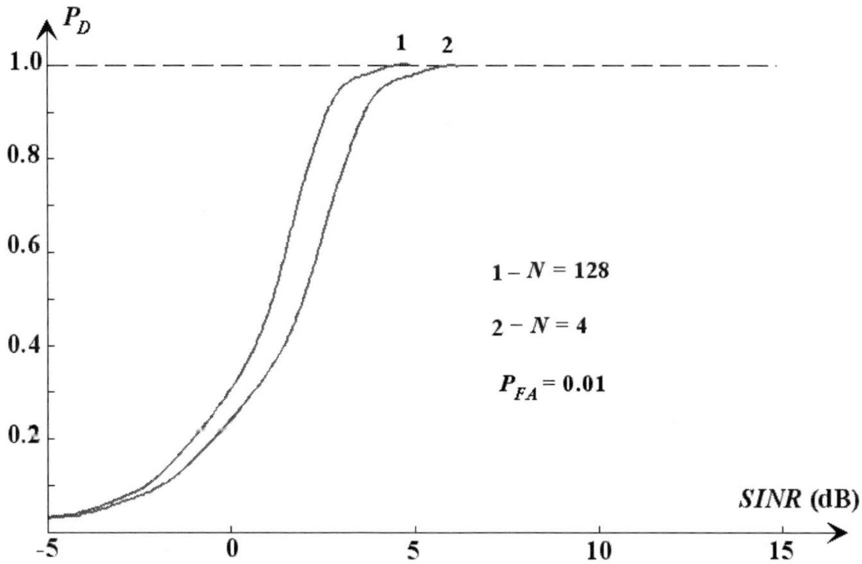

Figure 6.17. Impact of pulse number N on parametric Rao/*GD*.

So far we have assumed that the model order P of the multichannel AR process is known. As mentioned, various model selection techniques can be used to estimate P, and it is not unusual for these techniques to underestimate or overestimate the model order by a small number relative to the true model order P [52] and [55].

It would be of interest to find out how the parametric Rao/GD performs when an inaccurate model order estimate is used. This is shown in Fig. 6.16, where the performance of the Rao/GD using the true, an underestimated, and an overestimated model order is depicted. As we can see, using an inaccurate model order estimate degrades the detection performance, but the degradation is not significant, especially in the case of model order overestimation. Overestimation is a more robust error since the high-order coefficients can be estimated close to zero providing that the size of the signals that can be used for estimation is large enough. The above-mentioned behavior of the parametric Rao/GD is typical and has been consistently observed in other experiments with a similar setup.

Here, we only considered the case where the model order is incorrectly estimated by one unit. A larger performance variation is expected if there is a larger estimation error for P. Finally, Fig. 6.17 depicts the probability of detection P_D versus *SINR* for the parametric Rao/GD if $J = 4, P_{FA} = 0.01$ and N varies from $N = 4$ to $N = 128$. It is seen that the detection performance increases with N.

6.5. Quadriphase DS-CDMA Wireless Communication Systems

6.5.1. Brief Review

We study the probability of error P_{er} performance of band-limited quadriphase direct-sequence code-division multiple-access (QDS-CDMA) wireless communication systems. The considered wireless communication systems have asynchronous users, pulse shaping, direct-sequence quadric-phase-shift-keying (DS-QPSK) spreading and aperiodic spreading sequences or random spreading [56]. The uplink of the wideband CDMA system [57] that uses the square-root raised cosine (Sqrt-RC) pulse of 22% excess bandwidth, DS-QPSK spreading and long pseudo noise sequences is an example of QDS-CDMA system. The pseudo noise sequences are modeled as random, i.i.d. binary sequences [56] and [58].

We focus on the analysis of pulse shaping in wireless communication systems that is very important under designing process. The pulse shaping influences a spectral efficiency and reduces out-of-band radio emissions. Firstly, its relevance was mentioned in [59]. In the majority cases, the effect of bandwidth efficient pulse shaping by using the pulses of one-chip period duration, in particularly, the rectangular pulse, is neglected [60]–[64]. This approach simplifies the framework for the probability of error P_{er} analysis and can lead to overly optimistic estimates for the probability of error P_{er} performance of band-limited systems using the direct-sequence binary phase shift keying (DS-BPSK) spreading [65].

DS-CDMA systems with band-limited pulse shapes and DS-BPSK spreading are referred to as B-CDMA. Previous study of the probability of error P_{er} for QDS-CDMA in [66]–[68] assumes a simple and widely used standard Gaussian approximation [61] and [62] that models the multiple-access interference (MAI) generated by the signals of many users as a Gaussian random process. The difficulty with the standard Gaussian approximation is its tendency to break down and return the over-optimistic probability of error P_{er} results for many systems with either a small number of users or a small number of high-powered users that was observed both for systems with DS-BPSK or DS-QPSK spreading using rectangular pulses [58], [60]–[63], [69] and both for B-CDMA [70]–[73].

In contrast to [56], [67], and [68], an accurate probability of error P_{er} analysis of QDS - CDMA was done in [66]. Though its focus was on the design of optimum chip waveforms for QDS-CDMA, it used a more accurate method of the probability of error P_{er} analysis based on the improved Gaussian approximation or conditional Gaussian approximation for pulses time-limited to one-chip period [74].

We study the probability of error P_{er} performance for the case of the generalized detector (GD) employment based on GASP in noise [5], [6], [9]–[1]. We assume that the standard Gaussian approximation can serve as an accurate one under conditions typically satisfied in practice [75]. The improved Gaussian approximation can be reduced to the standard one for pulses with zero excess bandwidth.

It is shown that the probability of error P_{er} performance of QDS-CDMA with GD is superior or equivalent to the B-CDMA the probability of error P_{er} performance with GD case. This result is consistent with a similar finding in [64] which was restricted to pulses (such as the rectangular pulse) of one-chip period. An exact the probability of error P_{er} formula for QDS-CDMA with GD is derived according to the characteristic function method described in [63] for pulses of one-chip period and in [70] and [71] for general pulse shapes with DS-BPSK spreading. More tractable formula for the improved Gaussian approximation that simplifies greatly for pulses with small excess bandwidth is presented. The simplified improved Gaussian approximation that has a complexity on par with that of standard Gaussian approximation is derived. Numerical examples with Monte Carlo simulation show the accuracy of the standard Gaussian approximation for QDS-CDMA with GD wireless communication systems given pulse shapes of small excess bandwidth and spreading factors (the chip-rate to bit-rate ratio) of moderate to large values.

6.5.2. System Model

The QDS-CDMA system model consists of $K+1$ asynchronous users in an additive white Gaussian noise (AWGN) channel. This section describes the modeling of the received signal, coherent GD, and bit decision statistic based on DS-QPSK spreading with pulse shaping. An arbitrary user is designated as the desired user and indexed as user 0.

6.5.2.1. Received Signal

The received signal can be presented in the following form

$$x(t) = \sum_{k=0}^{K} s^{(k)}(t) + w(t) \tag{6.136}$$

For $-\infty < t < \infty$.The second term $w(t)$ is an AWGN process with a two-sided power spectral density of $0.5\mathcal{N}_0$. The first term is the sum of the received signals from each user.

The DS-QPSK signal of user k can be expressed as

$$s^{(k)}(t) = \sqrt{P_k}[s_I(t)\cos(\omega_c t + \theta_k) + s_Q \sin(\omega_c t + \theta_k)] \ , \tag{6.137}$$

where ω_c and $\theta_k \in [0,2\pi)$ represent, respectively, the carrier frequency in radians and carrier phase offset of user k relative to that of user 0 the in-phase (I) and quadrature phase(Q) baseband signals are, respectively

$$x_I(t) = \sum_{n=-\infty}^{\infty} b^{(k)}_{\lfloor n/N^{(k)} \rfloor} \alpha_n^{(I,k)} q(t - \tau_k - nT_c) \ ; \tag{6.138}$$

$$x_Q(t) = \sum_{n=-\infty}^{\infty} b^{(k)}_{\lfloor n/N^{(k)} \rfloor} \alpha_n^{(Q,k)} q(t - \tau_k - nT_c) \ . \tag{6.139}$$

The bits $b_i^{(k)} \in \{\pm 1\}$ of user k are transmitted at a rate of $1/T_b^{(k)}$ over both the I and Q branches. The floor function $\lfloor y \rfloor$ returns the integer portion of a real number y. The spreading factor $N^{(k)}$ is the chip-rate $1/T_c$ to bit-rate ratio. The terms $\alpha_n^{(I,k)}$ and $\alpha_n^{(Q,k)}$ represent, respectively, the chips of long pseudo noise sequences used to direct-sequence spread the bits in the in-phase and quadrature branches.

The pseudo noise sequence of each user and in each branch is modeled as i.i.d. random sequence with the probabilities given by [56] and [61]

$$P[\alpha_n^{(I,k)} = 1] = P[\alpha_n^{(I,k)} = -1] = P[\alpha_n^{(Q,k)} = 1] = P[\alpha_n^{(Q,k)} = -1] = 0.5 \ . \qquad (6.140)$$

The term P_k in (6.137) represents the received signal power for user k. The term $\tau_k \in [0, T_b^{(k)})$ represents the bit delay of user k relative to that of the receiver for user 0. For $k \in [1, K]$, the parameters θ_k and τ_k are modeled as independent and uniformly distributed random variables. At $k = 0$, the code and phase synchronization (for coherent detection) are assumed such that $\tau_0 = \theta_0 = 0$.The received baseband chip pulse $q(t)$ is assumed to be a real, even symmetric pulse with an energy of

$$\int_{-\infty}^{\infty} | q^2(t) | \, dt = E_b \qquad (6.141)$$

and with a time duration NT_c.

The bandwidth of the pulse can be expressed as

$$W = \frac{1+\alpha}{2T_c} \qquad (6.142)$$

where α represents the excess bandwidth of the pulse. We consider pulses which either satisfy the Nyquist criterion for zero intersymbol interference (ISI) or introduce ISI at a level negligible with respect to that of MAI. Furthermore, the effect of ICI and ISI is negligible to large processing gains [76].

6.5.2.2. Coherent GD

Let $N = N^{(0)}$ and $E_b = E_b^{(0)}$. Henceforth, N and E_b will be used to denote, respectively, the spreading factor and bit energy for user 0. The coherent GD for user 0 employed by QDS-CDMA wireless communication system is shown in Fig. 6.18. The received signal given by (6.136) is demodulated to form the I and Q baseband signals in the correlation channel of GD. The received signal in each branch of the GD correlation channel is passed through the receiver filter $Q^*(f)$ matched to the received chip pulse shape $Q(f)$ and direct-sequence spread. The outputs of two GD correlation channel branches are added and mixed with the GD autocorrelation channel owing to the GD compensation channel.

The resulting process is passed to a summation device to generate the bit decision statistics as

$$Z_i^{(0)'} = \sqrt{\sum_{l=iN}^{(i+1)N-1} \left\{ \left[\alpha_l^{(I,0)}\right]^2 + \left[\alpha_l^{(Q,0)}\right]^2 \right\}}$$

$$\times \left\{ \left\{ \int_{-\infty}^{\infty} \sum_{k=0}^{K} \left[s^{(k)}(t) q(-u) \right]^2 du \right\}^2 + \left\{ \int_{-\infty}^{\infty} \left[n_{l_{AF}}^2 - n_{l_{PF}}^2 \right] q^2(-u) du \right\}^2 \right\} . \quad (6.143)$$

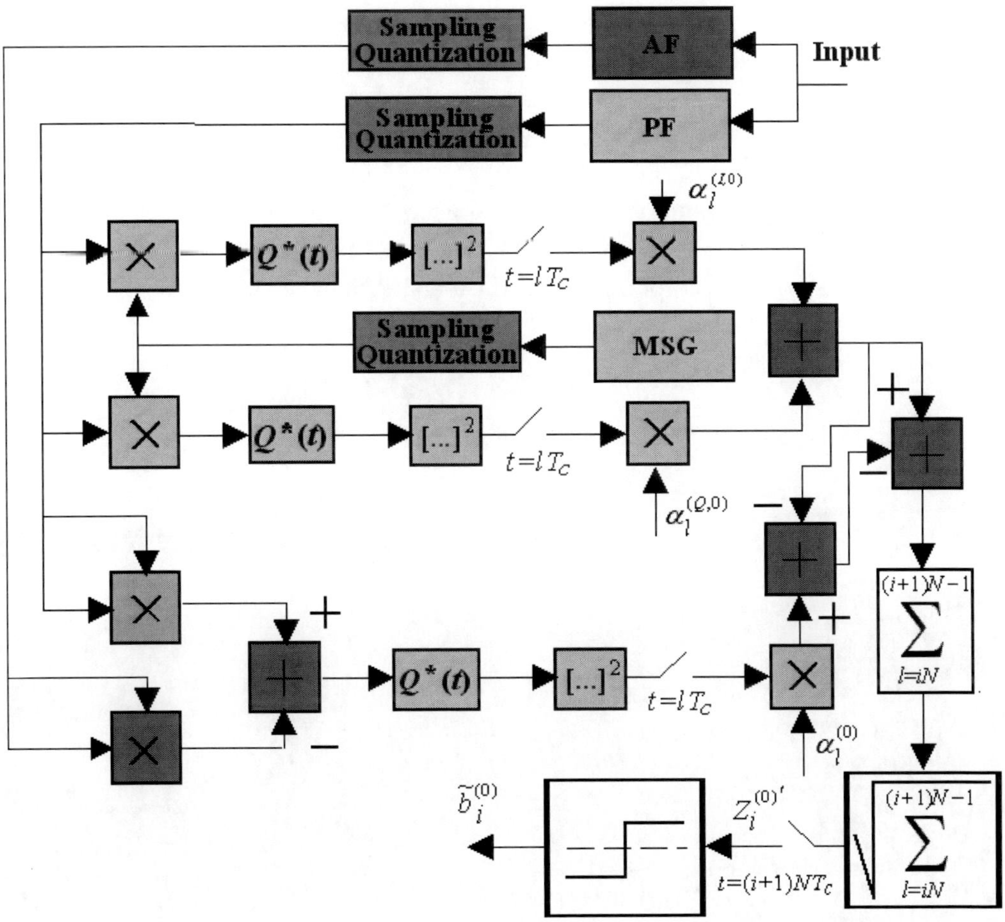

Figure 6.18. Coherent GD flowchart.

This is then fed to a hard decision device to produce the bit decision

$$\widetilde{b}_i^{(0)} = \text{sgn}\, Z_i^{(0)'} , \quad (6.144)$$

where sgn(...) is the *signum* function.

6.5.2.3. Bit Decision Statistics

Without loss of generality, the bit decision statistic $Z_0^{(0)'}$ for bit 0 of user 0, $b_0^{(0)}$ is considered. In anticipation of the characteristic function method of the following section, the statistic $Z_0^{(0)'}$ is rescaled and expressed as

$$Z_0^{(0)} = b_0^{(0)} + \mathcal{M} + [n_{I_{AF}}^2 - n_{I_{PF}}^2] , \tag{6.145}$$

where

$$Z_0^{(0)} = \frac{Z_0^{(0)'}}{\sqrt{P_0 N T_c}} . \tag{6.146}$$

The third term in (6.145) is the GD background noise defined in [6] and can be approximated by the Gaussian pdf with zero mean and variance equal to $4\sigma_n^4$, where σ_n^2 is given by (6.16a).

The received energy per bit for user 0 is defined in the following form:

$$E_b = P_0 N T_c . \tag{6.147}$$

The contribution of the MAI takes the form:

$$\mathcal{M} = \sum_{k=1}^{K} \mathcal{M}_k ; \tag{6.148}$$

$$\mathcal{M}_k = \sqrt{\frac{P_k}{P_0}} \frac{1}{2N}$$
$$\times \left[X(\mathbf{a}^I, \mathbf{d}^{I,k})\cos\theta_k + X(\mathbf{a}^I, \mathbf{d}^{Q,k})\sin\theta_k + X(\mathbf{a}^Q, \mathbf{d}^{I,k})\sin(-\theta_k) + X(\mathbf{a}^Q, \mathbf{d}^{Q,k})\cos\theta_k\right] , \tag{6.149}$$

where

$$X(\mathbf{a}^{\lambda_1}, \mathbf{d}^{\lambda_2,k}) = \sum_{l=0}^{N-1} \sum_{n=-\infty}^{\infty} \alpha_l^{(\lambda_1,0)} d_n^{(\lambda_2,k)} \rho[(l-n)T_c - T_k] \tag{6.150}$$

and $\lambda_1, \lambda_2 \in \{I, Q\}$ and

$$T_k = \text{mod}\{\tau_k, T_c\} \tag{6.151}$$

is the chip delay of user k with a uniform distribution over $T_k \in [0, T_c)$.

The function

$$\rho(\tau) = \mathcal{F}^{-1}(T_c^{-1} \, | Q(f) |^2) \tag{6.152}$$

is the inverse Fourier transform of $T_c^{-1} \, | Q(f) |^2$. It is the autocorrelation function of the chip pulse normalized by $\sqrt{T_c}$. For example, given a Sqrt-RC pulse for $q(t), \rho(\tau)$ is the raised cosine pulse scaled by T_c^{-1}.

6.5.3. MAI Analysis

This section derives an expression for the probability of error P_{er} and presents an accurate approximation for the distribution of the MAI component $\mathcal{M}$ in QDS-CDMA.

6.5.3.1 Characteristic Function of $\mathcal{M}$

The exact characteristic function for QDS-CDMA wireless communication system is derived according to the approach used in [71] and [77] for B-CDMA. First, the conditional characteristic function

$$\phi_{\mathcal{M}|\mathbf{a}^I,\mathbf{a}^Q,\mathbf{T},\mathbf{\Theta}}(\omega) = E\left[e^{j\omega\mathcal{M}} \,|\, \mathbf{a}^I, \mathbf{a}^Q, \mathbf{T}, \mathbf{\Theta}\right] \tag{6.153}$$

is derived by substituting $\mathcal{M}$ from (6.148) and (6.149). The terms $\mathbf{T}$ and $\mathbf{\Theta}$ are defined as

$$\mathbf{T} = [T_1, \ldots, T_K] \tag{6.154}$$

and

$$\mathbf{\Theta} = [\theta_1, \ldots, \theta_K] \ , \tag{6.155}$$

respectively.

Next, the conditioning of each random vector is removed step by step in the order of $\mathbf{T}, \mathbf{\Theta}$ and, finally, both $\mathbf{a}^I$ and $\mathbf{a}^Q$. Doing so yields

$$\phi_{\mathcal{M}}(\omega) = \frac{1}{2^{2N}} \sum_{p=1}^{2^N} \sum_{q=1}^{2^N} \phi_{\mathcal{M}|\mathbf{a}^I,\mathbf{a}^Q}(\omega) \ , \tag{6.156}$$

where $\mathcal{M}$ for $p, q \in \left[1, 2^N\right]$ represent, respectively, the 2^N distinct sequences both for $\mathbf{a}^I$ and $\mathbf{a}^Q$. Each $\{\pm 1\}$ sequence is of equal likelihood and of length N. The characteristic function of $\mathcal{M}$, conditioned on $\mathbf{a}^I$ and $\mathbf{a}^Q$, is given by the following equations

$$\phi_{\mathcal{M}|\mathbf{a}^I,\mathbf{a}^Q}(\omega) = \prod_{k=1}^{K} \phi_{\mathcal{M}_k|\mathbf{a}^I,\mathbf{a}^Q}(\omega) \ ; \tag{6.157}$$

$$\phi_{\mathcal{M}_k|\mathbf{a}^I,\mathbf{a}^Q}(\omega)=\frac{1}{2\pi T_c}\int_0^{2\pi}\int_0^{T_c}\phi_{\mathcal{M}_k|\mathbf{a}^I,\mathbf{a}^Q,T_k,\theta_k}(\omega)dT_k d\theta_k \ ; \tag{6.158}$$

$$\phi_{\mathcal{M}_k|\mathbf{a}^I,\mathbf{a}^Q,T_k,\theta_k}(\omega)=\prod_{n=-M}^{N+M-1}\cos\left\{\frac{\omega\varsigma_k}{N}\left[A^I\cos\theta_k-A^Q\sin\theta_k\right]\right\}$$

$$\times\prod_{n=-M}^{N+M-1}\cos\left\{\frac{\omega\varsigma_k}{N}\left[A^I\cos\theta_k+A^Q\sin\theta_k\right]\right\}, \tag{6.159}$$

where

$$\varsigma_k=\frac{1}{2}\sqrt{\frac{P_k}{P_0}}A^{\lambda_1}=\sum_{l=0}^{N-1}\alpha_l^{(0,\lambda_1)}\rho[(l-n)T_c-T_k] \tag{6.160}$$

And $\lambda_1\in\{I,Q\}$.

To evaluate (6.156), (6.157) must be averaged across $2^N\times2^N$ distinct sequences. Thus, $2^{2N}K$ double integrations are required in general. This is a daunting task for moderate to large values of *N*. For example, if $N=10$, roughly 10^6K double integrations are required. Thus, a more computationally efficient expression is needed for $\phi_{\mathcal{M}}(\omega)$ given moderate to large *N*, i.e. $N>10$.

6.5.3.2. Approximation for the Distribution of $\mathcal{M}|_{\mathbf{T},\mathbf{\Theta}}$

The approach in deriving an approximation for the distribution of $\mathcal{M}|_{\mathbf{T},\mathbf{\Theta}}$ follows that of [70] and [71]. The approximation assumes large bit finite *N*, where $\gamma=K/N$ is the constant. Under these conditions, it can be shown that

$$\mathcal{M}|_{\mathbf{T},\mathbf{\Theta}}\approx\mathcal{N}[0,NVar(X_0)]\ , \tag{6.161}$$

where

$$Var(X_0)=\frac{2}{N^2}\sum_{k=1}^{K}\varsigma_k^2\left\{B(0)+\sum_{i=1}^{1+\lfloor\alpha\rfloor}2B\left(\frac{i}{T_c}\right)\cos\left(\frac{2iT_k\pi}{T_c}\right)\right\}. \tag{6.162}$$

Two key observations follow. First, in contrast to that for DS-BPSK in [71], the effect of interference phase offsets disappear as $Var(X_0)$ in (6.162) is independent of $\mathbf{\Theta}$. Hence

$$Var(\mathcal{M}|_{\mathbf{T},\mathbf{\Theta}})=Var(\mathcal{M}|_{\mathbf{T}})\ . \tag{6.163}$$

Second, if $\alpha = 0$, $Var(X_0)$ reduces to

$$Var(X_0) = \frac{2B(0)}{N^2} \sum_{k=1}^{K} \varsigma_k^2 \tag{6.164}$$

and is independent of **T** as well. Hence,

$$Var(\mathcal{M}|_{\mathbf{T}}) = Var(\mathcal{M}) \quad \text{and} \quad \mathcal{M} \approx \mathcal{N}[0, NVar(X_0)] \,. \tag{6.165}$$

That is the standard Gaussian approximation for large *K* and *N* if $\alpha = 0$.

6.5.3.3. Approximation of $\phi_{\mathcal{M}}(\omega) : \hat{\phi}_{\mathcal{M}}(\omega)$

An approximation for $\phi_{\mathcal{M}}(\omega) : \hat{\phi}_{\mathcal{M}}(\omega)$, following the same steps outlined in [70], can be derived based on (6.161) and the simple form of the characteristic function of a normal random variable. The approximation can be expressed as

$$\hat{\phi}_{\mathcal{M}}(\omega) - \prod_{k=1}^{K} \psi_{\mathcal{M}_k}(\omega) \;; \tag{6.166}$$

$$\hat{\phi}_{\mathcal{M}_k}(\omega) = \frac{1}{T_c} \int_0^{T_c} \hat{\phi}_{\mathcal{M}_k|T_k}(\omega) dT_k \;; \tag{6.167}$$

$$\hat{\phi}_{\mathcal{M}_k|T_k}(\omega) = \exp\left\{-\xi_{\omega}^{(k)} B(0)\right\} \prod_{i=1}^{1+\lfloor \alpha \rfloor} \exp\left\{-2\xi_{\omega}^{(k)} B\left(\frac{i}{T_c}\right) \cos\left(\frac{2iT_k \pi}{T_c}\right)\right\}, \tag{6.168}$$

where

$$\xi_{\omega}^{(k)} = \frac{\omega^2 P_k}{4P_0 N} \,. \tag{6.169}$$

Two simplifications can be obtained for pulses with small α .First, if $\alpha < 1, \lfloor \alpha = 0 \rfloor$ and (6.168) reduces to

$$\hat{\phi}_{\mathcal{M}_k|T_k}(\omega) = \exp\left\{-\xi_{\omega}^{(k)} B(0)\right\} \exp\left\{-2\xi_{\omega}^{(k)} B\left(\frac{1}{T_c}\right) \cos\left(\frac{2iT_k \pi}{T_c}\right)\right\}. \tag{6.170}$$

Substituting (6.170) into (6.167) returns

$$\hat{\phi}_{\mathcal{M}_k}(\omega) = \exp\left\{-\xi_{\omega}^{(k)} B(0)\right\} I_0\left\{2\xi_{\omega}^{(k)} B(T_c^{-1})\right\}, \tag{6.171}$$

where

$$I_0(x) = \frac{1}{2\pi}\int_{-\pi}^{\pi}\exp\{x\cos\theta\}d\theta \tag{6.172}$$

is the modified Bessel function of the first kind of zero order [78]. Second, if $\alpha = 0$, then $\hat{\phi}_{\mathcal{M}_k}(\omega)$ reduces to

$$\hat{\phi}_{\mathcal{M}_k}(\omega) = \exp\left\{-\xi_{\omega}^{(k)}B(0)\right\} . \tag{6.173}$$

6.5.4. Bit Error Probability Performance Analysis

This section derives an expression for the probability of error P_{er} and presents an accurate approximation for the distribution of the MAI component $\mathcal{M}$ in QDS-CDMA.

6.5.4.1. Standard Gaussian Approximation

The standard Gaussian approximation models $\mathcal{M}$ as a Gaussian random variable and approximates the probability of error P_{er} as

$$P_{er}^{SGA} = \mathcal{Q}\left\{\sqrt{\frac{2\mathcal{N}_0^{\circ 2}}{E_b^2} + \frac{B(0)}{2N}\sum_{k=1}^{K}\frac{P_k}{P_0}}\right\} , \tag{6.174}$$

where

$$\mathcal{Q}(x) = \frac{1}{\sqrt{2\pi}}\int_{x}^{\infty}\exp\{-0.5y^2\}dy . \tag{6.175}$$

6.5.4.2. Characteristic Function Procedure

According to the characteristic function procedure [63], [70], and [71], the probability of error P_{er} can be expressed as

$$P_{er}^{CF_{exact}} = \frac{1}{2} - \frac{1}{\pi}\int_{0}^{\infty}\phi_{\mathcal{M}}(\omega)\phi_{n_{AF}^2 - n_{PF}^2}(\omega)\frac{\sin\omega}{\omega}d\omega , \tag{6.176}$$

where

$$\phi_{n_{AF}^2 - n_{PF}^2}(\omega) = \exp\left\{-\frac{2\omega^2\mathcal{N}_0^{\circ 2}}{E_b^2}\right\} \tag{6.177}$$

and $\phi_{\mathcal{M}}(\omega)$ is given by (6.156). This expression is exact. However, as pointed out earlier, the number of computations grows exponentially with *N*.

6.5.4.3. Characteristic Function Procedure Approximation

An approximation for the probability of error P_{er}, denoted as P_{er}^{CF}, can be obtained by replacing $\phi_{\mathcal{M}}(\omega)$ in (6.176) with $\hat{\phi}_{\mathcal{M}}(\omega)$. This yields an expression with significantly less complexity

$$P_{er}^{CF} = \frac{1}{2} - \frac{1}{\pi}\int_0^\infty \hat{\phi}_{\mathcal{M}}(\omega)\phi_{n_{AF}^2 - n_{PF}^2}(\omega)\frac{\sin\omega}{\omega}d\omega \ , \tag{6.178}$$

where the complexity in $\hat{\phi}_{\mathcal{M}}(\omega)$ is independent of N. The results of this procedure are equivalent to those obtained by the improved Gaussian approximation described below.

6.5.4.4. Improved Gaussian Approximation

The improved Gaussian approximation can be expressed as the mean of the probability of bit error P_{er} conditioned on $\mathbf{T}$ and evaluated by random sampling across $\mathbf{T}$ [66]

$$P_{er}^{IGA} = E\left\{Q\left[\sqrt{\frac{E_b'^2 + \Psi}{2\mathcal{N}_0'^2}}\right]\right\} , \tag{6.179}$$

where the conditional variance, $\Psi = Var(\mathcal{M}|_{\mathbf{T}})$ can be expressed as

$$\Psi = \sum_{k=1}^{K}\Psi_k \ , \tag{6.180}$$

where

$$\Psi_k = \frac{P_k}{2NP_0}\left\{B(0) + \sum_{i=1}^{1+\lfloor\alpha\rfloor} 2B\left(\frac{i}{T_c}\right)\cos\left(\frac{2iT_k\pi}{T_c}\right)\right\} . \tag{6.181}$$

A simplified expression is derived if $\alpha < 100\%$ using the approach of [74]. According to $\mathcal{M}|_{\mathbf{T},\mathbf{\Theta}}$ in (6.161), (6.181) can be expressed as

$$\Psi_k = C_1 + C_2 U_k \ , \tag{6.182}$$

where

$$\begin{cases} C_1 = \dfrac{B(0)P_k}{2NP_0} \quad \text{and} \quad C_2 = \dfrac{B(T_c^{-1})P_k}{2NP_0} \ ; \\ U_k = \cos(2\pi T_k T_c^{-1}) \ . \end{cases} \tag{6.183}$$

The Ψ_k is independent since the T_k is independent. Since $T_k \in [0, T_c)$ is uniformly distributed, the pdf of U_k can be presented in the following form:

$$f_{U_k}(u) = \frac{1}{\pi\sqrt{1-u^2}} , \tag{6.184}$$

where $|u| \le 1$ [79]. If $C_2 > 0$, the pdf of Ψ can be expressed as the $K-1$ convolutions

$$f_{\Psi}(z) = f_{\Psi_1}(z) \times f_{\Psi_2}(z) \times \cdots \times f_{\Psi_K}(z) , \tag{6.185}$$

where the pdf of Ψ_k, a linear function of U_k [79] takes the following form

$$f_{\Psi_k}(z) = \frac{1}{\pi\sqrt{C_2^2 - (z - C_1)^2}}, \qquad |z - C_1| \le C_2 . \tag{6.186}$$

Thus, the improved Gaussian approximation can be expressed as

$$P_{er}^{IGA} = \int_0^\infty \mathcal{Q}\left[\sqrt{\frac{E_b^2 + \Psi}{2\mathcal{N}_0^2}}\right] f_{\Psi}(z)dz . \tag{6.187}$$

If $\alpha = 0$, $C_2 = 0$

$$\Psi = \frac{B(0)}{2NP_0}\sum_{k=1}^{K} P_k \tag{6.188}$$

and the improved Gaussian approximation reduces to the standard Gaussian approximation given by (6.174).

6.5.4.5. Simplified Improved Gaussian Approximation

The simplified improved Gaussian approximation for QDS-CDMA is derived based on the simplified improved Gaussian approximation obtained for B-CDMA [72], 73]. It approximates the probability of error P_{er}^{IGA} in (6.187) by applying the procedure of expansion of differences as described in [69]. It can be expressed as the sum of three weighted $\mathcal{Q}$-functions

$$P_{er}^{SIGA} = \frac{2}{3}P(\mu) + \frac{1}{6}P\left[\mu + \sqrt{3}\sigma\right] + \frac{1}{6}P\left\{\max\left[0, \mu - \sqrt{3}\sigma\right]\right\}, \tag{6.189}$$

where $P(x) = \mathcal{Q}(x^{-0.5})$. The terms μ and σ represent, respectively, the mean and standard deviation of the variance of the interference including AWGN.

Two terms are defined as

$$\mu = \sum_{k=1}^{K} E[\Psi_k] + \frac{2\mathcal{N}_0^2}{E_b^2} \tag{6.190}$$

$$\sigma^2 = \sum_{k=1}^{K} E[\Psi_k^2] - \{E[\Psi_k]\}^2 . \tag{6.191}$$

Due to the earlier normalization of the bit statistic, $\mu^{-0.5}$ represents the *SNR*. It can be readily shown that two moments $E[\Psi_k]$ and $E[\Psi_k^2]$ for general pulses are, respectively

$$E[\Psi_k] = \frac{B(0)P_k}{2NP_0} ; \tag{6.192}$$

$$E[\Psi_k^2] = \frac{P_k^2}{4N^2P_0^2}\left\{ |B(0)|^2 + \sum_{i=1}^{1+\lfloor\alpha\rfloor} 2\left|B(iT_c^{-1})\right|^2 \right\} . \tag{6.193}$$

In the case of ideal Nyquist pulse, simplifications occur, i.e.

$$B(0) = 1,\ B(iT_c^{-1}) = 1,\ E[\Psi_k^2] = \{E[\Psi_k]\}^2,\ \sigma^2 = 0 . \tag{6.194}$$

Thus, consistent with the earlier results for P_{er}^{IGA}, the simplified improved Gaussian approximation expression reduces to $P_{er}^{SIGA} = P(\mu)$ that is equivalent to P_{er}^{SGA}.

6.5.4.6. Comparison of QDS-CDMA and B-CDMA

The probability of error P_{er} performance of QDS-CDMA is compared with that of B-CDMA. The comparison is made on their respective expressions for the variance of the MAI component conditioned on $\mathbf{\Theta}$ and $\mathbf{T}$ and the probability of error P_{er}^{SIGA} at identical conditions. For convenience, the corresponding B-CDMA expressions for the QDS-CDMA expression of $\Psi, \Psi^{(B)}$ and $P_{er}^{SIGA}, P_{er}^{B}$ are presented from [72]. The variance of the conditional MAI component can be expressed as

$$\Psi^{(B)} = \sum_{k=1}^{K} \Psi_k^{(B)} , \tag{6.195}$$

where

$$\Psi_k^{(B)} = \frac{P_k \cos^2\theta_k}{NP_0}\left\{ B(0) + \sum_{i=1}^{1+\lfloor\alpha\rfloor} 2B\left(\frac{i}{T_c}\right)\cos\left(\frac{2iT_k\pi}{T_c}\right) \right\} \tag{6.196}$$

such that

$$\Psi_k^{(B)} = 2\Psi_k \cos^2\theta_k \ . \tag{6.197}$$

The corresponding simplified improved Gaussian approximation expression for B-CDMA [72], P_{er}^B is identical to that for QDS-CDMA in (6.189), (6.190)–(6.193) except that (6.195) is replaced with

$$E\left\{\left[\Psi_k^{(B)}\right]^2\right\} = \frac{3P_k^2}{8N^2P_0^2}\left\{|B(0)|^2 + \sum_{i=1}^{1+\lfloor\alpha\rfloor} 2\left|B(iT_c^{-1})\right|^2\right\} \tag{6.198}$$

such that

$$E\left\{\left[\Psi_k^{(B)}\right]^2\right\} = 1.5E[\Psi_k^2] \ . \tag{6.199}$$

The means of the two MAI components over $\mathbf{\Theta}$ and $\mathbf{T}$ are equal:

$$E[\Psi] = E\left[\Psi^{(B)}\right] = \mu \ . \tag{6.200}$$

However, the variance of $\Psi^{(B)}$, σ_B^2 is greater than that of Ψ and

$$\sigma_B^2 - \sigma^2 = \frac{1}{2}\sum_{k=1}^{K} E[\Psi_k^2] > 0 \ . \tag{6.201}$$

The larger variation in the conditional MAI component variance (or power) for B-CDMA is not unexpected since the power level of the k-th interferer in B-CDMA is scaled by a $\cos^2\theta_k$ factor while that in QDS-CDMA remains constant irrespective of θ_k. The larger variance of the conditional MAI component in B-CDMA follows intuitively and was previously hinted at in [56]. For example, in a two user scenario of B-CDMA, the signal of an interferer and the conditional MAI component, in the best case, disappears if out-of-phase relative to other desired user at $\theta_1 = \{\pi/2, 3\pi/2\}$. In the worse case, it appears at its maximum power level if it is in phase at $\theta_1 = \{0, \pi\}$.

In contrast, in QDS-CDMA, the variance of the conditional MAI component with respect to phase is constant. The probability of bit error P_{er} for QDS-CDMA is shown to be less or equal to P_{er}^B, P_{er} for B-CDMA, in regions of large SNR and nearly identical to P_{er}^B in regions of small SNR and $\sigma_B^2 << SNR$. We can introduce the following difference

$$\Delta P_{er} = P_{er}^B - P_{er}^{SIGA} = \frac{1}{6}\left\{P\left[\mu+\sqrt{3}\sigma\right] + P\left[\mu-\sqrt{3}\sigma\right] - P\left[\mu+\sqrt{3}\sigma\right] - P\left[\mu-\sqrt{3}\sigma\right]\right\}. \tag{6.202}$$

The function given by (6.190) is a monotonically increasing function of x for $x>0$ as it represents the area in the tail end of the standard Gaussian pdf over the interval $[x^{-0.5},\ \infty)$. In regions of large μ, small *SNR*, and $\mu >> \sigma_B$, $P(x)$ is a slowly increasing function of x for large x. This implies that the four $P(\cdots)$ functions in (6.202) are approximately equal and that $\Delta P_{er} \approx 0$. On other hand, in regions of small μ, high *SNR*, $P(x)$ is an exponentially increasing function with x for small $x<<1$. This implies that the first $P(\cdots)$ function in (6.202) dominates the sum and that $\Delta P_{er} > 0$. Thus, in regions where μ is small we obtain $P_{er} > P_{er}^B$.

6.5.4.7 DS-CDMA versus DS-QPSK Spreading. Pulse Shaping Effect on Standard Gaussian Approximation Accuracy.

The effect of DS-BPSK spreading, DS-QPSK spreading, and pulse shaping on the accuracy of the standard Gaussian approximation is discussed. The inaccuracy of the stan-dard Gaussian approximation for B-CDMA can be explained intuitively based on the dependence of the conditional MAI variance on interferer phase offset. Implicit in the standard Gaussian approximation is the assumption that the MAI, like AWGN, is circularly symmetric and not depend on the phase offsets [57].

In B-CDMA, the effect of phase offsets on the conditional MA variance implies that the MAI process is not circularly symmetric under DS-BPSK spreading. Indeed, this was observed previously in [80] as well as in [66]. Pulse shaping and, more specifically, the excess bandwidth α of the pulse shape also play an important role on the accuracy of the standard Gaussian approximation.

It is well-known that the standard Gaussian approximation can be fairly inaccurate for time-limited pulses like the rectangular pulse [58], [60]–[63], [69]. In contrast, as was shown in Section 6.5.3, the improved Gaussian approximation in QDS-CDMA reduces to the standard Gaussian approximation if $\alpha = 0$ indicating a greater accuracy of standard Gaussian approximation for QDS-CDMA. This phenomenon can be explained by the effect of pulse shaping on the wide-sense cyclostationarity of the MAI process conditioned on the interferer chip delays [80].

A wide-sense cyclostationary process $X(t)$ with a period of T is a process whose auto-correlation function is periodic with T such that [81]–[83].

$$E[X(t+T)X(u+T)] = E[X(t)X(u)] \ . \tag{6.203}$$

Implicit in the standard Gaussian approximation is the assumption that the conditional MAI, like AWGN, is wide-sense stationary (WSS). However, given $\alpha > 0$, the conditional MAI process is wide-sense cyclostationary (WSCS) with a period of T_c under aperiodic spreading [81].

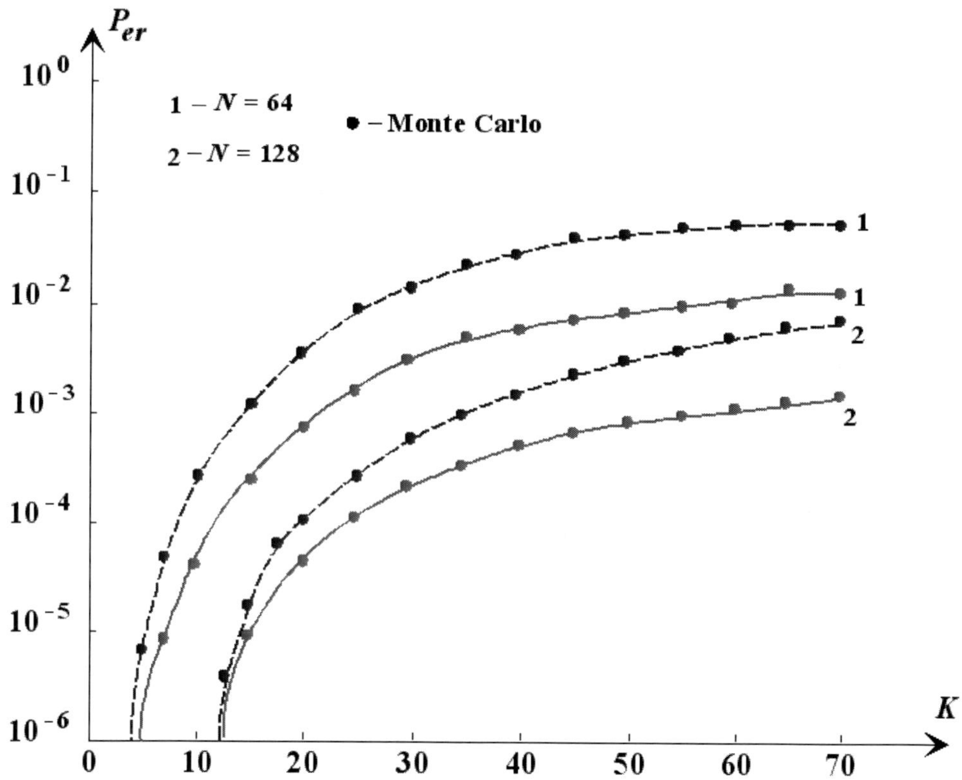

Figure 6.19. P_{er} versus K in QDS-CDMA. P_{er}^{IGA}, P_{er}^{SIGA}, P_{er}^{SGA} are matched and indistinguishable, $\alpha = 22\%$.

The level of cyclostationarity is intimately related to the excess band-width α while in contrast, as $\alpha \to 0$, the cyclostationarity disappears and the WSCS process reduces to a WSS process. This observation explains the larger deviation of the standard Gaussian approximation for time-limited pulses, like the rectangular pulse, with large α and its smaller deviation for band-limited pulse shapes.

6.5.5. Simulation Result Discussion

Numerical examples are presented to illustrate the accuracy of the standard Gaussian approximation with respect to the improved Gaussian approximation for performance analysis in QDS-CDMA. All the examples consider systems under uniform power conditions and MAI-limited conditions where the effect of AWGN is neglected by setting $N_0 = 0$. Estimates of the probability of bit-error P_{er} based on Monte Carlo simulations are also presented. These results denoted by $P_{er}^{MonteCarlo}$ requires of 100-1000 times the inverse of the estimated P_{er}.

Figure 6.19 plots the probability of bit error P_{er} for QDS-CDMA systems employing the GD with Sqrt-RC $\alpha = 22\%$ pulse where $N \in \{64{,}128\}$. It shows that $P_{er}^{IGA}, P_{er}^{SIGA}$, P_{er}^{SGA}, and $P_{er}^{MonteCarlo}$ are essentially identical. For example, with $N = 128$, the four estimates return $P_{er} = 6.3 \times 10^{-5}$ at $K = 20$. Also, comparison with QDS-CDMA system employing the Neyman-Pearson receiver is presented. Comparative analysis demonstrates superiority in favour of the GD.

Figure 6.20 plots the probability of bit error P_{er} versus K for the Sqrt-RC $\alpha = 100\%$ pulse with $N \in \{64,128\}$ and compares QDS-CDMA against B-CDMA. Again, it shows that P_{er}^{IGA}, P_{er}^{SIGA}, P_{er}^{SGA}, and $P_{er}^{MonteCarlo}$ are essentially identical and that the standard Gaussian approximation is still a very good approximation for based on GD with α as large as 100%. It also shows that as α increases, the probability of bit error P_{er} performance improves as would be expected.

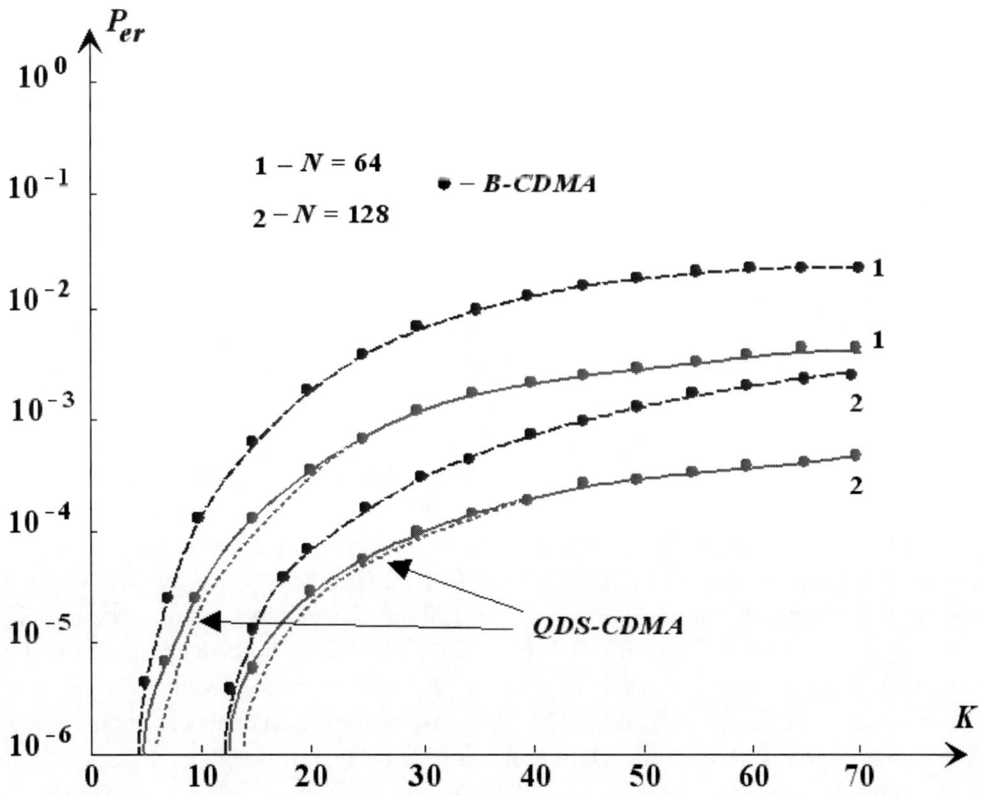

Figure 6.20. P_{er} versus K in QDS-CDMA. P_{er}^{IGA}, P_{er}^{SIGA}, P_{er}^{SGA} are very close to each other, $\alpha = 100\%$.

6.6. Optimal Waveforms for MIMO Systems

6.6.1 Brief Review

Recent advances in linear amplifier and arbitrary waveform generation technology, and the ever-increasing processing power, have spawned interest in the development of radars employed by wireless communication systems, which attempt to make full use of the spatial and temporal degrees of freedom available to the radar transmitter. These technological advances make it possible to consider the design of radar systems that allow the transmitter full flexibility to select the transmitted waveform within the given bandwidth and power constraints on a pulse-by-pulse and antenna-by-antenna basis. The flexibility to use a multiplicity of transmitted waveforms and of adaptively adjusting these waveforms offers significant performance advantages. For example, the technique based on employment of probing signal transmission in multiple-input multiple-output (MIMO) radar system is discussed in [84].

Fundamentally, the additional degrees of freedom afforded by the ability to vary the transmit waveform can be used to optimize a desired performance criterion. For example, the waveform can be adapted to the target signature to enhance detectability, to increase clutter or interference rejection, or to improve robustness to multipath. To put these works into perspective we note that the radar design is driven by the assumed models of the target and the interference-plus-noise environment. Targets are often modeled, as a rule, as point scatterers. However, since the resolution of radar systems increases, the better model is that of an extended target that is spread in range, azimuth, and Doppler. The target model can be deterministic or statistical: the former assumes that the target characteristics are fixed and known, possibly up to some unknown parameters, which can be estimated, while the latter treats the target as a random variable and attempts to characterize its statistics. Similarly, different models can be used for the interference environment, for example, the clutter, jamming, nearby targets, etc.

The work on optimum transmit-receive design in [85]–[87], for example, assumes a deterministic target model with a range spread, using a single transmit antenna, or an antenna with multiple polarization modes [88]. Optimal waveform design for single antenna radar is studied in [89]. A signal subspace framework that allowed the derivation of the optimal radar waveform for a given scenario and to evaluate the corresponding radar performance is discussed in [89]. Recently there has been considerable interest in radar systems employing the multiple antennas at both the transmitter and receiver and performing space-time processing on both, commonly referred to as MIMO radar. The concept of MIMO radar allows each transmitting antenna element to transmit an arbitrary waveform. This provides extra degrees of freedom compared to the traditional transmit beamforming approach. Joint optimization of waveforms and receiving filters in the MIMO radar systems for the case of extended target in clutter is discussed in [90]. The case of the signal dependent noise (clutter) for two scenarios, namely, the first is based on assumption that different antennas are able to find the uncorrelated aspects of the target and the second is based on the correlated target with dependence between the clutter and signal is considered in [91].

This work has focused almost entirely on the point target model, and assumes transmission of orthogonal signals on the different antennas. This make it possible to separate the signals arriving from the different transmit antenna at the receiver, and to perform any transmit array processing functions on the receive side "after the fact." For example, one can scan the transmit beam across the illuminated area within a single dwell time, or perform adaptive beamforming to reduce interference and improve resolution [92]–[98]. No-te however that the coherent transmitter array gain is lost when doing the transmit beamforming after, rather than during, transmission. Employing adaptive processing it is possible to improve clutter rejection in ways that are not possible in conventional radar [99] and [100].

MIMO radar can also provide angular diversity, which is useful in some scenarios [101]–[103]. MIMO radar systems with widely separated antennas provide spatial diversity by viewing the targets from different angles. A novel approach to accurately estimate properties, for instance, the position and velocity of multiple targets using such systems by employing sparse modeling is discussed in [104]. New metric to analyze the performance of the radar system and an adaptive mechanism for optimal energy allocation at the different transmit antennas are proposed in [104], also. In this case, the adaptive energy allocation mechanism significantly improves in performance over MIMO radar systems that transmit fixed equal energy across all the antennas.

In this section, we investigate the waveform designing problem for MIMO radar systems employing the GD. We study the case of an extended target, and without limiting ourselves to orthogonal waveforms [105]. Instead, we use a procedure discussed in [106] to design the optimal waveform, which maximizes the *SINR* at the GD output. The optimal waveform requires knowledge of both the target and clutter statistics. We also study a development of several suboptimal waveforms for radar systems employing the GD. This development requires knowledge of the target statistics only, clutter statistics only, or both.

In this section, we present in some detail a model for the radar signal, which incorporates the waveforms transmitted by the antenna array elements. We derive the optimal GD for the received signals assuming, that statistics of the radar return are completely known. We study how to maximize the *SINR* at the GD output and derive iterative algorithms for computing the *SINR* maximizing the transmit waveform. We also investigate several suboptimal waveform design algorithms. Theoretical study is strengthened by simulation results illustrating the performance gains achievable by adaptive waveform design compared with conventional radar waveforms and compare obtained results with radar systems employing the well-known generalized minimum variance distortionless response detector [106].

6.6.2. Problem Formulation

We consider MIMO radar employing M antennas at the transmitter and N antennas at the receiver. Assume that the radar operates at a bandwidth B, and all the baseband signals of interest are sampled at a rate f_s greater than the Nyquist rate, i.e., $f_s > 2B$. If the radar operates using a pulse repetition rate f_{prf}, we need $N_s = f_s / f_{prf}$ samples to fully represent the signal over a one pulse-to-pulse interval. Thus, all of the signals of interest will be represented by vectors of length N_f.

It is more convenient to use a frequency domain representation. All the signals discussed below are the Fourier transforms of the collected samples. Let $\mathbf{s}_k$ be the $N_f \times 1$ vector representing the signal transmitted at the k-th antenna, where $k = 1, \ldots, M$. These vectors can be stacked into a single transmit vector $\mathbf{s}$ of size $MN \times 1$, i.e.,

$$\mathbf{s}_n = [\mathbf{s}_1^T, \ldots, \mathbf{s}_M^T]^T, \tag{6.204}$$

The radar illuminates a set of l scatterers located at coordinates (x_i, y_i, z_i) where $i = 1,\ldots,l$ which have complex amplitudes h_i. These amplitudes can be assembled into $l \times 1$ vector **h.** The radar illumination at the location of the ith scatterer is denoted by g_i. Thus, the radar signal reflected from this scatterer towards the receiver takes the form $h_i g_i$. In the vector form we can write that the vector **g** is the $l \times 1$ illumination vector and the reflected signal is $\mathbf{h} \otimes \mathbf{g}$, where $\otimes$ denotes the element-by-element vector or matrix multiplication.

Assume each target can be considered of so many points. A point target is assumed between each pair of transmitter-receiver antennas so that the received signal component at the ith receiver due to the kth transmitter can be considered at the instant n as

$$y_{ik}(n) = g_{ik} \times h_k(n) \ , \tag{6.205}$$

where g_k is the path gain from the kth transmitter to the ith receiver.For narrow bandwidth waveforms, the point target model is often valid; see, for instance [107] and [108].

The received signal at the ith receiver is the superposition of all the signals originating from various transmitters plus the additive noise. Denote by $y_i(n)$ the received signal and by $w(n)$ the additive Gaussian noise at the ith receiver input and then $y_i(n)$ takes the following form

$$y_i(n) = \sum_{k=1}^{M} \mathbf{g}_{ik} \mathbf{h}_k(n) + w \ , \tag{6.206}$$

where

$$\mathbf{g}_l^T = [g_{l1}, g_{l2}, \ldots, g_{lM}] \ . \tag{6.207}$$

Denote by

$$\mathbf{y}_n = [y_1(n), \ldots, y_l(n)] \ . \tag{6.208}$$

As the collection of the received signals at the various receiving elements at the instant n, the received signal vector can be described by the following model:

$$\mathbf{y}_n = \mathbf{g}\,\mathbf{h}_n + \mathbf{w}_n \ , \tag{6.209}$$

where $M \times N$ matrix

$$\mathbf{g} = [\mathbf{g}_1, \mathbf{g}_2, \ldots, \mathbf{g}_N]^H \tag{6.210}$$

is the target scattering matrix similar to the channel matrix in [109] and [110]. Since entries of **g** are the path gains related to different aspects of the target, this matrix can be seen as the extended model of the target. It has been shown that **g** matrix entries obey to the Gaussian distribution law.

In order to estimate **g**, let the finite length signal $s_k(n)$ by the length $l \geq N$ be transmitted from each element. Due to the $N \times l$ transmitted matrix

$$\mathbf{S} = [\mathbf{s}_1, \mathbf{s}_2, \ldots, \mathbf{s}_l] \ , \tag{6.211}$$

The $N \times l$ received matrix **y** can be expanded as

$$\mathbf{y} = \mathbf{gS} + \mathbf{w} \ , \tag{6.212}$$

where **w** is the noise matrix defined by

$$\mathbf{w} = [\mathbf{w}_1, \mathbf{w}_2, \ldots, \mathbf{w}_l] \ . \tag{6.213}$$

The *i*th row of **y** indicates the received signal throughout *N* samples by the *l*th element.

Alternatively, the reflected signal vector can be written in the form of diagonal matrix $diag\{\mathbf{h}\}\mathbf{g}$ o r $diag\{\mathbf{g}\}\mathbf{h}$, $diag\{\mathbf{x}\}$ denotes a diagonal matrix whose diagonal elements are the elements of the vector **x**. The illumination vector is a linear function of the transmit signal and can be written as $\mathbf{g} = \mathbf{A}_M\mathbf{s}$, where $\mathbf{A}_M$ is a $l \times MN_f$ matrix whose elements are the gains from a particular transmit antenna to a scatterer location, at a specific frequency. Similarly, we define $\mathbf{A}_n$ as the $NN_f \times l$ matrix whose elements are the gains from a given scatterer location to a receive antenna, at a specific frequency.

The received signal is the reflected signal multiplied by this gain. Thus, the $NN_f \times 1$ vector **y** of received signals can be written in the following form:

$$\mathbf{y} = \mathbf{A}_N diag\{\mathbf{h}\}\mathbf{g} = \underbrace{\mathbf{A}_N diag\{\mathbf{h}\}\mathbf{A}_M}_{\mathbf{A}_s}\mathbf{s} = \mathbf{A}_s\mathbf{s} \tag{6.214}$$

or equivalently as

$$\mathbf{y} = \mathbf{A}_N diag\{\mathbf{g}\}\mathbf{h} = \underbrace{\mathbf{A}_N diag\{\mathbf{g}\}}_{\mathbf{A}_h}\mathbf{h} = \mathbf{A}_h\mathbf{s} \ . \tag{6.215}$$

These equations simply state that the received radar signal is a bilinear function of the transmitted signal and of the scatterer reflectivity. The gain matrices $\mathbf{A}_M$ and $\mathbf{A}_N$ can be written explicitly for a given radar-scatterer geometry, i.e., given the coordinates of all the transmitted and received antenna phase centers and the scatterer coordinates. If the antenna elements are not omnidirectional, their radiation patterns are also required. The detailed structure of $\mathbf{A}_M$ and $\mathbf{A}_N$ is discussed in [106] for the case where the transmitted and received arrays are collocated. The scatterer amplitudes are usually represented as a random process. In the following, we assume that **h** is a multivariate complex Gaussian vector with zero mean and covariance $\mathbf{R}_h$, i.e., $\mathbf{h} \approx \mathcal{CN}(0, \mathbf{R}_h)$.

The covariance matrix $\mathbf{R}_h$ is assumed to have a low rank decomposition

$$\mathbf{R}_h = \mathbf{V}_h \mathbf{V}_h^H \ , \tag{6.216}$$

where $\mathbf{V}_h$ is a $l \times N$ matrix, N being the rank of $\mathbf{R}_h$. The covariance of the received signal is then given by

$$\mathbf{R}_x = \mathbf{A}_h \mathbf{R}_h \mathbf{A}_h^H \ . \tag{6.217}$$

The reflection from a target is almost always accompanied by reflections from the surrounding environment, e.g. ground, ocean, etc., referred to as the clutter. In the following discussion, we need to distinguish between the radar signal returned from the target and from clutter. We denote the target-related quantities by the superscript t, namely, $\mathbf{x}_t$, $\mathbf{h}_t$, $\mathbf{R}_{xt}, \mathbf{V}_{xt}$. The clutter related quantities are denoted by the subscript c, namely $\mathbf{x}_c, \mathbf{h}_c$, $\mathbf{R}_{xc}, \mathbf{V}_{xc}$. The complete signal model for the received radar signal $\mathbf{y}$ is then given by the following form

$$\mathbf{y} = \mathbf{g}_t \mathbf{S} + \mathbf{g}_c \mathbf{S} + \mathbf{w} = \mathbf{x}_t + \mathbf{x}_c + \mathbf{w} \ , \tag{6.218}$$

where $\mathbf{x}_t$ is the target signal, $\mathbf{x}_c$ is the clutter signal; $\mathbf{w}$ is a noise vector where $\mathbf{w} \approx \mathcal{CN}(0, \sigma_n^2 \mathbf{I})$ and $\mathbf{I}$ is the identity matrix; $\mathbf{g}_t$ and $\mathbf{g}_c$ are assumed to be the Gaussian distributed matrices with zero mean and covariance $\mathbf{R}_{xt}$ and $\mathbf{R}_{xc}$, respectively.These covariance matrices are defined by

$$\begin{cases} \mathbf{R}_{xt} = E\{\mathbf{g}_t^H \mathbf{g}_t\}, \\ \mathbf{R}_{ct} = E\{\mathbf{g}_c^H \mathbf{g}_c\}, \end{cases} \tag{6.219}$$

where $E\{\cdot\}$ denotes the mean.

6.6.3. Detection by GD

Given the signal model above we formulate the target detection problem as the following Gauss-Gauss binary hypothesis-testing problem:

$$\begin{cases} \mathcal{H}_0 \Rightarrow \mathbf{y} \approx \mathcal{CN}(0, \mathbf{R}_{xc} + \sigma_n^2 \mathbf{I}) \ , \\ \mathcal{H}_1 \Rightarrow \mathbf{y} \approx \mathcal{CN}(0, \mathbf{R}_{xt} + \mathbf{R}_{xc} + \sigma_n^2 \mathbf{I}) \ , \end{cases} \tag{6.220}$$

where $\mathcal{CN}$ denotes the multivariate complex Gaussian distribution. The GD for the case where the statistics $(\mathbf{R}_{xt}, \mathbf{R}_{xc}, \sigma_n^2)$ are known has the following decision statistic:

$$Z = 2\mathbf{x}_t^H \mathbf{Q}\mathbf{y} - \mathbf{y}^H \mathbf{Q}\mathbf{y} + \xi_{AF}^H \mathbf{R}_{xcn} \xi_{AF} \ , \tag{6.221}$$

where the $nN_f \times 1$ vector ξ_{AF} represents the reference noise forming by the linear system at the GD front end (see (6.16)).

As we can see from (6.221)

$$\mathbf{Q} = \mathbf{R}_{xcn}^{-1} - [\mathbf{V}_{xt}\mathbf{V}_{xt}^{T} + \mathbf{R}_{xcn}]^{-1} . \tag{6.222}$$

Here

$$\mathbf{R}_{xcn} = \mathbf{R}_{xc} + \sigma_n^2\mathbf{I} \tag{6.223}$$

is the clutter-plus-noise covariance matrix. This can be equivalently written in the following form:

$$\mathbf{Q} = \mathbf{R}_{xcn}^{-1}\mathbf{V}_{xt}[\mathbf{V}_{xt}^{H}\mathbf{R}_{xcn}^{-1}\mathbf{V}_{xt} + \mathbf{I}]^{-1}\mathbf{V}_{xt}^{H}\mathbf{R}_{xcn}^{-1} = \mathbf{W}\mathbf{W}^{H} , \tag{6.224}$$

where

$$\mathbf{W} = \mathbf{R}_{xcn}^{-1}\underbrace{\mathbf{V}_{xt}[\mathbf{V}_{xt}^{H}\mathbf{R}_{zcn}^{-1}\mathbf{V}_{xt} + \mathbf{I}]^{-0.5}}_{\tilde{\mathbf{V}}_{xt}} . \tag{6.225}$$

Then the detection statistic of GD given by (6.221) can be rewritten in the following form

$$Z = 2\mathbf{x}_t^H\mathbf{W}\mathbf{W}^H\mathbf{y} - \mathbf{y}^H\mathbf{W}\mathbf{W}^H\mathbf{y} + \xi_{AF}^H\mathbf{R}_{xcn}^{-1}\xi_{AF} = 2\mathbf{x}_t^H\mathbf{W}\mathbf{W}^H\mathbf{y} - \| \mathbf{W}^H\mathbf{y} \|^2 + \xi_{AF}^H\mathbf{R}_{xcn}^{-1}\xi_{AF} . \tag{6.226}$$

In the case of a unit rank target covariance matrix, $\mathbf{V}_{xt}$ is a vector. To emphasize this fact we denote it by the lower case symbol $\mathbf{v}_{xt}$. In this case,

$$\mathbf{W} = \mathbf{R}_{xcn}^{-1}\mathbf{v}_{xt} , \tag{6.227}$$

where we discarded a scalar multiplier term that can be absorbed into the detection threshold. We may say that in this case the GD is constructed using the principles of the well-known minimum variance distortionless response (MVDR) detector [111].

The more general case where the rank is greater than unity is the generalized MVDR (GMVDR) detector discussed in [112]. The case of unit rank $\mathbf{R}_{xt}$ corresponds to a rigid target with fixed orientation relative to the radar. Motion relative to the radar will introduce the same random phase fluctuations to the returns from all of the scatterers comprising the target, so that the target scatters can be represented by $\mathbf{h}_t \exp(j\varphi)$, where $\mathbf{h}_t$ is a deterministic vector and φ is a random phase.

In this case,

$$\mathbf{R}_{ht} = \mathbf{v}_t\mathbf{v}_t^H , \tag{6.228}$$

where $\mathbf{v}_t$ equals to $\mathbf{h}_t$ up to an arbitrary unit magnitude complex scalar. A target that is not a rigid collection of point scatterers, or is one that is rotating relative to the radar will induce different random variations along the vector $\mathbf{h}_t$ and the corresponding covariance matrix $\mathbf{R}_{ht}$ defined above will have an effective rank $n > 1$.

6.6.4. SINR Analysis

Because both the target and clutter statistics $(\mathbf{R}_{xt}, \mathbf{R}_{xc})$ depend on the transmit signal **s**, the matrix **W** defining the GD structure is a function of the transmit waveform **s**. We want to select this waveform so, as to maximize the performance of the GD detector. This can be achieved by maximizing the *SINR* at the GD output.

It follows from (6.217) that the *SINR* is given by

$$SINR = \frac{\mathrm{tr}\{\mathbf{W}^H \mathbf{R}_{xt} \mathbf{W}\}}{\mathrm{tr}\{\xi_{AF}^H \mathbf{R}_{xcn}^{-1} \xi_{AF} - \xi_{PF}^H \mathbf{R}_{xcn}^{-1} \xi_{PF}\}} \tag{6.230}$$

or equivalently

$$SINR = \frac{\mathrm{tr}\{\widetilde{\mathbf{V}}_{xt}^H \mathbf{R}_{xcn}^{-1} \mathbf{R}_{xt} \mathbf{R}_{xcn}^{-1} \widetilde{\mathbf{V}}_{xt}\}}{\mathrm{tr}\{\xi_{AF}^H \mathbf{R}_{xcn}^{-1} \xi_{AF} - \xi_{PF}^H \mathbf{R}_{xcn}^{-1} \xi_{PF}\}}, \tag{6.231}$$

where $\mathrm{tr}\{\cdots\}$ is the trace operation of matrix. This *SINR* is a nonlinear function of the elements of **s** and, in general, we must resort to numerical optimization techniques to solve for the optimal transmit waveform **s**.

To gain some insight into this problem we consider next the case of a unit rank target covariance matrix. In the unit rank case, the matrix **W** given by (6.228) is a vector and the trace operation in (6.230) is not needed. We can then rewrite (6.230) in the following form

$$SINR = \frac{\mathbf{v}_{xt}^H \mathbf{R}_{xcn}^{-1} (\mathbf{v}_{xt} \mathbf{v}_{xt}^H) \mathbf{R}_{xcn}^{-1} \mathbf{v}_{xt}}{\xi_{AF}^H \mathbf{R}_{xcn}^{-1} \xi_{AF} - \xi_{PF}^H \mathbf{R}_{xcn}^{-1} \xi_{PF}}. \tag{6.232}$$

Because the numerator is the squared value of the $\mathbf{v}_{xt}^H \mathbf{R}_{xcn}^{-1} \mathbf{v}_{xt}$, we can write (6.232) in the following form:

$$SINR = \frac{\{\mathbf{v}_{xt}^H [\mathbf{V}_{xc} \mathbf{V}_{xc}^H + \sigma_n^2 \mathbf{I}]^{-1} \mathbf{v}_{xt}\}^2}{\xi_{AF}^H \mathbf{R}_{xcn}^{-1} \xi_{AF} - \xi_{PF}^H \mathbf{R}_{xcn}^{-1} \xi_{PF}}. \tag{6.233}$$

This can also be written as

$$SINR = \frac{(\mathbf{v}_{xt}^H)^2 \left\{ \mathbf{I} - \frac{1}{\sigma_n^2} \mathbf{V}_{xc} \left[\frac{1}{\sigma_n^2} \mathbf{V}_{xc}^H \mathbf{V}_{xc} + \mathbf{I} \right]^{-1} \mathbf{V}_{xc}^H \right\} (\mathbf{v}_{xt})^2}{\sigma_n^2 (\xi_{AF}^H \xi_{AF} - \xi_{PF}^H \xi_{PF})} . \tag{6.234}$$

If the clutter is much stronger than the noise, we can write this as

$$SINR = \frac{(\mathbf{v}_{xt}^H)^2 \mathbf{P}_{V_{xc}}^{\perp} (\mathbf{v}_{xt})^2}{4\sigma_n^4} , \tag{6.235}$$

where

$$\mathbf{P}_{V_{xc}}^{\perp} = \mathbf{I} - \mathbf{V}_{xc} (\mathbf{V}_{xc}^H \mathbf{V}_{xc})^{-1} \mathbf{V}_{xc}^H \tag{6.236}$$

is the orthogonal projection operator onto the column space of $\mathbf{V}_{xc}$.

The *SINR* expression above has the intuitive interpretation of being the energy of the target component, which is orthogonal to the clutter, i.e., what is left after completely removing the clutter component from the target, divided by the noise energy. In the following discussion, we consider only the unit-rank target covariance. We note, however, that all of our results can be extended to the general low rank case. An optimal transmit waveform **s** can be derived by maximizing the *SINR* over all possible choices of **s**. This optimization can be carried out using, for example, the numerical gradient descent algorithm. Consider the main statements of the numerical gradient descent algorithm.

6.6.5. Gradient Descent Algorithm

To solve the *SINR* maximization problem we need a more explicit expression of the *SINR* in terms of **s**. Recall that

$$\mathbf{x}_{st} = \mathbf{A}_{st}\mathbf{s} , \tag{6.237}$$

where

$$\mathbf{A}_{st} = \mathbf{A}_n diag\{\mathbf{h}_t\} \mathbf{A}_m . \tag{6.238}$$

In the unit-rank case $\mathbf{h}_t$ is deterministic up to a random scalar phase term, so $\mathbf{h}_t$ can be replaced by $\mathbf{v}_t$, where

$$\mathbf{R}_{ht} = \mathbf{v}_t \mathbf{v}_t^H . \tag{6.239}$$

In other words, for a unit-rank target covariance we have

$$\mathbf{A}_{st} = \mathbf{A}_n diag\{\mathbf{v}_t\}\mathbf{A}_m . \tag{6.240}$$

Recall that

$$\mathbf{R}_{xc} = \sum_{i=1}^{r} \mathbf{A}_{xc}[i]\mathbf{s}\mathbf{s}^H \mathbf{A}_{sc}^H[i] . \tag{6.241}$$

Inserting (6.241) into the *SINR* expression we obtain

$$SINR = \frac{\mathbf{v}_{xt}^H \mathbf{R}_{xcn}^{-1} \mathbf{v}_{xt}}{\xi_{AF}^H \xi_{AF} - \xi_{PF}^H \xi_{PF}} = \frac{\mathbf{s}^H \mathbf{A}_{st}^H \mathbf{R}_{xcn}^{-1} \mathbf{A}_{st}\mathbf{s}}{\xi_{AF}^H \xi_{AF} - \xi_{PF}^H \xi_{PF}} ; \tag{6.242}$$

$$\mathbf{R}_{xcn} = \sum_{i=1}^{r} \mathbf{A}_{sc}[i]\mathbf{s}\mathbf{s}^H \mathbf{A}_{sc}^H[i] + 4\sigma_n^4 \mathbf{I} . \tag{6.243}$$

Various numerical optimization techniques can be used to solve for the transmit waveform **s**, which maximizes the *SINR*. Here we consider the gradient descent method discussed in [113] that require knowledge of the derivatives of *SINR* with respect to the elements of **s**. The update equation of the gradient descent method is given by

$$\mathbf{s} \leftarrow \mathbf{s} + \mu \left[\frac{\partial (SINR)}{\partial \mathbf{s}} \right]^H , \tag{6.244}$$

where μ is a constant controlling the convergence rate of the algorithm. Taking the derivative of the *SINR* with respect to s_i we obtain

$$\frac{\partial (SINR)}{\partial s_i} = \frac{1}{\xi_{AF}^H \xi_{AF} - \xi_{PF}^H \xi_{PF}} \left[2\mathbf{s}^H \mathbf{A}_{st}^H \mathbf{R}_{xcn}^{-1} \mathbf{A}_{st}\mathbf{e}_i - 2\mathbf{s}^H \mathbf{A}_{st}^H \mathbf{R}_{xcn}^{-1} \frac{\partial \mathbf{R}_{xcn}}{\partial s_i} \mathbf{R}_{xcn}^{-1} \mathbf{A}_{st}\mathbf{s} \right] , \tag{6.245}$$

where s_i is the i-th entry of **s** and $\mathbf{e}_i$ is a vector of zeros with a 1 at the i-th position. Examples of calculating this type of derivatives are presented in [114] and [115].

Define $\mathbf{R}_{xc}$ in the following form:

$$\mathbf{R}_{xc} = \sum_{i=1}^{r} \mathbf{A}_h \mathbf{v}_c[i]\mathbf{v}_c^H[i]\mathbf{A}_h^H . \tag{6.246}$$

Note that

$$\frac{\partial \mathbf{R}_{xcn}}{\partial s_i} = \frac{\partial \mathbf{R}_{xc}}{\partial s_i} . \tag{6.247}$$

In this case, we obtain

$$\frac{\partial \mathbf{R}_{xcn}}{\partial s_i} = \sum_{n=1}^{r} \mathbf{A}_{sc}[n]\mathbf{e}_i \mathbf{s}^H \mathbf{A}_{sc}^H[n] \ . \tag{6.248}$$

Then (6.245) can be represented in the following form

$$\frac{\partial (SINR)}{\partial s_i} = \frac{2p_i - 2q_i}{\xi_{AF}^H \xi_{AF} - \xi_{PF}^H \xi_{PF}} \ , \tag{6.249}$$

Where p_i are the elements of the row vector

$$\mathbf{p} = \mathbf{s}^H \mathbf{A}_{st}^H \mathbf{R}_{xcn}^{-1} \mathbf{A}_{st} \tag{6.250}$$

and

$$\mathbf{q}_i = \mathbf{s}^H \mathbf{A}_{st}^H \mathbf{R}_{xcn}^{-1} \left[\sum_{n=1}^{r} \mathbf{A}_{sc}[n]\mathbf{e}_i \mathbf{s}^H \mathbf{A}_{sc}^H[n] \right] \mathbf{R}_{xcn}^{-1} \mathbf{A}_{st} \mathbf{s} \ . \tag{6.251}$$

Finally, collecting the derivatives into a row vector we obtain

$$\frac{\partial (SINR)}{\partial \mathbf{s}} = \frac{2(\mathbf{p} - \mathbf{q})}{\xi_{AF}^H \xi_{AF} - \xi_{PF}^H \xi_{PF}} \ . \tag{6.252}$$

Equations (6.244), (6.250)÷(6.252) define the gradient descent algorithm for computing the *SINR*-maximizing waveform **s**. To initialize the algorithm we let **s** be one of the suboptimal waveforms described in Section 6.6.6.

6.6.5.1. Iterative Optimization

A second optimization technique involves iterating between updating *SINR* in (6.251) assuming **s**, and consequently $\mathbf{R}_{xcn}$, is fixed, and then updating $\mathbf{R}_{xcn}$. More specifically, this algorithm proceeds as follows:

- Maximize

$$SINR = \frac{1}{\xi_{AF}^H \xi_{AF} - \xi_{PF}^H \xi_{PF}} \times \underbrace{\mathbf{s}^H \mathbf{A}_{st}^H \mathbf{R}_{xcn}^{-1} \mathbf{A}_{st} \mathbf{s}}_{\mathbf{Q}_1} \tag{6.253}$$

assuming $\mathbf{R}_{xcn}$ is fixed. This is accomplished by letting **s** be the eigenvector corresponding to the largest eigenvalue of $\mathbf{Q}_1$;

- Use the resulting **s** to update $\mathbf{R}_{xcn}$ via (20);
- Repeat until convergence. To initialize the algorithm we let **s** be one of the suboptimal waveforms described in Section 6.6.6.

6.6.6. Suboptimal Waveforms

The optimal waveform requires knowledge of the target and clutter statistics, namely, $\mathbf{R}_{ht}$ and $\mathbf{R}_{hc}$. The waveforms considered here provide suboptimal performance, but, with one exception, require less information about the target and clutter than the optimal waveform. The transmit signal will be normalized to unit energy in all the different waveforms we consider to allow for a fair comparison of the results. In other words, the transmitted signal vector **s** must obey the constraint $\| \mathbf{s} \|^2 = 1$.

6.6.6.1. Standard Waveform

As a reference waveform against which to compare the performance achieved using optimal and suboptimal waveforms, we selected a linear frequency-modulated (FM) waveform (chirp), which is commonly used in radar systems, fed to a single transmit antenna. Without loss of a generality, we can assume that the first antenna is active while the other antennas have zero input. Thus, we can write the standard transmit waveform **s** in the following form:

$$\mathbf{s} = \frac{1}{\sqrt{N_f}} \left[f(1), 0, \ldots, 0, f(N_f), 0, \ldots, 0 \right]^T , \tag{6.254}$$

where $f(n)$ are the values of the discrete Fourier transform of the signal.

It is possible, of course, to select other reference waveforms. For example, we can use a linear FM waveform fed to all antennas with an appropriate set of complex weights so, as to form a transmit beam in a desired direction. However, the resulting beam will illuminate only a portion of an extended target and the results will be highly dependent on the orientation of the beam relative to the target scatterers. We prefer instead to use for reference a system employing a single transmit antenna, so that the entire area of interest is uniformly illuminated.

6.6.6.2. Target-Based Waveforms

Consider the case, where we know the target statistics but not the clutter statistics. This case corresponds, for instance, to the situation where the radar is searching for speci-fic targets with known signatures. The target signature $\mathbf{v}_t$ represents the azimuth-frequency target scattering function. By designing **s**, we can control the azimuth-frequency radar illumination of the target.

In the following, we consider two ways to design the illuminating function. In the first, we focus all of the radar energy to illuminate the azimuth-frequency element of the target, i.e., the component of $\mathbf{v}_t$, which has the largest magnitude. The second approach is to match the illuminating function to the target scattering function. In other words, put more of the radar energy, where the scattering is strong, and less, where it is weak. The second approach will produce less target energy at the receiver than the first approach. In the case, where the clutter energy is uniformly distributed in azimuth-frequency, the first approach is expected to provide the better performance. However, given a non-uniform clutter distribution the second approach might be preferred.

Consider, for example, the case, where the azimuth-frequency cell, where the target scattering has its largest magnitude, happens to also have a strong clutter component. In this case, the better performance will be obtained by illuminating a part of the target scattering function, where the clutter is weak. Since we assume here that the clutter distribution is unknown, the second approach will be more robust than the first. Both of these approaches are made more precise next.

Maximizing Target Energy. Consider designing of the transmitted waveform **s** so that the received target energy is maximized. In other words, we want to maximize $\mathbf{W}^H\mathbf{R}_{xt}\mathbf{W}$ in the case, where

$$\mathbf{R}_{ht} = \mathbf{v}_t\mathbf{v}_t^H \tag{6.255}$$

is the unit rank, we have

$$\mathbf{W}^H\mathbf{R}_{xt}\mathbf{W} = \mathbf{W}^H(\mathbf{A}_h\mathbf{v}_t\mathbf{v}_t^H\mathbf{A}_h^H)\mathbf{W} = \| \mathbf{W}^H\mathbf{A}_h\mathbf{v}_t \|^2 \tag{6.256}$$

or

$$\mathbf{W}^H\mathbf{R}_{xt}\mathbf{W} = \mathbf{W}^H(\mathbf{A}_{st}\mathbf{s}\mathbf{s}^H\mathbf{A}_{st}^H)\mathbf{W} = \| \mathbf{W}^H\mathbf{A}_{st}\mathbf{s} \|^2 , \tag{6.257}$$

where

$$\mathbf{A}_{st} = \mathbf{A}_n diag\{\mathbf{v}_t\}\mathbf{A}_m . \tag{6.258}$$

Clearly $\mathbf{W}^H\mathbf{R}_{xt}\mathbf{W}$ will be maximized if **W** and **s** are chosen to be left and right eigenvectors of $\mathbf{A}_{st}$ corresponding to its largest eigenvalue. Without loss of generality, we assume that **W** is normalized to have unit norm, so as to keep the noise variance at the GD output constant. We can interpret this result as follows. The vector $\mathbf{v}_t$ represents the azimuth-frequency distribution of the target scattering function. The maximizing solution focuses the transmit energy on the azimuth-frequency portion of the target, which provides the strongest radar return.

Matching the Target Distribution. Consider the illumination vector $\mathbf{g} = \mathbf{A}_m\mathbf{s}$, which was introduced earlier. We now design **s**, so as to have the illumination function match the target scattering function that is to have $\mathbf{g} = \mathbf{v}_t$. In other words, let $\mathbf{A}_m\mathbf{s} = \mathbf{v}_t$. The target matching transmit waveform **s** is the least squares solution of this equation,

$$\mathbf{s} = (\mathbf{A}_m^H\mathbf{A}_m)^{-1}\mathbf{A}_m^H\mathbf{v}_t . \tag{6.259}$$

6.6.6.3. Minimizing Clutter Energy

Now, consider the case where we know the clutter statistics but not the target statistics. This will be the situation if we estimate the clutter covariance from measurement assumed to be target free, but do not have prior knowledge of the target characteristics. Consider designing $\mathbf{s}$ so that the received clutter energy is minimized, i.e., we want to minimize $\mathbf{W}^H\mathbf{R}_{xc}\mathbf{W}$. Recall that

$$\mathbf{R}_{xc} = \mathbf{A}_h\mathbf{R}_{hc}\mathbf{A}_h^H , \tag{6.260}$$

where

$$\mathbf{R}_{hc} = \mathbf{V}_c\mathbf{V}_c^H . \tag{6.261}$$

Let

$$\mathbf{V}_c = \{\mathbf{v}_c[1],\ldots,\mathbf{v}_c[r]\}, \tag{6.262}$$

where r is the rank of the clutter covariance matrix, and $\mathbf{v}_c[n]$ are the columns of $\mathbf{V}_c$. Note that

$$\mathbf{A}_h\mathbf{v}_c[i] = \mathbf{A}_{sc}[i]\mathbf{s}, \tag{6.263}$$

where

$$\mathbf{A}_{sc}[i] = \mathbf{A}_n diag\{\mathbf{v}_c[i]\}\mathbf{A}_m . \tag{6.264}$$

Referring to (6.246) we obtain

$$\mathbf{R}_{xc} = \sum_{i=1}^{r}\mathbf{A}_{sc}[i]\mathbf{s}\mathbf{s}^H\mathbf{A}_{sc}^H[i] . \tag{6.265}$$

We can write

$$\mathbf{W}^H\mathbf{R}_{xc}\mathbf{W} = \mathbf{W}^H\underbrace{\Big(\sum_{i=1}^{r}\mathbf{A}_{sc}[i]\mathbf{s}\mathbf{s}^H\mathbf{A}_{sc}^H[i]\Big)}_{\mathbf{Q}_{sc}}\mathbf{W} \tag{6.266}$$

or

$$\mathbf{W}^H\mathbf{R}_{xc}\mathbf{W} = \mathbf{s}^H\underbrace{\Big(\sum_{i=1}^{r}\mathbf{A}_{sc}[i]\mathbf{W}\mathbf{W}^H\mathbf{A}_{sc}^H[i]\Big)}_{\mathbf{Q}_{wc}}\mathbf{s} . \tag{6.267}$$

We want to minimize $\mathbf{W}^H\mathbf{R}_{xc}\mathbf{W}$ jointly over $\mathbf{s}$ and $\mathbf{W}$ subject to the norm constraints

$$\|\mathbf{s}\|^2 = 1 \text{ and } \|\mathbf{W}\|^2 = 1. \tag{6.268}$$

Note that if **W** is known then **s**, which minimizes the received clutter power, is the eigenvector of $\mathbf{Q}_{wc}$ corresponding to its smallest eigenvalue. If **s** is known then **W**, which minimizes the received clutter power, is the eigenvector of $\mathbf{Q}_{sc}$ corresponding to its smallest eigenvalue. This suggests an iterative solution procedure where we solve successively for **W** and **s**. More specifically, the following procedure is used:

a) Use an initial waveform **s**, for example, the linear FM signal;
b) Compute $\mathbf{Q}_{sc}$ corresponding to **s**;
c) Let **W** be eigenvector of $\mathbf{Q}_{sc}$ corresponding to its smallest eigenvalue;
d) Compute $\mathbf{Q}_{sc}$ corresponding to **W**;
e) Let **s** be eigenvector of $\mathbf{Q}_{wc}$ corresponding to its smallest eigenvalue;
f) Repeat steps b)-e) until convergence.

Maximize Signal-to-Interference Ratio. Finally, we consider an approach where the target-to-clutter ratio or signal-to-interference ratio (*SIR*) at the GD output is maximized rather than the *SINR*. As in the case of the optimal waveform design, both the clutter and the target statistics are known. The *SIR* is defined as

$$SIR = \frac{\mathbf{W}^H \mathbf{R}_{xt} \mathbf{W}}{\mathbf{W}^H \mathbf{R}_{xc} \mathbf{W}} . \tag{6.269}$$

Recall that the clutter energy can be alternatively written as $\mathbf{W}^H \mathbf{Q}_{xc} \mathbf{W}$ or $\mathbf{s}^H \mathbf{Q}_{xc} \mathbf{s}$. Similarly, the target energy can be written as $\mathbf{W}^H \mathbf{Q}_{st} \mathbf{W}$ or $\mathbf{s}^H \mathbf{Q}_{wt} \mathbf{s}$, where

$$\mathbf{Q}_{st} = \mathbf{A}_{st} \mathbf{s}\mathbf{s}^H \mathbf{A}_{st}^H \tag{6.270}$$

and

$$\mathbf{Q}_{wt} = \mathbf{A}_{st} \mathbf{W}\mathbf{W}^H \mathbf{A}_{st}^H . \tag{6.271}$$

Thus, the *SIR* can be written as

$$SIR = \frac{\mathbf{W}^H \mathbf{Q}_{st} \mathbf{W}}{\mathbf{W}^H \mathbf{Q}_{sc} \mathbf{W}} \tag{6.272}$$

or

$$SIR = \frac{\mathbf{s}^H \mathbf{Q}_{wt} \mathbf{s}}{\mathbf{s}^H \mathbf{Q}_{wc} \mathbf{s}} . \tag{6.273}$$

We want to maximize *SIR* jointly over all norm-constrained choices of $\mathbf{W}$ and $\mathbf{s}$. Note that if $\mathbf{s}$ is assumed known the value of $\mathbf{W}$, which maximizes the *SIR*, is the generalized eigenvector of $\{\mathbf{Q}_{st}, \mathbf{Q}_{sc}\}$ corresponding to the largest generalized eigenvalue. Similarly, if $\mathbf{W}$ is known, the value of $\mathbf{s}$, which maximizes *SIR*, is the generalized eigenvector of $\{\mathbf{Q}_{wt}, \mathbf{Q}_{wc}\}$ corresponding to the largest generalized eigenvalue. This suggests the following iterative solution procedure:

a) Use an initial waveform $\mathbf{s}$, for example, the linear FM signal;
b) Compute $\mathbf{Q}_{st}$ and $\mathbf{Q}_{sc}$ corresponding to $\mathbf{s}$.
c) Let $\mathbf{W}$ be the generalized eigenvector of $\{\mathbf{Q}_{st}, \mathbf{Q}_{sc}\}$ corresponding to its smallest eigenvalue;
d) Compute $\mathbf{Q}_{wt}$ and $\mathbf{Q}_{wc}$ corresponding to $\mathbf{W}$;
e) Let $\mathbf{s}$ be the generalized eigenvector of $\{\mathbf{Q}_{wt}, \mathbf{Q}_{wc}\}$ corresponding to its smallest eigenvalue;
f) Repeat steps b)-f) until convergence.

Note that the only difference between the *SIR* and the *SINR* is that the denominator of the *SINR* expression has an additional noise term $4\sigma_n^4 \mathbf{W}^H \mathbf{W}$. As the clutter-to-noise ratio increases, the *SIR* approaches the *SINR*. Therefore, we expect that the performance of the *SIR* maximizing waveform will be close to that of the optimal waveform for large values of the clutter-to-noise ratio.

6.6.7. Simulation and Discussion

In this section, we present some numerical examples illustrating the performance of the radar, when using the waveforms described earlier. It should be emphasized that these examples are presented only to understand the problem. To evaluate the actual performance tradeoffs of different waveform design methods requires studying them in the context of a specific radar system and a well-defined surveillance scenario, and requires addressing other issues, which are beyond the scope of the present chapter.

6.6.7.1. Conditions of Simulation

The clutter and target are modeled by a collection of scatterers placed on a grid in the azimuth/range plane. The grid consists of $N_a \times N_r$ points, where N_a is the number of azimuth sample points and N_r is the number of range sample points. Different clutter and target models correspond to different distributions of the complex scatterer amplitudes over this grid. For the examples presented here both the target and the clutter were represented by independent random scatterers whose amplitudes are complex Gaussian with zero mean and variance equal to *SNR* and the clutter-to-noise ratio (*CNR*), respectively.

In other words, we did not assume any particular structure for the target and clutter. We note however that it is straightforward to incorporate any desired clutter and target distribution into this type of simulation. The clutter scatterers were placed at all $N_a \times N_r$ grid points, while the target scatters were placed on a rectangular portion of the grid. The surveillance scenario is assumed stationary, i.e., the target and clutter scatters are not moving and neither is the radar. Therefore, both the target and clutter covariance matrices are assumed to have unit rank so that $\mathbf{R}_{ht} = \mathbf{v}_t \mathbf{v}_t^H$ and $\mathbf{R}_{hc} = \mathbf{v}_c \mathbf{v}_c^H$.

Given the clutter and target statistics we can now proceed to compute the optimal and suboptimal waveforms **s** as described earlier. The optimal waveform was computed using the gradient descent algorithm discussed in [113]. For each waveform **s,** we then compute the *SINR* at the GD output. The procedure above provides the GD *SINR* for a single randomly chosen scenario with a particular clutter and target. To get a more representative assessment of performance we repeat this procedure for many randomly selected scenarios and collect the probability density function (pdf) and cumulative density function (cdf) of the *SINR* at the GD output. Comparison of cdfs corresponding to the different waveform design methods provides valuable insight into their relative performance. A performance comparison of the GD and GMVDR detector [112] employed by radar system is presented.

6.6.7.2. Discussion

Next, we present results for three cases differing by the size of the target relative to footprint of the radar. We present the cdf of the *SINR* at the outputs of GD and GMVDR detector discussed in [112] for the following transmit waveforms: standard, target maximizing, matched to target, clutter minimizing, *SIR* maximizing, and the optimum computed using the gradient descent method [113]. The radar is assumed to employ a 10 element uniformly spaced linear array.

Case 1: Target is Small Compared with Radar Footprint. The results for this case are shown in Fig.6.21. Investigation of Fig.6.21 leads us to the following observations. The optimum waveform offers the best performance and provides a 10 dB advantage over the standard waveform. We carry out a comparison at the level of 0.9 for *SINR* cdf. This gain may be interpreted as a combination of an 8 dB transmit array gain, and a 2 dB temporal gain. The first is the gain of the transmit array due to the fact, that the transmit power is concentrated almost entirely on the part of the azimuth-frequency plane occupied by the target, whereas in the standard case the same power is spread evenly over the entire plane. The second is the gain due to the fact, that the transmit energy is concentrated at those frequencies where the target has the strongest scattering. In other words, we have gain due to both spatial and temporal effects. Note, however, that conventional beamforming would capture the 8 dB array gain. The cdfs of the *SINR* for the matched-to-target and maximum *SIR* waveform are almost identical, and are approximately 2 dB for the matched-to-target and 1 dB for the maximum *SIR* from the optimum. Both of these capture most of the performance gain of the optimal waveform in this case. The target maximizing waveform is next in order of performance. Note that its *SINR* has a larger variance than the *SINR*s of the previously mentioned waveforms. This is due to the variability for clutter at the azimuth-frequency cells where the target energy is largest. The clutter minimizing waveform performs poorly because it often focuses the energy where the target energy is low or nonexistent. In other words, illuminating area of low clutter may cause us to miss the target. Superiority of the GD SINR cdf over GMVDR one in the case of optimum waveform is for about 12 dB.

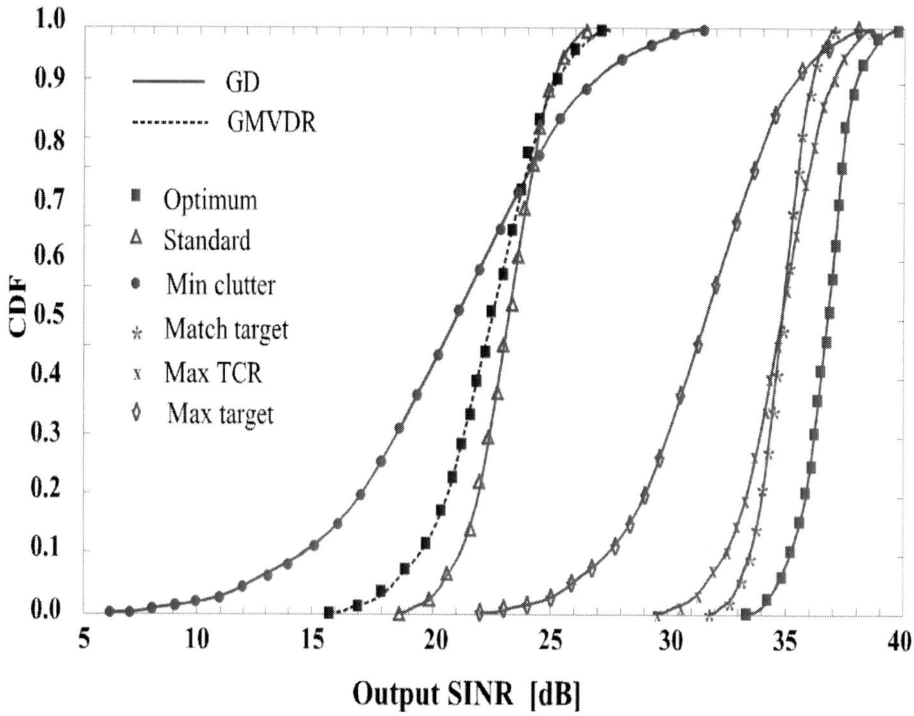

Figure 6.21. CDF of *SINR* at the GD output for different transmit waveforms. Comparison with GMVDR detector. Case 1.

Case 2: Target and Radar Footprint Have the Same Size. The results for this case are depicted in Fig.6.22. Examination of Fig.6.22 shows that the optimum waveform offers a 7 dB advantage over the standard waveform. Note that the target-matched waveform offers 1 dB improvement only. These waveforms spread the transmit power over the entire azimuth-frequency plane so they do not benefit from the transmit array gain. The target-to-clutter maximizing waveform gives close to optimal performance. As before, the clutter-based waveforms perform poorly. Superiority of the GD *SINR* cdf over GMVDR one in the case of optimum waveform is for about 14 dB.

Case 3: Target Size is Half of the Radar Footprint. The results for this case are depicted in Fig.6.23. As expected the results fall in between those of the two cases discussed above. Superiority of the GD *SINR* cdf over GMVDR one in the case of optimum waveform is for about 13 dB.

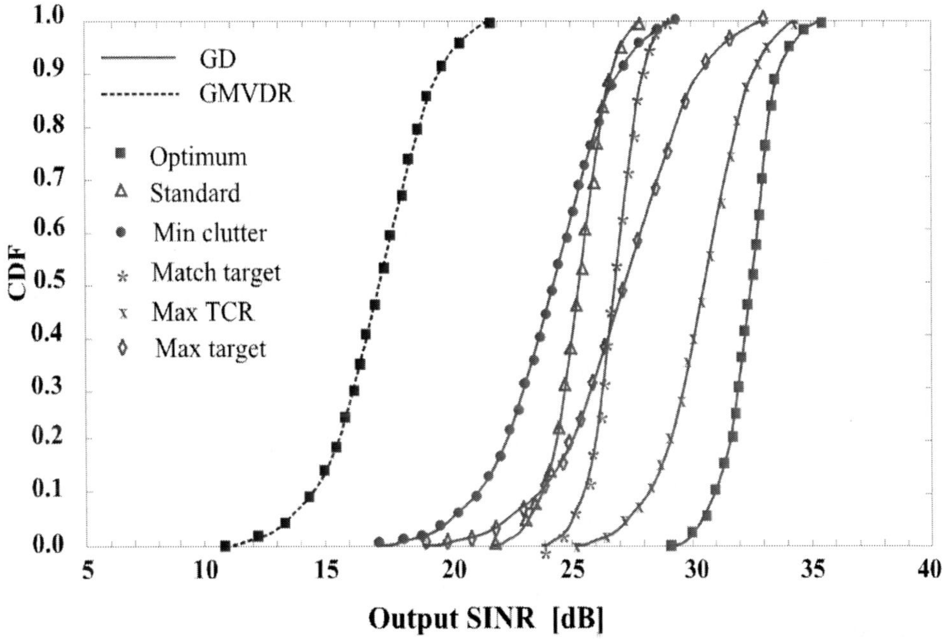

Figure 6.22. CDF of *SINR* at the GD output for different transmit waveforms. Comparison with GMVDR detector. Case 2.

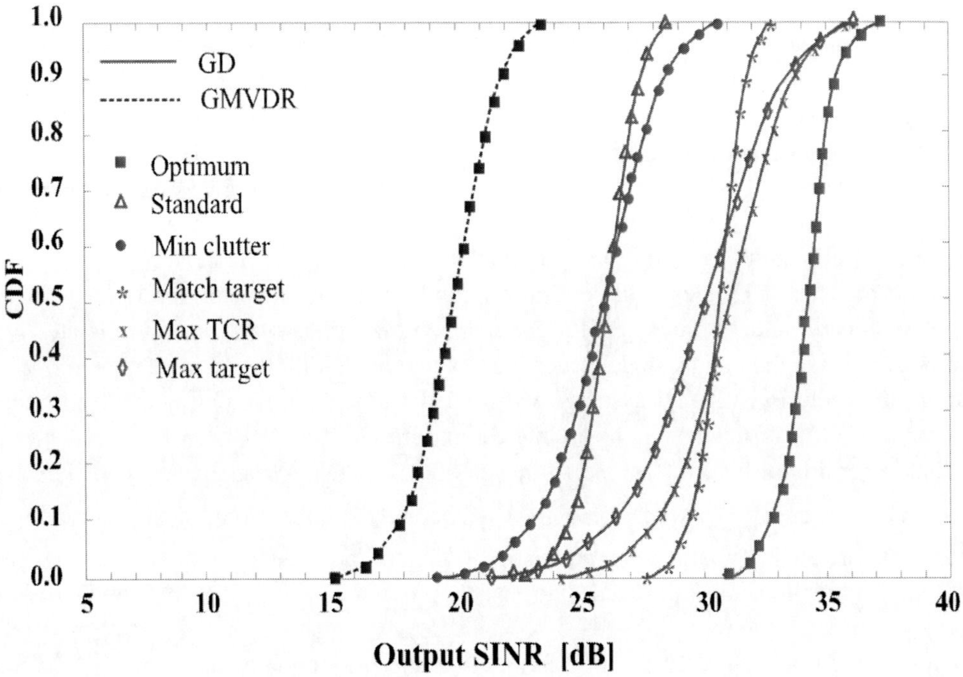

Figure 6.23. CDF of *SINR* at the GD output for different transmit waveforms. Comparison with GMVDR detector. Case 3.

6.7. Summary and Discussion

We investigated the GLRT GD and GD beamformer for spatially distributed signal sources. The analysis and computer simulation demonstrate that the GD performance is better in comparison with the conventional beamformer one that are designed for point sources. The performance difference increases as angular spread increases.

We have developed a parametric Rao/GD test for the multichannel adaptive signal detection problem by exploiting a multichannel AR model. We have derived the ML estimates of the parameters involved in the test. The parametric Rao/GD test is asymptotic PGD, and the asymptotic distributions of its test statistic under both hypotheses have been obtained in closed form. Computer simulations show that:

- Our asymptotic analysis provides fairly accurate prediction of the parametric Rao/GD test performance;
- Even with relatively limited training, the parametric Rao/GD is quite close to the ideal GD;
- The parametric Rao/GD outperforms the AMF detector, which does not exploit a parametric model;
- The performance of the parametric Rao/GD is affected by inaccurate model order estimation, but the resulting performance degradation is tolerable when the model order estimation error is small.

Our asymptotic analysis of the parametric Rao/GD is based on several assumptions made in Section 6.4.2, including that the disturbance can be modeled as an AR(P) process with known model order P, and that the training signals are i.i.d. When these assumptions are violated, we expect that the analysis will be less accurate, but may still be informative if the assumptions are not significantly violated. For example, we have noticed in simulation that when the disturbance is an MA process, the test threshold obtained by analysis assuming an AR model is still quite accurate. One possible reason is that AR models are fairly general parametric models, and under mild conditions, can be used to model or approximate a large class of stationary random processes, for example, an MA process can be approximated as an AR process with high enough model order [53]. Nevertheless, there is a need to find out how accurate our analysis is in real systems with real data, when assumptions made in Section 6.4.2 may not all be met. This will be an interesting future effort.

The probability of bit error P_{er} performance of QDS-CDMA was investigated. The exact probability of bit error P_{er} expression using the characteristic function procedure and several approximations, namely, the characteristic function procedure with an approximate characteristic function, the improved Gaussian approximation and the simplified improved Gaussian approximation were derived.

Under conditions typically satisfied in practice, the MAI component of the GD statistics and the GD probability of bit error P_{er} performance can be accurately approximated by the standard Gaussian approximation. The accuracy of the standard Gaussian approximation in QDS-CDMA stems primarily from two sources: DS-QPSK spreading and bandwidth-efficient pulse shaping, in addition to moderate to large spreading factors and aperiodic spreading. The former makes the conditional MAI process, conditional on interferer chip delays and phase offsets, circularly symmetric while the latter improves the approximation of the wide-sense cyclostationary MAI process as a wide-sense stationary process.

Second, the probability of bit error P_{er} performance of QDS-CDMA is shown to be superior or equivalent to that of B-CDMA. The performance difference diminishes as the number of users increase and as probability of bit error P_{er} performance increases. Great superiority of the GD employment in comparison with the matched filter and Neyman-Pearson receiver is demonstrated.

Techniques for designing transmit waveforms for MIMO radar systems employing the GD are presented. We have shown that by controlling the space-time (or azimuth-frequency) distribution of the transmitted signal it is possible to get significant improvements in the detection performance employing the GD. To achieve this advantage it is necessary to have knowledge of the clutter and/or target statistics.

In this chapter, we assumed that this statistical information is available. Statistics of specific targets of interest may be assumed known through measurements of their radar signatures. By tuning the transmit waveform to a given target type, using, for example, the matched target waveform described earlier, it is possible to enhance significantly the detectability of targets of that type. Knowledge of clutter statistics can be obtained by co-llecting data over multiple pulse periods. The use of GD by radar systems allows us to get a great advantage in the detection performance. A more complete discussion of issues re-lated to estimating target and clutter statistics and the impact of estimation accuracy on detection performance are beyond the scope of this chapter and are the subject of ongoing research.

ACKNOWLEDGMENT

This research was supported by the Kyungpook National University Research Grant, 2012.

REFERENCES

[1] Kay, S., 1998. *Fundamentals of Statistical Signal Processing : Detection Theory*. Upper Saddle River, NJ, Prentice-Hall, Inc.

[2] Van Trees, H., 2002. *Detection, Estimation and ModulationTheory.Parts I–III.*New York, Wiley & Sons, Inc.

[3] Helstrom, C., 1995. *Elements of Signal Detection and Estimation*. Englewood Cliffs, NJ, Prentice-Hall, Inc.

[4] Lehman, E., 1986. *Testing Statistical Hypotheses*. 2nd Ed. New York, Wiley & Sons, Inc.

[5] Tuzlukov. V.P., A new approach to signal detection theory, *Digital Signal Processing*, 1998, Vol. 8, No. 3, pp. 166–184.

[6] Tuzlukov, V.P., *Signal Detection Theory*. New York, Springer-Verlag, 2001.

[7] Maximov, M., Joint correlation of fluctuating noise at the outputs of frequency filters, *Radio Engineering*, 1956, No. 9, pp. 28–38.

[8] Chernyak, Y., Joint correlation of noise voltage at the outputs of amplifiers with non-overlapping responses, *Radio Physics and Electronics*, 1960, No. 4, pp. 551–561.

[9] Tuzlukov, V.P., *Signal Processing Noise*, Boca Raton, London, New York, Washington, D.C. : CRC Press, Taylor & Francis Group, 2002.

[10] Tuzlukov, V.P., *Signal and Image Processing in Navigational Systems*, Boca Raton, London New York, Washington, D.C. : CRC Press, Taylor & Francis Group, 2005.

[11] Tuzlukov, V.P., *Signal Processing in Radar Systems*, Boca Raton, London, New York, Washington, D.C. : CRC Press, Taylor & Francis Group, 2012.

[12] Gibson, J. and Melsa, J., 1975. *Introduction to Nonparametric Detection with Applications*. San Diego. Academic Press.

[13] Adachi, F., Freeney, M.T., Williamson, A.G., and D. Parson, Cross correlation between the envelopes of 900 MHz signals received at a mobile radiobase station site, *Proceedings of Institute of Electrical Engineers F,* 1991, Vol. 133, No. 6, pp. 506–512.

[14] Ikegami, F. and S. Yoshida, Analysis of multipath propagation structure in urban mobile radio environments, *IEEE Transactions on Antenna Propagation*, 1980, Vol. AP-28, No. 7, pp. 531–537.

[15] Lee, W. C.-Y., Finding the approximate angular probability density function of wave arrival by using a directional antenna, *IEEE Transactions on Antenna Propagation*, 1973, Vol. AP-21, No. 5, pp. 328–334.

[16] Eggers, P., Angular propagation descriptions relevant for base station adaptive antenna operations, *Wireless Personal Communications*, 1999, Vol. 11, No. 10, pp. 3–29.

[17] Pedersen, K.I., Mogensen, P.E., and B.H. Fleury, Power azimuth spectrum in outdoor environments, *Electronics Letters*, 1997, Vol. 33, No. 8, pp. 1583–1584.

[18] Salz, J. and J.H. Winters, Effect of fading corelation on adaptive arrays in digital mobile radio, *IEEE Transactions on Vehicle Technologies*, 1994, Vol. 43, No. 11, pp. 1049–1057.

[19] Tuzlukov, V., Implementation of generalized detector for distributed sources using sensor array, in *Proceedings of SPIE, Radar Sensor Technology XVI*, part of *SPIE Internation Symposium on Defense, Security & Sensing*, Editors, Kenneth Ranney and Armin Doery, April 23-25, 2012, Baltimore, Maryland, Volume 8361, pp. 83610V-1–83610V-12.

[20] Abramovich, Yu., Spencer, N., and Gorohov, Detection-estimation of distributed Gaussian sources, in *Proceedings Sensor Array Multichannel Signal Processing Workshop*, August 4-6, 2002, Posslyn, VA, USA, pp. 513–517.

[21] Besson, O. and P. Stoica, Decoupled estimation of DOA and angular spread for a spatially distributed source, *IEEE Transactions on Signal Processing*, 2000, Vol. SP-48, No. 7, pp. 1872–1882.

[22] Raich, R., Goldberg, 7J., and H. Messer, bearing estimation for a distributed source : Modeling, inherent accuracy limitations and algorithms, *IEEE Transactions on Signal Processing,* 2000, Vol. SP-48, No.2, pp. 429–441.

[23] Valaee, S., Champagne, B., and P. Kabal, Parametric localization of distributed sources, *IEEE Transactions on Signal Processing*, 1995, Vol. SP-43, No. 9, pp. 2144–2153.

[24] Raghavan, R.S., Pulsone, N., and D.J. McLauhlin, Performance of the GLRT for adaptive vector subspace detection, *IEEE Transactions on Aerospace Electronics Systems*, 1996, Vol. AES-32, No. 10, pp. 1473–1487.

[25] Scharf, LL. And B. Friedlander, Matched subspace detection, *IEEE Transactions on Signal Processing*, 1994, Vol. SP-42, No. 8, pp. 2146–2157.

[26] Tuzlukov, V.P., Implementation of generalized detector for distributed source using sensor arrays, in *Radar Sensor Technology XVI*, edited by Kenneth Ranney and Armin Doerry, Proceedings of SPIE Vol. 8361, SPIE, Belingham, WA, USA 2012, pp. 8361-0V-1–8361-0V-12.

[27] Box, G.E.P., Some theorems on quadratic forms applied in the study of analysis of variance problems, I. Effect of inequality of variance in the one-way classification, *Annual Mathematical Statistics*, 1954, Vol. 25, No. 2, pp.290–302.

[28] Lee, H.B., Eigenvalues and eigenvectors of covariance matrices for signals closely spaced in frequency, *IEEE Transactions on Signal Processing*, 1992, Vol. SP-40, No. 10, pp. 2518–2535.

[29] Ma, M.T., *Theory and Application of Antenna Array*, New York : Wiley & Sons, Inc., 1974.

[30] Ward, J., Space-time adaptive processing for airborne radar, *Technical Report 1015*, Lincoln Laboratory, MIT, December, 1994.

[31] Klemm, R., *Principles of Space-Time Adaptive Processing*, London, IEE Press, 2002.

[32] Shaw, G. and Manolakis, D., Signal processing for hyperspectral image exploitation, *IEEE Signal Processing Magazine*, 2002, Vol. 19, No. 1, pp. 12–16.

[33] Manolakis, D. and Shaw, G., Detection algorithms for hyperspectral imaging applications, *IEEE Signal Processing Magazine*, 2002, Vol. 19, No. 1, pp. 29–43.

[34] Paulraj, A. and Papadias, C.B. Space-time processing for wireless communications, *IEEE Signal Processing Magazine*, 1997, Vol. 14, No. 11, pp. 49–83.

[35] Justice, J.H., Array processing in exploration seismology, in *Array Signal Processing*, S.Haykin, Ed. Englewood Cliffs, NJ : Prentice-Hall, Inc., 1985.

[36] Reed, I.S., Mallett, J.D., and L.E.Brennan, Rapid convergence rate in adaptive arrays, *IEEE Transactions on Aerospace and Electronics Systems*, 1974, Vol. AES-10, No. 6, pp. 853–863.

[37] Kelly, E.J., An adaptive detection algorithm, *IEEE Transactions on Aerospace and Electronics Systems*, 1986, Vol. AES-22, No. 2, pp. 115–127.

[38] Cai, L. and Wang, H. On adaptive filtering with the CFAR feature and its performance sensitivity to non-Gaussian interference, in *Proceedings 24th Annual Conference on Information Sciences and Systems*, Princeton, NJ, USA, March 1990, pp. 558–563.

[39] Chen, W. and Reed, I.S., A new CFAR detection test for radar, *Digital Signal Processing*, 1991, Vol. 1, No. 4, pp. 198–214.

[40] Robey, F.C., Fuhrmann, D.R., Kelly, E.J., and Nitzberg, R., A CFAR adaptive matched filter detector, *IEEE Transactions on Aerospace and Electronics Systems*, 1992, Vol.AES-28, No. 1, pp. 208–216.

[41] Kraut, S. and Scharf, L.L., The CFAR adaptive subspace detector is a scale-invariant GLRT, *IEEE Transactions on Signal Processing*, 1999, Vol. SP-47, No. 9, pp. 2538–2541.

[42] Kraut, S., Scharf, L.L., and McWhorter, L.T., TAdaptive subspace detectors, *IEEE Transactions on Signal Processing*, 2001, Vol. SP-49, No. 1, pp. 1–16.

[43] Tuzlukov, V.P., Adaptive generalized detector for unknown power spectral densities of noise, in *Proceedings International Signal Processing Conference (ISPC'03),*Dallas, TX, USA, March 31-April 4, 2003.

[44] Tuzlukov, V.P., Adaptive beam-former generalized detector, in *Proceedings IASTED International Conference on Signal Processing, Pattern Recognition, Applications (SPPRA'03)*, Rhodes Island, Greece, June 30-july 2, 2003.

[45] Kim, Y.D., Yoon, W.S. and Tuzlukov, V.P., Adaptive beamformer generalized detector in wireless sensor networks, in *Proceedings IASTED International Conference on Parallel and Distributed Computing and Networks (PDCN 2004)*, Innsbruck, Austria, February 17-19, 2004, pp. 195–200.

[46] Rangaswamy, M. and Michels, J.H., A parametric multichannel detection algorithm for correlated non-Gaussian random processes, in *Proceedings 1997 IEEE National Radar Conference,* Syracuse, NY, USA, May 13-15, 1997, pp. 349–354.

[47] Roman, J.R., Rangaswamy, M., Davis, D.w., Zhang, Q., Himed, B., and Michels, J.H. Parametric adaptive matched filter for airborne radar applications, *IEEE Transactions on Aerospace and Electronics Systems*, 2000, Vol.AES-36, No. 4, pp. 677–692.

[48] Swindlehurst, A.I., and Stoica, P., Maximum likelihood methods in radar array signal processing, *IEEE Proceedings*, 1998, Vol. 86, No.2, pp.421–441.

[49] Li, J., Liu, G., Jiang, N., and Stoica, P., Moving targetfeature extraction for airborne high-range resolution phased-array radar, *IEEE Transactions on Signal Processing*, 2001, Vol. SP-49, No. 2, pp. 277–289.

[50] Tuzlukov, V.P., Multichannel adaptive generalized detector based on parametric Rao test, in *Signal and Data Processing of Small targets 2012*, edited by Oliver Drummond and Richard Teichgraeber, Proceedings of SPIE Vol. 8393, SPIE, Belingham, WA, USA 2012, pp. 8393-07-1–8393-07-12.

[51] Conte, E. and De Maio, A., Distributed target detection in compound-Gaussian noise with Rao and Wald tests, *IEEE Transactions on Aerospace and Electronics Systems*, 2003, Vol. AES-39, No. 4, pp. 568–582.

[52] De Maio, A., Alfano, G., and Conte, E., Polarization diversity detection in compound-Gaussian clutter, *IEEE Transactions on Aerospace and Electronics Systems*, 2004, Vol. AES-40, No. 1, pp. 114–131.

[53] Stoica, P. and Moses, R.L., *Spectral Analysis of Signals*, Englewood Cliffs, NJ : Prentice-Hall, Inc. 20056.

[54] Tsao, T., Himed, B., and Michels, J.H., Effects of interference rank estimation on the detection performance of rank reduced STAP algorithm, in *Proceedings 1998 IEEE Radar Conference (RADARCON'98)*, May 11-14, 1998, Dallas, TX, USA, pp. 147–152.

[55] Kay, S.M., *Modern Spectral Estimation : Theory and Application*, Englewood Cliffs, NJ : Prentice-Hall, 1988.

[56] Viterbi, A., *CDMA : Principles of Spread Spectrum Communication*, Addison-Wesley, Reading, MA, 1995.

[57] Adachi, F., Sawahashi, M., and Suda, H., Wideband DS-CDMA for next generation mobile communication systems, *IEEE Communication Magazine,* 1998, Vol.36, No. 9, pp. 56–69.

[58] Lehnert, J.S. and Pursley. M.B., Error probabilities for binary direct-sequence spread-spectrum communications with random signature sequences, *IEEE Transactions on Communications*, 1987, Vol. COM-35, No. 1, pp. 87–98.

[59] Imai, H. and Miyakawa, H., Pulse waveform in the asynchronous CDM communications, *Record of Joint Convention of Four Institutes with Electrical Engineering*, 1968, pp. 2017.

[60] Lok, T.M. and Lehnert, J.S., Error probabilities for generalized quadriphase DS/SSMA communication systems with random signature sequence, *IEEE Transactions on Communications*, 1996, Vol. COM-44, No. 7, pp. 876–885.

[61] Morrow Jr., R.K., Accurate CDMA BER calculations with low computational complexity, *IEEE Transactions on Communications*, 1998, Vol. COM-46, No. 11, pp. 1413–1417.

[62] Sunay, M.O. and McLane, P.J., Calculating error probabilities for DS-CDMA systems: When not to use the Gaussian approximation, in *Proceedings IEEE GLOBECOM Conference*, London, U.K., November 18-22, 1996, pp. 1744–1749.

[63] Geraniotis, E.A., and Chaffari, B., Performance of binary and quanternary direct-sequence spread-spectrum multiple-access systems with random signature sequences, *IEEE Transactions on Communications*, 1991, Vol. COM-39, No. 5, pp. 713–724.

[64] Torrieri, D.J., Performanceof direct-sequence systems with long pseudonoise sequences, *IEEE Journal of Selected Areas of Comunications*, 1992, Vol. 10, No. 5, pp. 770–781.

[65] Asano, Y., Daido, Y., and Holtzman, J.M., Performance evaluation for band-limited DS-CDMA communication system, in *Proceedings IEEE Vehicular Technology Conference,* Secaucus, NJ, USA, May 18-20, 1993, pp. 464–468.

[66] Cho, J.H. and Lehnert, J.S., Optimum chip waveforms for DS-SSMA communications with random spreading sequences and a matched filter receiver, in *Proceedings IEEE Wireless Communications and Networking Conference*, New Orleans, LA, USA, September 21-24, 1999.

[67] Shibata, T., Katayama, M., and Ogawa, A., Performance of asynchronous band-limited DS/ SSMA systems, *IEICE Transactions on Communications*, 1993, Vol. E76-B, No. 8, pp. 921–928.

[68] Huat, C.P., Performance comparison of DS-BPSK and DS-QPSK CDMA systems, in *Proceedings International Symposium on Spread-Spectrum Techniques Applications*, Sun City, South Africa, September 2-4, 1998, pp. 802–805.

[69] Holtzman, J.M., Simple, accurate, method to calculate spread-spectrum multiple-access error probabilities, *IEEE Transactions on Communications*, 1992, Vol. 40, No. 3, pp. 461–464.

[70] Yoon, Y.C., Accurate approximations for probability of bit error in asynchronous band-limited DS-CDMA systems with multiple data rates, in *Proceedings 20th Biennial Symposium on Communications*, Kingston, Canada, May 28-31, 2000, pp. 326–330.

[71] Yoon, Y.C., An improved Gaussian approximation for probability opf bit error analysis of asynchronoud band-limited DS-CDMA systems with BPASK spreading, *IEEE Journal of Selected Areas of Comunications: Wireless Communications Series*, 2002, Vol. 1, No. 7, pp. 373–382.

[72] Yoon, Y.C., Simple and accurate BER analysis in band limited DS-CDMA systems with multi-class services, in *Proceedings International Conference on 3rd Generation Wireless Communications*, San Francisco, CA, USA, June 14-16, 2000, pp. 505–512.

[73] Yoon, Y.C., SA simple and accurate method of probability of bit error analysis for asynchronous band-limited DS-CDMA systems, *IEEE Transactions on Communications*,2002, Vol. COM-50, No. 4, pp. 656–663.
[74] Morrow Jr., R.K. and Lehnert, J.S., Packet Throughput in slotted ALOHA DS/SSMA radio systems with random signature sequences, *IEEE Transactions on Communications,* 1992, Vol. 40, No. 7, pp. 1223–1230.
[75] Tuzlukov, V., Quadriphase DS-CDMA wireless communication systems employing the generalized detector, in *Wireless Sensing, Localization, and Processing VII*, edited by Sohai Dianat and Michael Zoltowski, Proceedings of SPIE, Vol. 8404, SPIE, Bellingham, WA, USA 2012, pp. 8404-0G-1–8404-0G-12.
[76] Lee, E.A. and Messerschmitt, D.G., *Digital Communications*, 2nd Ed. Kluwer Inc.: Norwel, MA, USA, 1994.
[77] Yoon, Y.C., Probability of bit error and the Gaussian approximation in asynchronous DS- CDMA systems with chip pulse shaping, in *Proceedings 20th Biannual Symposium on Communications*, Kingston, Canada, May 28-31, 2000, pp. 111–115.
[78] Haykin, S., *Communication Systems*, 5th Ed. Wiley & Sons, Inc.: New York, 2009.
[79] Leon-Garcia, A., *Probability and Random Processes for Electrical Engineering*, Addison-Wesley: Reading, MA, USA, 1989.
[80] Yoon, Y.C. and H. Leib, Maximizing SNR in improper complex noise and applications to CDMA, *IEEE Communications Letters*, 1997, Vol. 1, No. 1, pp. 5–8.
[81] Yoon, Y.C. and H. Leib, Chip-delay locked matched filters for DS-CDMA systems using long sequence spreading, *IEEE Transactions on Communications,* 2001, Vol. 49, No. 8, pp. 1468–1478.
[82] Gardner, W.A., An introduction to cyclostationary signals, in *Cyclostationarity in Communications and Signal Processing*, W.A. Gardner, Ed. IEEE Press : Piscataway, NJ, USA, 1994, pp. 1–90.
[83] Gardner, W.A. and L.E. Franks, Characterization of cyclostationary random signal processes, *IEEE Transactions on Information Theory*,1975,Vol. IT-21, No. 1, pp. 4–14.
[84] X. Luzhou, P. Stoica, K.W. Forsythe, and D.W. Bliss, Range compression and waveform optimization for MIMO Radar: A Cramer-Rao bound based study," *IEEE Transactions on Signal Processing*, Vol. 56, No. 1, 2008, pp. 218-232.
[85] Guerci, J.R. and S.U. Pillai, Theory and application of optimum transmit-receive radar, in *Proceedings IEEE 2000 Radar Conference*, May 4-8, Pasadena, CA, USA, 2000, pp. 705–710.
[86] Pillai, S.U., Youla, D.C., Oh, S., and J.R. Guerci, Optimum transmit-receiver design in the presence of signal-dependent interference and channel noise, in *Proceedings 33rd Asilomar Conference on Signals, Systems and Computers*, October 24-27, Pacific Grove, CA, USA, 1999, pp. 24–27.
[87] Pillai, S.U., Youla, D.C., Oh, S., and J.R. Guerci, Matched transmit-receiver design in the presence of signal dependent noise, *IEEE Transactions on Information Theory*, 2000, Vol. IT-46, No.2, pp. 577–584.
[88] Pillai, S.U., Oh, S., and J.R. Guerci, Multichannel matched transmit-receiver design in the presence of signal dependent interference and noise, in *Proceedings First IEEE Sensor Array and Multichannel Signal Processing Workshop*, March 16-17,2000, Cambridge, USA, pp. 385–389.

[89] Friedlander, B., A subspace framework for adaptive radar waveform design, in *Proceedings 33th Asilomar Conference on Signals, Systems and Computers*, Pacific Grove, CA, USA, October 30-November 2, 2005, pp. 1135–1139.

[90] Chen, C.-Y. and P.P. Vaidyanathan, MIMO radar waveform optimization with prior information of the extended target and clutter, *IEEE Transactions on Signal Processing*, 2008, Vol. 57, No. 9, pp. 3533–3544.

[91] Naghibi, T., Namvar, M., and F. Behnia, Optimal and robust waveform design for MIMO radars in the presence of clutter, *Signal Processing*, 2010, Vol. 90, No. 4, pp. 1103–1117.

[92] Bliss, D.W. and K.W. Forsythe, Multiple-input multiple-output (MIMO) radar: degrees of freedom and resolution, in *Proceedings 37th Asilomar Conference on Signals, Systems and Computers*, Pacific Grove, CA, USA, November 9-12, 2003, pp. 54–59.

[93] Forsythe, K.W., Bliss, D.W., and G. Fawcett, Multiple-input multiple-output (MIMO) radar: performance issues, *Proceedings 37th Asilomar Conference on Signals, Systems and Computers*, Pacific Grove, CA, USA, November 7-10, 2004, pp. 310–315.

[94] Forsythe, K.W.and D.W. Bliss, Waveform correlation and optimization issues for MIMO radar, in *Proceedings 33th Asilomar Conference on Signals, Systems and Computers*, Pacific Grove, CA, USA, October 30-November 2, 2005, pp. 1306–1310.

[95] Rabideau, D. and P. Parker, Ubiquitous MIMO multifunctional digital array radar, in *Proceedings 37th Asilomar Conference on Signals, Systems and Computers*, Pacific Grove, CA, USA, November 9-12, 2003, pp. 1057–1064.

[96] Robey, F.C., Coutts, S., Weikle, D., McHarg, J.C., and K. Cuomo, MIMO radar theory and experimental results, in *Proceedings 38th Asilomar Conference on Signals, Systems and Computers*, Pacific Grove, CA, USA, November 7-10, 2004, pp. 300–304.

[97] White, L.B. and P.S. Ray, Receiver design for MIMO tracking radars, in *Proceedings 2004 International Conference on Waveform Diversity and Design*, Edinborough, UK, November 8, 2004.

[98] Fuhrmann, D.R., and San Antonio, Transmit beamforming for MIMO radar systems using partial signal cross-correlation, in *Proceedings 38th Asilomar Conference on Signals, Systems and Computers*, Pacific Grove, CA, USA, November 7-10, 2004, pp. 295–299.

[99] Mecca, V., Ramakrishnan, D., and J. Krolik, MIMO radar space-time adaptive processing for multipath clutter mitigation, in *Proceedings IEEE Sensor Array and Multichannel Signal Processing Workshop*, Waltham, Massachusetts, USA, July 12-14, 2006, pp. 249–253.

[100] Xu, L., Li, L., and P. Stoica, Adaptive techniques for MIMO radar, in *Proceedings IEEE Sensor Array and Multichannel Signal Processing Workshop*, Waltham, Massachusetts, USA, July 12-14, 2006, pp. 249–253.

[101] Fishler, E., Haimovich, A., Blum, R., Chizik, D., Cimini, L., and R. Valenzuela, MIMO radar: an idea whose time has come, in *Proceedings IEEE Radar Conference*, Philadelphia, Pennsylvania, USA, April 26-29, 2004, pp. 71–78.

[102] Fishler, E., Haimovich, A., Blum, R., Cimini, L., Chizik, D., and R. Valenzuela, Statistical MIMO radar, in *Proceedings 12th Conference on Adaptive Sensor Array Processing*, Lexington, MA, USA, March 16-18, 2004.

[103] Fishler, E., Haimovich, A., Blum, R., Cimini, L., Chizik, D., and R. Valenzuela, Performance of MIMO radar systems: advantages of angular diversity, in *Proceedings*

38th Asilomar Conference on Signals, Systems and Computers, Pacific Grove, CA, USA, November 7-10, 2004, pp. 305–309.

[104] Gogineni, S. and A. Nehorai, Target estimation using sparse modeling for distributed MIMO radar, *IEEE Transactions on Signal Processing*, 2011, Vol. SP-59, No. 11, pp. 5315–5325.

[105] Tuzlukov, V., Optimal waveforms for MIMO radar systems employing the generalized detector, in *Signal Processing, Sensor Fusion, and Target Recognition XIX*, edited by Ivan Kadar, Proceedings of SPIE, Vol. 7697, SPIE, Bellingham, WA, 2010, pp. 7697-1G-1–7697 -1G-12.

[106] Friedlander, B., Waveform design for MIMO radars, *IEEE Transactions on Aerospace and Electronic Systems*, 2007, Vol. AES-43, No. 3, pp. 1227–1238.

[107] Van Trees, H.L., *Detection, Estimation, and Modulation Theory*, Part III, New York :Willey & Sons, Inc., 2002.

[108] Skolnik, M., *Introduction to Radar Systems*, 3^{td} Ed. New York : McGraw-Hill, 2008.

[109] Fishler, E., Haimovich, A., Blum, R., Cimini, L., Chizik, D., and R. Valenzuela, Spatial diversity in radars – models and detection performance, *IEEE Transactions on Signal Processing*, 2006, Vol. SP-54, No. 3, pp. 823–838.

[110] Larsson, E.G. and P. Stoica, *Space-Time Block Coding for Wireless Communications*, Cambridge, U.K.: Cambridge University Press.

[111] Haykin, S. *Array Signal Processing*, Englewood Cliffs, NJ: Prentice-Hall, 1985.

[112] Jin, Y. and B. Friedlander, Detection of distributed sources using sensor arrays, *IEEE Transactions on Signal Processing*, 2004, Vol. SP-52, No. 6, pp. 1537–1548.

[113] Ljung, L. and T. Soderstrom, *Theory and Practice of Recursive Identification*, Cambridge, MA, USA: MIT Press.

[114] Haykin, S., *Adaptive Filter Theory*, 3th Ed. Upper Saddle River, NJ, USA : Prentice-Hall, 2004.

[115] Moon. T.K. and W.C. Stirling, *Mathematical Methods and Algorithms for Signal Processing*, Upper Saddle River, NJ, USA : Prentice-Hall,

In: Communication Systems: New Research
Editor: Vyacheslav Tuzlukov

ISBN: 978-1-62618-654-5

Chapter 7

ULTRA-WIDE BAND COMMUNICATIONS IN THE EXTREMELY HIGH FREQUENCY (EHF) BAND

Yosef Pinhasi, Asher Yahalom and Gad A. Pinhasi
Faculty of Engineering, Ariel University of Samaria

ABSTRACT

The deficiency of wide frequency bands within the conventional spectrum, poses limits on the realization of broadband wireless communication links. The fact that the Extremely High Frequencies (EHF) band is relatively free of users, combined with recent developments of millimeter wave technologies, enables the utilization of frequencies above 30GHz for ultra-wide band communication links and radar applications.

One of the principal challenges in realizing modern wireless communication links in the EHF band are absorptive and dispersive effects emerging while millimeter waves propagate in the atmospheric medium. When millimeter-wave radiation passes through the atmosphere, it suffers from selective molecular absorptions peaking at 22GHz and 183GHz, where resonance absorption of water occurs, as well as at 60GHz and 119GHz, due to absorption resonances of oxygen. Between these frequencies, minimum attenuation is obtained at 35GHz (Ka-band), 94GHz (W-band), 130GHz and 220GHz, which are known as atmospheric transmission 'windows'. The transmission characteristics are also determined by weather conditions such as temperature, pressure and humidity. Attenuation due to fog, haze, clouds, rain and snow is one of the dominant causes of fading in wireless communication links operating in the EHF band.

In this chapter, we present a space-frequency theory for the propagation of an ultra-wide band radiation in a lossy dielectric media. Characterization of the atmospheric medium is via its refractivity leading to a transfer function, which describes the response of the medium in the frequency domain. This description enables the consideration of broadband signals taking into account inhomogeneous absorptive and dispersive effects of the medium.

Keywords: Extremely high frequencies, Millimeter waves, Atmospheric propagation, Satellite Networking

INTRODUCTION

The growing demand for broadband wireless communication links and the deficiency of wide frequency bands within the conventional spectrum, require utilization of higher microwave and millimeter-wave spectrum at the Extremely High Frequencies (EHF) above 30GHz (see table 1).

Table 1. The electromagnetic spectrum

BAND		IEEE	FREQUENCY	WAVELENGTH
Extremely Low Frequency	ELF		3 - 30Hz	
Super Low Frequency	SLF		30 - 300Hz	
Ultra Low Frequency	ULF		300 - 3,000 Hz	1,000 - 100 Km
Very Low Frequency	VLF		3 - 30 KHz	100 – 10 Km
Low Frequency	LF		30 - 300 KHz	10 - 1 Km
Medium Frequency	MF		300 - 3,000 KHz	1 - 0.1 Km
High Frequency	HF		3 - 30 MHz	100 - 10 m
Very High Frequency	VHF		30 - 300 MHz	10 – 1 m
Ultra High Frequency	UHF		300 - 3,000 MHz	1 - 0.1 m
		L	1 - 2 GHz	
		S	2 - 4 GHz	
Super High Frequency	SHF		3 - 30 GHz	10 - 1 cm
		C	4 - 8 GHz	
		X	8 - 12 GHz	
		Ku	12 - 18 GHz	
		K	18 - 26.5 GHz	
		Ka	26.5 - 40 GHz	
Extremely High Frequency	EHF		30 - 300 GHz	1 - 0.1 cm
		V	40 - 75 GHz	
		W	75 - 110 GHz	
Sub-millimeter (TeraHertz)	FIR		300 - 3,000 GHz	1 – 0.1 mm
Mid infra-red	MIR		3 – 30 THz	100 – 10 μm
Near infra-red	NIR		30 – 300 THz	10 – 1 μm

In addition to the fact that the EHF band (30-300GHz) covers a wide range, which is relatively free of spectrum users, it offers many advantages for wireless communication and RADAR systems as follows:

* Broad bandwidths for high data rate information transfer
* High directivity and spatial resolution
* Low transmission power (due to high antenna gain)
* Low probability of interference/interception (due to narrow antenna beam-widths)
* Small antenna and equipment size
* No multipath fadings (although fading can be caused by atmospheric conditions)

Some of the principal challenges in realizing modern wireless communication links at the EHF band are the effects emerging when the electromagnetic radiation propagates through the atmosphere. When millimeter-wave radiation passes through the atmosphere, it suffers from selective absorption in molecules of the gases composing the air [1-3]. The inhomogeneous transmission in a band of frequencies causes absorptive and dispersive effects in the amplitude and in phase of wide-band signals transmitted in the EHF band. The frequency response of the atmosphere plays a significant role as the data rate of a wireless digital radio channel is increased. The resulting amplitude and phase distortion leads to inter-symbol interference, and thus to an increase in the bit error rate (BER). These effects should be taken into account in the design of broadband communication systems, including careful consideration of appropriate modulation, equalization and multiplexing techniques.

An accurate study of the effects emerging when ultra-wide band signals are propagating in a lossy-dispersive medium, like the atmosphere, is done via presentation of the electromagnetic waves in the frequency domain. It is also found to be much more convenient to solve the propagation in the frequency domain, instead of using space-time field functions. We use a space-frequency approach to find the evolution of the field along its propagation path in the atmospheric medium [4-5].

SPACE-FREQUENCY MODEL OF A WIRELESS COMMUNICATION CHANNEL

The electromagnetic field is described by the space-time electric $E(\mathbf{r},t)$. $\mathbf{r}$ stands for the (x, y, z) coordinates, where (x, y) are the transverse coordinates and z is the axis of propagation. The Fourier transform of the field is:

$$\mathbf{E}(r,f)=\int_{-\infty}^{+\infty} E(r,t)\cdot e^{-j2\pi ft}dt \tag{1}$$

where f denotes the frequency. The response of the medium in which the field propagates is usually characterized by quantities given in the positive frequency domain. Thus, it is convenient to use the analytic representation of the electric field in the time domain [4-5]:

$$\widetilde{E}(r,t)=E(r,t)+j\hat{E}(r,t) \tag{2}$$

where:

$$\hat{E}(r,t)=\frac{1}{\pi}\int_{-\infty}^{+\infty}\frac{E(r,t')}{t-t'}dt' \tag{3}$$

is the Hilbert transform of $E(r,t)$, with a Fourier transform:

$$\widetilde{\mathbf{E}}(r,f)=\int_{-\infty}^{+\infty}\widetilde{E}(r,t)\cdot e^{-j2\pi ft}dt \tag{4}$$

the resulted presentation is a 'phasor-like' function defined for positive frequencies and related to $\mathbf{E}(r,f)$ via:

$$\widetilde{\mathbf{E}}(r,f)=\begin{cases} 2\mathbf{E}(r,f) & f>0 \\ 0 & f<0 \end{cases} \tag{5}$$

Since the electromagnetic signal is real (i.e. $E^*(r,t)=E(r,t)$), its Fourier transform satisfies $\mathbf{E}^*(r,f)=\mathbf{E}(r,-f)$. Consequently, the Fourier transform can be decomposed in terms of the phasor-like function according to:

$$\mathbf{E}(r,f)=\frac{1}{2}\widetilde{\mathbf{E}}(r,f)+\frac{1}{2}\widetilde{\mathbf{E}}^*(r,-f) \tag{6}$$

The time domain field is obtained by inverse Fourier transformation of Eq. (6):

$$E(r,t)=\int_{-\infty}^{+\infty}\mathbf{E}(r,f)\cdot e^{+j2\pi ft}df=\mathrm{Re}\int_{0}^{\infty}\widetilde{\mathbf{E}}(r,f)\cdot e^{j2\pi ft}df \tag{7}$$

A typical communication channel is shown in Figure 1. Now we assume that a carrier wave at f_0 is modulated by a wide-band signal $A_{in}(t)$:

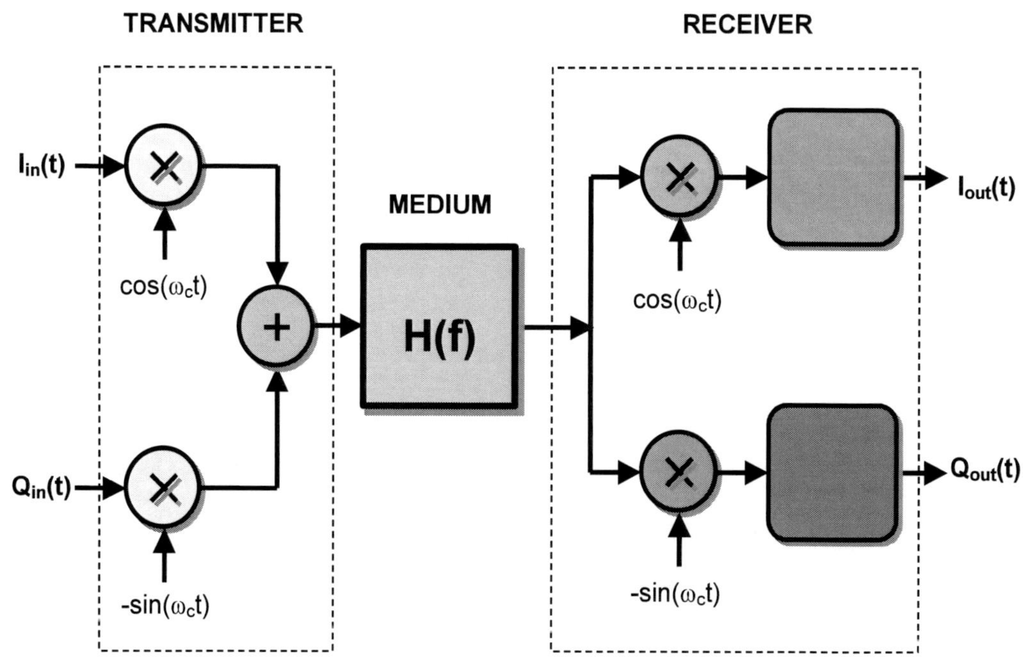

Figure 1. Wireless communication link.

$$E_{in}(t) = \mathrm{Re}\left\{A_{in}(t) \cdot e^{j2\pi f_0 t}\right\} \tag{8}$$

Here $A_{in}(t) = I_{in}(t) - jQ_{in}(t)$ is a complex envelope, representing the base-band signal fed into the modulator of the transmitter, where $I_{in}(t) = \mathrm{Re}\{A_{in}(t)\}$ and $Q_{in}(t) = -\mathrm{Im}\{A_{in}(t)\}$ are the in-phase and the quadrature information waveforms respectively (See Figure 1).

The Fourier transform of the transmitted field is:

$$\mathbf{E}_{in}(f) = \frac{1}{2}\mathbf{A}_{in}(f - f_0) + \frac{1}{2}\mathbf{A}_{in}^*[-(f + f_0)] \tag{9}$$

where $\mathbf{A}_{in}(f)$ is the Fourier transform of the complex envelope $A_{in}(t)$. After propagating along a path with a distance d, the field is:

$$\mathbf{E}_{out}(f) = \frac{1}{2}\mathbf{A}_{in}(f - f_0) \cdot H(f) + \frac{1}{2}\mathbf{A}_{in}^*[-(f + f_0)] \cdot H^*(-f) \tag{10}$$

Since the transfer function $H(f)$ is the Fourier transform of a real function, $H(f) = H^*(-f)$. The inverse Fourier transformation of Eq. (10), results in a received field given in the time domain by:

$$E_{out}(t) = \mathrm{Re}\left\{A_{out}(t) \cdot e^{j2\pi f_0 t}\right\} \tag{11}$$

where the complex envelope of the signal obtained at the receiver cite is given by:

$$A_{out}(t) = \int_{-\infty}^{+\infty} \mathbf{A}_{in}(f) \cdot H(f + f_0) \cdot e^{+j2\pi f t} df \tag{12}$$

The flow chart in Figure 2 summarizes the procedure carried out for solving Eq. (12).

A base-band digital signal $A_{in}(t)$, fed to the modulator at the transmitter site modulates a carrier at f_0. The transfer function $H(jf)$ is called for to calculate the complex transmittance of the medium. Finally, the de-modulated complex signal $A_{out}(t)$ and its related quadrature components $I_{out}(t)$ and $Q_{out}(t)$ obtained at the receiver output are found, employing an algorithm of Fast Fourier Transformation (FFT) for solving equation (12).

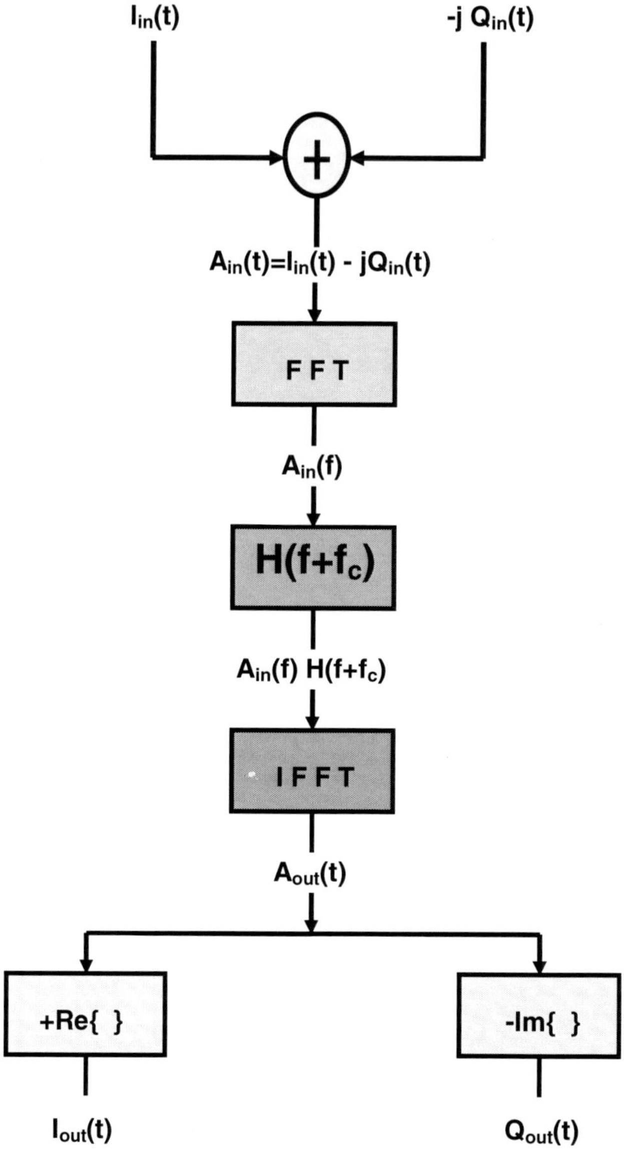

Figure 2. The procedure for calculation of de-modulated quadrature signals received in a wireless communication channel.

In the far field, transmission of a wave, radiated from a localized (point) isotropic source and propagating in a (homogeneous) medium is characterized in the frequency domain by the transfer function:

$$H(jf) = \frac{\mathbf{E}_{out}(jf)}{\mathbf{E}_{in}(jf)} \propto e^{-jk(f)\cdot d} \tag{13}$$

where, $k(f) = 2\pi f\sqrt{\mu\varepsilon}$ is a frequency dependent propagation factor and ε and μ are the permittivity and the permeability of the medium, respectively. The symbol d represents the distance from the source to the target. The transfer function $H(jf)$ describes the frequency response of the medium. In a dielectric medium the permeability is equal to that of the vacuum $\mu=\mu_0$ and the permittivity is given by $\varepsilon(f)=\varepsilon_r(f)\varepsilon_0$. If the medium introduces losses and dispersion, the relative dielectric constant $\varepsilon_r(f)$ is a complex, frequency dependent function. Knowing the dielectric coefficient of the atmosphere in the EHF regime enables calculation of the transfer function for a wireless communication link operating in millimeter wavelengths.

THE MILLIMETER WAVE PROPAGATION MODEL IN ATMOSPHERE

The atmosphere of Earth is a composition of gases surrounding the planet Earth that is retained by Earth's gravity. The atmosphere is divided into several principal layers as shown in Table 2. Pressure and density decrease in the atmosphere as height increases, while temperature may remain relatively constant or even increase with altitude in some regions.

Table 2. The layers of the atmosphere

70,000Km	Outer space				Magnetosphere
10,000Km				Plasmasphere	
800Km	Atmosphere	Exosphere	Ionosphere		
500Km		Thermosphere			
80Km		Mesosphere			
50Km		Stratosphere			
8-18Km		Troposphere			

The lower layer of the atmosphere is the Troposphere beginning at the surface and extends to between 8Km at the poles and 18 km at the equator, with some variation due to weather. The troposphere is mostly heated by transfer of energy from the surface, resulting in descending temperature with altitude. The tropospheric air is a mixture of gasses containing roughly 80% of the mass of the atmosphere. Dry air contains 78.084% Nitrogen (N_2), 20.948% Oxygen (O_2), 0.934% Argon (Ar), 0.0314% Carbon dioxide (CO_2), and small amounts of other gases. Humid atmosphere also contains a variable amount of water vapor, around 1-4%. The air also contains of dust, haze and other pollutant particles.

At lower frequencies, up to the UHF regime, the atmosphere is transparent to the electromagnetic radiation propagating in the medium. However, as the frequency is increased, absorption is revealed. When millimeter-wave radiation passes through the atmosphere, it suffers from selective molecular absorption due to molecular rotational resonances mainly in water H_2O and oxygen O_2. Several empirical and analytical models were suggested for estimating the millimeter and infrared wave transmission of the atmospheric medium. In this study, the transmission characteristics of the atmosphere at the EHF band, was calculated with the millimeter propagation model (MPM), developed by Liebe [6-9]. It is based on calculation of the local index of refraction presented by:

$$n(f)=\sqrt{\varepsilon_r(f)}=1+N(f)\cdot 10^{-6} \tag{14}$$

Where the complex refractivity given in ppm is:

$$N(f)=N_0+N'(f)-jN''(f) \tag{15}$$

The term in the refractivity that is constant in frequencies is given by a real quantity N_0. The frequency dependent term of the refractivity is a complex function consisting of a real $N'(f)$ and imaginary $N''(f)$ quantities. The propagation factor can be written in terms of the index of refraction:

$$\begin{aligned} k(f)&=\frac{2\pi f}{c}\cdot n(f)= \\ &=\underbrace{-j\frac{2\pi f}{c}\cdot N''(f)\cdot 10^{-6}}_{\alpha(f)}+\underbrace{\frac{2\pi f}{c}\cdot\left(1+N_0\cdot 10^{-6}\right)+\frac{2\pi f}{c}\cdot N'(f)\cdot 10^{-6}}_{\beta(f)} \end{aligned} \tag{16}$$

We identify the expression of the field attenuation coefficient [10]:

$$\alpha(f)=-\mathrm{Im}\{k(f)\}=-\frac{2\pi f}{c}\cdot\mathrm{Im}\{n(f)\}=\frac{2\pi f}{c}\cdot N''(f)\cdot 10^{-6} \tag{17}$$

and the wavenumber:

$$\begin{aligned} \beta(f)&=\mathrm{Re}\{k(f)\}=\frac{2\pi f}{c}\cdot\mathrm{Re}\{n(f)\}= \\ &=\frac{2\pi f}{c}\cdot\left(1+N_0\cdot 10^{-6}\right)+\underbrace{\frac{2\pi f}{c}\cdot N'(f)\cdot 10^{-6}}_{\Delta\beta(f)} \end{aligned} \tag{18}$$

The MPM model was used to calculate the transmission characteristics of the atmosphere at the EHF band, as shown in Figure 3. Curves are drawn for several values of relative-humidity (RH), assuming clear sky and no rain. Inspection of Figure 3.a reveals absorption peaks at 22GHz and 183GHz, where resonance absorption of water (H_2O) occurs, as well as absorption peaks at 60GHz and 119GHz, due to absorption resonances of oxygen (O_2). Between these frequencies, minimum attenuation is obtained at 35GHz (Ka-band), 94GHz (W-band), 130GHz and 220GHz, which are known as atmospheric transmission 'windows'. The behavior of the wavenumber increment $\Delta\beta(f)$ is shown in Figure 3.b demonstrating the expected phase changes in propagating signal.

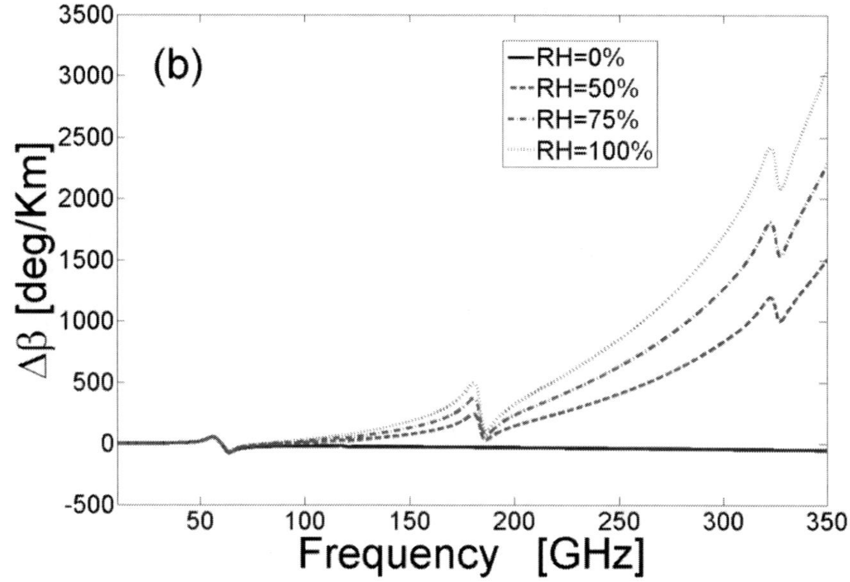

(b)

Figure 3. Millimeter wave a) Attenuation coefficient $20\log(e)\cdot\alpha(f)$ in [dB/Km] and b) $\Delta\beta(f)=\frac{2\pi f}{c}\cdot N'(f)\cdot 10^{-6}$ in [deg/Km] for various values of relative humidity (RH) at sea level.

Terrestrial Communication Link

Terrestrial wireless communication, operating in millimeter waves, is usually based on a line-of sight horizontal link operating in the lower layer of the troposphere. The power transfer function along the propagation path is given by:

$$|H(f)|^2=\left(\frac{d_0}{d_0+d}\right)^2\cdot e^{-2\alpha(f)d} \tag{19}$$

Figure 4 describes the power loss as a function of the distance d between the transmitting antenna and the receiving one (d_0 is some reference distance). Curves are drawn for frequencies of 35GHz and of 94GHz, where minimum attenuation ('transmission window') is obtained, as well for 60GHz and 119GHz where maximum absorption by oxygen molecules occurs. A curve for the path free space loss assuming no attenuation ($\alpha(f)=0$) is also shown for a reference.

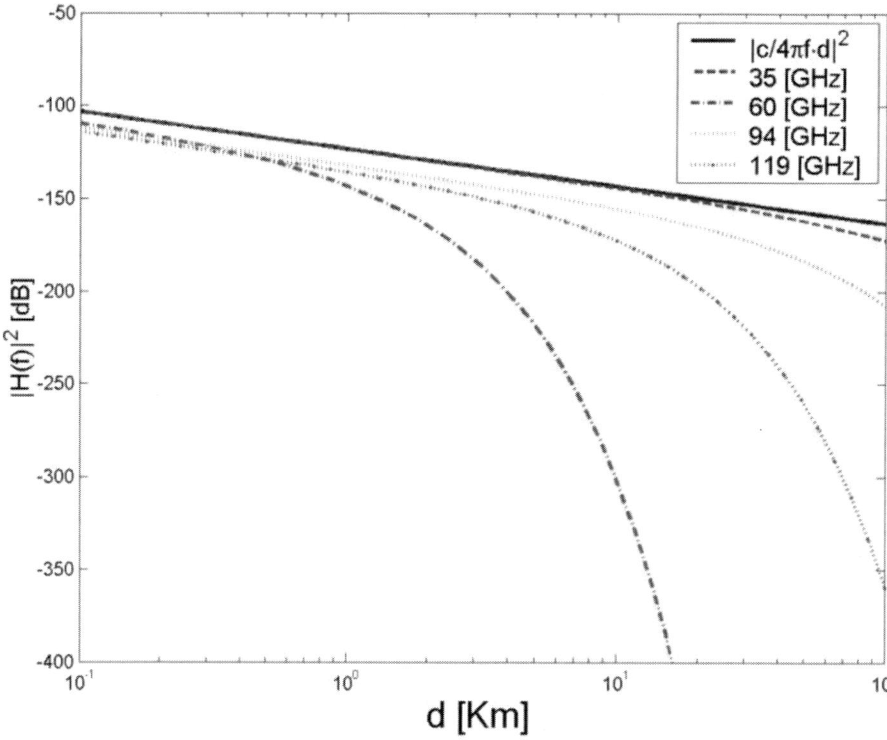

Figure 4. Power transfer at the "transmission window" frequencies of 35GHz and of 94GHz and at the oxygen absorption frequencies of 60GHz and of 119GHz.

Inspection of Figure 4 reveals that long range terrestrial communication links can be realized only at frequencies where minimum absorption is occurred, i.e. atmospheric windows in the Ka band (35GHz) and in the W band (94GHz). However, communication at 60GHz can be established only in short distances, due to the high atmospheric attenuation caused by the absorption of the oxygen molecules at this frequency region [11-14].

The transmission characteristics are determined by weather conditions such as temperature, pressure and humidity. The absorption is proportional to air density, and thus reduces with height. Attenuation due to fog, haze, clouds, rain and snow is one of the dominant causes of fading in wireless communication links operating in the EHF band. Raindrops and dust scatter millimeter wave radiation, resulting in amplitude fluctuations and phase randomness in the received signal. This further degrades the availability and performance of the communication links. Sufficient fade margins are essential for a reliable system. Figure 5 shows the power attenuation as a function of the atmospheric 'visibility' VS in the atmosphere. Typical values of VS are 1.1Km for haze, 0.27Km for fog, and 0.03Km for convective cloud. Severe attenuation in the propagation of millimeter wave radiation arises in a rainy weather as shown in Figure 6.

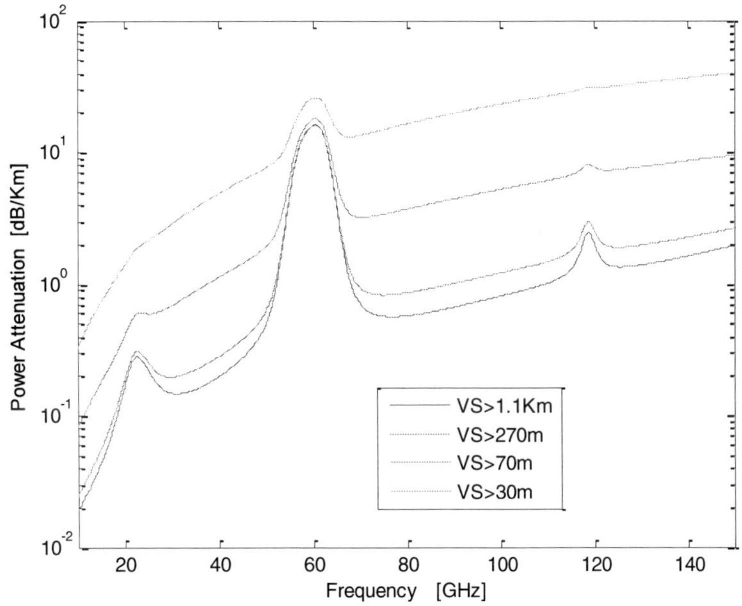

(a)

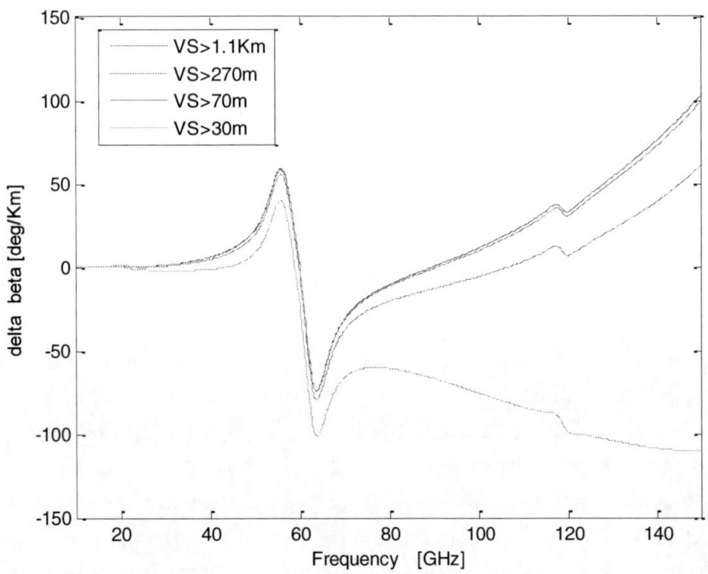

(b)

Figure 5. a) Attenuation coefficient $20\log(e)\cdot\alpha(f)$ in [dB/Km] and b) $\Delta\beta(f)$ in [deg/Km] for various values of visibilities.

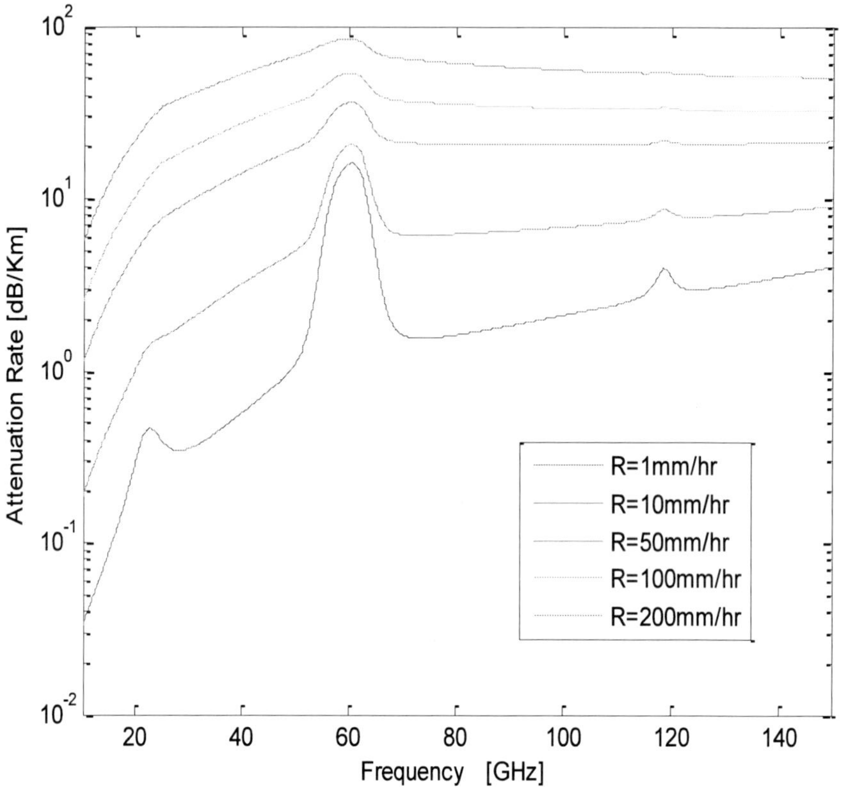

Figure 6. Attenuation curves for several values of rain dropping rate.

SATELLITE EHF CHANNEL

Among the practical advantages of using the EHF region for satellite communications systems is the ability to employ smaller transmitting and receiving antennas. This allows the use of a smaller satellite and a lighter launch vehicle. The main disadvantages of satellite communications at EHF arise from the increased free-space path loss, the increased water vapor absorption and rain fading as well as the low efficiency and relatively low power available from currently available sources [15-19]. For the higher part of the EHF spectrum, such as the W-band, no realistic model is given for the effects of clouds, gases and atmospheric precipitations. Moreover, physical layer issues, such as: phase noise, carrier recovery in the presence of large Doppler shifts and nonlinear distortions require deep investigations and involve stringent requirements on the baseband modem section [20].

Most of the attenuation is accumulated in the troposphere, which is the lowest portion of Earth's atmosphere, containing approximately 75% of the atmosphere's mass and 99% of its water vapor and aerosols. Fig. 7.b shows the accumulated attenuation at 94GHz. It can be shown that most of the attenuation is caused by the lowest 5-6Km layer of the troposphere, depending on humidity. In the worst cases it accumulates about 3.5 dB loss in addition to the free space loss present. For all practical purposes this should be taken as an upper limit for vertical losses.

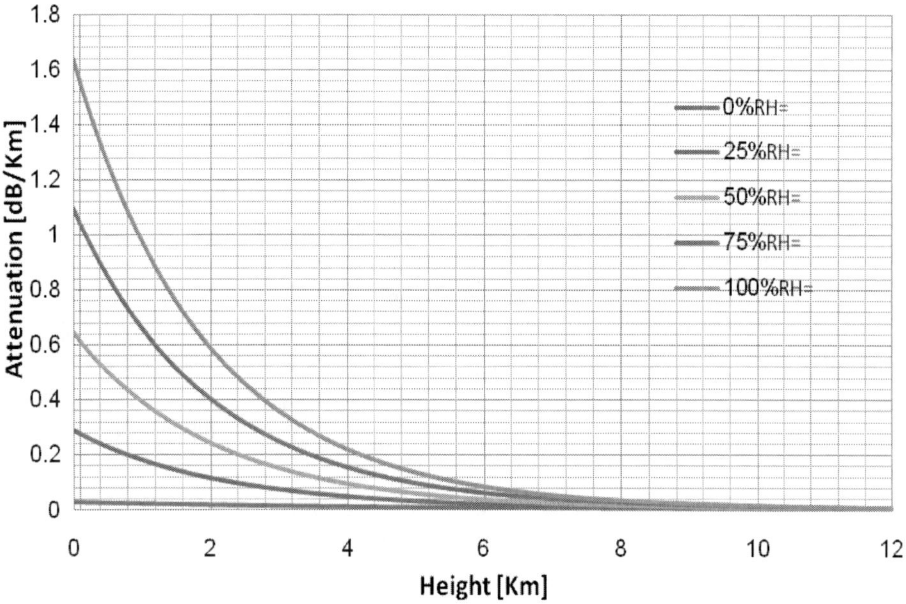

(a)

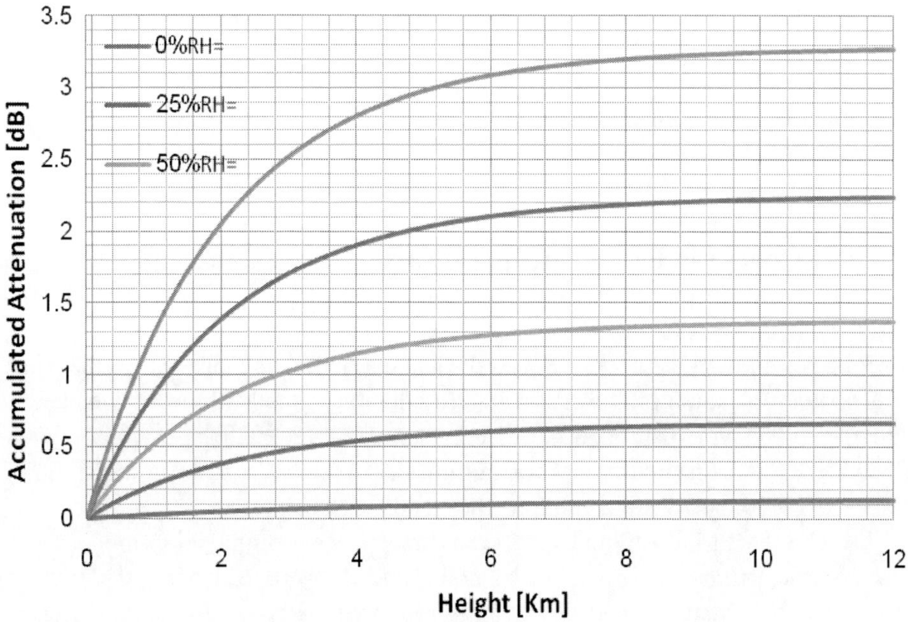

(b)

Figure 7. Attenuation at 94GHz as a function of height above sea level: a) local attenuation coefficient $20\log(e)\cdot\alpha(f)$ in [dB/Km] and b) accumulated attenuation in [dB].

SUMMARY

The inhomogeneous transmission in a band of frequencies causes absorptive and dispersive effects in the amplitude and in phase of wide-band signals transmitted in the EHF band. The frequency response of the atmosphere plays a significant role as the data rate of a wireless digital radio channel is increased. The resulting amplitude and phase distortion leads to inter-symbol interference, and thus to an increase in the bit error rate (BER). These effects should be taken into account in the design of broadband communication systems, including careful consideration of appropriate modulation, equalization and multiplexing techniques.

In this chapter we reviewed some of the key issues related to the establishment of an ultra-wide band wireless communication links in the millimeter wavelengths. The presented space-frequency model takes into account the inhomogeneous frequency behavior of the atmosphere. It enables calculation of broadband quadrature baseband components of a modulated signal, while propagating in a lossy-disperssive media. We considered EHF channel propagation aspects in the framework of terrestrial and satellite communications.

REFERENCES

[1] R. K. Crane: "*Propagation phenomena affecting satellite communication systems operating in the centimeter and millimeter wavelength bands", Proc. IEEE*, vol.59, no.2, Feb. 1971, pp. 173-188.

[2] L. J. Ippolito: "*Radio propagation for space communication systems", Proc. IEEE*, vol. 69, no. 6, Jun. 1981, pp. 697-727.

[3] H. J. Liebe: "Atmospheric EHF window transparencies near 35, 90,140 and 220 GHz", *IEEE Trans. on Antennas and Propagat.*, vol. 31, no.1, Jan. 1983, pp. 127-135.

[4] Y. Pinhasi, A. Yahalom, O. Harpaz, and G. Vilner: "Study of ultra wideband transmission in the extremely high frequency (EHF) band", *IEEE Trans. on Antennas and Propagat.*, vol. 52, no.11, Nov. 2004, pp. 2833-2842.

[5] Y. Pinhasi, and A. Yahalom "Spectral Characteristics of Gaseous Media and Their effects on Propagation of Ultra-Wideband Radiation in the Millimeter Wavelengths", *J. Non-Cryst. Sol.*, vol. 351, no. 33-36, 15 Sept. 2005, 2925-2928

[6] H. J. Liebe: "An updated model for millimeter wave propagation in moist air", *Radio Science*, vol. 20, no.5 ,May 1985, pp. 1069-1089.

[7] H. J. Liebe: "MPM – An atmospheric millimeter-wave propagation model", *Int. J. of Infrared and Millimeter waves*, vol.10, no.6, June 1989, pp. 631-650.

[8] H. J. Liebe, T. Manabe, and G. A. Hufford: "Millimeter-wave attenuation and delay rates due to fog / cloud conditions", *IEEE Trans. on Antennas and Propagat.,* vol. 37, no.12, Dec. 1989, pp. 1617-1623.

[9] H. J. Liebe, G. A. Hufford, and T. Manabe: "A model for the complex permittivity of water at frequencies below 1THz", *Int. J. of Infrared and Millimeter Waves,* vol. 12, no.7, July 1991, pp. 659-675.

[10] Y. Pinhasi, A. Yahalom, and G. A. Pinhasi: "Propagation Analysis of Ultra-Short Pulses in Resonant Dielectric Media", *J. Opt. Soc. Am. B*, vol. 26, no. 12, Dec. 2009, pp. 2404-2413.

[11] J.H. van Vleck, "The absorption of microwaves by oxygen", *Phys. Rev.*, 71, (1947), 413-424
[12] P. W. Rosenkranz, "Shape of the 5 mm oxygen band in the atmosphere", *IEEE Trans. On Antennas and Propagation*, 23 (1975), 498-506
[13] H.J. Liebe, P. W. Rosenkranz and G.A. Hufford, "Atmospheric 60GHz oxygen spectrum: new laboratory measurements and line parameters", *J. of Quantitative Spectroscopy and Radiative Transfer*, 48, (1992), 629-643
[14] F. Giannetti, M. Luise and R. Reggiani, "Mobile and personal communications in the 60GHz band: a survey", *Wireless Personal Communications*, 10, (1999), 207-243
[15] R.A. Bohlander, and R. W. McMillan: *"Atmospheric effects on near millimeter wave propagation", Proc. IEEE*, vol.73, no.1, Jan. 1985, pp. 49-60.
[16] ITU-R P.618, *"Propagation data and prediction methods required for the design of Earth-space telecommunication systems".*
[17] J.D. Laster, and W.L. Stutzman, "Frequency scaling of rain attenuation for satellite communication links", *IEEE Transactions on Antennas and Propagat.*, vol.43, no. 2, Nov 1995, pp.1207-1216.
[18] S. De Fina, M. Ruggieri, A. V. Bosisio, "Exploitation of the W-band for High-Capacity Satellite Communications", *IEEE Transactions on AES*, vol. 39, no. 1; January 2003, pp. 82-93.
[19] C. Sacchi, G. Gera, and C. Regazzoni, "W-band Physical Layer Design Issues in the Context of the DAVID-DCE Experiment", *Int. Jour. of Satellite Communications and Networking*, vol. 22, no. 2, March-April 2004, pp. 193-215.
[20] E. Cianca, T. Rossi, A. Yahalom, Y. Pinhasi, J. Farserotu, C. Sacchi: *"EHF for Satellite Communications: the New Broadband Frontier",* Proceedings of the IEEE | Vol. 99, No. 11, November 2011

ABOUT THE AUTHORS

Yosef Pinhasi is the Dean of the Faculty of Engineering at the Ariel University of Samaria. He was born in Israel on May 3, 1961, received the B.Sc., M.Sc. and Ph.D. degrees in electrical engineering from Tel-Aviv University, Israel in 1983, 1989 and 1995 respectively. He served as the head of the Department of Electrical and Electronic Engineering between the years 2004-2007.

Since 1990 he is working in the field of electromagnetic radiation, investigating mechanisms of its excitation and generation in high power radiation sources like microwave and millimeter wave electron devices, free-electron lasers (FELs) and masers. He developed a unified coupled-mode theory of electromagnetic field excitation and propagation in the frequency domain, enabling study of wideband interactions of electromagnetic waves in media in the linear and non-linear (saturation) operation regimes.

Prof. Pinhasi investigates utilization of electromagnetic waves in a wide range of frequencies for various applications such as communications, remote sensing and imaging. The space-frequency approach, which developed by him, is employed to study propagation of wide-band signals in absorptive and dispersive media in broadband communication links, and wireless indoor and outdoor networks as well as in remote sensing Radars operating in the millimeter and Tera-Hertz regimes. http://www.ariel.ac.il/sites/yosip/

Asher Yahalom is an Associate Professor in the Faculty of Engineering at the Ariel University Center of Samaria and the Academic director of the free electron laser user center which is located within the University Center campus. He was born in Israel on November 15, 1968, received the B.Sc., M.Sc. and Ph.D. degrees in mathematics and physics from the Hebrew University in Jerusalem, Israel in 1990, 1991 and 1996 respectively. Asher Yahalom was a postdoctoral fellow (1998) in the department of electrical engineering of Tel-Aviv University, Israel. In 1999 he joined the faculty of the Ariel University Center of Samaria. During the years 2005-2006 he was a Senior Academic Visitor in the University of Cambridge, Cambridge, UK. Details can be found in: http://www.ariel.ac.il/sites/ayahalom/

Gad Pinhasi is a senior lecturer in the Faculty of Engineering at the Ariel University Center of Samaria, at the Department of Chemical Engineering and Biotechnology. He was born in Israel on August 4, 1963, received the B.Sc., M.Sc. degree from Ben-Gurion University of the Negev and Ph.D. degree from Tel-Aviv University Israel, all in mechanical engineering, in 1990, 1993 and 2001 respectively.

Since 2001 he is working in numerical analysis and computational fluid dynamics, investigating mechanisms of two-phase flashing flow. He developed a novel model for predicting the thermodynamic and the dynamic state of the boiling liquid during a boiling liquid expanding vapor explosion (BLEVE) event. http://www.ariel.ac.il/sites/gadip/

In: Communication Systems: New Research
Editor: Vyacheslav Tuzlukov
ISBN: 978-1-62618-654-5

Chapter 8

MINIMUM SYMBOL ERROR RATE PULSE DESIGN WITH ROBUSTNESS TO TIMING JITTER

Jacob Gunther and Todd Moon*
Department of Electrical and Computer Engineering
Utah State University
Logan, Utah

Abstract

For a communication system employing linear modulation, an expression for the probability of symbol error is derived under the assumption that the receive matched filter is sampled with a timing offset. This expression is used as an objective to design filters for digital data transmission that are robust in the sense that they attain low symbol error rates in the presence of sampling jitter. Several simplifications lead to a convex objective function which is easily optimized subject to linear constraints for controlling attributes such as stopband attenuation. The simplified objective measures the inner boundary of the eye pattern at points away from the ideal sample times. Filters designed using this technique satisfy the Nyquist property and have equiripple behavior in the stopband. Comparisons with existing filter design techniques are presented.

1. Introduction

Linear pulse amplitude modulation is one of the most widely used formats for digital communication systems. Receivers for this type of modulation usually pass the received signal through a filter that is matched to the transmitter pulse and then sample the output of this filter at the symbol rate to produce symbol decision variables. Sample timing errors, also known as timing jitter, cause intersymbol interference and increases the probability of symbol decision error. We present an approach for designing transmit and receive filters to be used with linear digital modulation. A specific definition for *robustness* is adopted in the following: A filter is said to be robust if it provides for low symbol error rates in the

*E-mail address: jake.gunther@usu.edu, todd.moon@usu.edu

presence of sample timing errors at the output of the receive matched filter. The relationship between robustness, intersymbol interference, the Nyquist criterion, and eye patterns is discussed further ahead.

1.1. Background

Achieving low symbol error rates with digital communication systems employing linear modulation depends upon the co-occurrence of two special circumstances: (1) the combination of transmitter filtering, channel response, and receiver filtering must satisfy the Nyquist property; and (2) perfectly synchronized sampling must occur. For non-dispersive channels, the first condition can be achieved through careful design of the transmit filter and by using a matched filter at the receiver. However, even if perfect Nyquist filter responses are employed, the zero intersymbol interference (ISI) property is attained only if the receive filter output is sampled at the ideal sampling instants, $t = kT_s$, where T_s is the symbol period. Any distortion in sample times leads to ISI and results in an increase in the probability of error. While designing Nyquist and root-Nyquist filters has been and continues to be an active area for research [1–10], robustness to timing jitter has gained increasing attention in recent years.

Without loss of generality, the zero-ISI property is satisfied by any continuous-time filter whose impulse response passes through zero at all non-zero integer multiples of the symbol period and whose response is equal to one at time zero. Additionally, to have practical value the filter's frequency response must be bandlimited. Nyquist [11,12] showed that the magnitude frequency response of a bandlimited, zero ISI filter has odd symmetry about a frequency equal to half the symbol rate. The set of bandlimited filters satisfying this condition is infinite. Thus there is an opportunity to explore this set to find filters that have other desirable properties. Characteristics that have been used as the basis for comparison in the literature are passband energy compaction [9,10] and robustness to timing jitter [1,6,8].

The raised-cosine filter [13, 14] is perhaps the most widely known Nyquist filter, being described in most textbooks. Other filter shapes have also been proposed. For example, Beaulieu, Tan and Damen [6] showed that, relative to the raised-cosine filter, a filter having a flipped-exponential rolloff in the transition band has faster decaying tails in its impulse response and a more open eye pattern leading to lower probability of error in the presence of sample timing errors. Assalini and Tonello [8] examined the bit error rate versus sample timing error performance of filters having fliped-hyperbolic secant and flipped-inverse hyperbolic secant magnitude response in the transition band.

In a digital communication system employing linear modulation, it is the cascade combination of the transmit and receive filters that must satisfy the Nyquist criterion. Therefore, a given continuous-time Nyquist filter must be factored into a matched pair of square-root Nyquist filters which are then applied, one at the transmitter and one at the receiver. The square-root raised-cosine filter is available in many books [13, 14]. Additional work is required to find square-roots versions of the filters in [6, 8].

Often the transmit and receive filtering is performed in discrete-time. Therefore, finite-length discrete-time approximations of continuous-time square-root Nyquist filters are needed. Sampled representations of continuous-time square-root Nyquist filters loose no information provided the sample rate is sufficiently high and that an infinite number of uniformly spaced samples are preserved. However, as soon as these filters are truncated, they loose both the bandlimited property and the zero ISI property. Assalini and Tonello [8] explain that filters whose impulse response tails decay faster suffer less when truncated than those with slowly decaying impulse responses.

The main drawback of the Nyquist and square-root Nyquist filters described in [6, 8, 13, 14] is that in a finite-length discrete-time representation the only adjustable parameters are the excess-bandwidth and truncation length. There is little control over the truncation-induced spectral growth in the stopband.

A variety of Nyquist and root-Nyquist filter design algorithms have been developed to specifically control stopband spectral growth and residual ISI. From their inception, these filter design approaches work with finite-length discrete-time representations. Thus, the results are immediately useful in a discrete-time system implementation. Vaidyanathan and Nguyen [2] adopted a squared error objective to control both frequency-domain and time-domain properties of filters. The formulation leads to finding the minimal eigenvector of a symmetric positive definite matrix. So it is reasonably fast, and computer software is readily available. Building upon [2], Tuncer [7] explored ISI-free and approximately ISI-free designs. The latter trades ISI in the time domain for stopband attenuation in the frequency domain.

Another objective for filter design is maximizing the fraction of energy that is concentrated within the desired bandwidth subject to the zero-ISI constraint. To achieve this objective, Sood and Xiao [9] expand the root-Nyquist pulse in a basis of prolate spheroidal wave functions (PSWF). Then the energy objective leads to a quadratic program and the zero-ISI property leads to quadratic equality constraints. A solution to this hard problem is obtained using duality. Nigam, Singh and Chaturvedi [10] arrived at a simpler problem by expanding the Nyquist pulse, rather than the root-Nyquist pulse, in PSWF basis. This led to a more easily solved linearly constrained linear program.

Minimax or equiripple objectives are also popular for designing Nyquist and root-Nyquist filters. Vaidyanathan and Nguyen [2] suggested an iterative approach in which each iteration involves an eigenvector computation. Samueli [3] developed a linear programming algorithm for designing equiripple Nyquist filters and then applied a spectral factorization to obtain root-Nyquist versions. In [4] Samueli developed an iterative equiripple design algorithm where each step requires solving a system of linear equations. Harris [15] developed an algorithm that iteratively adjusts the band edge frequencies of an equiripple filter until it satisfies the Nyquist property.

The family of generalized raised-cosine filters proposed by Alagha and Kabal [5] offer greater flexibility to control the asymptotic decay of the filter impulse response so that residual ISI due to truncation is reduced. However, Tan and Beaulieu [16] showed that the bit error rate at offset sampling times is better when using the standard raised-cosine filter than when using the generalized versions.

1.2. Contributions and Overview

While robustness to timing errors is used as a basis to compare filters in [6,8], in none of the foregoing work is robustness explicitly coded into the design algorithm. We note that Tufts and Berger [1] approach the idea of robust pulse design through the addition of a flatness constraint. Besides having zero impulse response at the ideal sample times, they argue that a filter will be robust to sample timing errors if the derivative of the impulse response is zero at the ideal sample times. Enforcing flatness through derivative constraints prevents the ISI from growing too quickly as the actual sample times depart from the ideal. Pulse designs with robustness to timing jitter and other uncertainties is also considered by Davidson [17].

To address the filter design problem in the presence of timing jitter, this chapter adopts a new strategy. We dispense with Nyquist's criterion because it does not connect with the idea of timing jitter which occurs away from the ideal sample times. Even enforcing derivative constraints as in [1] does not allow precise control over the filter response at points away from the ideal sampling instants. Two things are needed: (1) a means to evaluate the continuous-time filter response at arbitrary times, and (2) an objective for optimization. The first need is addressed by expanding the filter response in a basis of continuous-time orthogonal functions. Given the coefficients in the expansion, the filter response can be evaluated at any point in time. Rather than a PSWF basis as in [9], time-shifted sinc functions are used in the sequel. Whereas the frequency support of the PSWF basis is concentrated below the excess bandwidth, the sinc functions are not. Therefore, in the sinc-based formulation, bandwidth limiting constraints must be employed. These translate into linear inequality constraints in subsequent optimization problems.

The ultimate performance metric in any digital communication system is probability of error. We adopt robustness, or probability of error in the presence of timing errors, as the objective for pulse design. As we will see, rather than seeking a perfect ISI canceling solution at the ideal sampling instants, this objective seeks for a filter response that maintains a low ISI across a user-specified band of time offsets. A more open eye pattern and improved probability of error results.

The following sections are organized as follows. Section 2. develops a model for the decision variables in a baseband digital communication system subject to sample timing errors. Based on this data model, Section 3. derives an expression for the probability of symbol decision error. Section 4. develops a suitable set of constraints and states the minimum symbol error rate filter design problem. Section 5. presents comparisons between filters designed according to the minimum error rate principle with other techniques in the literature.

1.3. Notation

Signals, systems and transforms that depend on a continuous variable have parenthesized arguments: $p(t)$. Discrete-time signals and systems have subscripted arguments: p_n. Linear convolution is denoted $p(t) * q(t)$ or $p_n * q_n$, and $p^*(t)$ is the complex conjugate of the function $p(t)$.

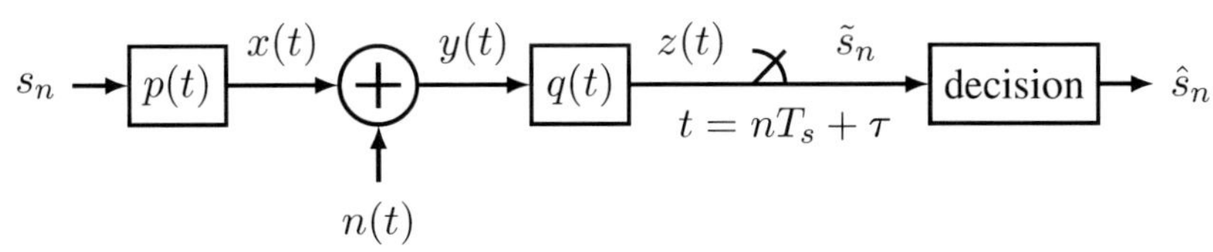

Figure 1. Continuous-time model of a baseband communication system employing linear modulation. Timing jitter τ is present in the samples $\tilde{s}_n$ at the output of the matched filter $q(t)$.

2. Data Model

The complex low-pass equivalent [13] of a bandpass communication system employing linear modulation is depicted in fig. 1. The transmitted signal consists of the pulse train

$$x(t) = \sum_n s_n p(t - nT_s), \tag{1}$$

where s_n are complex-valued symbols chosen from the M-ary set $\mathcal{C}$ and $p(t)$ is the impulse response of the transmit filter. Throughout this chapter, the receive filter is assumed to be matched to the transmit filter, $q(t) = p^*(-t)$, and, since $p(t)$ is real-valued, the conjugate can be dropped. The received signal is $y(t) = x(t) + n(t)$, where $n(t)$ is complex circular Gaussian white noise process. As this section deals mainly with a signal model in the presence of timing jitter, the additive noise in the channel is neglected here, but noise will be considered later when the probability of error is evaluated. Therefore, for the time being, suppose the received signal is $y(t) = x(t)$. In the absence of noise, the matched filter output admits the model

$$z(t) = \sum_n s_n r(t - nT_s), \tag{2}$$

where

$$r(t) = p(t) * q(t) = p(t) * p(-t) \tag{3}$$

is the combined response of transmit and receive filters. If the Nyquist criterion was enforced on $r(t)$, it would be required to satisfy

$$r(kT_s) = \delta_k = \begin{cases} 1, & k = 0, \\ 0, & k \neq 0, \end{cases} \tag{4}$$

and $p(t)$ and $q(t)$ would be referred to as root-Nyquist filters. Observe that Nyquist's criterion is silent about the response of $r(t)$ at points away from the ideal sample times. In this chapter, the Nyquist criterion is replaced by an objective that measures the probability of error when $r(t)$ is sampled with time offsets.

Ideally the matched filter output is sampled at $t = nT_s$, but a random process τ_n, referred to as timing jitter, distorts the sample times, and samples are obtained at $t =$

$nT_s + \tau_n$. Henceforth, the time dependence on τ is dropped with the understanding that τ is random, taking on new values at every sampling instant. The sample values $\tilde{s}_n = z(nT_s + \tau)$ are used for making symbol decisions according to

$$\hat{s}_n = \arg\min_{s \in C} |s - \tilde{s}_n|. \tag{5}$$

For convenience, a discrete-time parameterization for the filters $p(t), q(t)$, and $r(t)$ is used here. These filters have one-sided bandwidth of $(1+\beta)/2T_s$, where $\beta \in [0,1]$ is the excess bandwidth. Therefore, according to the sampling theorem [18], the continuous-time and discrete-time versions of these filters are related by

$$r(t) = \sum_{k=-\infty}^{\infty} r_k \varphi(t - kT), \qquad -\infty < t < \infty, \tag{6}$$

$$r_n = r(t)|_{t=nT}, \qquad n \in \{\cdots, -1, 0, 1, \cdots\}, \tag{7}$$

where $T = T_s/N$ is the sampling interval, $N \geq 2$ is an integer, and $\varphi(t)$ is the sinc function,

$$\varphi(t) = \frac{\sin(\pi t/T)}{\pi t/T}, \tag{8}$$

with Fourier transform

$$\Phi(F) = \begin{cases} T, & |F| \leq 1/2T, \\ 0, & \text{otherwise.} \end{cases} \tag{9}$$

A finite length approximation is obtained by truncating the sum in (6) to $2NL+1$ terms. Thus, in the following, we define

$$r(t) = \sum_{k=-NL}^{NL} r_k \varphi(t - kT), \qquad -JT_s \leq t \leq JT_s, \tag{10}$$

$$r_n = r(t)|_{t=nT}, \qquad n \in \{-NL, \cdots, NL\}, \tag{11}$$

where J is chosen large enough that the envelope of $r(t)$ in (10) has decayed to negligible values for $|t| > JT_s$. Thus, the ISI due to $r(t)$ in (10) spans $2J+1$ symbol periods. Substituting (10) into (2) lets the matched filter output be written in terms of r_n as

$$z(t) = \sum_k \sum_m s_m r_k \varphi(t - [k + mN]T) \tag{12}$$

$$= \sum_k \sum_m s_m r_{k-mN} \varphi(t - kT) \tag{13}$$

Define $\alpha - \tau/T_s = \tau/NT$ to be the normalized timing error, and let $z(t)$ be sampled with jitter at

$$t = nT_s + \tau = (n + \alpha)NT \tag{14}$$

to obtain the decision variables $\tilde{s}_n$ given by

$$\tilde{s}_n = \sum_k \left(\sum_m s_m r_{k-mN} \right) \varphi([(n+\alpha)N - k]T). \tag{15}$$

This form for the decision variables $\tilde{s}_n$ has an intuitive interpretation as resulting from a sequence of signal processing operations. The inner sum on m in (13) is a multirate convolution between an N-fold upsampled version of s_n with r_n. The outer sum on k shifts the time samples by αN samples through sinc interpolation, i.e. convolution with $\varphi([n+\alpha N]T)$, and finally downsamples the result by N. A more convenient form for $\tilde{s}_n$, involving matrix-vector products, can be obtained by inserting the sample times (14) into (12) and then changing variables using $i = n - m$. This leads to

$$\tilde{s}_n = \sum_{i=-J}^{J} \sum_{k=-NL}^{NL} s_{n-i} r_k \varphi([(i+\alpha)N - k]T), \tag{16}$$

which is now readily recognized as the matrix-vector product

$$\tilde{s}_n = \mathbf{r}^T \mathbf{F}(\alpha) \mathbf{s}_n, \tag{17}$$

where $\mathbf{F}(\alpha)$ is the $(2NL+1) \times (2J+1)$ block Toeplitz matrix having elements

$$[\mathbf{F}(\alpha)]_{(k,i)} = \varphi([(i+\alpha)N - k]T), \tag{18}$$

and

$$\mathbf{r} = [r_{-NL}, \cdots, r_0, \cdots, r_{NL}]^T, \tag{19}$$

$$\mathbf{s}_n = [s_{n+J}, \cdots, s_n, \cdots, s_{n-J}]^T. \tag{20}$$

It is useful to encode the filter design problem in terms of the square-root filter $p(t)$ and its samples p_n. Using (2), (3) and the sampled representations of $p(t)$ and $p(-t)$ we obtain decision variables $\tilde{s}_n$ written in terms of p_n as

$$\tilde{s}_n = \sum_k \sum_l \sum_m s_m p_{l-mN} p_{-[k-l]} \varphi([(n+\alpha)N - k]T). \tag{21}$$

This equation has the signal processing interpretation: s_n is upsampled by N, convolved with the transmit filter p_n, convolved with the matched filter p_{-n}, interpolated by αN, and finally downsampled by N. This processing is illustrated in fig. 2. From the perspective of the decision device looking back toward the information source, the systems in fig. 1 and 2 are equivalent in the sense that the that they both produce decision variables $\tilde{s}_n$ with the same signal, noise and timing jitter.

While the matched filter output z_n is oversampled by a factor of N, it is likely that none of these samples fall directly on the ideal zero-ISI sample times. Interpolators may be used to resample the signal at the desired times. As pictured in fig. 2, a timing recovery phase locked loop (PLL) can be used to control the interpolation and downsampling processes. Timing recovery is treated in [14, 19] where it is shown that, in the absence of noise, a

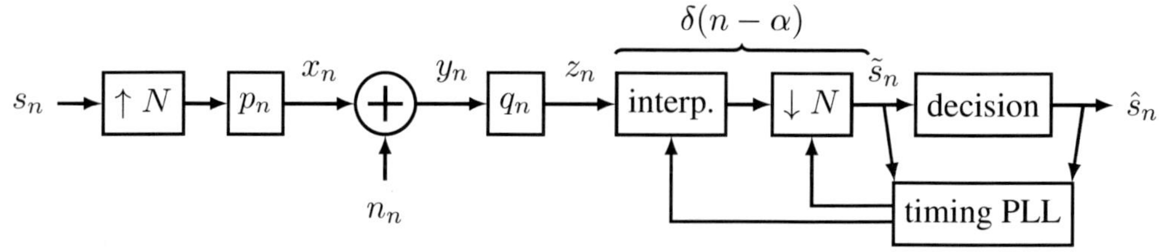

Figure 2. Discrete-time model of a baseband transmitter/receiver system including a symbol timing recovery loop. Timing jitter α is present in the interpolator outputs.

fairly simple PLL can perfectly synchronize with the received pulse train. Here "perfect synchronization" means having zero timing error after some initial transient. However, in practice the additive noise in the received signal passes through the timing recovery PLL and perturbs the interpolation process. This is one of the main sources of timing jitter. For the sake of this discussion, it is sufficient to think of the interpolator as a filter that implements a fractional sample delay, αN, which is random and varies with time.

3. Probability of Error

The developments in the previous section ignored noise. Let η_n be an i.i.d. discrete-time complex circular Gaussian process with zero mean and variance σ^2. To complete the model in (17), η_n should be added on the right hand side. The decision variables $\tilde{s}_n$ are affected by noise and by ISI due to both the pulse shape and timing error α. This section derives an expression for the probability of error due to all these factors. Minimization of the error probability with respect to $\mathbf{r}$ is considered in subsequent sections. Though expressions for error probability for ISI channels with Gaussian noise were previously derived in [20], an alternate route is pursued here.

The probability of error depends upon the distributions of the variables α, η_n, and $\mathbf{s}_n$. Under the assumption that these variables are stationary, expressions for the probability of error are time invariant. For convenience, we choose to evaluate the probability of error at time $n = 0$. Henceforth time dependence is dropped in the notation except when referring to the elements of the symbol vector $\mathbf{s}$ which are

$$\mathbf{s} = [s_{-J}, \cdots, s_{-1}, s_0, s_1, \cdots, s_J]^T. \tag{22}$$

Note that at time $n = 0$, the desired symbol is s_0. Any contribution to $\tilde{s}$ from the other symbols in $\mathbf{s}$ is ISI. With the notational updates described above, the model in (17) become

$$\tilde{s} = \mathbf{r}^T \mathbf{F}(\alpha)\mathbf{s} + \eta. \tag{23}$$

Suppose the symbols are drawn uniformly from the set $\mathcal{C} = \{c_1, c_2, \cdots, c_M\}$. We need to consider cases in which s_0 takes on different symbols in $\mathcal{C}$ and the notation

$$\mathbf{s}[c_i] = [s_{-J}, \cdots, s_{-1}, c_i, s_1, \cdots, s_J]^T \tag{24}$$

is used to indicate a vector of random variables in which the center variable s_0 takes on the specific value c_i. With this notation, assuming that $s_0 = c_i$ was transmitted, the probability of error is

$$P(\mathcal{E}|\mathbf{s}[c_i], \alpha) = P(\hat{s} \neq c_i|\mathbf{s}[c_i], \alpha), \tag{25}$$

where $\hat{s}$ is the result of the decision based on $\tilde{s}$ according to (5). Define the decision region $\mathcal{R}_i$ associated with c_i to be the set

$$\mathcal{R}_i = \{\tilde{s} \in \mathbb{C}||\tilde{s} - c_i| < |\tilde{s} - c_j|, \text{for all } j \neq i\}, \tag{26}$$

then the probability in (25) can be expressed as

$$P(\mathcal{E}|\mathbf{s}[c_i], \alpha) = P(\tilde{s} \in \cup_{j\neq i}\mathcal{R}_j|\mathbf{s}[c_i], \alpha) \tag{27}$$

$$= \sum_{j\neq i} P(\tilde{s} \in \mathcal{R}_j|\mathbf{s}[c_i], \alpha), \tag{28}$$

where the second equality follows from the disjointness of the decision regions. Because the shape of the regions $\mathcal{R}_i$ can be rather complicated, a simplification and bound is obtained by defining the half spaces

$$\mathcal{S}_{ji} = \{\tilde{s} \in \mathbb{C}||\tilde{s} - c_j| < |\tilde{s} - c_i|\}. \tag{29}$$

Then we have the bound

$$P(\mathcal{E}|\mathbf{s}[c_i], \alpha) \leq \sum_{j\neq i} P(\tilde{s} \in \mathcal{S}_{ji}|\mathbf{s}[c_i], \alpha). \tag{30}$$

This bound is a reasonably good approximation when the signal to noise ratio is high. Define $\mathcal{N}_i$ to be the indexes of the nearest neighbor symbols to c_i. As most of the error probability is due to deciding in favor of one of these neighbors, we have

$$P(\mathcal{E}|\mathbf{s}[c_i], \alpha) \approx \sum_{j\in\mathcal{N}_i} P(\tilde{s} \in \mathcal{S}_{ji}|\mathbf{s}[c_i], \alpha). \tag{31}$$

The error event can be expressed as

$$\{\tilde{s} \in \mathcal{S}_{ji}\} = \{\tilde{s} \in \mathbb{C}||\tilde{s} - c_j|^2 < |\tilde{s} - c_i|^2\} \tag{32}$$

$$= \left\{\tilde{s} \in \mathbb{C}\middle|\Re\left[\left(\tilde{s} - \frac{c_i + c_j}{2}\right)\left(\frac{c_i - c_j}{|c_i - c_j|}\right)^*\right] < 0\right\} \tag{33}$$

Using the model in (23), the error event in (33) can be expressed as

$$\left\{\eta \in \mathbb{C}\middle|\Re[\eta] > \Re\left[\left(\mathbf{r}^T\mathbf{F}(\alpha)\mathbf{s}[c_i] - \frac{c_i + c_j}{2}\right)\left(\frac{c_i - c_j}{|c_i - c_j|}\right)^*\right]\right\}. \tag{34}$$

Integrating over this set yields

$$P(\mathcal{E}|\mathbf{s}[c_i], \alpha) = \sum_{j\in\mathcal{N}_i} Q\left(\sqrt{\frac{2}{\sigma^2}}\Re\left[\left(\mathbf{r}^T\mathbf{F}(\alpha)\mathbf{s}[c_i] - \frac{c_i + c_j}{2}\right)\left(\frac{c_i - c_j}{|c_i - c_j|}\right)^*\right]\right), \tag{35}$$

where $Q(x) = \int_x^\infty \exp(-y^2/2)dy/\sqrt{2\pi}$ is the standard Gaussian tail probability.

The expression in (35) gives the conditional probability of error assuming a specific symbol sequence $\mathbf{s}[c_i]$ was transmitted and given a specific timing error α. The total (average) probability of error can be obtained as

$$P(\mathcal{E}|\alpha) = E_{c_i} E_{\mathbf{s}[c_i]} P(\mathcal{E}|\mathbf{s}[c_i], \alpha) \tag{36}$$

$$P(\mathcal{E}) = E_\alpha P(\mathcal{E}|\alpha). \tag{37}$$

Substituting from above yields a general expression for the total probability of error:

$$P(\mathcal{E}) = E_\alpha E_{c_i} E_{\mathbf{s}[c_i]} \sum_{j \in \mathcal{N}_i} Q\left(\sqrt{\frac{2}{\sigma^2}} \Re\left[\left(\mathbf{r}^T \mathbf{F}(\alpha)\mathbf{s}[c_i] - \frac{c_i + c_j}{2}\right)\left(\frac{c_i - c_j}{|c_i - c_j|}\right)^*\right]\right). \tag{38}$$

This expression is parameterized by the probability distribution on the timing error α and on the probability distribution on the symbols in the constellation $\mathcal{C}$.

The expectations over c_i and $\mathbf{s}[c_i]$ involve sums over all M^{2J+1} possible symbol vectors. Even for binary symbol alphabets where $M = 2$, these sums can be prohibitively expensive to compute. Lower complexity approaches are desirable. Furthermore, while the probability of error in (37) is not a convex function of $\mathbf{r}$, it can be transformed in several different ways into problems that are convex. Several convexity-revealing transformations, approximations and simplifications are considered below.

Because the Q function decays rapidly, most of the terms in expectations in (36) are insignificant. Indeed, a reasonable approximation is to retain only the terms that maximize the conditional probability of error in (35). Define

$$\mathcal{S}(\alpha) = \arg\max_{\mathbf{s}[c_i]} P(\mathcal{E}|\mathbf{s}[c_i], \alpha) \tag{39}$$

to be the set of symbol vectors $\mathbf{s}[c_i]$ that maximize $P(\mathcal{E}|\mathbf{s}[c_i], \alpha)$ for a given timing error α. The set $\mathcal{S}(\alpha)$ will depend upon the timing error α. To see this, note that the symbol sequence that maximizes the error probability for timing error α would be different than for an error of $-\alpha$. There is a direct correspondence between the symbol vectors in $\mathcal{S}(\alpha)$ and the eye pattern. In particular, given the filter $\mathbf{r}$, the inner boundary of the eye pattern may be generated by plotting $\tilde{s} = \mathbf{r}^T \mathbf{F}(\alpha)\mathbf{s}$ versus $\alpha \in [-0.5, 0.5]$ and for all $\mathbf{s} \in \mathcal{S}(\alpha)$. Examples will be shown in section 5..

Using the symbol vectors in $\mathcal{S}(\alpha)$ a simplified objective for filter design would be

$$P_{\max} = E_\alpha \left(\sum_{\mathbf{s}[c_i] \in \mathcal{S}(\alpha)} P(\mathcal{E}|\mathbf{s}[c_i], \alpha)\right). \tag{40}$$

The expectation over α in (40) overcomplicates the objective $P_{\max}$. A significant simplification may be obtained by recalling that the filter response is bandlimited. This prevents the impulse response from changing significantly over short time intervals. This has a direct impact on the eye pattern which is that the eye pattern may not fluctuate rapidly in a short interval. We can use this to advantage by focusing filter design on the error probability at only a few timing errors α. Let $\mathcal{A} = \{0, 0.5\alpha_{\max}, \alpha_{\max}\}$ be a set of three timing errors where the maximum offset $\alpha_{\max} \in [0, 0.5]$ is specified by the user. Because of the bandwidth limitation and the objective favoring low error probability, the eye pattern will tend to open wider for timing errors in the range $[-\alpha_{\max}, \alpha_{\max}]$. Though we have no proof of this, the examples shown in section 5. support this claim. Also, because of the even symmetry of the filter impulse response $\mathbf{r}$ (a constraint we enforce), only positive timing errors need be considered because the eye pattern will be symmetric.

Based on the foregoing considerations, given a user specified $\alpha_{\max}$, define $\mathcal{S} = \cup_{\alpha \in \mathcal{A}} \mathcal{S}(\alpha)$. An objective that could be optimized for filter design is

$$P_1(\mathbf{r}) = \sum_{\mathbf{s}[c_i] \in \mathcal{S}} P(\mathcal{E}|\mathbf{s}[c_i], \alpha_{\max}). \tag{41}$$

A special case of (35) is worth considering for its simplification. For the binary alphabet $\mathcal{C} = \{-1, +1\}$, the error probability in (35) reduces to

$$P(\mathcal{E}|\mathbf{s}[+1], \alpha) = Q\left(\sqrt{2}\, \mathbf{r}^T \mathbf{F}(\alpha) \mathbf{s}[+1]/\sigma\right). \tag{42}$$

Because of the symmetry of $\mathcal{C}$, $P(\mathcal{E}|\mathbf{s}[-1], \alpha) = P(\mathcal{E}|\mathbf{s}[+1], \alpha)$. Therefore, (42) is equal to $E_{c_i} P(\mathcal{E}|\mathbf{s}[c_i], \alpha)$. The symbol sequence that maximizes (42) can be determined in closed form. First define

$$\mathbf{r}(\alpha) = \mathbf{F}(\alpha)^T \mathbf{r}. \tag{43}$$

As Q is a monotonically decreasing function, maximizing (42) with respect to the $2J$ free variables in $\mathbf{s}[+1]$ is equivalent to minimizing the inner product

$$P_2(\mathbf{r}) = \mathbf{r}^T(\alpha)\mathbf{s}[+1] \tag{44}$$

$$= (+1) r_0(\alpha) + \sum_{i \neq 0} r_i(\alpha) s_i. \tag{45}$$

Minimizing the second term subject to the constraint $s_i \in \mathcal{C} = \{-1, +1\}$ leads to $s_i = -\text{sign}(r_i(\alpha))$. Substituting this into (45) yields

$$P_2(\mathbf{r}) = r_0(\alpha) - \sum_{i \neq 0} |r_i(\alpha)| \tag{46}$$

$$= 2r_0(\alpha) - \sum_{i=-J}^{J} |r_i(\alpha)| \tag{47}$$

$$= 2\mathbf{r}^T \mathbf{F}(\alpha) \mathbf{e}_0 - \|\mathbf{F}(\alpha)^T \mathbf{r}\|_1, \tag{48}$$

where $\mathbf{e}_0 = [0, \cdots, 0, 1, 0, \cdots, 0]^T$ is a $(2J + 1) \times 1$ vector of zeros with a 1 in the center position. In the binary case, plotting (48) versus α shows the inner boundary of the eye pattern. Note that the second term in (46)-(47) is the peak ISI. Thus $P_2(\mathbf{r})$ in (46) through (48) measures the maximum deviation away from $r_0(\alpha)$. Relating (46) back to (42) agrees with the intuition that the filter $\mathbf{r}$ that minimizes the maximum probability of error is related to the filter that maximizes (48), which is essentially equivalent to minimizing the peak ISI. Note that $P_2(\mathbf{r})$ in (48) is concave in $\mathbf{r}$ being the sum of affine and concave (negative 1-norm) functions [21].

4. Minimum Symbol Error Rate Filter Design

4.1. Constraints

4.1.1. Filter symmetry

The because $r(t)$ is the autocorrelation of $p(t)$, it should be symmetric. This can be enforced by requiring $r_n - r_{-n} = 0$ for $n = 1, 2, \cdots, NL$. Arranging these relations in matrix-vector form leads to NL linear equality constraints

$$\mathbf{B}\mathbf{r} = \mathbf{0}, \tag{49}$$

where the elements of $\mathbf{B}$ satisfy

$$[\mathbf{B}]_{i,j} = \delta_{i-j} - \delta_{i+j-2NL-2}, \tag{50}$$

$$i = 1, 2, \cdots, NL, \qquad j = 1, 2, \cdots, 2NL + 1. \tag{51}$$

4.1.2. Spectral mask

In (10) the filter $r(t)$ is represented using a finite length expansion in terms of $\varphi(t - nT)$. The functions $\varphi(t - nT)$ are bandlimited to $1/2T$ Hz, but $r(t)$ should ideally have all its energy concentrated below $(1 + \beta)/2T_s = (1 + \beta)/2NT < 1/2T$ Hz. This can be accomplished through constraints on the coefficients in $\mathbf{r}$.

Let $R_c(F)$ and $R_d(f)$ be the continuous-time and discrete-time Fourier transforms of $r(t)$ and r_n, respectively. These functions are related by

$$R_c(F) = R_d(FT)\Phi(F). \tag{52}$$

Note that because of even symmetry of r_n, $R_d(f)$ may be expressed as

$$R_d(f) = \sum_{n=-NL}^{NL} r_n \cos(2\pi f n) \tag{53}$$

$$= \mathbf{r}^T \mathbf{c}(f) \tag{54}$$

where $\mathbf{c}(f) = [\cos(-2\pi f NL), \cdots, \cos(-2\pi f), 1, \cos(2\pi f), \cdots, \cos(2\pi f NL)]^T$.

Consider the following simple spectral mask constraint

$$|R_c(F)| \leq \gamma, \qquad |F| \geq (1 + \beta)/2NT. \tag{55}$$

Because $\Phi(F)$ is already zero for frequencies above $1/2T$, we can focus on the frequency band $(1 + \beta)/2NT \leq |F| \leq 1/2T$. This is equivalent to the normalized frequency band $(1 + \beta)/2N \leq f \leq 1/2$. Sample this band at K points given by

$$f_i = \frac{1 + \beta}{2N} + i\delta, \qquad i = 0, 1, \cdots, K - 1, \qquad \delta = \frac{\frac{1}{2}\left(1 - \frac{1+\beta}{N}\right)}{K - 1}. \tag{56}$$

At each frequency, we require $-\gamma \leq \mathbf{r}^T \mathbf{c}(f_i) < \gamma$. Stacking these constraints leads to $2K$ linear inequality constraints

$$\mathbf{C}\mathbf{r} \leq \gamma \mathbf{1}, \tag{57}$$

where the first K rows of $\mathbf{C}$ are $\mathbf{c}(f_i)^T$ for $i = 0, 1, \cdots, K - 1$ and the last K rows are $-\mathbf{c}(f_i)^T$.

4.1.3. Impulse Response Constraints

The Nyquist property could easily be enforced on $\mathbf{r}$ using linear constraints. Even though this course is not pursued here, an inspection of the filters that are designed show that they do possess the Nyquist property.

The only impulse response constraint that we impose on $r(t)$ is $r(0) = 1$. This is easily accomplished by setting $r_0 = \mathbf{e}_0^T\mathbf{r} = 1$. This is a normalizing condition which assures that $p(t)$ has unit energy because $r(0) = \int p^2(t)dt$.

4.2. Filter Design

Assembling the objective in (48) with the symmetry, stopband and energy constraints above yields the following convex optimization problem.

$$\begin{aligned} &\text{maximize} \sum_{\alpha\in\mathcal{A}} 2\mathbf{r}^T\mathbf{F}(\alpha)\mathbf{e}_0 - \|\mathbf{F}(\alpha)^T\mathbf{r}\|_1 \\ &\text{subject to } \mathbf{Br} = \mathbf{0}, \quad \mathbf{e}_0^T\mathbf{r} = 1, \quad \mathbf{Cr} \leq \gamma\mathbf{1} \end{aligned} \tag{58}$$

5. Results

This section compares filters designed to minimize the probability of error in the presence of timing offsets with those presented in [2, 6, 8, 15]. Filters designed by the technique presented in this chapter, i.e. according to (58), are referred to as minimum symbol error rate (MSER) filters. In the following comparisons, the filters being compared are constrained to have the same length. The figure of merit used is the offset error rate which is defined to be the probability of error obtained in the presence of a given timing offset α. As spectral mask constraints must be satisfied in practice, another factor to be considered in the following comparisons is stopband attenuation. The filter shapes proposed in [6,8] offer no control over stopband attenuation when these infinite length filters are truncated. The design algorithms in [2, 15] directly control attenuation in the stop band. These differences makes a universal comparison difficult. Therefore, this section is divided into two parts: section 5.1. compares the proposed technique to those in [6, 8], section 5.2. compares with [2, 15].

5.1. Comparison with Raised-Cosine and Flipped-Exponential Filters

As the raised-cosine (RC) filter is the most widely recognized Nyquist filter, it will be used as a baseline for comparison. The flipped-exponential (FEXP) [6], flipped-hyperbolic secant (FSECH), and filpped-inverse hyperbolic secant (FARSECH) [8] filters have lower offset error rates than the RC filter. While FARSECH has the lowest error rate among these four filters [8], we will compare with the FEXP filter in addition to the RC filter. The offset error rate for FEXP is almost as good as that for FARSECH, and a closed form expression for the impulse response of FEXP is available. Initially, we focus on the MSER versus RC comparison.

An RC filter with 75% excess bandwidth was obtained by evaluating the impulse response given in [6]. The coefficients $\mathbf{r}$ for a MSER filter were obtained according to the optimization in (58). Each filter spanned $2L + 1 = 13$ symbol periods with $N = 4$ samples per symbol period. Because the truncated RC filter had an attenuation of -63 dB at the edge of the stopband ($f = (1+\beta)/(2N) = 0.21875$), the MSER filter frequency response was constrained to be below -63 dB over the same band. The maximum timing offset for the MSER filter was set to $\alpha_{\text{max}} = 0.1$. Frequency responses for these two filters are shown in fig. 3. The dashed vertical line in fig. 3 is placed at $f = 0.21875$ to mark the edge of the stopband. Consider the response of the RC filter. Its first sidelobe is 63 dB below the main lobe and the response decreases for higher frequencies. The MSER filter has an approximately equiripple response across the stopband with an attenuation of -63 dB. We have observed that equiripple behavior is typical for MSER filters. The MSER filters are not designed to have equi-amplitude stopband ripples. The optimization in (58) utilizes all available degrees of freedom to minimize a probability of error-related objective subject to spectral mask constraints. That MSER filters have equiripple behavior in the stopband indicates that some of the higher frequencies can be utilized to shape the time-domain response to improve the offset error rate.

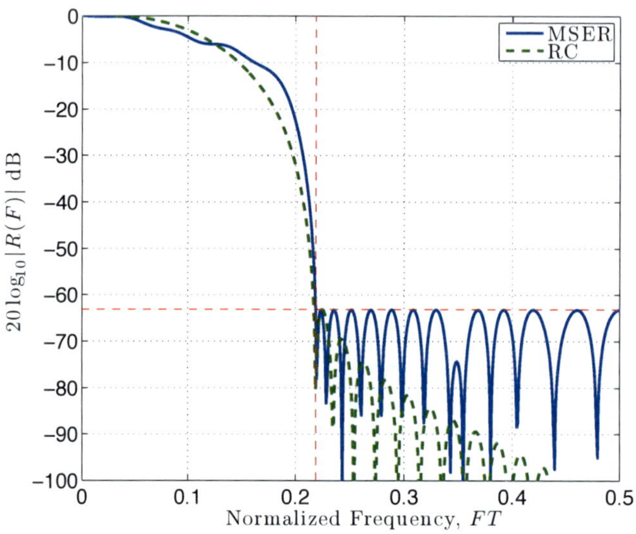

Figure 3. Frequency response of the MSER (solid) and RC (dashed) filters.

Because the pulses have been designed or truncated to a reasonably small length and because a binary symbol alphabet is being used, it is possible to calculate the offset error rate $P(\mathcal{E}|\alpha)$ exactly using (36). For the MSER and RC filters designed above, the error rates for an offset of $\alpha = 0.1$ are $5.96e-8$ for MSER and $1.53e-7$ for RC. This difference in error probability can be investigated by examining the corresponding eye patterns.

Eye patterns for MSER and RC filters appear in fig. 4. The dashed vertical lines correspond to a timing offset of $\alpha = \pm 0.1$ which is the maximum timing offset for the MSER designs. From an inspection of the eye patterns, it is difficult to determine which eye is more open. However, notice that the MSER filter does possess the Nyquist property even though this was not a design constraint.

As noted above, plotting (48) versus $\alpha \in [-0.5, 0.5]$ gives the inner boundary of the eye pattern. The small circles in the eye patterns in fig. 4 show these inner boundaries. In fig. 5 the inner boundaries for MSER and RC are shown together. We observe that the eye patterns are nearly equal over a wide range of timing offsets. The inset plot in fig. 5 shows a magnification of the eye pattern around $\alpha = 0.1$. It is clear that the MSER eye is more open. This agrees with the intuition that the more open eye should correspond with a lower probability of error. While this is indeed the case, there is more going on here than meets the “eye”.

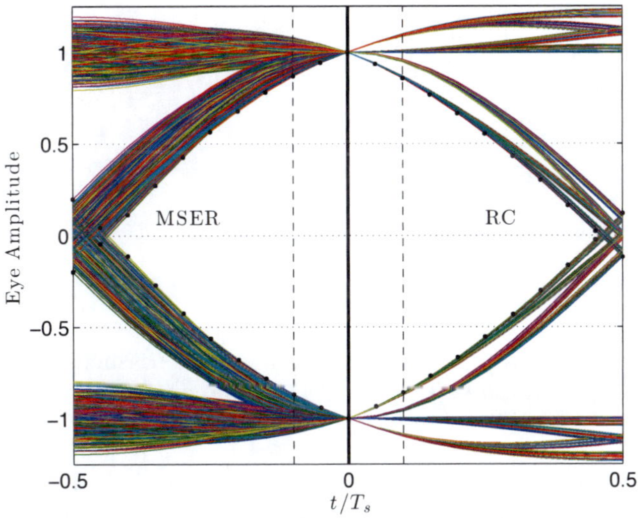

Figure 4. Eye patterns for the MSER (left hand side) and RC (right hand side) filters shown together in one split plot. The vertical dashed lines correspond to a timing offset of $\alpha = \pm 0.1$.

For a given time offset, the average probability of error behavior for a given filter is determined by the distribution of the traces across the eye pattern and not the single trace corresponding to the inner boundary. To investigate this idea, we computed cumulative distribution functions (CDF) for the eye patterns when sampled at time $\alpha = 0.1$ corresponding to the dashed vertical lines in fig. 4. Note that the CDFs are only for the positive portion of the eye. These CDFs appear in fig. 6. The stair step behavior of the RC CDF arises from the fact that the traces across the RC eye pattern are concentrated around a few points for offset $\alpha = 0.1$ (see fig. 4). The MSER CDF indicates that the traces across the MSER eye are fairly uniformly distributed. While fig. 5 suggests that the inner eye opening is about the same for the MSER and RC filters, the CDFs in fig. 6 reveals that MSER and RC filters distribute traces across the eye pattern very differently.

The significant difference in the offset probabilities of error between these two filters can be understood by comparing their CDFs. Notice that for low eye amplitudes the MSER CDF tends to lie below the RC CDF. At higher amplitudes, the reverse occurs. The reason that the MSER filter delivers a significantly lower offset probability of error is because the probability of error is so much higher for low signal amplitudes. For reference, fig. 6 also shows a plot of $\log_{10} Q(\sqrt{2A^2/N_0})$ which is the log of the probability of error for BPSK when A is the amplitude of the signal, i.e. A is the eye amplitude. This plot corresponds

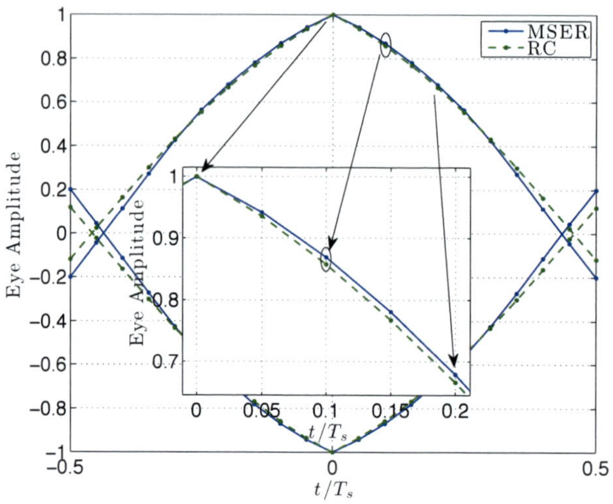

Figure 5. Inner boundaries of the MSER (solid) and RC (dashed) eye patterns. The inset plot magnifies the picture around the offset time $\alpha = 0.1$.

to E_b/N_0 of 12 dB when $A = 1$. Observe that the error probability is orders of magnitude higher in the lower amplitudes than in the higher amplitudes. The high density of low-amplitude traces across the RC eye pattern dominate its probability of error performance. Relative to the low-amplitude traces, the contribution of the high-amplitude traces to the probability of error is negligible. The density of low-amplitude traces across the MSER eye is much lower than for the RC eye. The probability of error performance for any filter is determined mainly by the way it distributes low-amplitude traces across the eye pattern. At low signal to noise ratios where there is less dynamic range in the probability of error across the range of eye amplitudes, there is also less difference in the performance gap. For example, at E_b/N_0 of 2 dB, the offset probabilities of error for the MSER and RC filters is 4.20×10^{-2} and 4.24×10^{-2}, respectively. The probability of error performance for a timing offset of $\alpha = 0.1$ across a range of signal to noise ratios is plotted in fig. 7.

The RC and MSER filters above have an excess bandwidth of 75%. Decreasing the excess bandwidth in the RC filter decreases the stopband attenuation (increases the gain). Raised cosine filters were obtained for β =50% and 25% excess bandwidths. The corresponding stop band attenuations were γ =-55 dB and -45 dB, respectively. MSER filters were designed with the same excess bandwidths and stopband attenuations. The offset (by $\alpha = 0.1$) probability of error was evaluated at $E_b/N_0 = $ 12dB for each of these filters and are given in Table 1. At each excess bandwidth setting, the probability of error of the MSER filter improves upon that for the RC filter.

Table 1 also compares the MSER filter with the FEXP filter [6] at the same signal to noise ratio and excess bandwidths as above. The frequency characteristics of the FEXP filter degrade more severely than the RC filter when its impulse response is truncated. At an excess bandwidth of 75% for example, the first sidelobe is only 45 dB down from the mainlobe and the rolloff of the sidelobes in the stopband is much slower than for the RC

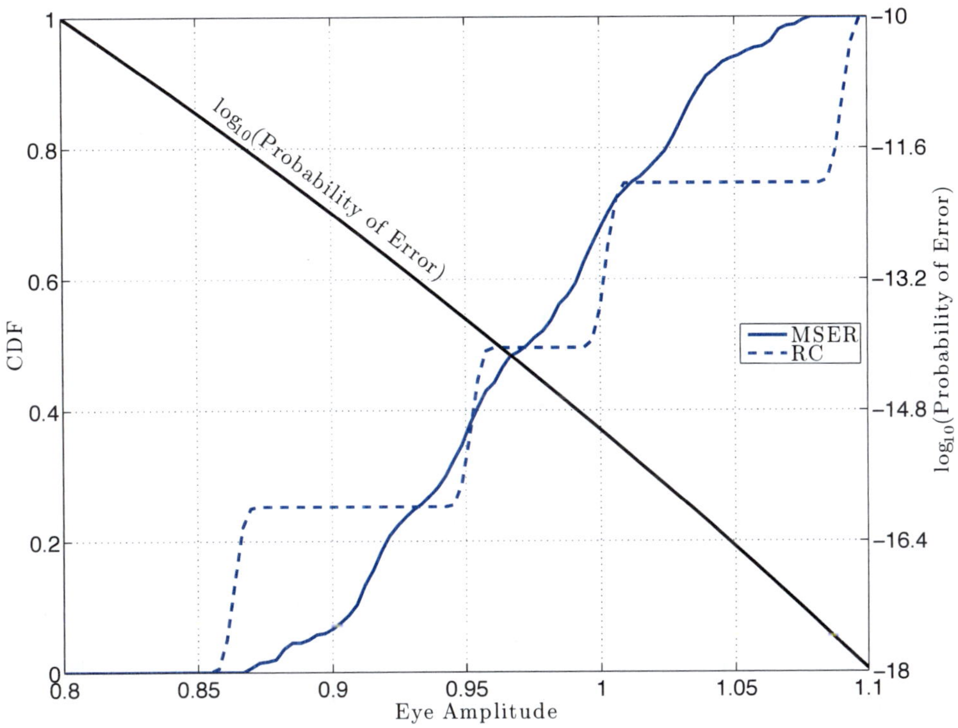

Figure 6. Cumulative distribution functions for the MSER (solid) and RC (dashed) filters having 75% excess bandwidth. The probability of error for a given eye amplitude is also shown.

filter. In addition, the magnitude frequency response of a truncated FEXP filter is down only 36.1 dB at the stopband edge frequency $f = (1 + \beta)/2N = 0.21875$. To correct the magnitude response at this frequency, the excess bandwidth for FEXP was decreased until the magnitude response at $f = 0.21875$ was at the same level as the first sidelobe. The corrected excess bandwidths were $0.96\beta, 0.93\beta$, and 0.89β at $\beta = 75\%, 50\%$, and 25%, respectively. With these corrections, the FEXP filter achieved the desired excess bandwidth. MSER filters were designed to achieve the same stopband attenuations as obtained by the FEXP filters. The offset probability of error was evaluated for each of the filters and appears in Table 1 along with the specified excess bandwidth and stopband attenuation.

5.2. Comparison with Nyquist Filter Design Techniques

The main disadvantage of the RC and FEXP filters is lack of control over the stopband attenuation obtained upon truncating these filters to be finite in length. The Nyquist filter design techniques presented in [2, 15] do provide control over stop band attenuation. For example, in the Nyquist eigenfilter approach [2] a parameter in the objective function trades off suppressing error in the passband versus attenuation in the stopband. This parameter can be tuned to obtain the desired level of stopband attenuation. In the iterative technique described in [15], each iteration designs a standard equiripple filter. The stopband attenuation

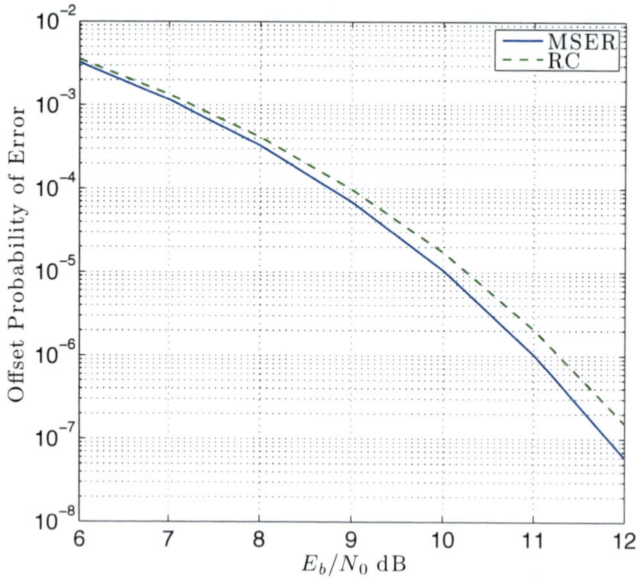

Figure 7. Offset probability of error performance for the MSER (above) and RC (below) filters. The timing offset is $\alpha = 0.1$. The excess bandwidth is 75%.

of equiripple filters can be specified directly.

This section compares filters designed according to [2] (denoted "Eigen" in the figures) and [15] (denoted "Equi" in the figures) to MSER filters with $\alpha_{\text{max}} = 0.1$. The length of each filter spanned $13 = 2L + 1$ symbol periods with $N = 4$ samples per symbol period. The stopband attenuation was set to -90 dB. This is much lower than the attenuation achieved by the RC or FEXP filters in the previous section. The frequency responses are shown in fig. 8, and the inner boundaries of the associated eye patterns are shown in fig. 9. The MSER eye is much more open than the eyes for either eigenfilter or equiripple filter. A better understanding of the distribution of the traces across the associated eye patterns can be obtained by examining the CDFs in fig. 10. Note that the CDFs are obtained by slicing the eye pattern at a time offset of $\alpha = 0.1$. The eigenfilter and equiripple filter

Table 1. Probability of Error for $E_b/N_0 = 12\ dB$, Timing Offset of $\alpha_{max} = 0.1$, BER=9.0060e-9 at the Optimum Sampling Time

Filter	(β = excess bandwidth, γ = stopband attenuation)		
	$(0.25, -45\text{dB})$	$(0.50, -55\text{dB})$	$(0.75, -63\text{dB})$
MSER	1.8023e-06	2.6882e-07	5.9552e-08
RC	2.7580e-06	5.3473e-07	1.5299e-07
	$(0.25, -33\text{dB})$	$(0.50, -43\text{dB})$	$(0.75, -45\text{dB})$
MSER	6.2642e-07	1.9693e-07	4.0141e-08
FEXP	1.6047e-06	2.2579e-07	4.9821e-08

distribute traces approximately uniformly. Not only is the MSER eye more open, but its traces appear to be concentrated around four amplitudes. Given the greater concentration of low-amplitude traces in the eigenfilter and equiripple filters, it is not surprising that they have higher offset error rates tabulated in Table 2. Probability of error curves for the three filters are shown in fig. 11 for a time offset of $\alpha = 0.1$. At high signal to noise ratios, the probability of error for MSER is five times better than for the other filters.

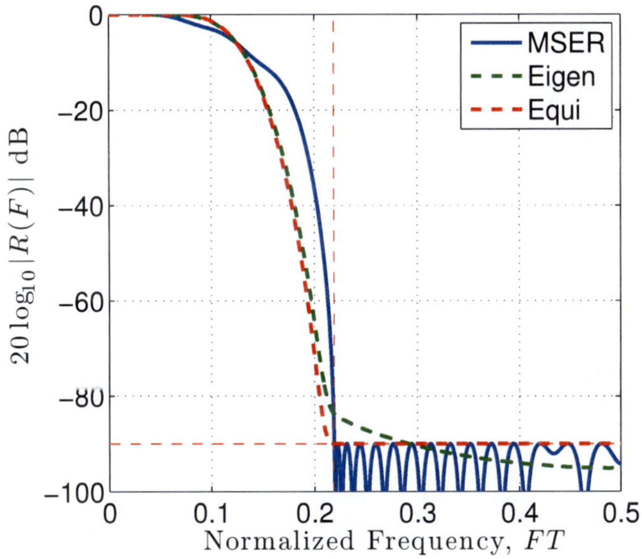

Figure 8. Filter frequency responses for MSER (blue-solid), Nyquist eigenfilter (green-dashed), and Nyquist equiripple (red-dashed) filters with 75% excess bandwidth.

These results highlight the utility of optimizing the offset error rate through the optimization problem described in (58). The eigenfilter minimizes error energy in pass and stop frequency bands subject to Nyquist constraints. The equiripple Nyquist filter minimizes the maximum error in pass and stop bands and iteratively enforces the Nyquist property. Neither of these design approaches is sensitive to the impulse response away from the ideal sample times. Therefore, they tend to behave poorly when sampled with an offset. The MSER filter design approach directly optimizes the impulse response to achieve good performance when subject to sampling offsets.

Table 2. Probability of Error for $E_b/N_0 = 12$ dB, Timing Offset of $\alpha_{max} = 0.1$, BER=9.0060e-9 at the Optimum Sampling Time

Filter	(β = excess bandwidth, γ = stopband attenuation)		
	$(0.25, -45\text{dB})$	$(0.50, -75\text{dB})$	$(0.75, -90\text{dB})$
MSER	1.8282e-06	5.6620e-07	1.5005e-07
Eigen	2.7250e-06	1.4131e-06	7.8197e-07
Equi	2.8452e-06	1.4322e-06	8.9575e-07

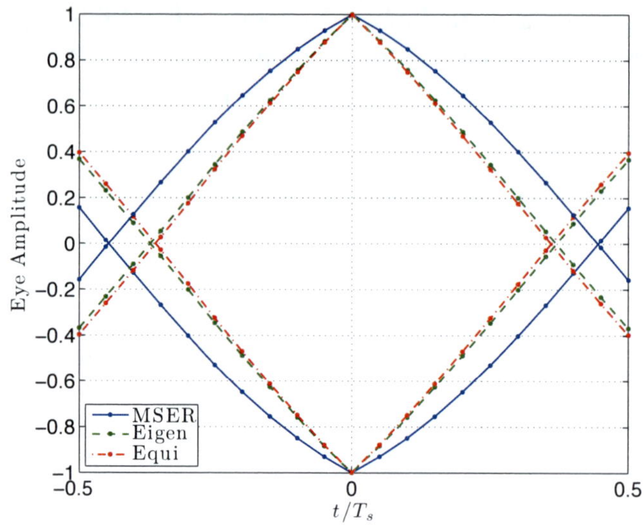

Figure 9. Inner boundary of the eye patterns for MSER (solid), Nyquist eigenfilter (dashed), and Nyquist equiripple (dash-dot) filters with 75% excess bandwidth.

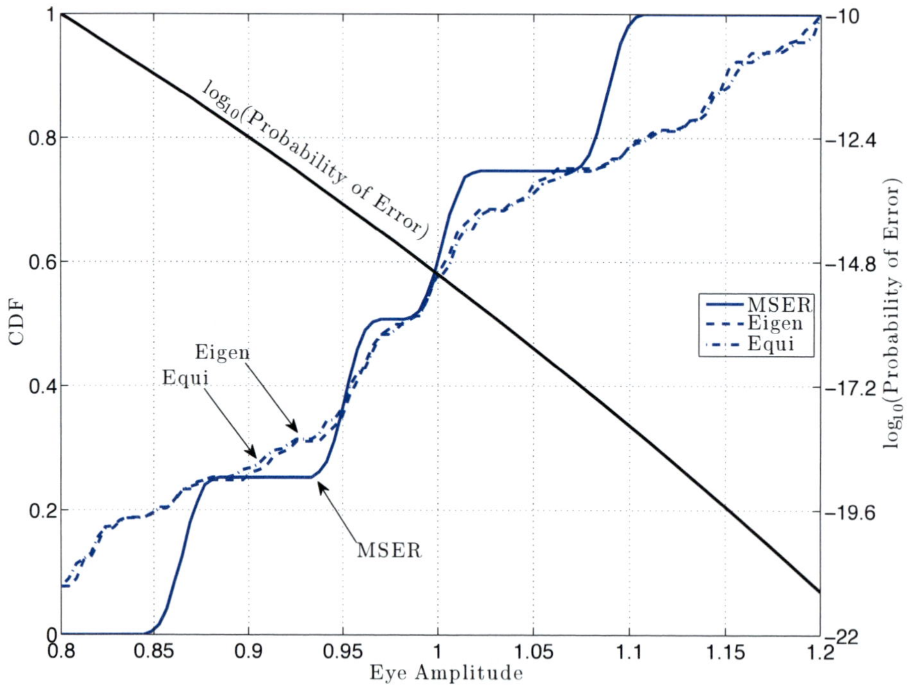

Figure 10. Cumulative distribution functions for the MSER (solid) and Nyquist eigenfilter (dashed) and Nyquist equiripple (dash-dot) filters having 75% excess bandwidth. The probability of error for a given eye amplitude is also shown.

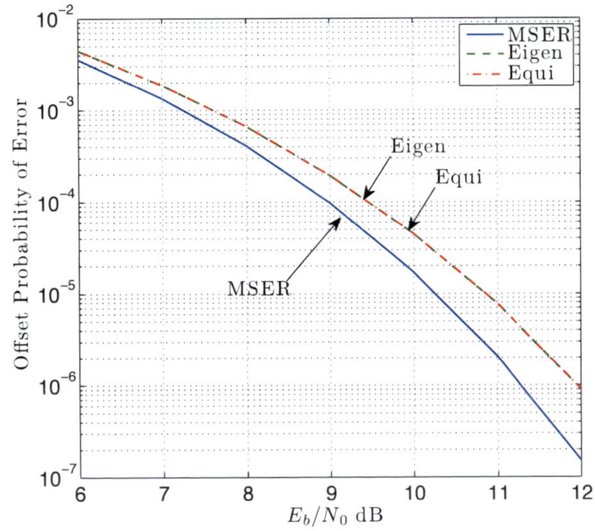

Figure 11. Offset probability of error performance for the MSER (solid), Nyquist eigenfilter (dashed) and Nyquist equiripple (dash-dot) filters. The timing offset is $\alpha = 0.1$. The excess bandwidth is 75%.

Conclusions

This chapter proposed using the offset probability of error as an objective for designing filters for digital data transmission. Using several simplifications, a convex optimization problem was obtained. The simplified objective measures the inner boundary of the eye pattern at points away from the ideal sample time. Symmetry and bandlimiting constraints were added to control the shape and stopband attenuation of the filters designed. Filters designed by this technique obeyed the Nyquist property at the ideal sample time. They exhibited equiripple behavior in the stopband. When compared to truncated raised cosine and flipped exponential filters with the same length and stopband attenuation, the minimum symbol error rate filters were more robust to timing jitter because they achieved lower error rates at offset sample times. The same is true for comparisons with Nyquist eigenfilters and Nyquist equiripple filters.

References

[1] D. Tufts and T. Berger, "Optimum pulse amplitude modulation-II: Inclusion of timing jitter," *IEEE Trans. on Information Theory*, vol. 13, pp. 209–216, Apr. 1967.

[2] P. P. Vaidyanathan and T. Q. Nguyen, "Eigenfilters: A new approach to least-squares FIR filter design and applications including Nyquist filters," *IEEE Trans. Circuits and Systems*, vol. CAS-34, pp. 11–23, Jan. 1987.

[3] H. Samueli, “On the design of optimal equiripple FIR digital filters for data transmission applications,” *IEEE Trans. on Circuits and Systems*, vol. 35, pp. 1542–1546, Dec. 1988.

[4] H. Samueli, “On the design of FIR digital data transmission filters with arbitrary magnitude specifications,” *IEEE Trans. Circuits and Systems*, vol. 38, pp. 1563–1567, Dec. 1991.

[5] N. S. Alagha and P. Kabal, “Generalized raised-cosine filters,” *IEEE Trans. on Communications*, vol. 47, pp. 989–997, July 1999.

[6] N. C. Beaulieu, C. C. Tan, and M. O. Damen, “A "better than" Nyquist pulse,” *IEEE Communications Letters*, vol. 5, pp. 367–368, Sept. 2001.

[7] T. E. Tuncer, “Optimum Nyquist filters with controlled time and frequency characteristics,” *Electronics Letters*, vol. 38, pp. 1292–1293, Oct. 2002.

[8] A. Assalini and A. M. Tonello, “Improved Nyquist pulses,” *IEEE Communications Letters*, vol. 8, pp. 87–89, Feb. 2004.

[9] R. Sood and H. Xiao, “Root Nyquist pulses with an energy criterion,” in *IEEE International Conference on Communications, ICC ’07*, pp. 2711–2716, 2007.

[10] G. Nigam, R. Singh, and A. K. Chaturvedi, “Finite duration root Nyquist pulses with maximum in-band fractional energy,” *IEEE Communications Letters*, vol. 14, pp. 797–299, Sept. 2010.

[11] H. Nyquist, “Certain topics in telegraph transmission theory,” *Trans. of the A.I.E.E.*, pp. 617–644, Feb. 1928.

[12] H. Nyquist, “Certain topics in telegraph transmission theory,” *Proceedings of the IEEE*, vol. 90, pp. 280–305, Feb. 2002. (Reprint of 1928 article.).

[13] J. G. Proakis and M. Salehi, *Digital Communications*. McGraw Hill, 2008.

[14] M. Rice, *Digital Communications: A Discrete-Time Approach*. Prentice Hall, 2009.

[15] f. j. harris, *Multirate Signal Processing for Communication Systems*. Prentice Hall, 2004.

[16] C. C. Tan and N. C. Beaulieu, “Transmission properties of conjugate-root pulses,” *IEEE Trans. on Communications*, vol. 52, pp. 553–558, Apr. 2004.

[17] T. N. Davidson, “Efficient design of waveforms for robust pulse amplitude modulation,” *IEEE Trans. on Signal Processing*, vol. 49, pp. 3098–3111, Dec. 2001.

[18] A. V. Oppenheim and R. W. Schafer, *Discrete-Time Signal Processing*. Prentice Hall, third edition ed., 2010.

[19] B. Farhang-Boroujeny, *Signal Processing Techniques for Software Radios*. `http://www.lulu.com`: Lulu publishing house, 2nd edition ed., 2010.

[20] N. C. Beaulieu, "The evaluation of error probabilities for intersymbol interference and cochannel interference," *IEEE Trans. on Communications*, vol. 39, pp. 1740–1749, Dec. 1991.

[21] S. Boyd and L. Vandenberghe, *Convex Optimization*. Cambridge University Press, 2004.

In: Communication Systems: New Research
Editor: Vyacheslav Tuzlukov
ISBN: 978-1-62618-654-5

Chapter 9

SPACE TIME CODING IN MULTIPLE INPUT MULTIPLE OUTPUT SYSTEMS: CHALLENGES AND APPLICATIONS

Lina Mroueh *
Institut supérieur d'électronique de Paris, Paris

Abstract

In a point-to-point communication, the use of multiple transmitter and receiver antennas enables an increased data throughput through spatial multiplexing and an increased range by exploiting the spatial diversity. The design of space time coding schemes that fully achieve the available diversity and the multiplexing gain in a MIMO system has been extensively addressed in literature yielding to the design of the optimal family of codes called perfect space time codes constructed from cyclic division algebra. These codes, originally designed for flat fading channels, received a lot of attention in industry in the last few years. However, the recent standards that use multiple antenna terminals are based on more realistic assumptions involving the use of outer codes and multi-taps channels. This chapter will give a literature overview on the design criteria of space time coding technique and their application in industrial systems.

Keywords: MIMO systems, space time coding, diversity, multiplexing gain

1. Introduction

One of the main challenge of the next generations of wireless communication systems is to offer with a high reliability a high data rate. The introduction of multiple antennas at the transmitter and the receiver side, commonly known as multiple input multiple output

*E-mail address: lina.mroueh@isep.fr

(MIMO) systems, offers a high data rate through spatial multiplexing and an increased reliability by exploiting the spatial diversity.

These MIMO systems have been widely studied in literature over the last few years aiming to conceive convenient transmission schemes that take advantages from the MIMO benefits. These works consider mainly the cases when no outer codes (*e.g.* convolutional code, turbo code, LDPC, ...) are used at the transmitter side. The design of such schemes depend critically on the availability of the channel state information (CSI) at the transmitter side. When full CSI is available at the transmitter, optimal power allocation can be performed on the different transmit antennas in order to maximize the capacity of the MMO system [1]. However, the full CSIT is not always feasible as it requires a large amount of feedback. The no CSIT assumption is more considered in a practical scenario. For this case, two approaches have been studied in the literature. The first approach proposed by Tarokh *et al.* in [2] is more tailored to the Rayleigh fading distribution and consists to minimize the error probability over all the fading distribution. The second approach proposed by Zheng and Tse in [3] is more general and characterizes at high signal-to-noise ratio (SNR) the dual benefits in term of diversity and spatial multiplexing using the diversity multiplexing tradeoff (DMT) framework.

More recently, Oggier *et al.* in [4] proposed a family of optimal space time codes known as perfect space time codes that fulfill the design criteria of Tarokh *et al.* in [2]. Moreover, it has been shown that these codes are the optimal codes over flat fading MIMO channel (when no outer codes are used) since they achieve full rate and full diversity, preserve the mutual information, achieve the Diversity Multiplexing Tradeoff (DMT) [3] and have a non vanishing determinant [5,6].

Unlike the simplified flat fading MIMO channel without outer codes, industrial transmission schemes are based on more realistic assumptions involving the use of outer codes such as the convolutional code and frequency selective channels [7–9]. The error performances of the perfect space time code in such scenario have been studied in [10] where the authors studied the impact of concatenating space time codes with good outer code in the case of bit interleaved coded modulation and multiple input multiple output system (BICM-MIMO) over a flat or a frequency selective channel.

This chapter will give first a literature overview on the design of transmission schemes for the case of MIMO flat fading channel when no outer code is used. Then the construction of the family of perfect space time codes from the cyclic division algebra is presented. Finally, based on our contribution in [10], we focus on the application of these MIMO codes in a standard context. Throughout this chapter and for the sake of the clarity of the presentation, we will skip all the details related to the mathematical derivations. However, we emphasize on the resulting interpretations and their impact on the MIMO transmission scheme design. The interested reader is invited to refer the indicated references for more details. The rest of the chapter will be organized as following. In Section 2., we review from literature some basic principles on MIMO systems that will be essential for the development of this chapter. Section 3. is dedicated to the code construction schemes over the MIMO channel and the evaluation of their error performance over a flat fading MIMO channel. Then, we present in Section 4. the performance of the optimal perfect space time codes in a practical scenario. Finally, we give in Section 5. some concluding remarks and perspectives.

Notation: The notation used in this chapter is as follows. Boldface lower case letters **v** denote vectors, boldface capital letters $\mathbf{M}$ denote matrices. $\mathbf{M}^\dagger$ and $\mathbf{M}^T$ denotes respectively the matrix transposition and conjugated transposition operations. $\|\mathbf{v}\|$ stands for the Euclidean norm of vector $\mathbf{v}$. The Frobenius norm of matrix $\mathbf{H}$ is denoted by $\|\mathbf{H}\|_F^2 = \mathrm{Tr}\{\mathbf{H}^\dagger \mathbf{H}\}$ where $\mathrm{Tr}\{\mathbf{A}\}$ refers to the trace of matrix $\mathbf{A}$. The $N \times N$ identity matrix is represented by $\mathbf{I}_N$. $\mathcal{CN}$ represents the complex Gaussian random variable. Finally, $\mathbb{E}_X\{.\}$ is the mathematical expectation of random variable X.

2. Basic Principles on MIMO Flat Fading Channel

The multiple input multiple output (MIMO) systems consist simply to transmit and to receive data from/to different locations over different uncorrelated fading paths. One of the main advantage of the MIMO system is the possibility to recover data over the independent uncorrelated fading paths if one or more paths are deeply faded. This gain is known as diversity gain in the MIMO terminology. Another advantage of the MIMO channel is the gain in data rate due to the spatial multiplexing of data over the different antennas. This gain corresponds to the multiplexing gain in the MIMO terminology.

The main challenge in conceiving a MIMO system is to code data over the different antennas in a convenient way to benefit from MIMO gains.

2.1. From SISO to MIMO channel

2.1.1. Modeling the SISO channel

For a single input single output SISO system, represented in Figure 1, the transmitted signal x is modified by the channel in an unpredicted way due to the multiplicative random fading h and the additive Gaussian noise z with variance N_0.

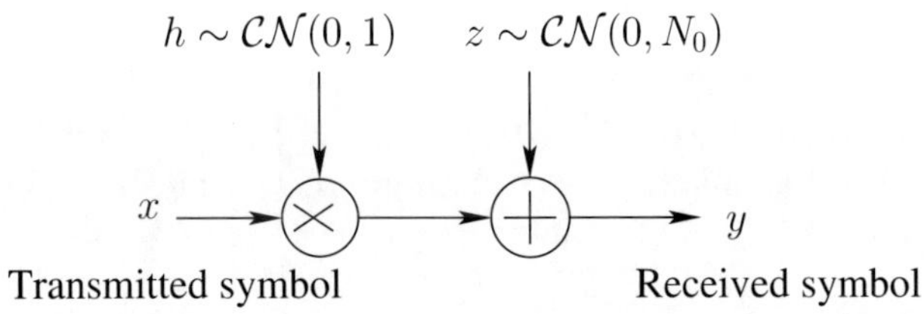

Figure 1: SISO channel.

The received signal is therefore

$$y = hx + z.$$

In wireless communication, the fading coefficient is often modeled as a complex Gaussian random variable $h = re^{j\theta} \sim \mathcal{CN}(0,1)$. In this case, r follows a Rayleigh distribution with $p(r) = \frac{1}{2\pi}e^{-\frac{r^2}{2}}$ and θ is uniformly distributed in $[0\ 2\pi]$.

2.1.2. Wireless MIMO system

The MIMO system is illustrated in Figure 2. In this case, each antenna j with $j = 1 \ldots n_r$ receives the noisy linear combination of all the transmitted signals. The received signal is,

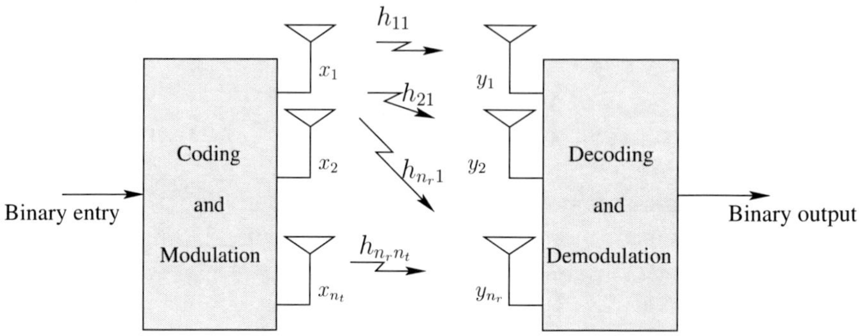

Figure 2: Multiple Input Multiple Output system.

$$\begin{bmatrix} y_1 \\ y_2 \\ \vdots \\ y_{n_r} \end{bmatrix} = \begin{bmatrix} h_{11} & h_{12} & \ldots & h_{1n_t} \\ h_{21} & h_{22} & \ldots & h_{2n_t} \\ \vdots & & & \\ h_{n_r1} & h_{n_r2} & \ldots & h_{n_rn_t} \end{bmatrix} \begin{bmatrix} x_1 \\ x_2 \\ \vdots \\ x_{n_t} \end{bmatrix} + \begin{bmatrix} \omega_1 \\ \omega_2 \\ \vdots \\ \omega_{n_r} \end{bmatrix}$$

Or equivalently, using the matricial notation is,

$$\mathbf{y} = \mathbf{H}\mathbf{x} + \boldsymbol{\omega},$$

where $\mathbf{x} = (x_i)_{1 \le i \le n_t}$ is the transmitted signal subjected to the power constraint $\mathrm{Tr}\left[\mathbb{E}[\mathbf{x}\mathbf{x}^\dagger]\right] \le P$, the channel matrix $\mathbf{H} = (h_{i,j})_{1 \le i \le n_r, 1 \le j \le n_t}$ and $\boldsymbol{\omega}$ is the additive noise vector. We assume here that the channel is flat fading which means that the channel remains constant during all the duration of the transmission.

2.2. Channel capacity and multiplexing gain

The capacity of a channel $C(\mathrm{SNR})$ represents the maximal number of bits that can be transmitted in one time slot where $\mathrm{SNR} = \frac{P}{N_0}$ is the signal to noise ratio. For a MIMO channel, the capacity was derived in [1] and is summarized in Theorem 1.

Theorem 1 (Instantaneous channel capacity). *The maximal rate that can be transmitted over a $n_t \times n_r$ MIMO channel, known as the capacity of MIMO channel is equal to,*

$$C(\mathrm{SNR}) = \max_{\mathbf{Q}:\mathrm{Tr}(\mathbf{Q}) \le P} \log_2 \det\left(\mathbf{I} + \frac{1}{N_0}\mathbf{H}\mathbf{Q}\mathbf{H}^\dagger\right),$$

where $\mathbf{Q} = \mathbb{E}\left[\mathbf{x}\mathbf{x}^\dagger\right]$ *is the covariance matrix of the transmitted signal with limited power P such that* $\mathrm{Tr}(\mathbf{Q}) \le P$.

The covariance matrix of the transmitted signal depends on the availability of the channel knowledge $\mathbf{H}$ at the transmitter side.

Definition 1 (Multiplexing gain). *The multiplexing gain is defined as,*

$$r = \lim_{\text{SNR}\to\infty} \frac{\mathbb{E}[C(\text{SNR})]}{\log_2 \text{SNR}} = \min(n_t, n_r),$$

where $\mathbb{E}[C(\text{SNR})]$ *represents the ergodic capacity obtained by averaging over the fading distribution. The multiplexing gain indicates the maximal number of symbols that can be simultaneously transmitted over a channel.*

2.3. Shannon theorem and implications

Theorem 2 (Shannon theorem). *The error-free transmission is possible as long as the transmitted rate R does not exceed the channel's capacity C.*

The main implication of the Shannon theorem is that one can predict the error if the channel's capacity is known at the transmitter side. The knowledge of this capacity at the transmitter side depends on the availability of the channel state information (CSI) at the transmitter side. This CSI is feasible at the receiver side as the receiver can perfectly estimate the wireless channel gains using pilots sequences. However, the situation becomes more complicated at the transmitter side as this CSI is not always feasible at the transmitter side unless a large amount of feedback from the receiver side to the transmitter is considered. The transmission strategy depends hence critically on this CSIT availability.

2.3.1. Full CSIT case: rate adaptation

When full CSIT is available at the transmitter side, the transmitted rate can be adapted to the maximal capacity.

For a given channel realization, the maximal capacity of the MIMO channel can be achieved using the so-called water-filling strategy [1]. This strategy uses the available CSI to jointly diagonalize as shown in Figure 3 at the transmitter and the receiver side the channel $\mathbf{H} = \mathbf{U}\mathbf{\Lambda}\mathbf{V}^\dagger$ by creating as many parallel channels as $n_{\min} = \min(n_t, n_r)$. The transmitted data vector $\tilde{\mathbf{x}}$ that contains $n_{\min}$ symbols is projected on the right eigenvector space $\mathbf{V}$ *i.e.*,

$$\mathbf{x} = \mathbf{V}\tilde{\mathbf{x}},$$

and is subjected to the following power constraint $\text{Tr}\left[\mathbb{E}[\mathbf{x}\mathbf{x}^\dagger]\right] \leq P$. The received vector is projected on the left eigenvector space $\mathbf{U}$, *i.e.*,

$$\begin{aligned}
\tilde{\mathbf{y}} = \mathbf{U}^\dagger \mathbf{y} &= \mathbf{U}^\dagger(\mathbf{H}\mathbf{x} + \omega) = \mathbf{U}^\dagger\mathbf{U}\mathbf{\Lambda}\mathbf{V}^\dagger\mathbf{V}\,\tilde{\mathbf{x}} + \mathbf{U}^\dagger\omega, \\
&= \mathbf{\Lambda}\tilde{\mathbf{x}} + \mathbf{z},
\end{aligned}$$

where $\mathbf{z} = \mathbf{U}^\dagger\omega \sim \mathcal{CN}(0, N_0\mathbf{I})$ is a Gaussian complex vector as $\mathbf{U}^\dagger$ is unitary. Let $\mathbf{P}_x$ denote the diagonal covariance matrix $\mathbf{P}_x = \mathbb{E}[\tilde{\mathbf{x}}\tilde{\mathbf{x}}^\dagger]$. The capacity of this equivalent MIMO channel is then

$$C = \max_{P^*:\sum_i P_i^* \leq P} \sum_{i=1}^{n_{min}} \log\left(1 + \frac{P_i^*\lambda_i^2}{N_0}\right).$$

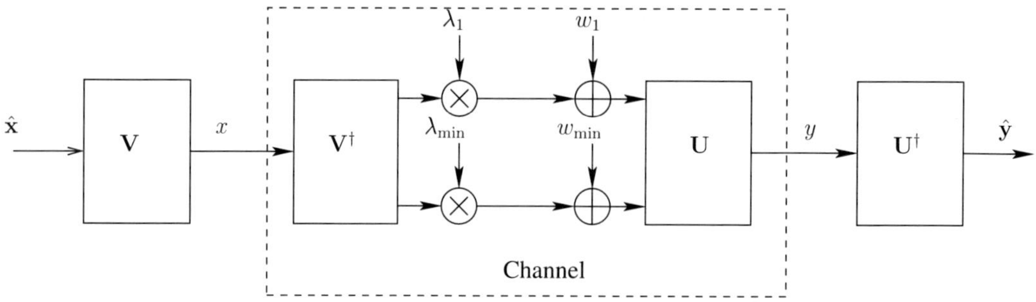

Figure 3: Water-filling: precoder and post-processing.

The solution of this convex optimization problem is,

$$P_i^* = \max\left(0, \mu - \frac{N_0}{\lambda_i^2}\right), \tag{1}$$

and the value of μ is determined using the power constraint $\sum P_i^* = P$. The above steps are summarized in Algorithm .

algorithm

[. htbp]

1: Perform SVD for the channel $\mathbf{H} = \mathbf{U}\mathbf{\Lambda}\mathbf{V}^\dagger$.

2: Compute the power allocated over parallel channel: $\mathbf{P}_x = \mathbb{E}[\tilde{\mathbf{x}}\tilde{\mathbf{x}}^\dagger] = \mu\mathbf{I} - N_0\mathbf{\Lambda}^{-2}$.

3: Calculate optimal water level μ : $\mathrm{Tr}[\mathbf{P}_x] = P$.

4: Compute the covariance matrix: $\mathbf{K}_x = \mathbb{E}[\mathbf{x}\mathbf{x}^\dagger] = \mathbf{V}\mathbf{P}_x\mathbf{V}^\dagger$.

Water-filling algorithm.

end

2.3.2. No CSIT case: diversity techniques

Unlike the CSIT case, optimal power allocation across the antennas cannot be performed in the absence of channel knowledge. In this case, a blind uniform power allocation is used instead, *i.e.*,

$$P_i = \frac{P}{n_t}. \tag{2}$$

The instantaneous channel capacity is then

$$C = \log\det\left(\mathbf{I} + \frac{\mathrm{SNR}}{n_t}\mathbf{H}\mathbf{H}^\dagger\right) = \sum_{i=1}^{m} \log_2\left(1 + \frac{\mathrm{SNR}}{n_t}\lambda_i\right) \tag{3}$$

where λ_i are the eigenvalues of the Wishart channel matrix $\mathbf{H}\mathbf{H}^\dagger$ and $q = \min(n_t, n_r)$. It can be easily observed from (3) that the uniform power allocation does not penalize the maximal multiplexing gain of $\min(n_t, n_r)$ that can be obtained. Using a convenient scheme it is always possible to transmit $\min(n_t, n_r)$ symbols.

Moreover, the transmitter cannot predict whether the channel is in deep fading or not if no CSI is available at the transmitter side. In this case, there is a non-zero probability that the transmitted rate exceeds the channel capacity and the two events defined in Definitions 2 and 3 should be considered.

Definition 2 (Outage event). *The outage event occurs when the MIMO channel cannot support the transmitted rate R, which means that R exceeds the capacity of the MIMO channel C, i.e.,*

$$O = \left\{ \mathbf{H} : \log\det(\mathbf{I} + \frac{\mathrm{SNR}}{n_t}\mathbf{H}\mathbf{H}^{\dagger}) < R \right\}$$

The outage probability is defined as,

$$P_{out}(R) = P(O)$$

Definition 3 (Error event). *The MIMO system is in error, when the decoded message is different than the transmitted message. The error event is therefore defined as,*

$$\varepsilon = \{\hat{\mathbf{X}} \neq \mathbf{X}\}$$

and the error probability is such that,

$$P_{error} = P(\varepsilon)$$

As a consequence of the Shannon theorem, when the outage event occurs, the system is in error almost surely, which means that the outage region is necessarily included in the error region, *i.e.*,

$$O \subseteq \varepsilon$$

For a given MIMO system operating at a rate R and under a power constraint P, the error probability is always lower-bounded by the outage probability as shown in Figure 2.3.2., *i.e.*,

$$\mathrm{P_{error}} \geq \mathrm{P_{out}}$$

The main objective is to design a coding scheme having an error probability that approaches the outage probability.

2.4. Design challenges without CSIT

In order to exploit fully the available diversity and the multiplexing gain of the MIMO channel, two approaches described in Paragraphs 2.4.1. and 2.4.2. have been considered in the literature. For these two approaches, the transmitted signal are coded across the time (during T time-slots) and the space (over the n_t antennas) using the so-called space-time coding. The transmitted codeword matrix $\mathbf{X} \in \mathbb{C}^{n_t \times T}$ is carved from a codebook denoted $\mathcal{X}_p$ having a size $|\mathcal{X}_p|$ and a rate R bits per channel use (bpcu) that is equal to

$$R = \frac{1}{T}\log_2|\mathcal{X}_p|.$$

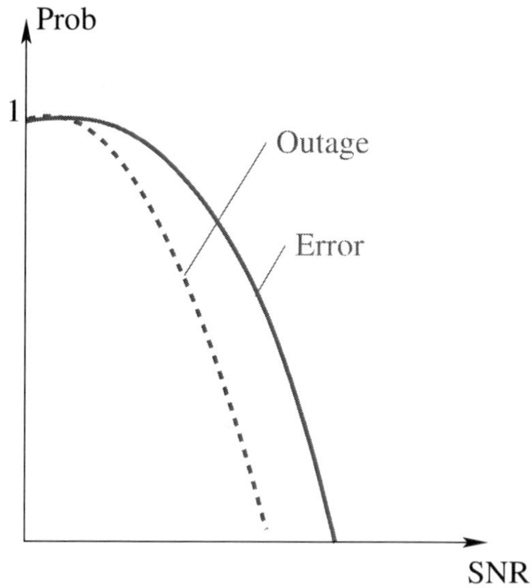

Figure 4: Relationship between outage and error event.

2.4.1. Fixed rate code construction

The first approach proposed by Tarokh *et al.* in [2] considers the case of fixed data rate R that does not scale as SNR. This approach is more tailored to the Rayleigh fading distribution and consists in the minimization of the error probability over all the fading distribution to approach the outage probability behavior in the high SNR regime.

For a fixed rate R and using the eigenvalue distribution of the Wishart matrix $\mathbf{HH}^\dagger$, it is well known from that the outage probability scales in the high SNR regime as,

$$\mathrm{P}_{\mathrm{out}}(\mathrm{SNR}) \doteq \mathrm{SNR}^{-d}, \qquad \text{where } d = n_t \times n_r \tag{4}$$

The slope d of the outage probability $\mathrm{P}_{\mathrm{out}}(\mathrm{SNR})$ is called diversity gain. Notice that the number of independent paths for a $n_t \times n_r$ MIMO channel is also equal to $d = n_t n_r$. It can be then deduced that the maximal diversity gain of a MIMO system corresponds simply to the number of independent paths.

For a given channel realization, the error probability is upper-bounded using the union bound [2, 15] such that,

$$\mathrm{P}_{\mathrm{error}|\mathbf{H}} \leq \frac{1}{|\mathcal{X}_p|} \sum_{i,j:i\neq j} \mathrm{Prob}\{\mathbf{X}_i \rightarrow \mathbf{X}_j|\mathbf{H}\}$$

where $\mathrm{Prob}\{\mathbf{X}_i \rightarrow \mathbf{X}_j|\mathbf{H}\}$ denotes the pairwise error probability (PEP) between two codewords $\mathbf{X}_i$ and $\mathbf{X}_j$. The PEP refers to the probability that a certain true codeword $\mathbf{X}_i \in \mathcal{X}_p$ is mistaken with another codeword $\mathbf{X}_i$ assuming that these two codewords are the only codewords of the codebook. At high SNR, the most significant error event corresponds to the case of neighboring $\mathbf{X}_i$ and $\mathbf{X}_j$, i.e.,

$$\mathrm{P}_{\mathrm{error}|\mathbf{H}} \dot{\leq} \frac{1}{|\mathcal{X}_p|} \max_{i,j:i\neq j} \mathrm{Prob}\{\mathbf{X}_i \rightarrow \mathbf{X}_j|\mathbf{H}\}.$$

Finally, the upper-bound on the error probability can be deduced by averaging over all the channel distribution. The goal of the space time code design is to minimize the worst case of error probability to approach the outage probability behavior in the high SNR regime.

2.4.2. Approximately universal code construction

While the first approach is more tailored to the Rayleigh fading distribution, Zheng and Tse proposed in [3] a powerful approach based on the high SNR characterization of the dual benefits in term of diversity and spatial multiplexing using the diversity multiplexing tradeoff (DMT) framework.

Definition 4. *Given a point-to-point MIMO system, the gains in terms of diversity gain d*

$$-d = \lim_{\mathrm{SNR}\to\infty} \frac{\log P_{out}(R,\mathrm{SNR})}{\log \mathrm{SNR}}$$

and spatial multiplexing gain r

$$r = \lim_{\mathrm{SNR}\to\infty} \frac{C(\mathrm{SNR})}{\log \mathrm{SNR}}$$

can be simultaneously obtained. But, there is a fundamental tradeoff $d_{\mathrm{out}}(r)$ between these two gains provided by any coding scheme.

Theorem 3 (Outage DMT of the MIMO channel). *The DMT of $n_t \times n_r$ Rayleigh channel is a piecewise-linear function connecting the points $\big(r, d(r)\big)$ where $r = 0,\ldots,\min(n_t,n_r)$ and*

$$d(r) = (n_t - r)(n_r - r). \tag{5}$$

Definition 5. *A coding scheme $X_p(\mathrm{SNR})$ with data rate R bits per channel use achieves a multiplexing gain r and diversity gain d if the data rate R is such that*

$$\lim_{\mathrm{SNR}\to\infty} \frac{R(\mathrm{SNR})}{\log \mathrm{SNR}} = r,$$

and the average error probability $P_e(\mathrm{SNR})$ with maximum likelihood-decoding is such that

$$-d = \lim_{\mathrm{SNR}\to\infty} \frac{\log P_e(\mathrm{SNR})}{\log \mathrm{SNR}}.$$

For a given multiplexing gain r, the largest diversity supported by any coding scheme is denoted by $d_{X_p}(r)$.

For any coding scheme with rate scaling as $r\log \mathrm{SNR}$, the DMT of the code $d_{X_p}(r)$ is upper bounded by $d_{\mathrm{out}}(r)$, i.e,

$$d_{X_p}(r) \le d_{\mathrm{out}}(r). \tag{6}$$

3. Code Construction for Flat Fading Channel

In this section, we consider the case of a flat fading MIMO channel where the wireless channel remains constant during all the duration of the transmission. The data bits streams are assumed to be transmitted over this wireless channel without using any outer codes.

3.1. Flat fading channel model

We consider first the case of flat fading channel model depicted in Figure 5 that is given by

$$\mathbf{Y} = \sqrt{\frac{\mathrm{SNR}}{n_t}}\,\mathbf{H}\mathbf{X} + \mathbf{Z} \tag{7}$$

where $\mathbf{X} \in \mathbb{C}^{n_t \times T}$ is a space time code drawn from code $\mathcal{X}_p$ of rate R per channel use, $\mathbf{Y} \in \mathbb{C}^{n_r \times T}$ is the received signal, $\mathbf{Z} \in \mathbb{C}^{n_r \times T} \sim \mathcal{CN}(0,1)$ is the additive noise and $\mathbf{H} \in \mathbb{C}^{n_r \times n_t}$ is the channel matrix with i.i.d complex Gaussian $\mathcal{CN}(0,1)$ entries. The scaling factor θ is chosen to ensure the power constraint,

$$\mathbb{E}\left[\|\mathbf{X}\|_{\mathrm{F}}^2\right] = T. \tag{8}$$

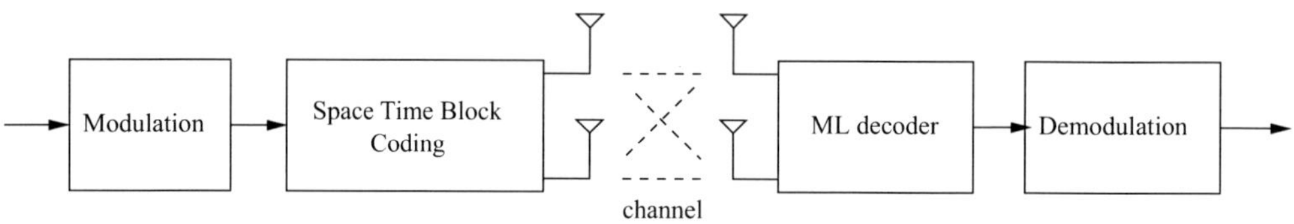

Figure 5: MIMO system.

3.2. Optimal Maximum Likelihood (ML) MIMO decoder

For the SISO channel considered in Paragraph 2.1.1., the maximum likelihood decoder should find the QAM constellation that minimizes,

$$\hat{x} = \arg \min_{x \in \mathrm{QAM}} |y - hx|^2$$

The maximum likelihood decoder is also extended to the MIMO case. In this case, the receiver should find the matrix $\hat{\mathbf{X}}$ in the family of space time code that minimizes the following Frobenius norm,

$$\hat{\mathbf{X}} = \arg \min_{\mathbf{C} \in \mathcal{X}_p} \mathrm{Tr}[(\mathbf{Y} - \mathbf{HC})(\mathbf{Y} - \mathbf{HC})^\dagger]$$

Exhaustive research among all the possible space time codewords can be performed. Algorithm with less complexity such as sphere decoder [11] and Schnorr-Euchner [12] are used in practice.

3.3. Space time code properties with fixed rate

In this section, we focus on space time codes design when the rate of the code is independent of SNR, *i.e.* $R(\mathrm{SNR}) = R$.

3.3.1. Error probability upper-bound

In this case, minimizing the average error probability over the distribution of the fading channel is studied. The average PEP for the $n_t \times n_r$ MIMO channel has been derived in [2].

Theorem 4 (PEP upper-bound). *Assuming that a maximum likelihood decoder is used, the worst PEP is bounded by,*

$$\mathrm{PEP} \leq c\ \mathrm{SNR}^{-d}, \tag{9}$$

where d is the diversity given by,

$$d = n_r \mathrm{rank}\{\Delta\mathbf{X}\Delta\mathbf{X}^\dagger\},$$

and c is the coding gain and is equal to,

$$c = 4^d \left(\min_{\Delta\mathbf{X}\neq 0} \det\left\{\Delta\mathbf{X}\Delta\mathbf{X}^\dagger\right\}\right)^{-n_r}.$$

The upper-bound of $c\,\mathrm{SNR}^{-d}$ *is minimized if:*

(i)- The diversity order $d = n_r \mathrm{rank}(\boldsymbol{\Delta X \Delta X}^\dagger)$ *is maximized;*

(ii)- The coding gain $c = 4^d \left(\min_{\Delta\mathbf{X}\neq 0} \det\left\{\Delta\mathbf{X}\Delta\mathbf{X}^\dagger\right\}\right)^{-n_r}$ *is minimized.*

3.3.2. Code design criteria

As a consequence of the outage probability lower-bound, the maximal diversity order that can be achieved by a coding scheme is equal to $n_t \times n_r$. As shown in Theorem 4, the diversity order of the coding scheme when using an ML decoder is $n_r \times \mathrm{rank}(\boldsymbol{\Delta X \Delta X}^\dagger)$. This means that the diversity corresponding to the multiple receive antennas of n_r, known as receive diversity, can be achieved regardless the used coding scheme. However, in order to achieve full transmit diversity, the rank of $\Delta\mathbf{X}\Delta\mathbf{X}^\dagger$ should be equal to n_t.

The gain in diversity is illustrated in Figure 6 where it is shown that significant gains can be observed when increasing the slope of the error probability. Moreover, the minimization of the coding gain in Theorem 4 results on a left shift of the error curve in Figure 6.

An optimal linear space time code should satisfy thus the following design criteria:

- Full rate symbol = $\min(n_t, n_r)$
- Full diversity, *i.e.* $\mathrm{rank}\{\Delta\mathbf{X}\Delta\mathbf{X}^\dagger\} = n_t$.
- Minimized coding gain $c = 4^d \left(\min_{\Delta\mathbf{X}\neq 0} \det\left\{\Delta\mathbf{X}\Delta\mathbf{X}^\dagger\right\}\right)^{-n_r}$.

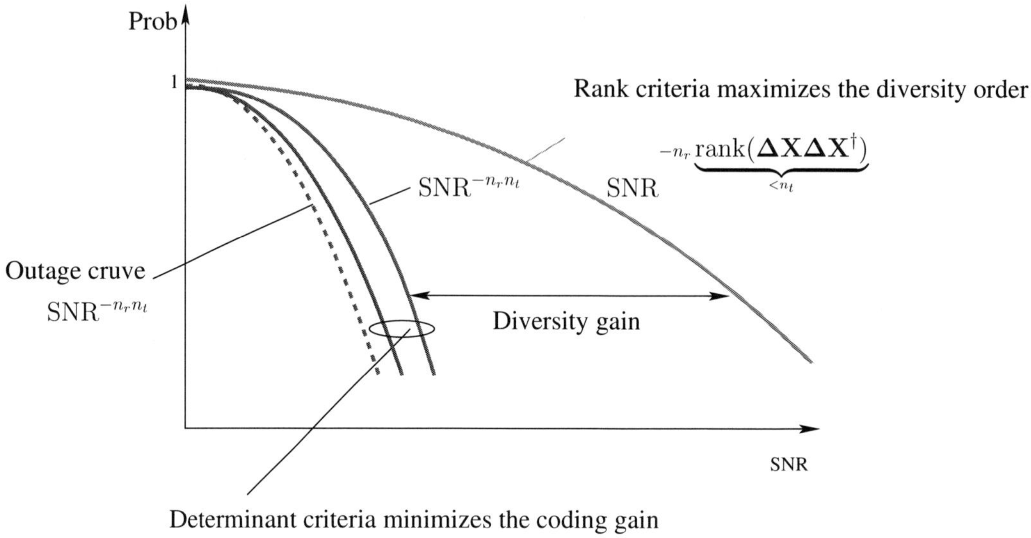

Figure 6: Rank and determinant criteria.

3.3.3. Examples of space-time codes

To illustrate this, we consider a MIMO channel having $n_t = 2$ antennas at the transmitter side and $n_r = 2$ antennas at the receiver side. We assume that the channel is known only at the receiver side but not at the transmitter side. The binary information $b_1 b_2 \ldots$ are first modulated using a 2^m-QAM constellation $x_1 x_2 x_3 x_4 \ldots$. Then, the stream of data symbols is coded using a $n_t \times T$ space time coding before being transmitted on the MIMO channel as described in the channel model in (7).

The QAM symbols are carved from a normalized 2^m-QAM constellation that are scaled to match $\frac{1}{\sqrt{E_s}}\left[2\mathbb{Z}[i] + (1+i)\right]$ with

$$E_s = \frac{2(2^m - 1)}{3}.$$

It can be easily checked that the minimal distance between two neighboring symbols is $d_{\min} = \frac{2}{\sqrt{E_s}}$.

A. Spatial division multiplexing:

This method is also known as Vertical Bell Labs Space Time code (VBLAST) [13]. In this case, two different symbols are transmitted on the different antennas without any coding at each time slot. The corresponding codeword is therefore,

$$\mathbf{X} = \frac{1}{\sqrt{2}} \begin{bmatrix} x_1 \\ x_2 \end{bmatrix}$$

Note that the normalization factor $\frac{1}{\sqrt{2}}$ is required in order to satisfy the power constraint in (8). For this code, two symbols are transmitted during one time slot and hence the symbol rate is 2 and is equal to the maximal multiplexing gain $\min(n_t, n_r) = 2$. The diversity of this code is,

$$d = n_r \text{rank}\{\Delta\mathbf{X}\Delta\mathbf{X}^\dagger\} \text{ with } \Delta\mathbf{X} \neq \mathbf{0}.$$

The difference codeword matrix is,

$$\Delta\mathbf{X} = \mathbf{X} - \mathbf{X}' = \frac{1}{\sqrt{2}} \begin{bmatrix} x_1 - x_1' \\ x_2 - x_2' \end{bmatrix}$$

is a rank one matrix. The diversity of this code is then only equal to 2. Finally, the minimal determinant of the SDM code is,

$$\min_{\Delta\mathbf{X}\neq\mathbf{0}} \det(\Delta\mathbf{X}\Delta\mathbf{X}^\dagger) = \min_{x_1\neq x_1'} \frac{1}{2}|x_1 - x_1'|^2 = \frac{1}{2}d_{\min}^2 = \frac{2}{E_s}$$

and the PEP is upper-bounded by, The maximal error probability is therefore,

$$\boxed{\text{PEP} \leq 4E_s^2\,\text{SNR}^{-2}.}$$

B. Repetition code :

For the repetition code described in [15], each data symbol is repeated during two time slots and the corresponding codeword is,

$$\mathbf{X} = \begin{bmatrix} x_i & 0 \\ 0 & x_i \end{bmatrix}.$$

Note that in this case, the power constraint in (8) is satisfied without need to multiply by a normalization factor as for the SDM case. It can be observed that for the repetition code only one symbol is transmitted during $T = 2$ time slots and then the symbol rate is $1/2$ which is less than the maximal multiplexing gain of 2. The diversity of this code can be computed from Theorem 4 as,

$$d = n_r \text{rank}\{\Delta\mathbf{X}\Delta\mathbf{X}^\dagger\} \text{ with } \Delta\mathbf{X} \neq \mathbf{0}.$$

The difference codeword matrix is,

$$\Delta\mathbf{X} = \mathbf{X} - \mathbf{X}' = \begin{bmatrix} x_1 - x_1' & 0 \\ 0 & x_1 - x_1' \end{bmatrix}.$$

and its determinant is

$$\det(\Delta\mathbf{X}\Delta\mathbf{X}^\dagger) = |x_1 - x_1'|^4 \neq 0 \text{ if } x_1 \neq x_1',$$

The non zero matrix is therefore full rank and the diversity that can be extracted is 4. The code extracts all the diversity of the 2×2 MIMO system but not all the degrees of freedom. The minimum of $\det(\Delta\mathbf{X}\Delta\mathbf{X}^\dagger)$ is,

$$\min_{\Delta\mathbf{X}\neq\mathbf{0}} \det(\Delta\mathbf{X}\Delta\mathbf{X}^\dagger) = \min_{x_1\neq x_1'} |x_1 - x_1'|^4 = d_{\min}^4 = \frac{2^4}{E_s^2}.$$

The error probability is then upper-bounded by,

$$\mathrm{PEP}_{\max} = c\,\mathrm{SNR}^{-4},$$

where

$$c = 4^{d}\big(\min_{\Delta\mathbf{X}\neq\mathbf{0}} \det(\Delta\mathbf{X}\Delta\mathbf{X}^{\dagger})\big)^{-n_r} = E_s^4.$$

The maximal error probability is therefore,

$$\boxed{\mathrm{PEP} \leq E_s^4\,\mathrm{SNR}^{-4}.}$$

C. Alamouti code

The Alamouti code structure in [14] is given by,

$$\mathbf{X} = \frac{1}{\sqrt{2}}\begin{bmatrix} x_1 & -x_2^* \\ x_2 & x_1^* \end{bmatrix}$$

Notice that the normalization factor of $\frac{1}{\sqrt{2}}$ is also required for the Alamouti case to satisfy the power constraint in (8). For the Alamouti code, two different symbols are transmitted during 2 time slots. The symbol rate of this code is then equal to 1 symbol per time slot which is not optimal in term of its multiplexing gain for the 2×2 MIMO case. However, for the 2×1 MISO configuration, this code is full rate. The diversity of this code can be computed by,

$$d = n_r \operatorname{rank}\{\Delta\mathbf{X}\Delta\mathbf{X}^{\dagger}\} \text{ with } \Delta\mathbf{X} \neq \mathbf{0}.$$

where

$$\Delta\mathbf{X} = \mathbf{X} - \mathbf{X}' = \frac{1}{\sqrt{2}}\begin{bmatrix} x_1 - x_1' & -(x_2 - x_2')^* \\ x_2 - x_2' & (x_1 - x_1')^* \end{bmatrix}.$$

The determinant of $\Delta\mathbf{X}\Delta\mathbf{X}^{\dagger}$ is,

$$\det(\Delta\mathbf{X}\Delta\mathbf{X}^{\dagger}) = \frac{1}{2}(|x_1 - x_1'|^2 + |x_2 - x_2'|^2)^2 \neq 0 \text{ if } (x_1, x_2) \neq (x_1', x_2')$$

This means that $\operatorname{rank}\{\Delta\mathbf{X}\Delta\mathbf{X}^{\dagger}\} = 2$ for $\Delta\mathbf{X} \neq \mathbf{0}$. The diversity achieved by this code is,

$$d = 2\times 2 = 4.$$

For 2×2 MIMO system, the Alamouti code extracts the diversity gain but not the full multiplexing gain. The minimum of $\det(\Delta\mathbf{X}\Delta\mathbf{X}^{\dagger})$ is,

$$\min_{\Delta\mathbf{X}\neq\mathbf{0}} \det(\Delta\mathbf{X}\Delta\mathbf{X}^{\dagger}) = \frac{1}{2}\min_{x_1\neq x_1', x_2\neq x_2'} (|x_1 - x_1'|^2 + |x_2 - x_2'|^2)^2 = \frac{1}{2} d_{\min}^4 = \frac{2^3}{E_s^2}.$$

The PEP is then upper-bounded by,

$$\mathrm{PEP} \leq c\,\mathrm{SNR}^{-4},$$

where

$$c = 4^{d}\big(\min_{\Delta\mathbf{X}\neq\mathbf{0}} \det(\Delta\mathbf{X}\Delta\mathbf{X}^{\dagger})\big)^{-n_r} = 4E_s^4.$$

The maximal error probability is therefore,

$$\boxed{\mathrm{PEP} \leq 4E_s^4\,\mathrm{SNR}^{-4}.}$$

D. Golden code

This code was introduced in [5]. The codeword matrix of the Golden code $\mathbf{X} = \frac{1}{\sqrt{2}} \times \boldsymbol{C}$ where,

$$\boldsymbol{C} = \frac{1}{\sqrt{5}} \begin{bmatrix} \alpha(x_1 + \theta x_2) & \bar{\alpha}(x_3 + \bar{\theta} x_4) \\ i\alpha(x_3 + \theta x_4) & \bar{\alpha}(x_1 + \bar{\theta} x_2) \end{bmatrix},$$

and $\theta = \frac{1+\sqrt{5}}{2}$, $\bar{\theta} = \frac{1-\sqrt{5}}{2}$, $\alpha = 1 + i - i\theta$ and $\bar{\alpha} = 1 + i - i\bar{\theta}$.
We can verify that $\theta\bar{\theta} = -1$ and $\alpha\bar{\alpha} = 2 + i$. This means that using the Golden code the combined symbols sent at each antenna has energy equal to 1 and therefore a normalization factor of $\frac{1}{\sqrt{2}}$ is required to satisfy (8). Using the Golden code structure, it can be seen that four symbols are transmitted during $T = 2$ time slots. The symbol rate is therefore equal to 2. The diversity of this code can be computed by,

$$d = n_r \text{rank}\{\Delta\mathbf{X}\Delta\mathbf{X}^\dagger\} \text{ with } \Delta\mathbf{X} \neq \mathbf{0}.$$

The difference codeword matrix is given by,

$$\Delta\boldsymbol{C} = \boldsymbol{C} - \boldsymbol{C}' = \frac{1}{\sqrt{5}} \begin{bmatrix} \alpha((x_1 - x_1') + \theta(x_2 - x_2')) & \bar{\alpha}((x_3 - x_3') + \bar{\theta}(x_4 - x_4')) \\ i\alpha((x_3 - x_3') + \theta(x_4 - x_4')) & \bar{\alpha}((x_1 - x_1') + \bar{\theta}(x_2 - x_2')) \end{bmatrix}.$$

The minimal determinant of this matrix is $\min_{\Delta\boldsymbol{C}\neq 0} \det(\Delta\boldsymbol{C}) = \frac{2+1i}{(\sqrt{5})^2}$ and

$$\min_{\Delta\boldsymbol{C}\neq 0} \det(\Delta\boldsymbol{C}\Delta\boldsymbol{C}^\dagger) = \frac{|2+1i|^2}{25} = \frac{1}{5}.$$

The matrix is therefore full rank and the maximal diversity is 4. The Golden code extracts the full diversity of the 2×2 MIMO system and the full multiplexing gain. The error probability is upper-bounded by,

$$\text{PEP} \leq c\,\text{SNR}^{-4},$$

where

$$c = 4^d \left(\min_{\Delta\mathbf{X}\neq\mathbf{0}} \det(\Delta\mathbf{X}\Delta\mathbf{X}^\dagger) \right)^{-n_r} = 100\, E_s^4.$$

The maximal error probability is,

$$\boxed{\text{PEP} \leq 100\, E_s^4\, \text{SNR}^{-4}.}$$

D. Codes comparison:

In order to have a fair comparison, the four above schemes should be compared using the same bit rate, for example 4 bits per time slot. As we can see from the above discussion, these codes have different symbol rate. In order to achieve the same bit rate per channel use, different 2^m-QAM constellation should be used to map the symbols of the different space time coding schemes. The VBLAST code and the Golden code have a symbol rate of 2. The bit rate when symbols are mapped using 2^m-QAM constellation is $2m$ and is equal to 4 if $m = 2$. Symbols of these two schemes should be then mapped using QPSK constellation.

The repetition code has a symbol rate of $1/2$. The bit rate of this code when using 2^m-QAM constellation is $1/2m$ and is equal to 4 if $m = 8$. This corresponds then to a 256QAM constellation. The Alamouti code has a symbol rate of 1. Using a 2^mQAM constellation, the bit rate of m is equal to 4 if $m = 4$ which corresponds to a 16QAM constellation.

Table 1 summarizes the PEP upper-bound of the schemes with their corresponding constellation. As it can be seen from Table 1, the upper-bound on the pairwise error probability expression depends on the choice of the constellation. constellation.

Space Time Code	**Constellation**	**PEP upper-bound**
VBLAST	QPSK	$16\ \mathrm{SNR}^{-2}$
Repetition code	256QAM	$170^4\ \mathrm{SNR}^{-4}$
Alamouti code	16QAM	$4\ 10^4\ \mathrm{SNR}^{-4}$
Golden code	QPSK	$400\ \mathrm{SNR}^{-4}$

Table 1. Upper-bound on PEP for a bit rate of 2 bpcu.

The error performances of these space time code are illustrated in Figure 7. At very low

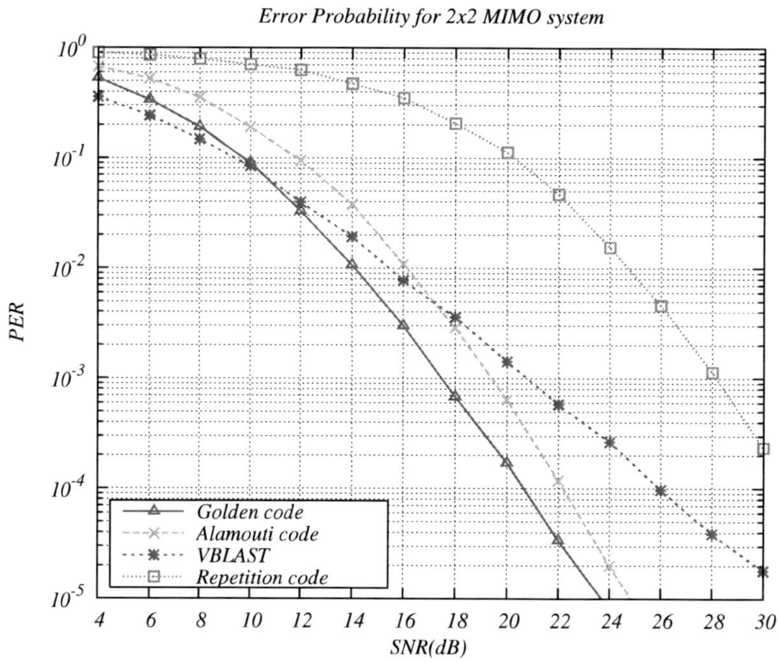

Figure 7: Comparison between error probability for the 2×2 MIMO scheme with a bit rate of 2 bpcu.

SNR, the SDM has better coding gain then both Alamouti schemes and the Golden code (Table 1) and has a better PER than the Golden code and the Alamouti code. However, for the high SNR regime, the diversity gain dominates the coding gain. As the Alamouti scheme is not a full rate scheme, using a 16QAM constellation instead of QPSK constellation in order to achieve the same spectral efficiency induces a loss in term of coding gain. The gain of the Golden code in diversity and in coding gain can be also observed in Figure 7. The repetition code has a full diversity of 4. However its coding gain in Table 1 is very high as a 256QAM constellation is used to compensate the loss in the rate symbol. This high coding gain dominates the error probability in the low SNR regime and the repetition code performs badly compared to the SDM and the other codes (Figure 7). In the high SNR regime, the gain diversity of the code becomes more dominant and the slope of the error probability becomes higher and compensate the loss in SNR compared to the SDM case.

3.4. Approximately universal code over a flat fading channel

The approximately universal code design provides a structured code design criterion that achieve the DMT [6]. This design criteria is derived from the performance of the code over the worst-channel case that is not in outage. Universal codes achieve reliable communication over MIMO channel realization that are not in outage. In the following, we review from [5, 6] the non-vanishing design criteria and the DMT-achieving construction of the perfect space time code in [4].

3.4.1. Sufficient condition for DMT achievability

Theorem 5. *A coding scheme $X_p(\mathrm{SNR})$ is approximately universal over the MIMO channel if and only if, for every pair of distinct codewords*

$$\mu_1^2\mu_2^2\ldots\mu_n^2 \geq \frac{c}{2^{R(\mathrm{SNR})+o(\log \mathrm{SNR})}}, \quad c > 0 \tag{10}$$

where $n = \mathrm{rank}\{\mathbf{H}\mathbf{H}^\dagger\} = \min(n_t, n_r)$ and $\mu_1^2 \leq \ldots \leq \mu_n^2$ are the m eigen-values of the codewords difference matrix $\Delta\mathbf{X}\Delta\mathbf{X}^\dagger$.

3.4.2. Non Vanishing Determinant (NVD) code definition

The non-vanishing determinant (NVD) criteria is a particular form of the approximately universal condition defined in Theorem 5. Before defining the NVD code structure in [4], we first define the normalized space time code $\mathbf{X}$ by,

$$\mathbf{X} = \theta\bar{\mathbf{X}}, \tag{11}$$

where $\bar{\mathbf{X}} \in \bar{X}_p(\mathrm{SNR})$ refers to the normalized space time code and θ is the scaling factor that ensures the power constraint in (8).

Definition 6 (NVD codes). *A coding scheme $\bar{X}_p(\mathrm{SNR})$ is called a rate-n NVD code if $X_p(\mathrm{SNR})$ satisfies the following properties*

- *Each entry $\bar{x}_{i,j}$ of the $X_p(\mathrm{SNR})$ is a linear combination of symbols from $\mathcal{A}(\mathrm{SNR})$, where $\mathcal{A}(\mathrm{SNR})$ is a universal code over the scalar channel with data rate $R_{\mathcal{A}}(\mathrm{SNR})$ bits PCU. The quadrature amplitude modulation (QAM) constellation such as QPSK, 16QAM, 64QAM,... or HEX constellation are usually used as scalar universal codes.*
- *The average number of symbols transmitted by $X_p(\mathrm{SNR})$ is equal to n symbols per channel use.*

- *The following NVD property is satisfied*

$$\det\left(\Delta\bar{\mathbf{X}}\Delta\bar{\mathbf{X}}^{\dagger}\right) \dot{\geq} \mathrm{SNR}^{0} \tag{12}$$

3.4.3. Scaling factor θ for NVD codes

For a non-vanishing determinant code, the average number of symbols transmitted by $\mathcal{X}_p(\mathrm{SNR})$ is $n = \min(n_t, n_r)$ symbols per channel use and can be defined as

$$n = \frac{1}{T}\log_{|\mathcal{A}|}|\mathcal{X}_p|, \tag{13}$$

where $|\mathcal{X}_p|$ denotes the total number of possible codewords in $\mathcal{X}_p$ and $|\mathcal{A}|$ denotes the total number of constellation symbols in $\mathcal{A}$. Equivalently,

$$|\mathcal{X}_p| = |\mathcal{A}|^{nT}. \tag{14}$$

Let $\mathcal{A}$ be the 2^mQAM constellation[1] with size $|\mathcal{A}| = 2^m$, such that

$$\mathcal{A} = \left\{a + ib, \quad |a|, |b| \leq m-1 \quad a, b \text{ are odd}\right\}.$$

The rate of the space time code $R = r\log\mathrm{SNR}$ can be related to $|\mathcal{X}_p|$ by $R = \frac{1}{T}\log_2|\mathcal{X}_p| = r\log\mathrm{SNR}$. or equivalently

$$|\mathcal{X}_p| = \mathrm{SNR}^{Tr}. \tag{15}$$

By combining (14) and (15), it follows that,

$$|\mathcal{A}| = \mathrm{SNR}^{\frac{r}{n}}. \tag{16}$$

As each entry $\bar{x}_{i,j}$ is a linear combination of symbols s_l carved from a 2^m-QAM constellation, i.e

$$\bar{x}_{i,j} = \sum_{l=1}^{n} a_l s_l, \quad a_l \in \mathbb{C} \text{ and } \|\mathbf{a}\|^2 = \|[a_1 \ldots a_n]\|^2 = 1$$

then it can be easily checked that ,

$$\mathbb{E}[|\bar{x}_{i,j}|^2] = \|a\|^2\,\mathbb{E}[|s|^2] = E_s = \frac{2(|\mathcal{A}|-1)}{3} \doteq \mathrm{SNR}^{\frac{r}{n}}.$$

Using the normalization constraint in (8), it follows that

$$\theta^2 \doteq \mathrm{SNR}^{-\frac{r}{n}}. \tag{17}$$

3.4.4. DMT of the code

Lemma 1. *NVD codes achieve the DMT for $n_t \times n_r$ MIMO configuration when $n_t \leq n_r$, and for full rate codes ($n = n_t$).*

Proof. The determinant of the non normalized space time code is such that

$$\det\left(\Delta\mathbf{X}\Delta\mathbf{X}^{\dagger}\right) = \theta^{2n_t}\det\left(\Delta\bar{\mathbf{X}}\Delta\bar{\mathbf{X}}^{\dagger}\right).$$

By replacing the scaling factor θ^2 by its value in (17) with $n = n_t$, the NVD property is then satisfied *i.e.*,

$$\det\left(\Delta\mathbf{X}\Delta\mathbf{X}^{\dagger}\right) \dot{\geq} \mathrm{SNR}^{-r}\,\mathrm{SNR}^{0} \doteq \frac{c}{2^{R(\mathrm{SNR})+o(\mathrm{SNR})}}.$$

□

[1] The same value of scaling factor is valid using HEX constellation.

3.4.5. NVD code construction: Perfect space time codes

Perfect space time codes are full rate codes ($n = n_t$) constructed from cyclic division algebras (CDA) defined as following. Let $\mathbb{L} = \mathbb{Q}(i, \theta)$ be a cyclic extension of degree n_t on the base field $\mathbb{Q}(i)$. The generator of Galois group $\mathrm{Gal}(\mathbb{L}/\mathbb{Q}(i))$ is denoted by σ, and assume that $\mathrm{Gal}(\mathbb{L}/\mathbb{Q}(i)) = \{\sigma^0, \ldots, \sigma^{n_t-1}\}$. Let $\gamma \in \mathbb{Q}(i)$ be such that $\gamma, \gamma^2, \ldots, \gamma^{n_t-1}$ are non-norm elements in $\mathbb{L}$. The CDA of degree n_t is given by

$$C = \left(\mathbb{L}/\mathbb{Q}(i), \sigma, \gamma\right).$$

Each element $\bar{\mathbf{X}}$ of C is given by,

$$\bar{\mathbf{X}} = \begin{pmatrix} x_1 & x_2 & \ldots & x_{n_t} \\ \gamma\sigma(x_{n_t}) & \sigma(x_1) & \ldots & \sigma(x_{n_t-1}) \\ \vdots & & & \vdots \\ \gamma\sigma^{n_t-1}(x_2) & \gamma\sigma^{n_t-1}(x_3) & \ldots & \sigma^{n_t-1}(x_1) \end{pmatrix} \tag{18}$$

where $x_i \in \mathbb{I} \subset O_{\mathbb{L}}$ is a linear combination of symbols carved from a QAM or Hex constellation, $O_{\mathbb{L}}$ being the ring of the integers, and $\mathbb{I}$ is an properly chosen ideal that preserves the constellation shaping. As perfect space time codes are linear codes constructed from a CDA, then

$$\min_{\Delta\bar{\mathbf{X}} \neq 0} \det\{\Delta\bar{\mathbf{X}}\Delta\bar{\mathbf{X}}^\dagger\} \geq \delta,$$

where δ is the inverse of the discriminant of $\mathbb{Q}(\theta)$ (refer to [4] for more details), and is independent of the constellation size. The NVD property in (12) can be easily verified, such that the determinant of the non-normalized space time code is such that,

$$\det\left(\Delta\mathbf{X}\Delta\mathbf{X}^\dagger\right) \geq \frac{\delta}{2^{R(\mathrm{SNR})+o(\mathrm{SNR})}}.$$

4. Practical Insights on the Use of MIMO Diversity Techniques

Unlike the simplified flat fading channel model considered in previous sections, the industrial transmission schemes are based on more realistic assumptions that include the use of the bit interleaved coded modulation scheme and a multi-tap channel. This section will review the result in [10] where the impact of concatenating multidimensional coding scheme with outer codes is studied.

4.1. BICM system model

The block diagram of the considered system based on the IEEE 802.11n transmission scheme is depicted in Figure 8. During the transmission, the binary information elements $\underline{b}$ are first encoded by a binary code of rate R_c *e.g.* a convolutional code C with free distance d_{free} and then interleaved by a bit interleaver π. The coded and interleaved sequence $\underline{c}$ is fed into the 2^m-QAM gray mapper and is mapped into a sequence of symbols $\underline{x}$. The resulting symbols are coded by a space time block codeword (STBC) X_p that associates to each $1 \times sr$ symbol vector a $n_t \times s$ matrix where s is the coding duration of the STBC known also as

STBC spreading factor and r is the number of symbols transmitted during one time slot *i.e.*, the rate of the STBC[2]. The coded space-time codes are finally transmitted on a multiple antenna channel $\mathbf{H}$ with n_t transmit antennas and n_r receive antennas.

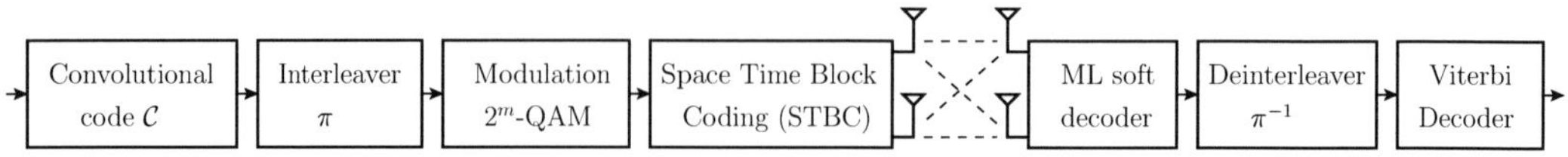

Figure 8: BICM MIMO system.

Channel model

The two cases where the channel is either flat fading or selective in frequency are addressed. The flat fading channel models the case when the channel remains constant during all the duration of coding. For the flat fading channels, N space time block codewords are transmitted. Each STBC has a coding duration equal to s time slots. The second considered model is the frequency selective fading which models the case when the bandwidth of the transmitted signal exceeds the coherence bandwidth of the channel. For a frequency selective channel with a delay profile of L taps, a MIMO-OFDM system is considered. The channel is therefore decomposed into N parallel channels, $\mathbf{H}(f_k) = \mathbf{H}(e^{j2\pi f_k})$ with $k = 0 \dots N-1$ that are statistically correlated where f_k is the channel frequency subcarrier. We assume that the channel is flat over each subcarrier f_k and that each space-time codeword is transmitted over one subcarrier f_k where $k = 0 \dots N-1$. Finally, to simplify the notation for both cases of flat fading and frequency selective channel, we refer by k the indexing of the space time codewords that will be used also to index the channel realization over the k-th STBC, *i.e.*, $\mathbf{H}(k) = \mathbf{H}(sk+j)$ with $j = 0 \dots s-1$.

Bit interleaver

Before detailing the interleaver structure, we recall that each block STBC is associated to a vector of $1 \times sr$ symbol vector or equivalently to a bit vector of $1 \times msr$.
The bit interleaver can be modeled as π: $k^{'} \to (k,i)$ where $k^{'}$ denotes the original ordering of the coded bits $c_{k^{'}}$, k denotes the index of the STBC matrix and i indicates the position of the bits $c_{k^{'}}$ in the corresponding $1 \times msr$ bit vector associated to the k-th STBC.

At the receiver, the received coded space time codeword is given by

$$\mathbf{Y}(k)^{[n_r \times s]} = \mathbf{H}(k)^{[n_r \times n_t]}\mathbf{C}(k)^{[n_t \times s]} + \mathbf{Z}(k)^{[n_r \times s]}, \tag{19}$$

where $\mathbf{Z}(k) \sim \mathcal{CN}(0, N_0 \mathbf{I}_{n_r})$ is the complex additive Gaussian noise.

ML soft decoder

The maximum likelihood (ML) soft decoder generates for each coded bit $c_{k,i}$ two metrics: $\lambda^i_{c_k=0}$ and $\lambda^i_{c_k=1}$. Theses metrics correspond to the log-MAP (Maximum A-Posteriori) com-

[2]We recall from [3] that $r \leq \min(n_t, n_r)$

puted over one codeword (refer to [7] for more details on λ-metrics), and are given by:

$$\begin{aligned}\lambda^i(c_k) &= \log \sum_{\mathbf{C}\in\mathcal{X}^i_{c_k}} p(\mathbf{Y}(k)|\mathbf{H}(k),\mathbf{C}),\\ &= \log \sum_{\mathbf{C}\in\mathcal{X}^i_{c_k}} \exp^{-\|\mathbf{Y}(k)-\mathbf{H}(k)\mathbf{C}(k)\|^2_{\mathrm{F}}}.\end{aligned}$$

These metrics can be well approximated by

$$\lambda^i(c_k) \approx \min_{\mathbf{C}\in\mathcal{X}^i_{c_k}} \|\mathbf{Y}(k)-\mathbf{H}(k)\mathbf{C}\|^2_{\mathrm{F}}, \tag{20}$$

where $\mathcal{X}^i_b$ denotes the constellation subset $\mathcal{X}^i_b = \{\mathbf{C}\in\mathcal{X}_p : l^i(\mathbf{C}) = b\}$ and $l^i(\mathbf{C})$ is the ith bit in the corresponding $1\times msr$ bit vector associated to $\mathbf{C}$. The problem in (20) is a variant of the well known closest problem and can be solved using low complexity algorithms such as accelerated decoder in [16]. Finally, λ metrics associated to the interleaved bits are de-interleaved and are used by the Viterbi decoder to decode the information bits by finding the shortest path in the trellis according to,

$$\underline{\hat{c}} = \arg\min_{c} \sum_{k'} \lambda(c^i_k). \tag{21}$$

4.2. MIMO gains for flat fading MIMO BICM channels

The first practical scenario is the case of MIMO-BICM system and a transmission over a flat fading channel. For this scenario, an upper-bound on the pairwise error probability was derived in [10] for both cases when MIMO symbols are coded with a perfect space time code and the case when there is no space time coding for the MIMO symbols *i.e.* the case of spatial division scheme (SDM).

In the first case when perfect space time codes are used, the PEP is upper-bounded at high SNR by,

$$\mathrm{PEP} \leq \left(d_{\mathrm{free}}\delta\right)^{-n_r}\left(E_s\right)^{n_t n_r}\mathrm{SNR}^{-n_t n_r}. \tag{22}$$

On the other hand, when no space time code is used (SDM scheme), the PEP can be upper-bounded by

$$\mathrm{PEP} \leq \left(d_{\mathrm{free}}\right)^{-n_r}\left(E_s\right)^{n_r}\mathrm{SNR}^{-n_r}. \tag{23}$$

From these PEP upper-bound, it can be easily seen that the perfect codes extract the full MIMO diversity $n_t n_r$. However, without MIMO coding (the spatial division schemes), the diversity is only n_r. To illustrate this, we consider the case of 2×2 MIMO system with QPSK constellation. The values of the upper-bounds in (22) and (23) can be written in the form $\gamma_{\mathrm{as}}\,\mathrm{SNR}^{-d}$ and are summarized in Table 3 for different values of free distance.

As it can be seen from Table 3, the asymptotical upper-bound gain γ_{as} of the SDM is lower than the one of the GC. However in the high SNR regime, this asymptotical upper-bound gain will not be significant and the diversity gain of the GC becomes more dominant. Two main remarks can be deduced: the first one is that at the high SNR the GC with a weak convolutional code gives better performance than the SDM with a strong outer code. The

Coding scheme	PEP upper-bound
GC without CC	$400\,\mathrm{SNR}^{-4}$
SDM without CC	$4\,\mathrm{SNR}^{-2}$
GC with $d_{\text{free}} = 5$	$16\,\mathrm{SNR}^{-4}$
SDM with $d_{\text{free}} = 5$	$0.16\,\mathrm{SNR}^{-2}$
GC with $d_{\text{free}} = 10$	$4\,\mathrm{SNR}^{-4}$
SDM with $d_{\text{free}} = 10$	$0.04\,\mathrm{SNR}^{-2}$

Table 2. Theoretical upper-bound for a 2×2 flat fading MIMO-BICM with QPSK symbols.

second remark is that the gain of the GC vs the SDM schemes can be observed at a moderate PEP range when no convolutional code is used. This will not be the case in a MIMO-BICM system and especially for higher values of d_{free}. Reaching very low target PER will be very hard to simulate in a real simulation context where reasonable target error rate are often addressed.

4.3. MIMO gains for frequency selective MIMO BICM channels

The second scenario studied in [10] is the case of BICM-MIMO system with frequency selective fading channel. When MIMO symbols over each subcarrier are coded using perfect space time codes, the PEP is upper-bounded at high SNR by,

$$\mathrm{PEP} \leq \mathcal{G}\,\mathrm{SNR}^{-n_r n_t \min(L,D)}, \tag{24}$$

where $\mathcal{G} = (\alpha\,\delta)^{-n_r d_{\text{free}}} (E_s)^{n_t n_r \min(L,D)}$ and $D \leq d_{\text{free}}$ denotes the number of different subcarriers on which erroneous bits are received. The interleaver design maximizes the parameter D. The parameter α is a constant that depends on the covariance matrix.

For the case, when no space time code is used, the PEP is bounded such that,

$$\mathrm{P}(\underline{c} \rightarrow \underline{\hat{c}}) \leq (\alpha)^{-n_r d_{\text{free}}} (E_s)^{n_r \min(L,D)}\,\mathrm{SNR}^{-n_r \min(L,D)}, \tag{25}$$

where α depends on the covariance matrix.

The numerical value of the upper-bounds in (24) and (25) are illustrated in Table 3 for the case of 2×2 MIMO with QPSK symbols over a frequency fading channel with $L = 6$ where the values of $D_1 \leq 5$ and $D_2 \leq 10$ depend on the used interleaver. The same remarks

Coding scheme	PEP upper-bound
GC with $d_{\text{free}} = 5$	$(5\alpha^{-1})^{10}\,(\mathrm{SNR}/2)^{-4\min(6,D_1)}$
SDM with $d_{\text{free}} = 5$	$\alpha^{-10}\,(\mathrm{SNR}/2)^{-2\min(6,D_1)}$
GC with $d_{\text{free}} = 10$	$(5\alpha^{-1})^{20}\,(\mathrm{SNR}/2)^{-4\min(6,D_2)}$
SDM with $d_{\text{free}} = 10$	$\alpha^{-20}\,(\mathrm{SNR}/2)^{-2\min(6,D_2)}$

Table 3. Theoretical upper-bound for a 2×2 MIMO with QPSK symbols over a frequency fading channel with $L = 4$.

as for the flat fading channel concerning the difficulty of observing the additional diversity

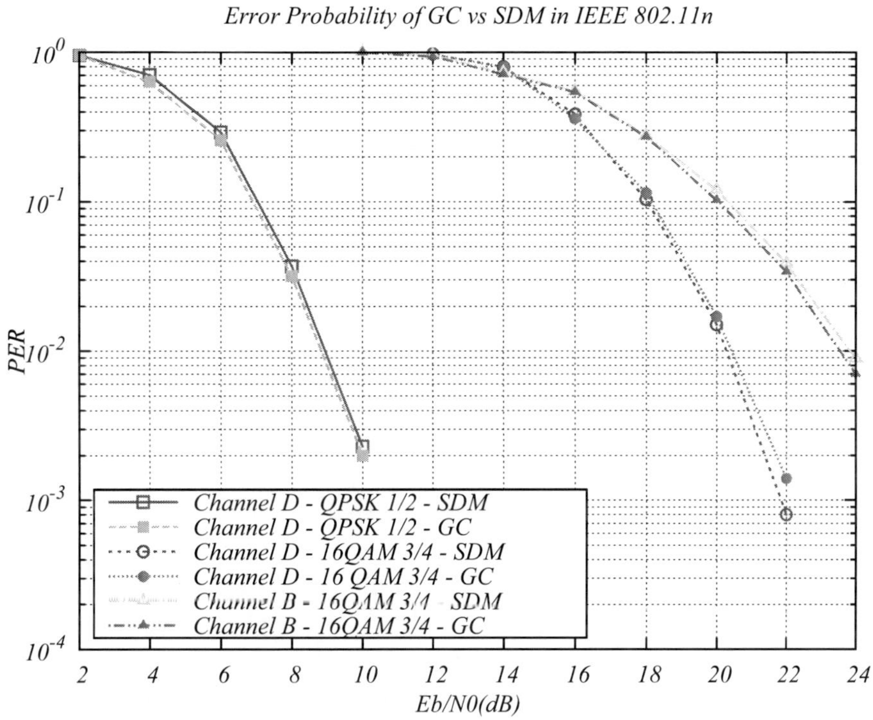

Figure 9: Golden Code vs SDM in IEEE 802.11n context.

gain of the GC can be observed also here. Moreover, this will be become more difficult to observe than the flat fading channel as in this case both schemes gain in diversity.

The performance of the Golden code versus SDM has been evaluated in the IEEE 802.11n context in terms of packet error rate (PER) versus SNR, for a packet length of 1000-bits. In the following, SNR gain will be related to a PER of 10^{-2}. The packet error rates in Figure 9 are evaluated over channel D using QPSK and 16QAM constellation. The channel D is characterized by a $50ns$ rms delay spread and 18 taps, and then by significant frequency diversity. In the IEEE 802.11n context, the convolutional code $[133\ 171]$ with a coding rate of $R_c = 1/2$ is used with $d_{\text{free}} = 10$. No additional gain is observed at a PER $= 10^{-2}$. The channel B in the IEEE 802.11n standard can be assimilated to a flat fading channel, for which the additional gain using a convolutional code with high free distance ($d_{\text{free}} = 10$) cannot be observed at reasonable PER.

4.4. Practical limits of space time codes use in a standard context

Recent standards that use MIMO system such that IEEE 802.11n and IEEE 802.16e aim to increase the throughput and the reliability of the system. However, increasing the reliability comes often at the expense of increased complexity at both transmitter and receiver side. Scarifying the complexity order can be done if promising gains at reasonable PER range are observed. Although theoretically, using the upper-bound on the PEP at high SNR, one expects that the error probability of these codes should be shifted more to the left compared with the SDM case, practical assumptions are more realistic, and address generally a moderate SNR regime and moderate range of PER.

For the flat fading case, when no outer code is used, the huge gain observed by the Golden code over all other known code make it promising to be used in such systems. However, when a complete chain as BICM-MIMO-OFDM system is considered, the situation become considerably different. As we show in a BICM-MIMO-OFDM system, the diversity of BICM-OFDM system can be extracted when no space time code is used. Additional diversity can be provided by using perfect space time codes over each subcarriers. This additional diversity comes at the expense of an increased lattice decoder, *i.e.* instead of using a $2n_t \times 2n_t$ ML soft decoder a $2n_t^2 \times 2n_t^2$ is required. Moreover, the impact of this additional diversity cannot be unfortunately observed at moderate range of PER.

5. Concluding Remarks

In this chapter, we reviewed the principle basis on the conception of multiple input multiple output systems. The knowledge of the channel knowledge at the transmitter side is crucial for this design. The case of a flat fading MIMO channel when the channel remains constant during all the duration of the transmission was firstly addressed. For the full CSIT case, we showed how the optimal power allocation over the eigen-channel modes optimizes the capacity of the channel. When no CSIT is available, we presented the diversity techniques used to benefit from the MIMO gains in term of diversity and multiplexing gain. The conception of these diversity techniques relies on two approaches. The first approach is based on the minimization of the pairwise error probability upper bound and yields to the well known space time code design criteria by Tarokh *et al.* (the rank and the determinant criteria) when the fading paths are assumed to be Rayleigh distributed. The second approach relies on the characterization of the diversity multiplexing tradeoff and is independent of the fading distribution. The construction of the perfect space time code family from the cycle division algebra that fully achieves these two gains is then detailed and the gain of the 2×2 MIMO Golden code are illustrated. Finally, we reviewed from [10] the impact of the concatenation of these multi-dimensional space time codes with powerful outer codes. The upper-bound on the PEP derived in [10] when using perfect space time codes is much lower than the one with space time codes. This should be traduced normally by a shift to the left of the packet error rate probability in the log-log scale. For the 2×2 MIMO channel, we show that these gains cannot be observed in the IEEE 802.11n context at a reasonable PER or SNR.

References

[1] I. E. Telatar, "Capacity of multi-antenna gaussian channels," *IEuropean Transactions on Telecommunications and Related Technologies*, vol. 10, no. 6, pp. 585–596, 1999.

[2] V. Tarokh, N. Seshadri and A. R. Calderbank, "Space time codes for high data rate wireless communication : performance criterion and code construction," *IEEE transactions on information theory*, vol. 44, no. 2, pp. 744–765, march 1998.

[3] L. Zheng and D. Tse, "Diversity and multiplexing : A fundamental tradeoff in multiple antenna channels," *IEEE Transactions on Information Theory*, vol. 49, no. 5, pp. 1073 – 1096, may 2003.

[4] F. Oggier, G. Rekaya, J.-C. Belfiore and E. Viterbo, "Perfect space time block codes," *IEEE transactions on information theory*, vol. 52, no. 9, septembre 2006.

[5] J. C. Belfiore, G. Rekaya and E. Viterbo, "The Golden Code: A 2×2 full rate space time code with non vanishing determinants," *IEEE Transactions on information theory*, vol. 51, no. 2, pp. 1432 – 1436, april 2005.

[6] P. Elia, K. R. Kumar, S. A. Pawar, P. V. Kumar and H.-F. Lu, "Explicit space time codes achieving the diversity multiplexing gain tradeoff," *IEEE Transactions on information theory*, vol. 52, no. 9, pp. 3869 – 3884, sept. 2006.

[7] G. Caire, G. Taricco and E. Biglieri, "Bit-Interleaved Coded Modulation," *IEEE transactions on information theory*, vol. 44, no. 3, may 1998.

[8] N. Gresset, L. Brunel and J. Boutros, "Space-time coding techniques with bit interleaved coded modulation over block-fading MIMO channels," *IEEE Transactions on Information Theory*, vol. 54, no. 5, pp. 2156–2178, may 2008.

[9] E. Akay and E. Ayanoglu, "Achieving Full Frequency and Space Diversity in Wireless Systems via BICM, OFDM, STBC, and Viterbi Decoding," *IEEE transactions on communications*, vol. 54, no. 12, pp. dec 2006.

[10] L. Mroueh, S. Rouquette-Léveil and J-C. Belfiore, "Application of perfect space time codes: PEP bounds and some practical insights," in *IEEE transaction on communication*, vol. 60, no. 3, pp 747-755, march 2012.

[11] E. Viterbo and J. Boutros, "A universal lattice code decoder for fading channels", *IEEE Transactions on Information Theory*, vol. 45, no. 5, pp. 1639-1642

[12] E. Agrell, T. Eriksson, A. Vardy and K. Zeger, "Closest point search in lattices," *IEEE transaction on Information Theory*, vol. 48, no. 8, pp. 2201-2214, 2002

[13] G. J. Foschini, "Layered space-time architecture for wireless communication in a fading environment when using multiple antennas," *Bell Lab. Tech. Journal*,vol. 1, no. 2, pp. 41–59, 1996

[14] S.M. Alamouti, "A simple transmit diversity technique for wireless communications,"IEEE Journal on Selected Areas in Communications, vol.16, no.8, pp.1451-1458, Oct 1998

[15] D. Tse and P. Viswanath, *Fundamentals of wireless communications.* Cambridge University Press, 2005.

[16] J. Boutros, N. Gresset, L. Brunel and M. Fossorier, "Soft-input soft-output lattice sphere decoder for linear channels," *IEEE Globecom*, dec 2003.

In: Communication Systems: New Research
Editor: Vyacheslav Tuzlukov
ISBN: 978-1-62618-654-5

Chapter 10

TRANSMISSION OF JPEG2000 IMAGES OVER BROADBAND WIRELESS NETWORKS USING UNEQUAL POWER ALLOCATION

Moein Shayegannia[1]*, Sami Muhaidat[2], Atousa Hajshirmohammadi[1] and Hamada Alshaer[2]
[1] Simon Fraser University, BC, Canada
[2]Khalifah University of Science Technology and Research, Abu-Dhabi, UAE & University of Surrey, UK

ABSTRACT

This chapter introduces an overview of our recent results on unequal power allocation for the transmission of JPEG2000 coded images over frequency selective channels. During transmission of the JPEG2000 codestream over frequency selective channels, the negative effects of inter symbol interference and inter carrier interference are eliminated by designing two separate systems for downlink and uplink data transmission: an orthogonal frequency division multiplexing system (downlink) and a single carrier frequency domain equalization system (uplink). The performance of these two systems is analyzed under deployment of multiple transmitter antennas, to achieve spatial diversity by incorporating space time block coding technique into the transmitter model. In addition, we explore addition of a linear constellation pre-coder matrix to the OFDM system to achieve multipath diversity during transmission of JPEG2000 images. During the transmission, the channel state information is assumed to be available at the receiver and can be feedback to the transmitter. Also, transmission of the header information of the JPEG2000 image codestream is assumed to be error free. To demonstrate the simulation results, the peak signal to noise ratio and the bit error rate profiles of the received images are used as the measure of metric. The results demonstrate a performance evaluation of our proposed systems for JPEG2000 image transmission with competing strategies, and measure transmission bandwidth efficiency and peak signal to noise ratio profile of the received images in our proposed systems.

* Corresponding Author - Email: moeins@sfu.ca.

Keywords: JPEG2000; Frequency Selective Channel; Unequal Power Allocation; Orthogonal Frequency Division Multiplexing; Single Carrier Frequency Domain Equalization

INTRODUCTION

The increasing demand on wireless image transmission requires high data rate and reliable transfer of multimedia streams over severe channel conditions. However, the hostile behavior of wireless medium and large number of users impose challenges in maintaining a high quality of the received images, conserving the available bandwidth, and preserving the total power for transmission. In addition, applications such as medical imagery require not only high quality of the reconstructed images, but also fast coding, delivery, and decoding of these images. To address these challenges, we need to alleviate the sources of disturbance in wireless channels, such as fading, interference, shadowing, path loss, and multipath, by providing techniques which can adapt to different channel conditions, while keeping the complexity of the system low.

To lessen the error impact, it is advantageous to use a source coding technique with some degree of protection against errors. JPEG2000 is one of the recent and advanced source coding techniques for image coding [1]. This standard is able to generate an error-resilient and scalable bitstream which allows progressive decoding of the received bitstream at different quality and resolution levels. The error resilient tool is able to detect but cannot correct the errors in the code-stream. Once an error is detected within a code block, the data associated with the erroneous coding pass and with the future coding pass of that code block are discarded, and resynchronization mechanism of the received data is performed [2]. The JPEG2000 coded bitstream has a hierarchical structure in which some bits hold more important information compared to others, thus it is natural to have higher protection over the more important bits. Different techniques such as Unequal Error Protection (UEP) and Unequal Power Allocation (UPA) are introduced in the literature to enhance the transmission of scalable coded bitstreams. The UEP methods apply Forward Error Correction (FEC) coding with different coding rates to different portions of the bitstream, based on the importance of each portion. The UPA techniques distribute the total available power for transmission of an image unequally over the bitstream in such a way that more power is allocated to the more important bits.

The pioneering works in this area mostly address implementation of the UEP technique to protect images against disturbances in wireless transmission. However, this technique leads to a lower data rate and excessive usage of bandwidth in practice. Recently, focus on protecting images via power allocation techniques for wireless transmission has been increased in the literature. However, most of the available manuscripts have assumed a basic transmission medium of non-frequency selective, which is impractical. Among those that consider frequency selectivity of the channel, a comprehensive investigation on the performance of the UPA or UEP techniques under different transmission scenarios, such as different channel memory lengths, various Signal to Noise Ratio (SNR) values, and multiple transmission antennas has not been analyzed. Furthermore, the literature on transmission of multimedia over frequency selective channels contains employment of the Orthogonal Frequency Division Multiplexing (OFDM) in the transmitter and receiver to combat Inter

Symbol Interference (ISI) and Inter Carrier Interference (ICI). Due to the large Peak to Average Power Ratio (PAPR) imposed by the OFDM system, this technique is employed at the base station for down link data transmission. However, for a communication system, exploration of multimedia data transmission in Single Carrier Frequency Domain Equalization (SCFDE) system, which is used to eliminate ISI and ICI in frequency selective channels in the uplink transmission, is mandatory [3].

The main purpose of this chapter is to investigate the impact of an optimized UPA technique on the quality of the received JPEG2000 images over wireless channels which suffer from frequency selectivity. We provide Monte-Carlo simulations for different channel memory lengths, on a range of different available SNR values, with single and multiple transmitter antennas. To combat the negative effects of ISI and ICI, we employ the UPA algorithm in an OFDM system for downlink data transmission, and in an SCFDE system for uplink data transmission. We compare the performance of these two systems on the quality of the received images and on the bandwidth efficiency during transmission. We utilize the UPA technique proposed in [4] for transmission of JPEG200 images over non-frequency selective channels.

This chapter is organized as follows. First section introduces the topic and the motivation behind this chapter. The second section of this chapter provides a brief review of important background materials related to this chapter. This section explains the JPEG2000 Image coder, and elaborates on the structure of this standard. Then, the available protection techniques in the literature for wireless image transmission such as UEP techniques or UPA schemes are discussed. Next, an overview about the Space Time Block Coding (STBC) technique, which is used in this chapter to obtain spatial diversity, is presented. The next section reviews a SCFDE based system for the uplink transmission of JPEG2000 images with UPA algorithm in broadband networks. The fifth section compares efficiency of the OFDM and SCFDE systems on transmitting JPEG2000 images over frequency selective channel. Finally, the last section concludes this chapter by providing a brief summary of all the sections in this chapter.

Background and Related Work

JPEG2000 Structure

The transformation technique used in JPEG2000 is the Discrete Wavelet Transform (DWT). In the JPEG2000 image coder, the first operation is to (optionally) partitiona source image into a number of rectangular non-overlapping blocks called tiles. Then DWT is applied to each tile which transforms the samples into spatial frequency subbands at different levels of resolution. The first level of decomposition consists of four subbands LL1, LH1, HL1, HH1 [5]. The LL1 subband is the lowest resolution of the tile and is a down-sampled low-resolution representation of the original tile-component. The LL1 subband can be further decomposed by applying DWT. This process can be repeated to obtain different resolution levels. Then, each resolution of each tile component is further partitioned into precincts. Within every subband, each precinct contributes one packet to the code-stream of the image. Precincts are not a partition of image data and do not impact sample data transformation or coding. Precincts are used to reconstruct the resolution [6]. Each subband is partitioned into small blocks, called code-blocks, where quantization and bit-plane coding are initiated. The packets furnished by the precincts identify their header and body from the contributions of the code-blocks belonging to the relevant precinct. Starting from the Most Significant Bit (MSB), the coder scans through the bit-planes of each coding pass within three coding passes. Each of the coding passes collects the relevant information about the bit-plane data. Based on the significance of a particular bit location and its neighboring, location of each bit in the coding passes is decided. The encoder uses this information to generate a compressed bitstream or a code-stream [1]. Figure 1 illustrates a 3 layer decomposition of a source image using DWT and its partitioning into four resolution levels, subbands, precincts, and code-blocks. This figure also presents how each code-block is decomposed into a number of bit-planes.

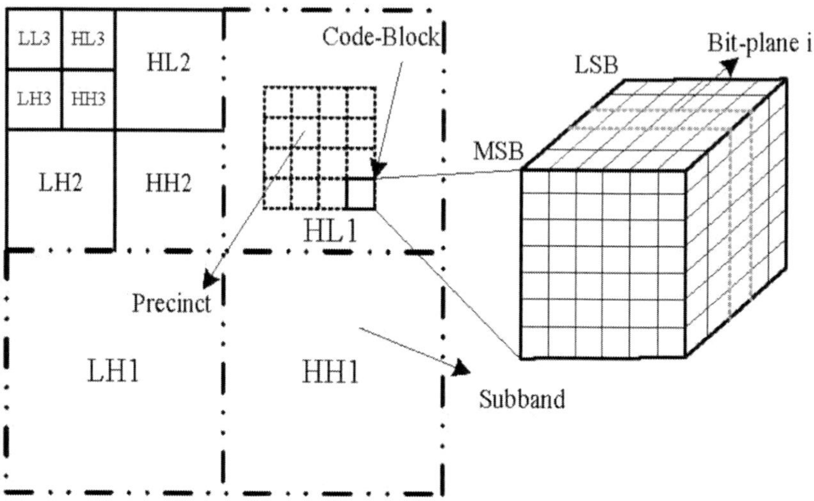

Figure 1. Components of a JPEG2000 transformed image.

To increase the robustness of the JPEG2000 bitstream against error propagation along the code-stream, error resilient feature is introduced in the standard. Small size code-blocks are independently coded and included with resynchronization markers. As a result, errors do not propagate beyond the code-block whose bit-stream is corrupted, and the markers keep the synchronization between the encoder and decoder in case of occurrence of bit errors. JPEG2000 standard also provides a mechanism to combine all the packet headers within the main header. This adds an advantage to the decoding process of the received data stream, if the main header can be transmitted in an error-free medium. It is necessary to correctly decode the header of a packet in order to extract the code-block contributions to the body of the packet [5], [1].

Overview of Error Protection Techniques for Wireless Image Transmission

Transmission of JPEG2000 images using UEP techniques over slow fading non-frequency selective channels has been widely investigated in [6], [7], [8], and [9]. In [6], UEP is achieved in JPEG2000 code-streams by using Reed Solomon (RS) block codes. In [7], Banister et al. make use of Viterbi algorithm to jointly optimize source rate and channel rate for the purpose of UEP. In [8], the authors employ product coded streams which consist of Turbo-codes and RS codes to obtain UEP. In [9], the authors obtain UEP using RS channel coding for the header and convolutional coding for the body of the image bitstream.

Protection of JPEG2000 images over slow fading non-frequency selective channels using UPA schemes has been reported in [10] and [11]. Atzori employs an optimized UPA scheme in [10] based on increasing JPEG2000 image quality as well as RS channel coding to protect coded bitstream. In [11], Torki et al. proposes an optimized UPA scheme based on minimizing the total end to end distortion of JPEG2000 images, which proved its effectiveness in improvement of the PSNR performance and at a lower complexity in comparison with the existing UEP techniques.

Evaluation of UEP for JPEG2000 image transmission over frequency selective channels has been investigated in [12], [13], [14], [15], [16], [17], [18], [19], [20], [21], [22], and [23]. Houas et al. utilize rate compatible punctured convolutional codes to enable UEP for transmission of the JPEG2000 images in OFDM systems [12]. In [13], the layer structure of the JPEG2000 is exploited by protecting the data in top layer with an FEC code, and the performance is analyzed in an OFDM transmission system for different channel delays. In [14], authors achieve UEP using an optimal joint source-channel coding and consider progressive image transmission over differentially space-time coded OFDM system. Sethakaset et al. propose an UEP scheme in the spatial domain through enhanced beamforming algorithms over closed loop multiple-input multiple-output OFDM system [15]. In [16], the authors propose an unequal density of pilot placement for OFDM systems in time-varying channel to achieve UEP for important multimedia transmission. They also present a dynamic image data assignment scheme for UEP by utilizing visual saliency to relocate image data in an OFDM symbol for the UEP transmission. in [17] an unequal density of pilot placement for OFDM systems in time-varying channels and a dynamic image data assignment scheme is proposed to achieve UEP for important multimedia transmission. UEP for the Set Partitioning In Hierarchical Trees (SPIHT) bitstream is obtained in [18] by designing a multi-rate turbo coder which adopts multiple generators to enable it cover wide range of code-rates in an Alamouti STBC OFDM system. Salah *et al.* propose to combine convolutional coding with unequal cyclic time guard in an OFDM system to obtain UEP for image transmission [19]. In [20], a practical joint source-channel coding scheme for OFDM system is proposed and spatial diversity is exploited using space time block coding. [21] studies transmission of progressively coded image bitstreams using channel coding in a 2-D time-frequency resource block in an OFDM network. The authors in [22] use a single carrier frequency domain multiple access system to compare discrete fourier transform based systems with discrete cosine transform systems for an efficient image transmission. In [23], the authors propose to transmit wavelet based images using an adaptive OFDM system which allocates the modulation bit of each subcarrier according to the channel condition.

Space Time Block Coding for Transmit Diversity

Space Time Block Coding (STBC) is a technique used in wireless communication to transmit multiple copies of the transmitting data across multiple antennas at different time slots. This technique enhances the reliability of the wireless link, benefits the receiver through spatial diversity, and leads to a higher quality of the received signal. Implementation of the STBC technique occurs as the last step in the transmitter side of our proposed system. The space time block coding for two transmit antennas is defined as [24]:

$$\mathbf{C} = \begin{pmatrix} \mathbf{z}_1 & \mathbf{z}_2 \\ -\mathbf{z}_2^* & \mathbf{z}_1^* \end{pmatrix} \tag{1}$$

where $\mathbf{z}_1$ and $\mathbf{z}_2$ are transmitted during the first time slot, from the first and the second antenna, respectively. Also $-\mathbf{z}_2^*$ and $\mathbf{z}_1^*$ are transmitted during the second time slot, from the first and the second antenna, respectively. $(.)^*$ also denotes complex conjugation.

The coded blocks of data are transmitted over the two channels during two consecutive time slots. The Channel Impulse Response (CIR) is assumed to remain constant during the two time slots. To keep the total transmit power constant in the multiple transmit antenna system, half of the power value used in the single antenna transmission scenario is provided to each transmitting antenna at each time slot. The received signal is thus formulated as follows:

$$\begin{aligned} \boldsymbol{r}_1 &= \boldsymbol{h}_1\boldsymbol{z}_1 + \boldsymbol{h}_2\boldsymbol{z}_2 + \boldsymbol{v}_1 \ , \\ \boldsymbol{r}_2 &= -\boldsymbol{h}_1\boldsymbol{z}_2^* + \boldsymbol{h}_2\boldsymbol{z}_1^* + \boldsymbol{v}_2 \ , \end{aligned} \tag{2}$$

where $\boldsymbol{r}_1$ is the received signal at the first time slot, and $\boldsymbol{r}_2$ is the received signal at the second time slot. $\boldsymbol{h}_1$ and $\boldsymbol{h}_2$ are the slow fading representation of the multi-tap channel; $\boldsymbol{v}_1$ and $\boldsymbol{v}_2$ indicate the Additive White Gaussian Noise (AWGN) of the channel at the first and the second time slot, respectively. At the receiver side, a linear combining is executed on the both received signals as follows [24]:

$$\begin{aligned} \hat{\mathbf{z}}_1 &= \boldsymbol{h}_1^*\boldsymbol{r}_1 + \boldsymbol{h}_2\boldsymbol{r}_2^* \ , \\ \hat{\mathbf{z}}_2 &= \boldsymbol{h}_2^*\boldsymbol{r}_1 - \boldsymbol{h}_1\boldsymbol{r}_2^* \ , \end{aligned} \tag{3}$$

where $\hat{\mathbf{z}}_1$ and $\hat{\mathbf{z}}_2$ are the recovered versions of $\mathbf{z}_1$ and $\mathbf{z}_2$. In the last step, the receiver recovers the information bits by a Minimum Mean Square Error (MMSE) or a maximum likelihood decoder.

TRANSMISSION OF JPEG2000 IMAGES OVER OFDM NETWORKS

Most of the literature concentration on protecting Images during transmission is with UEP techniques, which lower the transmission data rate and leads to excessive usage of power for the coding bits.Yet a shortage of research exists in the literature on the UPA technique for wireless image transmission over frequency selective channels. In [25], a UPA technique is adapted for OFDM systems to transmit JPEG2000 imamges over broadband networks. In their paper, Shayegannia *et. al* demonstrate effectiveness of the UPA technique proposed in [11] for transmission of JPEG2000 images with UPA over multicarrier systems, with different channel memory lengths.

TRANSMISSION OF JPEG2000 IMAGES WITH UPA OVER SINGLE CARRIER TRANSMISSION SYSTEMS

OFDM is a multicarrier communication technique that has attracted the interests of the academic and industrial communities in the resent years. In wire line communications, OFDM has been employed by several standards such as the High-bit-rate Digital Subcarrier Line (HDSL) and Very-high-speed Digital Subcarrier Line. In wireless systems, it has been used by the Digital Audio Broadcasting (DAB), Digital Video Broadcasting for Terrestrial Television (DVB-T) and Wireless Local Area Networks (WLANs), which are based on the IEEE 802.11 standards

Unfortunately, OFDM has the drawback of high Peak to Average Power Ratio (PAPR). As an attractive alternative to Multi-Carrier (MC) techniques, Cyclic Prefixed Single-Carrier (CP-SC) transmission techniques have recently attracted considerable attention due to their low PAPR. In practice, OFDM is implemented in the base station of a cellular network for downlink data transmissions, and Single Carrier Frequency Domain Equalization (SCFDE) is adapted at the end user station for uplink data transmissions [2].

To the best knowledge of this author, most, if not all, of the literature reports on multimedia data transmission over frequency selective channels, with UEP or UPA, utilize the OFDM technique to combat the negative impact of ISI and ICI. This chapter reviews our proposal in [26] which integrates the UPA algorithm with SCFDE to provide protection for uplink transmission of JPEG2000 coded bitstreams in block fading frequency selective channels. Moreover, our SCFDE system obtains spatial and multipath diversity during transmission by using Alamouti's Space Time Block Coding (STBC) scheme [24] and MMSE frequency domain equalization, respectively.

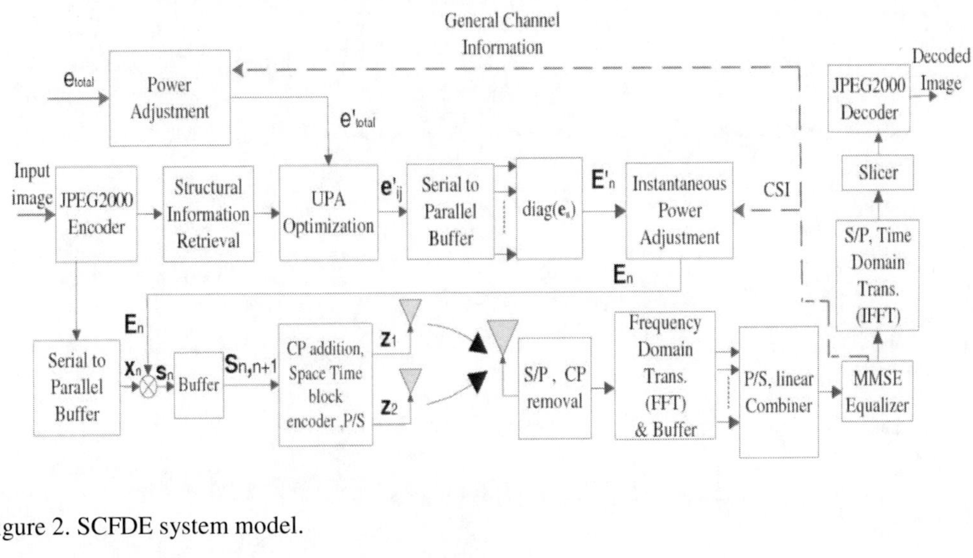

Figure 2. SCFDE system model.

System Model

The overall system block diagram is presented in Figure 2. The functionality of the JPEG2000 encoder, Structural Information Retrieval, and the UPA optimization block are detailed in [25]. In brief, these blocks convert an input image into a code stream, and allocate unequal amount of power values to the coding passes based on their importance in reconstructing the received image. The importance of each coding pass is determined using a distortion model described in [25]. Once the optimum power of each coding pass, and the bits within that coding pass is allotted, the bitstream and the power assigned to each bit are categorized into N_B blocks. Before the transmission, every block of data is appended with a cyclic prefix, and every two consecutive blocks, i.e, n^{th} and $(n+1)^{\text{th}}$, are Alamouti coded, where $(n = 1, 2, \ldots, N_B)$. The two transmit antennas model exploit spatial diversity which increases the reliability of the wireless link and improves the quality of the received image. This model improves the quality of the received image without increasing the transmit power. Thus, the two transmit antenna system is capable of conserving more energy than the single transmit antenna system.

The channel impulse response between the jth transmitter antenna and the receiver, for the nth transmission block is given by $\mathbf{h}_{n,j} = [h_0, h_1, \ldots, h_{m-1}]^T$, where m is the channel memory length. The beginning of each data block, with a size of $N \times 1$, is appended with the last m samples of the same block, to eliminate the impact of ISI and to form the channel matrix circulant. At the receiver side, the first m samples of every block are discarded and only N samples are processed. Forming the channel into a circulant matrix accounts for the addition and removal of the cyclic prefixes for every n^{th} transmission block.

FDE-STBC was suggested in [3] as an extension of Alamouti scheme [24] to frequency selective channels by imposing the Alamouti orthogonal structure at a block level, rather than the original symbol-level implementation used for frequency-flat fading channels. Following [3], the transmitted blocks over the first and the second antenna are generated according to the encoding rule presented in Figure 3.

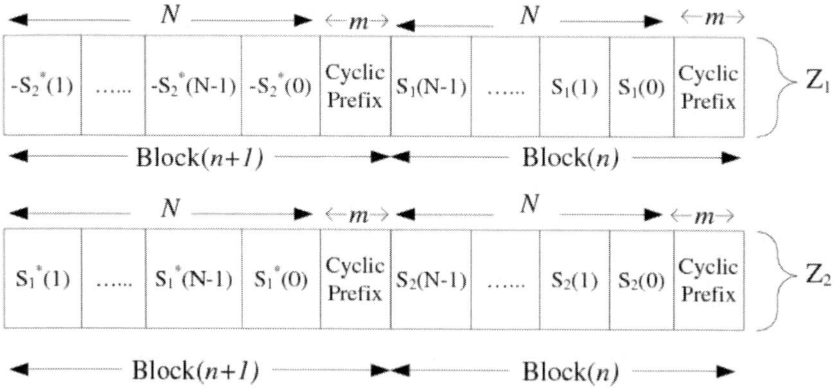

Figure 3. Transmission block format for SCFDE-STBC [3].

Denote in this figure, $\mathrm{S_n} = \sqrt{\mathrm{E_n}}\mathrm{x_n}$, and $\mathrm{S_n} = [\mathrm{s_0}, \mathrm{s_1}, \ldots, \mathrm{s_{N-1}}]$, and the nth transmitted block from antenna j by $\mathrm{S_{n,j}(i)}$. At times $\mathrm{n} = 0, 2, 4, \ldots$, pairs of length-N blocks $\mathrm{S_{n,1}(i)}$ and

$S_{n,2}(i)$ (for $i = 0, 1, \ldots, N-1$) are generated by an information source with the following transmit diversity scheme [3]:

S_(n+1,1) (i)=-S_(n,2)^* ((-i)_N) , i=0,1,…,N-1
S_(n+1,2) (i)=S_(n,1)^* ((-i)_N) , n=0,2,4,… (4)

where $(.)^*$ and $(.)_N$ denote complex conjugation and modulo-N operations, respectively. The received signal in the first and the second time slots at the receiver side are given by:

r_(n_1)=H_(n,1) √(E_(n,1)) x_(n,1)+H_(n,2) √(E_(n,2)) x_(n,2)+v_(n_1),
r_(n_2)=⟦-H⟧_(n,1) √(E_(n,2)) x_(n,2)^*+H_(n,2) √(E_(n,1)) x_(n,1)^*+v_(n_2), (5)

where $\mathbf{r}_{n_1}$ and $\mathbf{r}_{n_2}$ are the received signal of the nth transmission block at the first and second time slot, respectively. $\mathbf{H}_{n,1}$ and $\mathbf{H}_{n,2}$ are circulant matrices, with entries $[\mathbf{H}_{n,j}]_{ik} = [\mathbf{h}_{n,j}]_{(i-k)\, mod\, N}$, $[\mathbf{h}_{n,j}]_i = 0$ for $i > m-1, j = 1, 2$, and they represent the CIR of the first and the second antenna, respectively. $\mathbf{v}_{n_1}$ and $\mathbf{v}_{n_2}$ symbolize the AWGN vector of mean zero and variance of $\frac{1}{2}$ per dimension for the first and second time slot, respectively. $\mathbf{E}_{n,1}$ and $\mathbf{E}_{n,2}$ are the actual power values of the nth block of data ($\mathbf{x}_{n,1}$ and $\mathbf{x}_{n,2}$) transmitted from the first and second transmission antenna, respectively. Equation (9) formulates the calculation of the actual power of each bit from the assigned valued by the UPA optimization algorithm. The available power for transmission at any of the two antennas is half of the values in the single transmission antenna case, in order to keep the total transmit power constant.

The received signal is converted into frequency domain by multiplying by the FFT matrix as follows [3]:

$$\begin{aligned} \mathbf{c}_{n_1} &= \Lambda_{n,1}\mathbf{Q}\sqrt{\mathbf{E}_{n,1}}\mathbf{x}_{n,1} + \Lambda_{n,2}\mathbf{Q}\sqrt{\mathbf{E}_{n,2}}\mathbf{x}_{n,2} + \mathbf{Q}\mathbf{v}_{n_1}, \\ \mathbf{c}_{n_2} &= -\Lambda_{n,1}\mathbf{Q}\sqrt{\mathbf{E}_{n,2}}\mathbf{x}^*_{n,2} + \Lambda_{n,2}\mathbf{Q}\sqrt{\mathbf{E}_{n,1}}\mathbf{x}^*_{n,1} + \mathbf{Q}\mathbf{v}_{n_2}, \end{aligned} \tag{6}$$

where $\mathbf{c}_{n_1} = \mathbf{Q}\mathbf{r}_{n_1}$ and $\mathbf{c}_{n_2} = \mathbf{Q}\mathbf{r}_{n_2}$.The next step is to perform a linear combining as follows:

$$\begin{aligned} \hat{\mathbf{r}}_{n,1} &= \Lambda^*_{n,1}\mathbf{c}_{n_1} + \Lambda_{n,2}\mathbf{c}^*_{n_2}, \\ \hat{\mathbf{r}}_{n,2} &= \Lambda^*_{n,2}\mathbf{c}_{n_1} - \Lambda_{n,1}\mathbf{c}^*_{n_2}. \end{aligned} \tag{7}$$

The received data in (7) are time reversed according to the encoding rule in Figure 3. Then, the receiver uses a diagonal matrix ($\mathbf{W}$) of size $N \times N$ to perform an MMSE equalization on the resulting signals in (7) which are in frequency domain. The elements of the equalizer matrix are given by [3]:

$$[\mathbf{W}_n]_{ii} = \frac{1}{\left|[\Lambda_{n,1}]_{ii}\right|^2 + \left|[\Lambda_{n,2}]_{ii}\right|^2 + \frac{1}{SNR}}, \tag{8}$$

where SNR stands for Signal to Noise Ratio. This equalizer can be used for the signals received at both time intervals. The output of the MMSE equalizer will be $\hat{\mathbf{y}}_n = \mathbf{W}_n \hat{\mathbf{r}}_n$, where $\hat{\mathbf{r}}_n$ is the output of the linear combiner. Next, The output of the equalizer is transferred into time domain by applying the IFFT to recover the original data given by $\hat{\mathbf{s}}_n = \mathbf{Q}^H \hat{\mathbf{y}}_n$, where $(.)^H$ denotes complex conjugate transpose operation. At the end, the receiver applies a hard decision slicer on the recovered data, and passes it to the JPEG2000 decoder to generate the received image.

Instantaneous Power Adjustment Unit

The UPA algorithm implemented in this single carrier system assumes an AWGN channel at the time of power allocation to the coding passes [11]. The Instantaneous Power Adjustment unit compensates for the effect of fading before transmission by using the instantaneous values of the channel impulse response. The compensation for the effect of fading in a multi-transmission antenna channel is given by:

$$\mathbf{E}_{n,j} = \frac{\acute{\mathbf{E}}_{n,j}}{\alpha + \sum_0^{m-1} |h_k|^2}, \tag{9}$$

where h_k is the kth channel tap for the nth transmission block and m is the channel memory length or the total number of taps. $\acute{\mathbf{E}}_n$ is the assigned power to the bits within the nth transmission block by the UPA optimization algorithm, and $\mathbf{E}_n$ represents the corresponding power of the bits after the effect of channel fading is taken into consideration.

Simulation Results

Kakadu software is used to encode JPEG2000 images of Lena and Peppers at a size of 512×512, and at a rate of 0.25 bpp. The settings for the JPEG2000 codec are 64×64 code-blocks, 128×128 precincts, and one level of decomposition for the sake of simplicity in analysis. The header information is assumed to be transmitted error-free. Binary Phase Shift Keying (BPSK) is used to modulate the bitstream produced by the JPEG2000 encoder. We analyze the performance of the system by measuring the average PSNR and the average BER profile of the received image transmitted through multi-tap frequency selective channels. The images are transmitted through a block fading frequency selective channel, and assume that each tap of the channel has a Rayleigh fading distribution with AWGN, and is independent of the other taps. OFDM and SCFDE techniques are employed with our UPA scheme to eliminate the effect of ISI and ICI, while an optimal amount power is allotted to each transmitting bit. The number of subcarriers (N) used in the transmitter is 16. The value of N does not have any impact on the performance of the system. However, at higher number of subcarriers, the size of the IFFT and the channel matrix grows linearly, thus the computational complexity of the system increases.

Based on the previous experiments in [11], the value of α in (9) is set to be 0.01. The performance of the UPA algorithm with SCFDE technique is analyzed by calculating the PSNR of the received image transmitted through block fading frequency selective channels.

Figure 4 to Figure 7 present the PSNR performance of the single carrier system with multiple transmission antennas when combined with the UPA optimization algorithm for image transmission through frequency selective channels, at different channel taps. The benchmark in the PSNR performance evaluation in this section is the scenario in which the UPA algorithm is used alone for image transmission over block fading non-frequency selective channels. Figure 4 reveals a degradation in the PSNR of the received "Lena" when the number of channel taps increases. For example, at an SNR value of 10 dB, integration of UPA and SCFDE lowers the PSNR of the system in a 2-tap channel for about 6 dB in comparison with the single tap channel. The latter is also lowered by about 2.2 dB compared to the case where only UPA is used for transmission over frequency flat channels. Figure 5 illustrates similar trends in the PSNR profile of the received "Peppers". The reason for the lower PSNR performance at higher channel taps in the UPA-SCFDE system is that our UPA algorithm initially assumes an AWGN channel condition during the optimization process. To compensate for the channel fading, (9) is applied in the algorithm before the transmission. This equation eliminates the effect of the instantaneous channel fading; however, the SCFDE technique requires the multi-tap fading behavior of the channel to obtain multipath diversity and enhance the system performance as the number of channel taps increases.

Table 1. Transmission scenarios in the UPA-SCFDE system (SNR=16 dB)

Scenario	Transmission Scenario	PSNR (dB)	
		Lena	Peppers
(a)	Original Image	----	-------
(b)	Using UPA over non-frequency selective channel	27.99	27.22
(c)	Using UPA & SCFDE over 1-tap frequency selective channel with 2 transmission antenna	26.95	26.73
(d)	Using UPA& SCFDE over 1-tap frequency selective channel with 1 transmission antenna	24.95	24.32
(e)	Using UPA& SCFDE over 2-tap frequency selective channel with 2 transmission antenna	20.01	19.56
(f)	Using UPA& SCFDE over 2-tap frequency selective channel with 1 transmission antenna	17.15	15.57

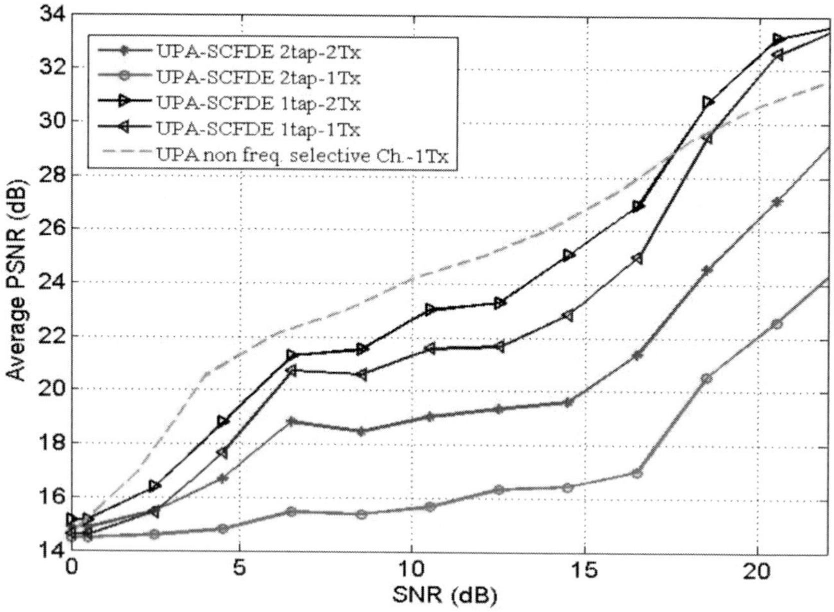

Figure 4. Comparison of the PSNR of the received Lena ($\mathbf{512 \times 512, 0.25}$ bpp) in the UPA-SCFDE system.

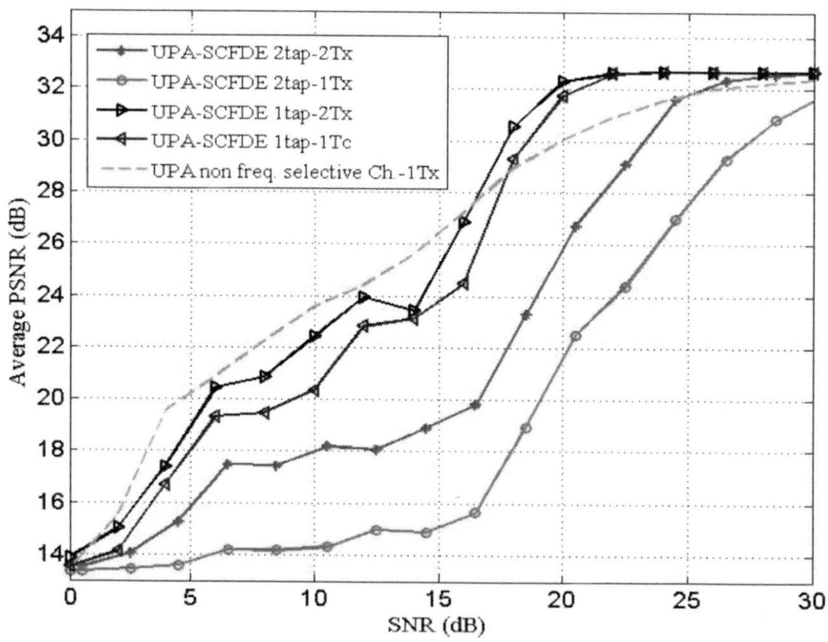

Figure 5. Comparison of the PSNR of the received Peppers ($\mathbf{512 \times 512, 0.25}$ bpp) in the UPA-SCFDE system.

Despite the loss in the PSNR profile at higher channel taps, our proposed system achieves spatial diversity when two transmit antennas are employed at the transmitter side and Alamouti's STBC is utilized. Figure 4 and Figure 5 demonstrate that at an SNR value of 10 dB, utilization of two transmitter antennas contributes to an increment of up to 2.5 dB in the PSNR of the received image transmitted through 2-tap frequency selective channel. Thus, we are able to receive the image with a higher quality at the receiver, or conserve energy by bounding the quality of the received image. Also at SNR values greater than 18 dB, the proposed system has a superior performance in a single tap channel in comparison to the benchmark scenario.

Figure 6. Visual comparison of "Lena" at 0.25 bpp, transmitted at SNR=**16** dB over the scenarios in Table 1.

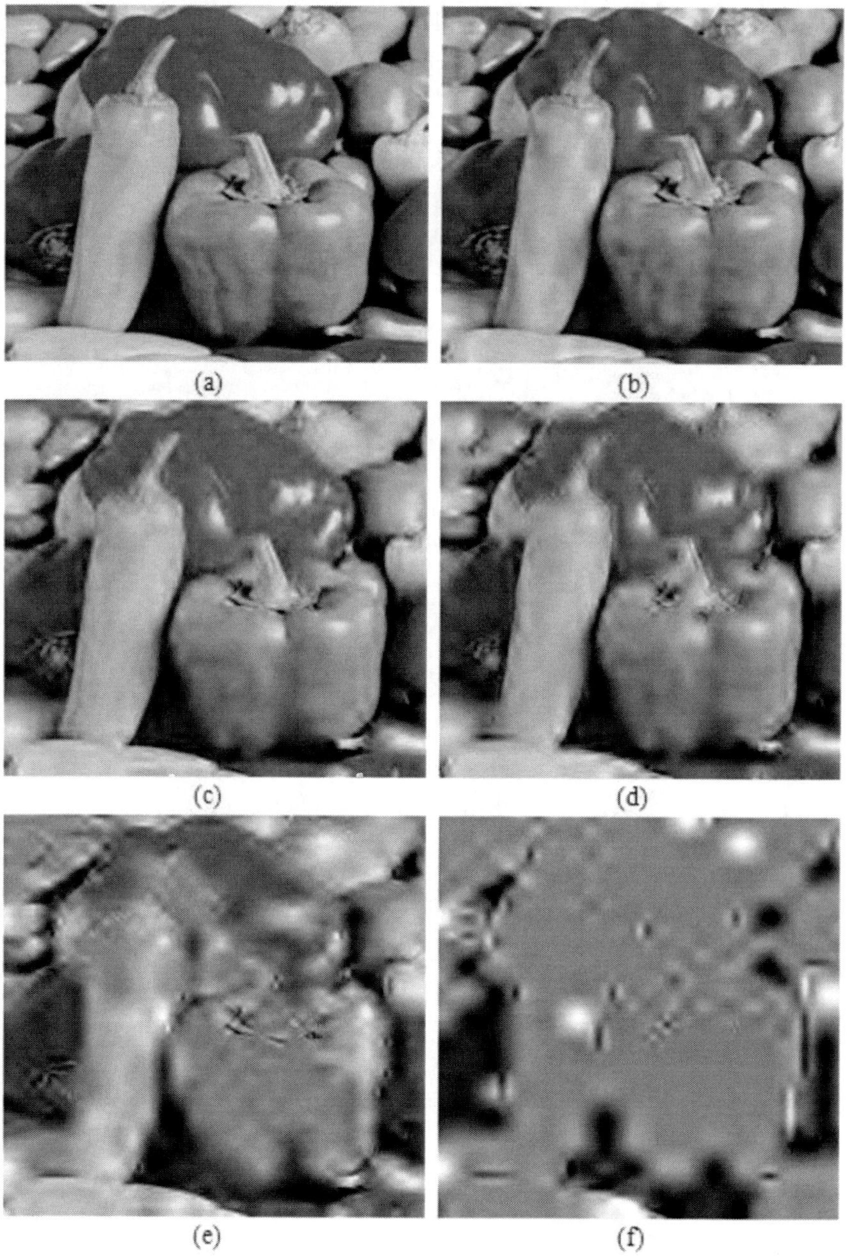

Figure 7. Visual comparison of "Peppers" at 0.25 bpp, transmitted at SNR=**16** dB over the scenarios in Table 1.

Figure 6 and Figure 7 present a visual comparison for transmission of "Lena" and "Peppers" (respectively) in the UPA-SCFDE system over the channel scenarios tabulated in Table 1. Note that the visual quality of the received image is enhanced when the number of transmitter antennas increases. In addition, increasing the number of channel taps lowers the quality of the received image.

Comparison of OFDM and MMSE-SCFDE over Frequency Selective Channels

Many researchers consider progressive image transmission in OFDM systems, and few evaluate the transmission of images in single carrier systems in the literature. A comparison between OFDM and SCFDE needs to be undertaken to investigate their different behaviors in transmission of JPEG2000 images in single antenna and multiple-input multiple-output systems, and how much these two techniques are capable of conserving bandwidth at different source coding rates. Also, an important criterion in designing a system for wireless image transmission is to maintain a low complexity in order to avoid time delay in coding and transmission, and reduce the fabrication cost.

This section investigates and compares the PSNR performance of the Linear Constellation Pre-Coding (LCP) OFDM and MMSE-SCFDE systems for the transmission of JPEG2000 images over frequency selective channels. For the sake of comparison, we consider EPA to distribute the available power over the coding passes. Finally, this section examines the amount of bandwidth conservation that these two systems can provide either in the case of single transmission antenna or multiple transmission antennas, by measuring the PSNR of the received image at different bit budgets.

System Model

Figure 8 and Figure 9 present the block diagrams which use SCFDE and OFDM, respectively. Yue *et al.* propose a transmission model for integrating STBC as an inner code with OFDM to achieve diversity gain in [27]. This section uses the proposals of [3] and [27] to integrate STBC with MMSE-SCFDE and LCP-OFDM, respectively, for wireless image transmission. In [25], the inability of the OFDM technique to achieve multipath diversity is demonstrated, and the PSNR performance of a received image from a frequency selective channel in an OFDM system deteriorates as the number of channel taps increases. A similar trend for a SCFDE system is demonstrated in the previous section, where a lower PSNR values at higher number of channel taps is measured. Both of these studies included integration of OFDM and SCDDE with an optimized unequal power allocation algorithm. This section presents allocation of an equal amount of power on codestream of an image to establish a fair comparison between LCP-OFDM and SCFDE for wireless image transmission.

The LCP technique can be used to obtain multipath diversity in an OFDM system. The multipath diversity enhances the quality of the received signal at higher number of channel taps [28]. In this technique, the symbols are pre-coded at the transmitter with a linear constellation matrix. The pre-coder matrix ($\boldsymbol{\theta}$) is unitary, as a result it neither increases power nor reduces the transmission rate. This matrix is represented by a Vandermonde matrix as follows [28]:

$$\boldsymbol{\theta} = \frac{1}{\beta}\begin{pmatrix} 1 & \alpha_1 & \dots & \alpha_1^{N-1} \\ 1 & \alpha_2 & \dots & \alpha_2^{N-1} \\ \vdots & \vdots & & \vdots \\ 1 & \alpha_N & \dots & \alpha_N^{N-1} \end{pmatrix}, \tag{10}$$

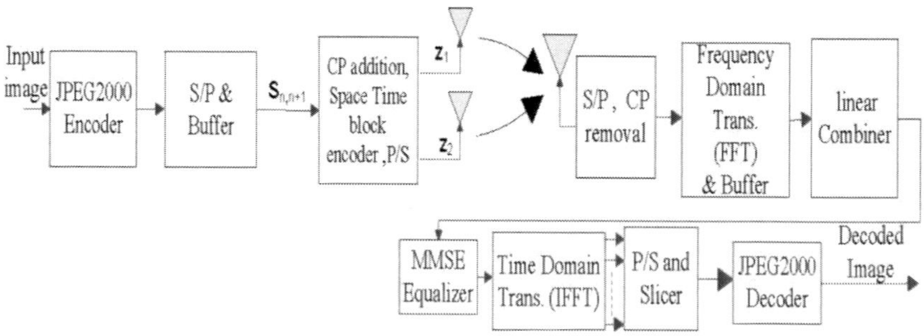

Figure 8. SCFDE system model with EPA.

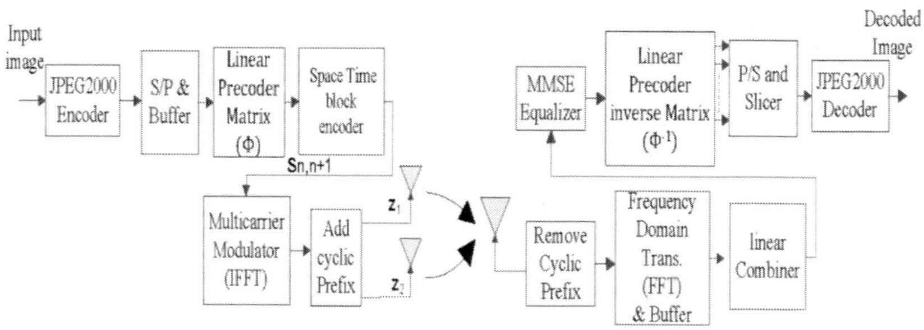

Figure 9. OFDM system model with EPA.

where N is the number of subcarriers, and β is a normalization factor to impose the power constraint $\text{tr}(\boldsymbol{\theta}\boldsymbol{\theta}^H) = N$, with tr(.) denoting matrix trace. $\{\alpha_N\}_{n=1}^{N}$ are the roots of $x^n - \sqrt{-1} = 0$, if n is a power two. In our analysis, we consider power of 2 values for N, albeit in [28] design of the LCP matrix for different values of N is explained. At the receiver side, inverse of the LCP matrix $(\boldsymbol{\theta}^{-1})$ is used to compensate for the pre-coding. Detailed algebraic derivation of constructing this matrix is available in [29].

On the transmitter side of the systems, the JPEG2000 encoder transforms the format of an input image into JPEG2000 format and generates a scalable coded bitstream. The OFDM system performs a multicarrier modulation (i.e, multiplying by the IFFT matrix) at the transmitter; however, the transmitter model in the SCFDE system does not include any Fourier transformation. This approach not only reduces the computational complexity in the SCFDE transmitter terminal, but also leads to a lower PAPR as compared with OFDM. The multicarrier implementation of OFDM technique leads to a large PAPR and high sensitivity to carrier frequency offsets. To tackle these issues, OFDM can be implemented in the base station of a cellular network for down-link data transmissions, and SCFDE is adapted at the end user station for uplink data transmissions [2].

We assume that the channel is block fading frequency selective, and the CIR remains constant over the two time slots during which the space time block codes are transmitted. The transmission model for SCFDE system has been already explained in this chapter; in the following section the transmission model for LCP-OFDM system with two transmit antennas is discussed.

Orthogonal Frequency Domain Multiplexing Transmission Model

The OFDM system, initially, pre-codes the transmitting data with a linear constellation matrix to exploit multipath diversity. The resulting symbols are passed to the STBC block to generate different copies of the signal before the transmission. The output of the STBC block is depicted in Figure 3. Before the transmission, the OFDM transmitter performs a Fourier transformation and sends the frequency domain symbols to the circulant structured channel in two different time slots.

$$\begin{aligned} \mathbf{r}_{n_1} &= \mathbf{H}_{n,1}\mathbf{Q}^H\boldsymbol{\theta}\mathbf{x}_{n,1} + \mathbf{H}_{n,2}\mathbf{Q}^H\boldsymbol{\theta}\mathbf{x}_{n,2} + \mathbf{v}_{n_1}, \\ \mathbf{r}_{n_2} &= -\mathbf{H}_{n,1}\mathbf{Q}^H\boldsymbol{\theta}\mathbf{x}^*_{n,2} + \mathbf{H}_{n,2}\mathbf{Q}^H\boldsymbol{\theta}\mathbf{x}^*_{n,1} + \mathbf{v}_{n_2}. \end{aligned} \tag{04}$$

In (11), $\mathbf{r}_{n_1}$ and $\mathbf{r}_{n_2}$ are the received signals at the receiver from the OFDM system in the first and the second time slot, respectively. $\mathbf{H}_{n,1}$ and $\mathbf{H}_{n,2}$ represent the circulant channel for the nth transmission block from the first and second antenna, respectively. $\mathbf{x}_{n,1}$ and $\mathbf{x}_{n,2}$ denote the nth block of transmitting data from the first and the second antenna, respectively. $\mathbf{v}_{n_1}$ and $\mathbf{v}_{n_2}$ are the AWGN vectors. $\mathbf{Q}$ and $\boldsymbol{\theta}$ represent the FFT and the LCP matrices, respectively, both with the sizes of $N \times N$, where N is the number of subcarriers.

The receiver performs a serial to parallel conversion, discards the redundant cyclic prefix data, and executes a Fourier time domain transformation (i.e, multiplying by FFT matrix). Then, the receiver combines the received signals linearly, similar to (7). The resulting signals are concatenated into a single block, $\hat{\mathbf{r}}_n$. Then, the MMSE equalizer in (8)) is used in the time domain, opposite to the SCFDE system which performs it in frequency domain. The output of the MMSE equalizer will be $\hat{\mathbf{y}}_n = \mathbf{W}_n\hat{\mathbf{r}}_n$. Afterward, the receiver compensates for the pre-coding in the transmitter by applying the inverse LCP matrix to recover the original data given by $\hat{\mathbf{x}}_n = \boldsymbol{\theta}^{-1}\hat{\mathbf{y}}_n$. At the end, a hard decision slicer is applied on the recovered data and passed to the JPEG2000 decoder to generate the received image.

Simulation Results

The simulation parameters are characterized similar to the ones in the previous section. The results provide an overview over the comparison of the SCFDE and OFDM systems for transmitting JPEG2000 images over frequency selective channels with equal power allocation. The review comprises investigation of the PSNR profile of the received image transmitted through block fading frequency selective channels, with different number of transmit antennas and for various source encoding rates.

Figure 10 compares the PSNR performance of the systems presented in Figure 8 and Figure 9, in a multi-tap channel and with one and two numbers of transmission antennas. The two systems present similar improvement trends in the case of 1-tap channel when the number of antennas is increased from one to two. In other words, both systems achieve spatial diversity with a gain of up to 10 dB, by employing two transmitter antennas and applying STBC to the transmitting codestream. For example, at an SNR value of 20 dB, the PSNR performance enhances up to 12 dB in a 1-tap channel when the number of transmitter antennas increases from one to two in the SCFDE and the OFDM systems. The OFDM system achieves a higher spatial diversity gain of up to 12 dB at an SNR value of 20 dB in a 2-tap channel condition, and this gain decreases to 5 dB for the SCFDE system. In addition, Figure 27 demonstrates that the higher is the number of channel taps in an OFDM system, the lower is the PSNR value of the received image. However, the SCFDE system exhibits an increase as high as 8 dB in the PSNR value of the received image when the number of channel taps increases from one to two. That is, in a severer channel condition, the SCFDE achieves multipath diversity, which leads to a better quality of the received image. This type of diversity, however, is not achievable in our OFDM system.

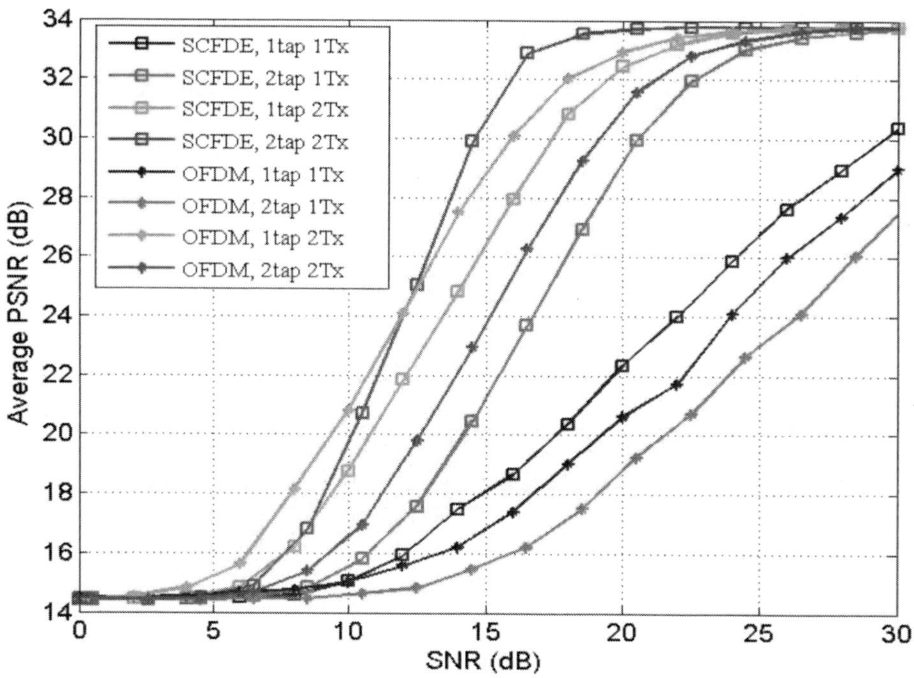

Figure 10. PSNR performance comparison of the received Lena ($512 \times 512, 0.25$ bpp) between SCFDE and OFDM systems with EPA.

To achieve multipath diversity in an OFDM system as well, a linear constellation pre-coder is employed at the transmitter side. Figure 11 demonstrates that the LCP-OFDM system achieves multipath diversity. Note that the LCP-OFDM system has an improved PSNR performance even for 1-tap channel. As the number of channel taps increases to two, the LCP-OFDM improves the PSNR performance to a higher value than the OFDM does. Even though the LCP-OFDM system increases the PSNR value of the received image passed through a 2-tap channel, up to 3 dB at SNR value of 20 dB, yet it is not able to enhance the performance of a 2-tap channel over that of a 1-tap as the SCFDE does. However, at SNR values higher than 30 dB, the PSNR performance of the LCP-OFDM for a 2-tap channel converges to that of the 1-tap channel.

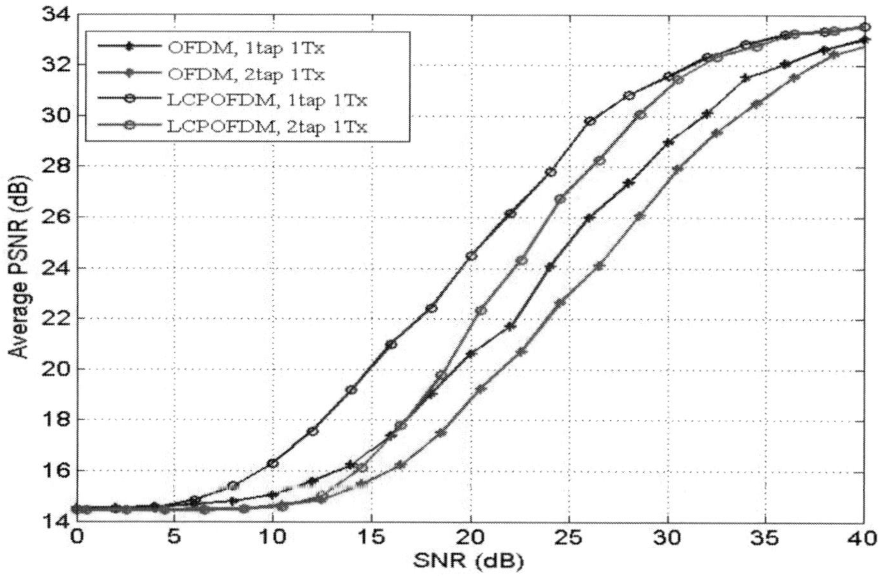

Figure 11. PSNR performance comparison of the received Lena ($\mathbf{512 \times 512, 0.25}$ bpp) between OFDM and LCP-OFDM techniques with EPA.

Figure 12 and Figure 13 evaluate the PSNR performance of the SCFDE and the OFDM system for different bit budgets, respectively. To establish a fair comparison between the two systems in this part, the images are transmitted only over a single tap channel to generate a PSNR profile without multipath diversity. The bit budget is varied for each SNR value by changing the average bits per pixel from 0.1 bpp to 0.3 bpp in steps of 0.05. For example at 0.1 bpp, the bit budget (i.e, the total number of bits of the codestream) is 21,000 bits, and it increases to 73,120 at 0.3 bpp. Thus, lowering the code rate is advantageous in conserving the bandwidth; however, it might deteriorate the image quality. Figure 12 illustrates that in our SCFDE system, at SNR values of less than 15 dB, the quality of the received image does not improve for higher coding rates. However, degradation of the received PSNR values at higher SNR values is notable when the coding rate decreases. Similar observations are valid when two transmitter antennas are utilized, except that we can conserve on bandwidth in this case by losing less of the quality in the received image. For example, when two transmitter antennas are used, we can save in bandwidth by lowering the coding rate from 0.3 bpp to 0.15 bpp, while keeping the PSNR of the received image above 30 dB -an acceptable quality for the received image- for SNR values higher than 30 dB. Figure 13 presents similar trend for the OFDM system when one and two antennas are utilized at the transmitter side. An interesting observation in (a) is that the PSNR performance at code rates between 0.1 bpp and 0.2 bpp outperforms the one at 0.25 bpp, when one transmitter antenna is used for SNR values between 10 dB and 20 dB. Depending on the application, insignificant loss in the quality can be sacrificed for the bandwidth conservation.

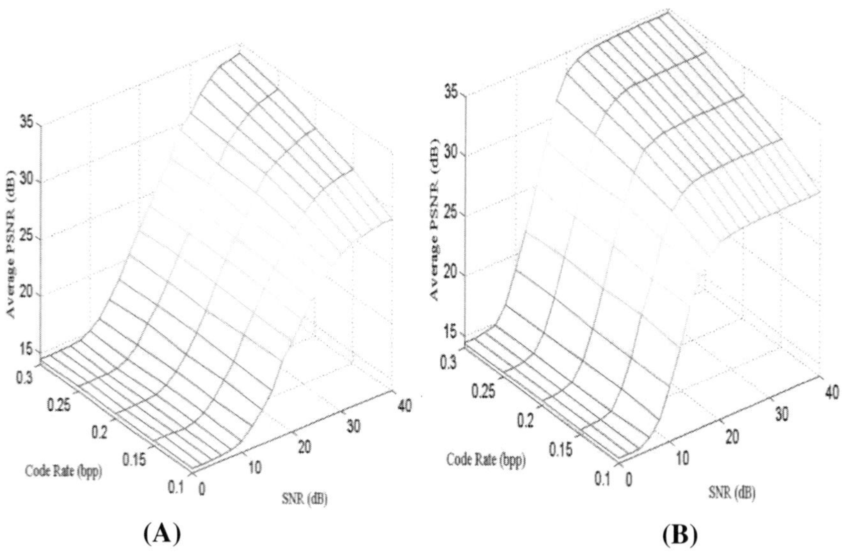

Figure 12. PSNR performance comparison for different coding rates in EPA-SCFDE system (a) 1 transmitter antenna (b) 2 transmitter antennas.

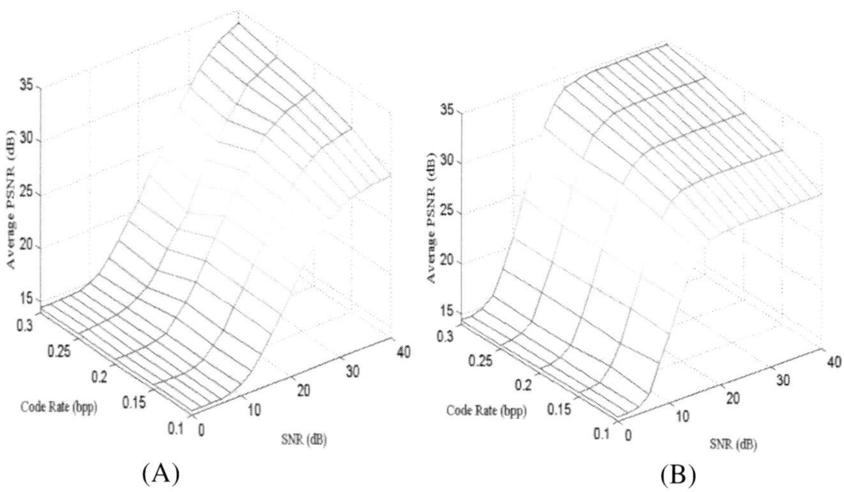

Figure 13. PSNR performance comparison for different coding rates in EPA-OFDM system (a) 1 transmitter antenna (b) 2 transmitter antennas.

Reducing the coding rate not only saves the available bandwidth, but also reduces the file size and lowers the delay time in coding, transmitting, and decoding of the images. In medical applications, where seconds play important roles in survival, fast coding, transmitting, and decoding of an X-RAY or FMRI image is critical.

Figure 14 presents a visual comparison for transmission of "Lena" through block fading frequency selective channel with EPA over the scenarios tabulated in Table 2. To simplify the comparison, transmission through 1-tap channel is only considered. Note that the visual quality of the received image increases at higher number of transmission antennas. Similar visual quality of the received image at 0.15 bpp and 0.3 bpp, within the OFDM and the

SCFDE systems for two transmitter antennas, implies that we can save in bandwidth by lowering bit budget without jeopardizing image quality.

Table 2. Comparison scenarios in between the SCFDE and the OFDM system with EPA (SNR=16 dB)for transmission of Lena

Scenario	Transmission Scenario	PSNR (dB)
(a)	Original image of Lena	------
(b)	Using OFDM, with 1 Transmitter antenna, at a rate of 0.3 bpp	18.48
(c)	Using SCFDE, with 1 Transmitter antenna, at a rate of 0.3 bpp	19.24
(d)	Using OFDM, with 2 Transmitter antenna, at a rate of 0.3 bpp	31.03
(e)	Using SCFDE, with 2 Transmitter antenna, at a rate of 0.3 bpp	28.62
(f)	Using OFDM, with 1 Transmitter antenna, at a rate of 0.15 bpp	18.21
(g)	Using SCFDE, with 1 Transmitter antenna, at a rate of 0.15 bpp	18.70
(h)	Using OFDM, with 2 Transmitter antenna, at a rate of 0.15 bpp	30.32
(i)	Using SCFDE, with 2 Transmitter antenna, at a rate of 0.15 bpp	27.45

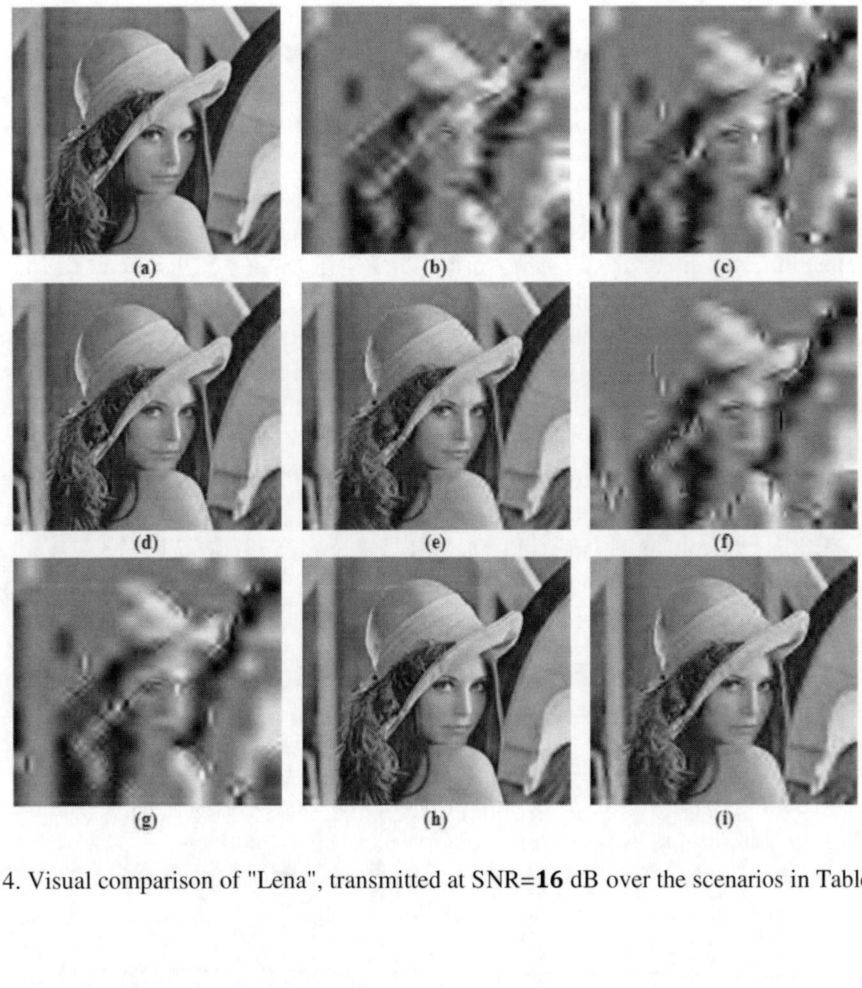

Figure 14. Visual comparison of "Lena", transmitted at SNR=**16** dB over the scenarios in Table 2.

SUMMARY AND CONCLUSION

Transmission of still images over wireless channels with optimal protection techniques such as UEP or UPA techniques has attracted much attention in the field of multimedia communications. However, implementation of an UPA scheme over frequency selective channels requires an in-depth investigation of the received PSNR profile of an image for a range of SNR values, different channel memory lengths, multiple transmitter antennas, and various source coding rate.

Second section of this chapter presents an overview of JPEG2000 structure, a brief literature survey error protection technique for JPEG2000 image transmission, and formulation of space time block coding. The main concentration of the literature is on applying unequal error protection techniques on the coding passes of JPEG2000 code stream in order to achieve a high quality of the received image. In order to cover the shortage of analysis on unequal power allocation for image transmission, this chapter presents a comprehensive over view of utilizing UPA technique for JPEG2000 image transmission over frequency selective channels.

Third section reviews integration of a UPA algorithm with the SCFDE system to enable transmission of JPEG2000 images over frequency selective channels. The SCFDE based system incorporates the STBC technique to achieve spatial diversity. The simulation results are validated with extensive Monte Carlo simulations, and they demonstrate the effectiveness of the SCFDE based UPA system in eliminating the negative impacts of ISI and ICI in a frequency selective channel is demonstd. Even though the system achieves spatial diversity at higher number of transmitter antennas, but it suffers from loss of multipath diversity when the number of channel taps increases.

Fourth section compares the performance of the OFDM and the SCFDE techniques for JPEG2000 image transmission over frequency selective channels, where the EPA technique is utilized. The comparison is based on measuring the PSNR values of the received images in a range of SNR values, with multi-tap channel condition, different number of transmitter antennas, and different source coding rates. The results demonstrate that both systems achieve spatial diversity; however, only the SCFDE obtains multipath diversity as the number of channel taps increases. To tackle this shortcoming of the OFDM system, an LCP technique is introduced to the OFDM system. The simulation results present that the LCP-OFDM system achieves multipath diversity during the transmission of JPEG2000 images over frequency selective channel. In addition, the simulation results for the PSNR profile at different coding rates reveal that the SCFDE and the OFDM system can conserve bandwidth during transmission of JPEG2000 images, if lower bit budget is utilized and certain loss in the received image quality can be tolerated.

REFERENCES

[1] D. S. Taubman and M. W. Marcellin, *JPEG2000: Image Compression Fundamentals, Standards and practices.*: Kluwer Academic Publishers, 2002.

[2] F. Pancaldi et al., "Single-Carrier Frequency Domain Equalization," *IEEE Signal Processing Magazine*, vol. 25, no. 5, pp. 37-56, 2008.

[3] N. Al-Dhahir, "Single Carrier Frequency Domain Equalization for Space-Time Block-Codded Transmissions over Frequency-Selective Fading Channels," *IEEE Communications Letters*, vol. 5, no. 7, pp. 304-306, 2001.

[4] M. Torki, "JPEG2000 Transmission Over Wireless Channels Using Unequal Power Allocation," Engineering Science, Simon Fraser University, Burnaby, Canada, Master's Thesis 2009.

[5] T. Acharya and P. S. Tsai, *JPEG2000 Standard for Image Compression: Concepts, Algorithems and VLSI Architecture*. New Jersey: John Wiley and Sons Incorporation, 2004.

[6] A. Natu and A. S. Taubman, "Unequal Protection of JPEG2000 Code-Streams in Wireless Channels," in *IEEE Global Telecommunications Conference (GLOBCOM)*, vol. 1, 2002, pp. 534-538.

[7] B. A. Banister, B. Belzer, and T. R. Fischer, "Robust Image Transmission using JPEG2000 and Turbo-codes," in *Proceedings of the 2000 International Conference on Image Processing (ICIP)*, 2004, pp. 375-378.

[8] N. Thomas, N. V. Boulgouris, and M. G. Strintzis, "Optimized Transmission of JPEG2000 Streams over Wireless Channels," *IEEE Transactions on Image Processing*, vol. 15, no. 1, pp. 54-67, 2006.

[9] Y. Wei, Z. Sahinoglu, and A. Vetro, "Energy Efficient JPEG2000 Image Transmission over Wireless Sensor Networks," in *IEEE Global Telecommunications Conference (GLOCOM)*, 2004, pp. 2738-2743.

[10] L. Atzori, "Transmission of JPEG2000 Images over Wireless Channels with Unequal Power Distribution," *IEEE Transactions on Consumer Electronics*, vol. 49, no. 4, pp. 883-888, 2003.

[11] M. Torki and A. Hajshirmohammadi, "Unequal Power Allocation for Transmission of JPEG2000 Images over Wireless Channels," in *IEEE Global Telecommunications Conference*, 2008, pp. 1-5.

[12] H. Houas, I. Fijalkow, and C. Baras, "Resource Allocation for the Transmission of Scalable Images on OFDM Systems," in *IEEE International Conference on Communications*, 2009, pp. 1-5.

[13] K. Munadi, M. Kurosaki, K. Nishikawa, and H. Kiya, "Error Protection for JPEG2000 Encoded Images and Its Evaluation Over OFDM Channel," in *Proceedings of the International Symposium on Circuits and Systems*, vol. 2, 2003, pp. 432-435.

[14] Y. Sun, Z. Xiong, and X. Wang, "Prospective Image Transmission over Differentially Space-Time Coded OFDM Systems: Research Articles," *Journal of Wireless Communications and Mobile Computing*, vol. 6, no. 8, pp. 1057-1075, December 2006.

[15] U. Sethakaset and S. Sumei, "Robust JPEG2000 Image Transmission over Closed-Loop MIMO-OFDM with Limited Feedback," in *IEEE International Symposuim on Personal, Indoor, and Mobile Radio Communications (PIMRC)*, 2008, pp. 1-5.

[16] C. Hsien Yun and C. Wei Ho, "Unequal Error Protection for OFDM Systems in Time-Varying Channels," in *Wireless Communications and Networking Conference (WCNC)*, 2011, pp. 1664-1669.

[17] H. Y. Chung and W. H. Chung, "Unequal Error Protection for OFDM Systems in Time-Varying Channels," in *IEEE Wireless Communications and Networking Conference*, 2011, pp. 1664-1669.

[18] Y. Qiao, Y. Wang, S. Yu, and D. Wang, "Robust Progressive Image Transmission Based on Multi-Rate Turbo Code and Space-Time OFDM," in *Proceedings of the Ninth International Symp. on Consumer Electronics*, 2005, pp. 128-133.

[19] M. M. Salah, A. A. Elrahman, and M. M. Mokhtar, "Performance Enhancement of Image transmission Using UEP Overlayed UTG_OFDM Scheme," in *National Radio Science Conference*, 2008, pp. 1-11.

[20] J. Song and K. J. R. Liu, "Robust Progressive Image Transmission Over OFDM Systems Using Space-Time Block Code," *IEEE Transactions on Multimedia*, vol. 4, no. 3, pp. 394-406, September 2002.

[21] L. Toni, Y. S. Chan, P. C. Cosman, and L. B. Milstein, "Channel Coding for Progressive Image in a 2-D Time-Frequency OFDM Block with Channel Estimation Error," *IEEE Transactions on Image Processing*, vol. 18, no. 11, pp. 2476-2490, 2009.

[22] A. Elbehery, S. A. S. Abdelwahab, M. A. Elnabi, E. S. Hassan, and S. Elarabi, "Efficient Image Transmission Over the Single Carrier Frequency Division Multiple Access System Using Chaotic Interleaving," in *29th National Radio Science Conference (NRSC)*, 2012, pp. 411-420.

[23] H. S. Chu, B. S. Park, C. K. An, J. S. Kang, and H. G. Son, "Wireless Image Transmission based on Adaptive OFDM System," in *International Forum on Strategic Technologies (IFOST)*, 2007, pp. 623-626.

[24] S. Alamouti, "A Simple Transmit Diversity Technique for Wireless Communications," *IEEE Journal on Selected Areas in Communications*, vol. 16, no. 8, pp. 1451-1548, 1998.

[25] M. Shayegannia, A. Hajshirmohammadi, S. Muhaidat, and M. Torki, "Transmission of JPEG2000 Images over Frequency Selective Channels with Unequal Power Allocation," *IET Transaction on Image Processing*, 2012, accepted for publication.

[26] M. Shayegannia, S. Muhaidat, and A. Hajshirmohammadi, "Transmission of JPEG2000 Images in an Uplink Cellular Network with UPA and SCFDE: A System Description," in *8th International Conference on Wireless and Mobile Communications*, Venice, 2012.

[27] J. Yue and J. D. Gibson, "Performance of OFDM Systems with Space-Time Coding," in *IEEE Wireless Comm. and Networking Conference (WCNC)*, vol. 1, 2002, pp. 280-284.

[28] Z. Liu, Y. Xin, and G. B. Giannakis, "Linear Constellation Precoding for OFDM with Maximum Multipath Divesity and Coding Gains," *IEEE Transactions on Communications*, vol. 51, no. 3, pp. 416-427, 2003.

[29] Y. Xin, Z. Wang, and G. B. Giannakis, "Space-Time Diversity Systems Based on Linear Constellation Precoding," *IEEE Transactions on Wireless Communications*, vol. 2, no. 2, pp. 294-309, March 2003.

In: Communication Systems: New Research
Editor: Vyacheslav Tuzlukov
ISBN:978-1-62618-654-5

Chapter 11

GENERALIZED HYPERGEOMETRIC FUNCTIONS WITH APPLICATIONS TO PERFORMANCE ANALYSIS OF WIRELESS COMMUNICATION SYSTEMS

J. F. Paris *
University of Málaga, Spain

Abstract

The fundamental performance metrics of wireless communications systems, i.e. capacity, error probability and outage probability can be frequently obtained in closed-form by the common classical special functions (e.g. Bessel's functions, Hypergeometric's functions or Meijer's functions), the Gaussian Q-function, the Marcum Q-function and the Nuttal Q-function [1].

In order to obtain new analytical closed-form results for the basic performance metrics, this set of special functions have been recently extended to include either less common classical special functions (e.g. Lauricella functions [2] or the Fox's H function [3]) or new generalizations of classical functions (e.g. the upper Fox's H function [4] or the generalized Fox's H function [5]). From a computational point of view, certain Lauricella functions (e.g. F_D, Φ_1, Φ_2 or Φ_3) are easier to compute than Fox type functions because either they have Euler type single integral representations with elementary integrands or their univariate Laplace transform is elementary being easily inverted numerically. Other Lauricella functions like F_A are more difficult to compute than the F_D function or its confluent forms; however, certain important metrics for wireless communications systems are obtained in closed-form by the function F_A.

In this chapter, general results on the applications of the classical Lauricella functions F_A, F_D, Φ_1, Φ_2 and Φ_3 to the performance of wireless communication systems are obtained. Then, new special functions which include these classical Lauricella functions as particular cases are proposed. This new type of functions, named $\mathcal{A}$, $\bar{\mathcal{A}}$, $\mathcal{B}$, $\mathcal{B}^+$ and $\mathcal{B}^-$, are nearly as easy to compute as the Lauricella functions listed above; at the same time, this new family of special functions will allow us to express in exact closed-form a variety of important analytically open problems in wireless communications.

Keywords: Wireless Communications, Fading Channels, Performance Analysis, Capacity, Error Probability, Outage Probability, Generalized Hypergeometric Functions

*E-mail address: paris@ic.uma.es

1. Introduction

For many decades, communication theorists have analyzed the performance of single channel and multichannel receivers in a fading environment. Among performance measures, the basic metrics are ergodic capacity, outage probability, capacity with outage and error probability.

Flat-fading channels are assumed along this chapter[1]. Let γ be the equivalent Signal-to-Noise Ratio (SNR) of a wireless communication system in a flat-fading channel as defined in [1], i.e. if the system is Single-Input Single-Output (SISO) then γ essentially represents the SNR statistics of the fading channel, if the system is Single-Input Multiple-Output (SIMO) with diversity combining then γ represents the SNR at the combiner output and, in general, if the system is Multiple-Output Multiple-Output (MIMO) then γ represents the SNR of the end-to-end equivalent SISO system. The system *ergodic capacity* [bit/s/Hz] is defined as

$$C = \mathrm{E}[\log_2(1+\gamma)] = \int_0^\infty \log_2(1+\gamma) f_\gamma(\gamma) d\gamma, \tag{1}$$

where f_γ is the Probability Density Function (PDF) of the random variable γ. The *outage probability* is defined as

$$P_o(\eta) = \Pr[\gamma \leq \eta] = F_\gamma(\eta), \tag{2}$$

where F_γ is the Cumulative Distribution Function (CDF) of the random variable γ and η is a predefined threshold. The *capacity with outage* [bit/s/Hz] is defined as

$$C_o(\eta) = (1 - P_o(\eta)) \log_2(1+\eta), \tag{3}$$

where η is again a predefined threshold. Finally, the *error probability* is defined as

$$P_e = \mathrm{E}[P_e(\gamma)] = \int_0^\infty P_e(\gamma) f_\gamma(\gamma) d\gamma, \tag{4}$$

where $P_e(\gamma)$ is the conditional error probability. For a variety of modulation/detection schemes the conditional Bit Error Probability (BEP) is given by a finite linear combination of Gaussian Q functions [6], i.e. $P_e(\gamma)$ is a finite linear combination of Gaussian error probabilities with the form $Q(\sqrt{p\gamma})$ where p is a real positive constant.

The classical Lauricella functions F_A, F_D, Φ_1, Φ_2 and Φ_3 allow us to obtain closed-form results for the last three metrics defined in (2)-(4) while closed-form results for ergodic capacity are usually obtained by the Fox's H function [3]. Specifically, Lauricella functions are very appropriate to obtain the CDF F_γ and the Gaussian error probability $\mathrm{E}[Q(\sqrt{p\gamma})]$ of many equivalent SNR distributions found in applications.

The remainder of this chapter has two main blocks. In the first block, which includes Sections 2 and 3, the applications of the classical Lauricella function F_D, its confluent forms Φ_1, Φ_2, Φ_3, and the Lauricella function F_A to wireless communication theory are presented. General results which allow us to evaluate performance metrics like the error probability, the outage probability or the capacity with outage in a variety of fading channels are derived. In the second block, which comprises Sections 4 and 5, new functions, named $\mathcal{A}$, $\bar{\mathcal{A}}$, $\mathcal{B}$, $\mathcal{B}^+$ and $\mathcal{B}^-$, which include the classical Lauricella functions F_A, F_D, Φ_1, Φ_2 and Φ_3 as particular cases are proposed and studied.

[1]Note that many results for flat-fading channels are also useful for frequency-selective channels.

2. Fundamentals of Certain Lauricella Functions

This Section briefly summarizes the definitions and key properties of the Lauricella functions F_A, F_D, Φ_1, Φ_2 and Φ_3. We only consider those properties which are key from the point of view of wireless communication theory. The reader should consult [2] and [7] for a comprehensive list of properties concerning Lauricella functions.

2.1. Definitions

The Lauricella functions F_A, F_D, Φ_1, Φ_2 and Φ_3 are usually defined from their series representations. These functions can be extended beyond its region of convergence if necessary using the principle of analytic continuation.

The Lauricella functions F_A and F_D are defined respectively as[2]

$$F_A^{(n)}\left[a, b_1, \ldots, b_n; c_1, \ldots, c_n; x_1, \ldots, x_n\right] = \sum_{m_1,\ldots,m_n=0}^{\infty} \frac{(a)_{m_1+\ldots+m_n} (b_1)_{m_1} \cdots (b_n)_{m_n}}{(c_1)_{m_1} \cdots (c_n)_{m_n}} \frac{x_1^{m_1}}{m_1!} \cdots \frac{x_n^{m_n}}{m_n!}, \quad |x_1| + \ldots + |x_n| < 1; \tag{5}$$

$$F_D^{(n)}\left[a, b_1, \ldots, b_n; c; x_1, \ldots, x_n\right] = \sum_{m_1,\ldots,m_n=0}^{\infty} \frac{(a)_{m_1+\ldots+m_n} (b_1)_{m_1} \cdots (b_n)_{m_n}}{(c)_{m_1+\ldots+m_n}} \frac{x_1^{m_1}}{m_1!} \cdots \frac{x_n^{m_n}}{m_n!}, \quad \max\left\{|x_1|, \ldots, |x_n|\right\} < 1. \tag{6}$$

Otherwise, the confluent Lauricella functions Φ_1, Φ_2 and Φ_3 are defined from F_D by the following confluence processes[3]

$$\left.\begin{aligned}
\Phi_1^{(n)}\left[a, b_1, \ldots, b_{n-1}; c; x_1, \ldots, x_n\right] &= \lim_{\lambda\to\infty} F_D^{(n)}\left[a, b_1, \ldots, b_{n-1}, \lambda; c; x_1, \ldots, x_{n-1}, \frac{x_n}{\lambda}\right], \\
\Phi_2^{(n)}\left[b_1, \ldots, b_n; c; x_1, \ldots, x_n\right] &= \lim_{\lambda\to\infty} F_D^{(n)}\left[\lambda, b_1, \ldots, b_n; c; \frac{x_1}{\lambda}, \ldots, \frac{x_n}{\lambda}\right], \\
\Phi_3^{(n)}\left[b_1, \ldots, b_{n-1}; c; x_1, \ldots, x_n\right] &= \lim_{\lambda\to\infty} F_D^{(n)}\left[\lambda, b_1, \ldots, b_{n-1}, \lambda; c; \frac{x_1}{\lambda}, \ldots, \frac{x_{n-1}}{\lambda}, \frac{x_n}{\lambda^2}\right].
\end{aligned}\right\} \tag{7}$$

The functions Φ_2 and Φ_3 are directly connected by the following confluence process

$$\Phi_3^{(n)}\left[b_1, \ldots, b_{n-1}; c; x_1, \ldots, x_n\right] = \lim_{\lambda\to\infty} \Phi_2^{(n)}\left[b_1, \ldots, b_{n-1}, \lambda; c; x_1, \ldots, x_{n-1}, \frac{x_n}{\lambda}\right]. \tag{8}$$

---

[2]For $n = 2$ these functions are also known as Apell's hypergeometric functions.

[3]For $n = 2$ these functions are also known as Humbert's confluent hypergeometric functions or bivariate confluent hypergeometric functions.

From (6) and (7) the following series representations are obtained

$$\left.\begin{aligned}
&\Phi_1^{(n)}\left[a, b_1, \ldots, b_{n-1}; c; x_1, \ldots, x_n\right] = \\
&\qquad \sum_{m_1,\ldots,m_n=0}^{\infty} \frac{(a)_{m_1+\ldots+m_n}(b_1)_{m_1}\cdots(b_{n-1})_{m_{n-1}}}{(c)_{m_1+\ldots+m_n}} \frac{x_1^{m_1}}{m_1!}\cdots\frac{x_n^{m_n}}{m_n!}, \\
&\qquad\qquad \max\left\{|x_1|, \ldots, |x_{n-1}|\right\} < 1, |x_n| < \infty, \\
&\Phi_2^{(n)}\left[b_1, \ldots, b_n; c; x_1, \ldots, x_n\right] = \\
&\qquad \sum_{m_1,\ldots,m_n=0}^{\infty} \frac{(b_1)_{m_1}\cdots(b_n)_{m_n}}{(c)_{m_1+\ldots+m_n}} \frac{x_1^{m_1}}{m_1!}\cdots\frac{x_n^{m_n}}{m_n!}, \\
&\qquad\qquad \max\left\{|x_1|, \ldots, |x_n|\right\} < \infty, \\
&\Phi_3^{(n)}\left[b_1, \ldots, b_{n-1}; c; x_1, \ldots, x_n\right] = \\
&\qquad \sum_{m_1,\ldots,m_n=0}^{\infty} \frac{(b_1)_{m_1}\cdots(b_{n-1})_{m_{n-1}}}{(c)_{m_1+\ldots+m_n}} \frac{x_1^{m_1}}{m_1!}\cdots\frac{x_n^{m_n}}{m_n!}, \\
&\qquad\qquad \max\left\{|x_1|, \ldots, |x_n|\right\} < \infty.
\end{aligned}\right\} \tag{9}$$

2.2. Some Key Properties

Starting with the Lauricella function F_A, the most interesting property for wireless communication theory is its Laplace type representation

$$\begin{aligned}
&F_A^{(n)}\left[a, b_1, \ldots, b_n; c_1, \ldots, c_n; x_1, \ldots, x_n\right] = \\
&\qquad \frac{1}{\Gamma(a)} \int_0^{\infty} e^{-t} t^{a-1} {}_1F_1(b_1; c_1; x_1 t) \ldots {}_1F_1(b_n; c_n; x_n t)\, dt, \\
&\qquad\qquad \operatorname{Re}\left\{x_1 + \ldots + x_n\right\} < 1,\ \operatorname{Re}\left\{a\right\} > 0,
\end{aligned} \tag{10}$$

where Γ is the gamma function and ${}_1F_1$ is the confluent hypergeometric function. The representation given in (10) is interesting from a theoretical point of view because both the Cumulative Distribution Function (CDF) of certain fading distributions and the Gaussian Q function can be expressed in terms of ${}_1F_1$ functions. More details on such applications of (10) will be given in Section 3. The integral representation given in (10) is also interesting from a numerical point of view since this type of integrals can be efficiently computed by Gauss-Laguerre quadrature.

The Lauricella function F_D has the following Euler type integral representation

$$\begin{aligned}
&F_D^{(n)}\left[a, b_1, \ldots, b_n; c; x_1, \ldots, x_n\right] = \\
&\qquad \frac{\Gamma(c)}{\Gamma(a)\Gamma(c-a)} \int_0^1 t^{a-1}(1-t)^{c-a-1}(1-tx_1)^{-b_1}\ldots(1-tx_n)^{-b_n}\, dt, \\
&\qquad\qquad \operatorname{Re}\left\{c\right\} > \operatorname{Re}\left\{a\right\} > 0.
\end{aligned} \tag{11}$$

The integral representation given in (11) is interesting from a numerical point of view because it is a one-dimensional integral with an elementary integrand which can be efficiently computed by standard numerical techniques. From a theoretical point of view, expression (11) has been used in literature to compute error probabilities by an indirect method. As it is shown in next Section, all these type of calculations are alternatively obtained in a simple and direct way by using the connection between the Lauricella function F_D and its confluent form Φ_2

$$\begin{aligned}&\int_0^\infty e^{-st}t^{a-1}\Phi_2^{(n)}\left[b_1,\ldots,b_n;c;x_1t,\ldots,x_nt\right]dt=\\&\qquad\frac{\Gamma(a)}{s^a}F_D^{(n)}\left[a,b_1,\ldots,b_n;c;\frac{x_1}{s},\ldots,\frac{x_n}{s}\right],\\&\qquad\qquad\mathrm{Re}\{a\}>0,\ \ \mathrm{Re}\{s\}>\max\{0,\mathrm{Re}\{x_1\},\ldots,\mathrm{Re}\{x_n\}\},\end{aligned}\tag{12}$$

i.e. the Lauricella function F_D with linear arguments in s^{-1} is essentially the Laplace transform of its confluent form Φ_2 with linear arguments in t. Definition (6) and the binomial expansion yields

$$F_D^{(n)}\left[a,b_1,\ldots,b_n;a;x_1,\ldots,x_n\right]=\prod_{r-1}^{n}(1-x_r)^{-b_r};\tag{13}$$

thus, from (12) and (13) we obtain ($t>0$)

$$\begin{aligned}&\Phi_2^{(n)}\left[b_1,\ldots,b_n;a;x_1t,\ldots,x_nt\right]=\frac{\Gamma(a)}{t^{a-1}}\frac{1}{2\pi j}\int_{\varepsilon-j\infty}^{\varepsilon+j\infty}\frac{1}{s^a}\prod_{r=1}^{n}\left(1-\frac{x_r}{s}\right)^{-b_r}e^{st}ds,\\&\qquad\qquad\mathrm{Re}\{a\}>0,\varepsilon>\max\{0,\mathrm{Re}\{x_1\},\ldots,\mathrm{Re}\{x_n\}\}.\end{aligned}\tag{14}$$

As shown in next Section, the inverse Laplace transform representation in (14) is very useful to obtain analytical expressions for the CDF of a variety of fading distributions. In addition, it can be also used to numerically compute the function Φ_2 since a number of efficient numerical methods are available to calculate the one-dimensional inverse Laplace transform [8].

As it occurs with F_D and Φ_2, the functions Φ_1 and Φ_3 constitute a Laplace transform pair. Applying an appropriate confluence process to both sides of (12) by (7) and (8) yields

$$\begin{aligned}&\int_0^\infty e^{-st}t^{a-1}\Phi_3^{(n)}\left[b_1,\ldots,b_{n-1};c;x_1t,\ldots,x_nt\right]dt=\\&\qquad\frac{\Gamma(a)}{s^a}\Phi_1^{(n)}\left[a,b_1,\ldots,b_{n-1};c;\frac{x_1}{s},\ldots,\frac{x_n}{s}\right],\\&\qquad\qquad\mathrm{Re}\{a\}>0,\ \ \mathrm{Re}\{s\}>\max\{0,\mathrm{Re}\{x_1\},\ldots,\mathrm{Re}\{x_{n-1}\}\}.\end{aligned}\tag{15}$$

The same confluence process applied to (14) yields ($t>0$)

$$\begin{aligned}&\Phi_3^{(n)}\left[b_1,\ldots,b_{n-1};a;x_1t,\ldots,x_nt\right]=\frac{\Gamma(a)}{t^{a-1}}\frac{1}{2\pi j}\int_{\varepsilon-j\infty}^{\varepsilon+j\infty}\frac{1}{s^a}\prod_{r=1}^{n-1}\left(1-\frac{x_r}{s}\right)^{-b_r}e^{\frac{x_n}{s}}e^{st}ds,\\&\qquad\qquad\mathrm{Re}\{a\}>0,\varepsilon>\max\{0,\mathrm{Re}\{x_1\},\ldots,\mathrm{Re}\{x_{n-1}\}\}.\end{aligned}\tag{16}$$

In next Section expression (16) will allow us to find the CDF of certain fading distributions, being also appropriate to compute Φ_3 by numerical methods for the inversion of Laplace transforms [8]. Fortunately, Φ_1 has a one-dimensional Euler type integral representation, which follows from (11) after applying the appropriate confluence process and is very useful to compute Φ_1 by standard numerical methods

$$\Phi_1^{(n)}\left[a, b_1, \ldots, b_{n-1}; c; x_1, \ldots, x_n\right] = \frac{\Gamma(c)}{\Gamma(a)\Gamma(c-a)} \int_0^1 t^{a-1}(1-t)^{c-a-1}\left(1-tx_1\right)^{-b_1} \ldots \left(1-tx_{n-1}\right)^{-b_{n-1}} e^{tx_n} dt, \quad \operatorname{Re}\{c\} > \operatorname{Re}\{a\} > 0. \tag{17}$$

3. Applications of Certain Lauricella Functions to Wireless Communications Theory

This Section show the applications of the classical functions F_A, F_D, Φ_1, Φ_2 and Φ_3 to wireless communications theory. General results are derived which allow us to obtain several performance metrics in a straightforward manner.

3.1. Applications of the Lauricella Function F_A

This function is very useful to obtain error probabilities by a direct application of (10).

Lemma 1 *If the PDF of the equivalent SNR of a wireless communication system can be expressed as*

$$f_\gamma(\gamma) = e^{-d\gamma}\gamma^{a-1}{}_1F_1(b_1; c_1; x_1\gamma) \ldots {}_1F_1(b_m; c_m; x_m\gamma),$$

with d a real positive number and its conditional error probability can be essentially expressed as[4]

$$P_e(\gamma) = e^{-d'\gamma}\gamma^{a'}{}_1F_1(b_{m+1}; c_{m+1}; x_{m+1}\gamma) \ldots {}_1F_1(b_{m+p}; c_{m+p}; x_{m+p}\gamma),$$

where d' is a real nonnegative number; then, the error probability is given by

$$P_e = \mathrm{E}\left[P_e(\gamma)\right] = \frac{\Gamma(a+a')}{(d+d')^{a+a'}} F_A^{(m+p)}\left[a+a', b_1, \ldots, b_{m+p}; c_1, \ldots, c_{m+p}; \frac{x_1}{d+d'}, \ldots, \frac{x_{m+p}}{d+d'}\right], \quad \operatorname{Re}\{x_1 + \ldots + x_{m+p}\} < d+d',\ \operatorname{Re}\{a+a'\} > 0. \tag{18}$$

Proof Use (10) after scaling the integral. □

The question which arises from Lemma 1 is whether both the system SNR and the conditional error probability can be expressed in the desired form for practical wireless

[4] For example, the conditional error probability could be a linear combination of terms with this form.

communication systems. Regarding first the system SNR, there are certain SISO fading channels with the PDF of the SNR in such desired form, e.g. Rayleigh, Nakagami-m and Rician shadowed [1]. Moreover, it is straightforward to show that the PDF of the maximum of independent non-identically distributed Rayleigh or Nakagami-m power random variables has again this desired form. Otherwise, it is well-known that the conditional error probabilities of a variety of binary modulation formats can be expressed in terms of the incomplete gamma function [1, eq. (8.100)], and after using [9, eq. 9.236-4] in terms of the ${}_1F_1$ function, with the desired form given in Lemma 1. In fact, the Gaussian conditional error probability $Q(\sqrt{p\gamma})$ with p a modulation-dependent real positive constant is included in the generalized formula given in [1, eq. (8.100)]. Moreover, the natural powers of $Q(\sqrt{p\gamma})$ can be also expressed in the desired form for the conditional error probability required by Lemma 1.

In summary, generalized binary error probabilities, bit and symbol Gaussian error probabilities, and error probabilities involving natural powers of $Q(\sqrt{p\gamma})$ can be obtained by Lemma 1 for a variety of wireless communication systems like SISO with Rician Shadowed fading and selection combining or transmit antenna selection with Rayleigh or Nakagami-m fading. A particular example of this type of result in literature is [10], where generalized binary error probabilities are obtained for transmit antenna selection in Nakagami-m fading channels.

3.2. Applications of the Lauricella Functions F_D and Φ_2

This pair of functions are useful to obtain outage probabilities or capacities with outage in terms of the Φ_2 function and error probabilities in terms of the F_D function.

Some comments on notation are in order. For an arbitrary function $\phi(x)$ we denote the Laplace transform as $\mathcal{L}[\phi(x); s]$. As in [1], we define the MGF of a nonnegative random variable γ as $\mathcal{M}_\gamma(s) = E[e^{s\gamma}] = \mathcal{L}[f_\gamma(\gamma); -s]$, where $s \in \mathbb{C}$ and $f_\gamma(\gamma)$ is the probability density function (PDF) of γ.

To grasp the generality of the results derived in this section, we may define the Gaussian error probability in a more fundamental form. Integral transforms theory provides a convenient framework for our purposes. In [11] a new integral transform, called *Shannon transform* was considered in relation to the ergodic capacity analysis. Regarding the problem treated here, we may consider the following integral transform.

Definition 1 (Gaussian $\mathcal{Q}$-transform) *Given a nonnegative absolutely continuous random variable γ with PDF $f_\gamma(\gamma)$, we define*

$$\mathcal{Q}_\gamma(p) \equiv \mathcal{Q}[f_\gamma(\gamma); p] = \mathrm{E}[Q(\sqrt{p\gamma})], \tag{19}$$

where p is a real nonnegative number.

Some straightforward properties of the Gaussian $\mathcal{Q}$-transform, including an inversion formula, are collected in the following result.

Lemma 2 *Let us consider a nonnegative absolutely continuous random variable γ. Then, the corresponding Gaussian $\mathcal{Q}$-transform is always defined in the set $\mathbb{R}^+ \equiv [0, \infty)$, and the following properties hold:*

(i) $\mathcal{Q}_\gamma(p)$ is a continuous and decreasing function in $\mathbb{R}^+$ such that $\mathcal{Q}_\gamma(0) = \frac{1}{2}$ and $\lim_{p\to\infty} \mathcal{Q}_\gamma(p) = 0$.
(ii) If $\mathcal{Q}_\gamma(p) = \Phi(p)$ then

$$f_\gamma(\gamma) = \sqrt{2\pi}\frac{d}{d\gamma}\left\{\sqrt{\gamma}\int_{\varepsilon-j\infty}^{\varepsilon+j\infty}\frac{\Phi(p)}{\sqrt{p}}e^{\frac{p\gamma}{2}}dp\right\},$$

for any $\varepsilon > 0$.

Proof The existence of the Gaussian $\mathcal{Q}$-transform is clear after taking into account that $Q(\sqrt{p\gamma}) \leq \frac{1}{2}$ for every $p > 0$.

(i) Continuity follows form the fact that the kernel $Q(\sqrt{p\gamma})$ is continuous and bounded in $\mathbb{R}^+$. If $p_1 < p_2$ then $Q(\sqrt{p_1\gamma}) > Q(\sqrt{p_2\gamma})$ for all γ in $\mathbb{R}^+$, thus, $\mathcal{Q}_\gamma(p_1) > \mathcal{Q}_\gamma(p_2)$. The property $\lim_{p\to\infty} \mathcal{Q}_\gamma(p) = 0$ follows after applying the monotone convergence theorem.

(ii) After integrating by parts $Q_\gamma(p)$ we can write

$$\Phi(p) = \frac{\sqrt{p}}{2\sqrt{2\pi}}\int_0^\infty e^{-\frac{pt}{2}}F_\gamma(t)t^{-1/2}dt. \tag{20}$$

The integral in (20) is recognized as a Laplace transform converging for $\mathrm{Re}[p] > 0$. Thus, after applying the inverse Laplace transform to this equality, the desired result is obtained.□

The following definition will be very useful for our purposes[5].

Definition 2 (Monomial and Posynomial MGF) *A nonnegative random variable γ has a posynomial MGF if its MGF has the form*

$$\mathcal{M}_\gamma(s) = \sum_{k=1}^{K} c_k \prod_{i=1}^{n_k}\left(1 - \frac{s}{a_{k,i}}\right)^{-b_{k,i}}, \tag{21}$$

where the involved parameters satisfy the following conditions

$$\begin{cases} \mathrm{Re}[a_{k,i}] > 0, \quad i = 1, 2, \ldots, n_k \text{ and } k = 1, 2, \ldots, K, \\ \sum_{i=1}^{n_k} b_{k,i} > 0, \quad k = 1, 2, \ldots, K, \\ \sum_{k=1}^{K} c_k = 1. \end{cases} \tag{22}$$

For the special case $K = 1$, the MGF will be called monomial.

In the interest of brevity, we will say that a random variable or a distribution is monomial or posynomial if its associated MGF is monomial or posynomial, respectively. In addition, the coefficients in Definition 2 will be referred as *characteristic coefficients* and the conditions given in (22) will be called *compatibility conditions*. The necessity of these

[5] This definition and its subsequent results were first presented in [12].

Table 1. Some Monomial MGFs.

Fading Distribution	MGF $\mathcal{M}_\gamma(s)$	References
Rayleigh	$(1-\bar{\gamma}s)^{-1}$; $0 \leqslant \bar{\gamma}$.	[13],[1]
Nakagami-q (Hoyt)	$\left(1-\frac{2q^2}{1+q^2}\bar{\gamma}s\right)^{-1/2}\left(1-\frac{2}{1+q^2}\bar{\gamma}s\right)^{-1/2}$; $0 \leqslant \bar{\gamma}, 0 < q \leqslant 1$.	[14],[1]
Nakagami-m	$\left(1-\frac{\gamma}{m}s\right)^{-m}$; $\frac{1}{2} \leqslant m$.	[15],[1]
Rician Shadowed*	$\left(1-\frac{\bar{\gamma}}{1+K}s\right)^{m-1}\left(1-\left(1+\frac{K}{m}\right)\frac{\bar{\gamma}}{1+K}s\right)^{-m}$; $0 \leqslant \bar{\gamma}, 0 \leqslant K, 0 < m$.	[16],[1]
η-μ Physical Model	$\left(1-s\frac{\bar{\gamma}}{n(h+H)}\right)^{-n/2}\left(1-s\frac{\bar{\gamma}}{n(h-H)}\right)^{-n/2}$ $0 \leqslant \bar{\gamma}, n = 1, 2, \ldots$ Format 1 $\Rightarrow h = \frac{2+\eta^{-1}+\eta}{4}, H = \frac{\eta^{-1}-\eta}{4}, 0 < \eta < \infty$ Format 2 $\Rightarrow h = \frac{1}{1-\eta^2}, H = \frac{\eta}{1-\eta^2}, -1 < \eta < 1$	[17]

* Note that $K \doteq \frac{\Omega}{2b_0}$, using the same notation as in [16],[1].

conditions will be revealed along this section[6]. As shown in Table 1, identifying γ with the system SNR, we observe that many common fading distribution have monomial MGF.

The key result of this section is conceptually summarized as follows: monomial distributions and certain expressions involving the Lauricella $F_D^{(n)}$ function are connected by the Gaussian $\mathcal{Q}$-transform. The extension of this idea to posynomial distributions is straightforward. From an operational point of view, once the underlaying fading distribution is identified as posynomial, obtaining the Gaussian $\mathcal{Q}$-transform reduces to extracting the characteristic coefficients from the MGF. The mathematically precise statements are given below.

It is most common to find monomial distributions in practical problems, thus, we start presenting the results for the monomial case.

Theorem 1 *Let γ be a monomial random variable, i.e.*

$$\mathcal{M}_\gamma(s) = \prod_{i=1}^{n}\left(1-\frac{s}{a_i}\right)^{-b_i}, \tag{23}$$

where $\{a_i\}_{i=1}^n$ and $\{b_i\}_{i=1}^n$ satisfy the compatibility conditions given in (22). Then

[6]We anticipate that the first and second compatibility conditions will be used in the proof of Theorem 1, while the third condition is necessary to satisfy $\mathcal{M}_\gamma(0) = 1$.

(i) The CDF of γ *is expressed as*

$$F_\gamma(\gamma) = \left\{ \frac{\prod_{i=1}^{n} (a_i)^{b_i}}{\Gamma\left(1 + \sum_{i=1}^{n} b_i\right)} \right\} \gamma^{\sum_{i=1}^{n} b_i} \Phi_2^{(n)}\left(b_1, \ldots, b_n; 1 + \sum_{i=1}^{n} b_i; -a_1\gamma, \ldots, -a_n\gamma\right), \tag{24}$$

where Γ *is the gamma function and* $\Phi_2^{(n)}$ *is the confluent Lauricella function.*
(ii) The Gaussian $\mathcal{Q}$*-transform of* γ *is given by*

$$\mathcal{Q}_\gamma(p) = \mathrm{E}[Q(\sqrt{p\gamma})] = \frac{1}{2\sqrt{\pi}} \left\{ \prod_{i=1}^{n} (a_i)^{b_i} \right\} \left\{ \frac{\Gamma\left(\frac{1}{2} + \sum_{i=1}^{n} b_i\right)}{\Gamma\left(1 + \sum_{i=1}^{n} b_i\right)} \right\} \left(\frac{2}{p}\right)^{\sum_{i=1}^{n} b_i}$$
$$\times F_D^{(n)}\left(\frac{1}{2} + \sum_{i=1}^{n} b_i, b_1, \ldots, b_n; 1 + \sum_{i=1}^{n} b_i; -\frac{2a_1}{p}, \ldots, -\frac{2a_n}{p}\right), \tag{25}$$

where $F_D^{(n)}$ *is the Lauricella function.*

Proof (i) Since $\mathcal{L}[f_\gamma(t); s] = \prod_{i=1}^{n}\left(1 + \frac{s}{a_i}\right)^{-b_i}$, we can write

$$\mathcal{L}[F_\gamma(t); s] = \frac{1}{s}\mathcal{L}[f_\gamma(t); s] = \left\{ \frac{\prod_{i=1}^{n} (a_i)^{b_i}}{\Gamma\left(1 + \sum_{i=1}^{n} b_i\right)} \right\} \left\{ \frac{\Gamma\left(1 + \sum_{i=1}^{n} b_i\right)}{s^{1+\sum_{i=1}^{n} b_i}} \right\} \prod_{i=1}^{n}\left(1 - \frac{(-a_i)}{s}\right)^{-b_i}, \tag{26}$$

for $\mathrm{Re}[s] > 0$. Taking into account the first and second compatibility conditions given in (22) and identifying (14) with (26), the expression (24) is obtained.

(ii) Introducing (24) in (20) yields

$$\mathcal{Q}_\gamma(p) = \frac{\sqrt{p}}{2\sqrt{2\pi}} \left\{ \frac{\prod_{i=1}^{n} (a_i)^{b_i}}{\Gamma\left(1 + \sum_{i=1}^{n} b_i\right)} \right\} \times$$
$$\int_0^\infty e^{-\frac{p}{2}t} t^{-1/2+\sum_{i=1}^{n} b_i} \Phi_2^{(n)}\left(b_1, \ldots, b_n, 1 + \sum_{i=1}^{n} b_i; -a_1 t, \ldots, -a_n t\right) dt. \tag{27}$$

Taking into account the first and second compatibility conditions given in (22) and identifying (12) with (27), the expression (25) is obtained. □

Extending this last result to posynomial random variables is straightforward.

Corollary 1 *If* γ *is a posynomial random variable such that its MGF is given by (21) then*

(i) The CDF of γ is expressed as

$$F_\gamma(\gamma) = \sum_{k=1}^{K} c_k \left\{ \frac{\prod_{i=1}^{n} (a_{k,i})^{b_{k,i}}}{\Gamma\left(1+\sum_{i=1}^{n} b_{k,i}\right)} \right\} \gamma^{\sum_{i=1}^{n} b_{k,i}} \times \Phi_2^{(n_k)}\left(b_{k,1},\ldots,b_{k,n_k}; 1+\sum_{i=1}^{n_k} b_{k,i}; -a_{k,1}\gamma,\ldots,-a_{k,n_k}\gamma\right). \tag{28}$$

(ii) The Gaussian $\mathcal{Q}$-transform of γ is given by

$$\mathcal{Q}_\gamma(p) = \frac{1}{2\sqrt{\pi}} \sum_{k=1}^{K} c_k \left\{ \prod_{i=1}^{n_k} (a_{k,i})^{b_{k,i}} \right\} \left\{ \frac{\Gamma\left(\frac{1}{2}+\sum_{i=1}^{n_k} b_{k,i}\right)}{\Gamma\left(1+\sum_{i=1}^{n_k} b_{k,i}\right)} \right\} \left(\frac{2}{p}\right)^{\sum_{i=1}^{n_k} b_{k,i}} \times F_D^{(n_k)}\left(\frac{1}{2}+\sum_{i=1}^{n_k} b_{k,i}, b_{k,1},\ldots,b_{k,n_k}; 1+\sum_{i=1}^{n_k} b_{k,i}; -\frac{2a_{k,1}}{p},\ldots,-\frac{2a_{k,n_k}}{p}\right). \tag{29}$$

Proof Repeat the steps of the proof of Theorem 1. □

Theorem 1 and Corollary 1 have important theoretical and practical consequences. Next result provides an integral representation for the Gaussian $\mathcal{Q}$-transform of posynomial random variables that is convenient for numerical evaluation.

Corollary 2 *If γ is a posynomial random variable such that its MGF is given by (21) then*

$$\mathcal{Q}_\gamma(p) = \frac{1}{2\pi} \sum_{k=1}^{K} c_k \left\{ \prod_{i=1}^{n_k} (a_{k,i})^{b_{k,i}} \right\} \left(\frac{2}{p}\right)^{\sum_{i=1}^{n_k} b_{k,i}} \times \int_0^1 u^{-\frac{1}{2}+\sum_{i=1}^{n_k} b_{k,i}} (1-u)^{-\frac{1}{2}} \left(1+\frac{2a_{k,1}}{p}u\right)^{-b_{k,1}} \cdots \left(1+\frac{2a_{k,n}}{p}u\right)^{-b_{k,n_k}} du. \tag{30}$$

Proof Check that the compatibility conditions allow us to use the Euler type integral representation given in (11) for the Lauricella functions in (29). □

From this last Corollary it is straightforward to obtain an asymptotic approximation for posynomial random variables.

Corollary 3 *Let γ be a posynomial random variable such that its MGF is given by (21). Then*

$$\mathcal{Q}_\gamma(p) \sim \frac{1}{2\sqrt{\pi}} \sum_{k=1}^{K} c_k \left\{ \prod_{i=1}^{n_k} (2a_{k,i})^{b_{k,i}} \right\} \frac{\Gamma\left(\frac{1}{2}+\sum_{i=1}^{n_k} b_{k,i}\right)}{\Gamma\left(1+\sum_{i=1}^{n_k} b_{k,i}\right)} \frac{1}{p^{\sum_{i=1}^{n_k} b_{k,i}}}, \tag{31}$$

for large values of p.

Proof Substitute the integrand in (30) by its asymptotic approximation and use the basic properties of the beta function [9, p. 908-910]. □

As expected, from (31) it is inferred that the underlaying diversity order of $\mathcal{Q}_\gamma(p)$ is

$$\lim_{p\to\infty} -\frac{\log\left(\mathcal{Q}_\gamma(p)\right)}{\log(p)} = \min_k \left\{ \sum_{i=1}^{n_k} b_{k,i} \right\}, \tag{32}$$

assuming γ is a posynomial random variable.

A particular example of result which can be obtained by Theorem 1 and published in literature is [18]. The Gaussian error probability for Rician-Nakagami fading channels is obtained in [18] in terms of the F_D function; however, instead of using the fact that the MGF of the fading distribution is monomial and then exploiting (12) (i.e. the steps followed in Theorem 1), the authors follow the usual indirect method found in literature based on an *ad hoc* change of variable from the fading MGF to get an Euler type representation like (11) and then identifying such representation with the F_D function. With the help of Theorem 1 and its corollaries all these type of results are obtained in a direct way after identifying the MGF of the equivalent SNR of the wireless communication system as either monomial or posynomial. Following the approach of Theorem 1 based on (12), the author has been able to obtain the exact and closed-form CDFs for the $\eta - \mu$ and the Rician shadowed fading distributions [19]-[20].

3.3. Applications of the Lauricella Functions Φ_1 AND Φ_3

This pair of functions are also useful to obtain outage probabilities or outage capacities in terms of the Φ_3 function and error probabilities in terms of the Φ_1 function, when the equivalent SNR of the wireless communication system has an special form.

The main application of the functions Φ_1 and Φ_3 is summarized in the following theorem which is analog to Theorem 1 for the Φ_2-F_D pair of functions.

Theorem 2 *Let γ be a random variable with MGF*

$$\mathcal{M}[f_\gamma(t);s] = \exp\left(\frac{1}{1-s/a_1}\right)\prod_{i=1}^{n}\left(1-\frac{s}{a_i}\right)^{-b_i}, \tag{33}$$

where in general $\{a_i\}_{i=1}^n$ and $\{b_i\}_{i=1}^n$ are complex numbers with $\max_{i=1,\ldots,n}\{\mathrm{Re}[a_i]\} > 0$. Then

(i) If $\mathrm{Re}\left[\sum_{i=1}^{n} b_i\right] > -1$ the CDF of γ is expressed as

$$F_\gamma(\gamma) = \frac{\left(\prod_{i=1}^{n} a_i^{b_i}\right)}{\Gamma\left(1+\sum_{i=1}^{n} b_i\right)}\gamma^{\sum_{i=1}^{n} b_i} e^{-a_1\gamma}\times$$
$$\Phi_3^{(n+1)}\left(1, b_2, \ldots, b_n; 1+\sum_{i=1}^{n} b_i; a_1\gamma, (a_1-a_2)\gamma, \ldots, (a_1-a_n)\gamma, a_1\gamma\right), \tag{34}$$

where Γ *is the gamma function and* Φ_3 *is the confluent Lauricella function.*

(ii) If $\mathrm{Re}\left[\sum_{i=1}^{n} b_i\right] > -\frac{1}{2}$ *the Gaussian* Q*-transform of* γ *is given by*

$$Q_\gamma(p) = \mathrm{E}[Q(\sqrt{p\gamma})] = \frac{\sqrt{p}}{2\sqrt{2\pi}} \frac{\prod_{i=1}^{n} (a_i)^{b_i}}{\Gamma\left(1 + \sum_{i=1}^{n} b_i\right)} \frac{\Gamma\left(\frac{1}{2} + \sum_{i=1}^{n} b_i\right)}{\left(\frac{p}{2} + a_1\right)^{1/2 + \sum_{i=1}^{n} b_i}} \times$$

$$\Phi_1^{(n+1)}\left(\frac{1}{2} + \sum_{i=1}^{n} b_i, 1, b_2, \ldots, b_n; 1 + \sum_{i=1}^{n} b_i; \frac{a_1}{\frac{p}{2} + a_1}, \frac{a_1 - a_2}{\frac{p}{2} + a_1}, \ldots, \frac{a_1 - a_n}{\frac{p}{2} + a_1}, \frac{a_1}{\frac{p}{2} + a_1}\right), \tag{35}$$

where Φ_1 *is the confluent Lauricella function.*

Proof Note that

$$\mathcal{L}[e^{a_1\gamma} F_\gamma(\gamma); s] = \mathcal{L}[F_\gamma(\gamma); s - a_1] = \frac{1}{s - a_1} \mathcal{L}[f_\gamma(\gamma); s - a_1] =$$

$$\left(\prod_{i=1}^{n} a_i^{b_i}\right) s^{-b_1} \exp\left(\frac{a_1}{s}\right) \frac{1}{s - a_1} \prod_{i=2}^{n} (a_i - a_1 + s)^{-b_i} -$$

$$\frac{\left(\prod_{i=1}^{n} a_i^{b_i}\right)}{\Gamma\left(1 + \sum_{i=1}^{n} b_i\right)} \frac{\Gamma\left(1 + \sum_{i=1}^{n} b_i\right)}{s^{1 + \sum_{i=1}^{n} b_i}} \exp\left(\frac{a_1}{s}\right)\left(1 - \frac{a_1}{s}\right)^{-1} \prod_{i=2}^{n}\left(1 - \frac{a_1 - a_i}{s}\right)^{-b_i},$$

identifying with (16) and checking the restrictions on the parameters yields (34). Then, if we apply (15) to Gaussian error probability which follows once (34) is substituted in (20), expression (35) is obtained after checking the restrictions on the parameters. □

An example of result found in literature which can be obtained by the utilization of Theorem 2 is [21]. In [21] the Gaussian error probability is obtained, by the indirect method which uses an *ad hoc* change of variable, when the fading SNR is modelled as a κ-μ random variable.

4. The Functions $\mathcal{A}$, $\bar{\mathcal{A}}$, $\mathcal{B}$, $\mathcal{B}^+$ AND $\mathcal{B}^-$

In this Section new special functions which include the classical Lauricella functions as particular cases are proposed. This new type of functions, named $\mathcal{A}$, $\bar{\mathcal{A}}$, $\mathcal{B}$, $\mathcal{B}^+$ and $\mathcal{B}^-$, are nearly as easy to compute as the Lauricella functions described above and easier to compute than the Fox type functions; at the same time, this new family of special functions will allow us to express in exact closed-form a variety of important analytically open problems in the field of wireless communications.

4.1. Definitions

The first set of functions, which generalizes the F_A, F_D and Φ_1 Lauricella functions, are defined below. In the following formal definitions all the variables and parameters are understood as complex numbers by default.

Definition 3 (Generalized Lauricella Functions $\mathcal{A}$ and $\bar{\mathcal{A}}$) *The generalized Lauricella function $\mathcal{A}$ is formally defined as follows*

$$\mathcal{A}\left(x;\alpha,\mathbf{A}\right)=\frac{1}{\Gamma\left(\alpha\right)}\int_0^{\infty}e^{-xt}t^{\alpha-1}\prod_{r=1}^{m}\Phi_3^{(n)}\left[a_{r,1},\ldots,a_{r,n-1};a_{r,n};a_{r,n+1}t,\ldots,a_{r,2n}t\right]dt,\tag{36}$$

where the $m\times 2n$ matrix parameter $\mathbf{A}$ is arranged as

$$\mathbf{A}=\begin{bmatrix} a_{1,1} & \cdots & a_{1,n-1} & a_{1,n} & a_{1,n+1} & \cdots & a_{1,2n} \\ \vdots & \ddots & \vdots & \vdots & \vdots & \ddots & \vdots \\ a_{m,1} & \cdots & a_{m,n-1} & a_{m,n} & a_{m,n+1} & \cdots & a_{m,2n} \end{bmatrix}.\tag{37}$$

The incomplete generalized Lauricella function $\bar{\mathcal{A}}$ is formally defined as follows

$$\bar{\mathcal{A}}\left(x,y;\alpha,\mathbf{A}\right)=\frac{1}{\Gamma\left(\alpha\right)}\int_0^{y}e^{-xt}t^{\alpha-1}\prod_{r=1}^{m}\Phi_3^{(n)}\left[a_{r,1},\ldots,a_{r,n-1};a_{r,n};a_{r,n+1}t,\ldots,a_{r,2n}t\right]dt,\tag{38}$$

where y is a real variable and the $m\times 2n$ matrix parameter $\mathbf{A}$ is arranged as in (37).

The second set of functions, which generalizes the Φ_2 and Φ_3 Lauricella functions, are defined below. Again, in the formal definitions given below all the variables and parameters are understood as complex numbers by default.

Definition 4 (Generalized Lauricella Functions $\mathcal{B}$, $\mathcal{B}^+$ and $\mathcal{B}^-$) *The bilateral generalized Lauricella function $\mathcal{B}$ is formally defined as follows*

$$\mathcal{B}\left(x;\beta,\mathbf{b}^+,\mathbf{b}^-\right)=$$
$$\frac{\Gamma\left(\beta\right)}{2\pi j}\int_C\frac{e^{sx}}{s^{\beta}}\prod_{r=1}^{n}\left(1-\frac{a_r^+}{s}\right)^{-b_r^+}e^{c_r^+\left(1-\frac{a_r^+}{s}\right)^{-1}}\prod_{q=1}^{m}\left(1-\frac{a_q^-}{s}\right)^{-b_q^-}e^{c_q^-\left(1-\frac{a_q^-}{s}\right)^{-1}}ds,\tag{39}$$

where the $3n$-dimensional vector parameter $\mathbf{b}^+$ and the $3m$-dimensional vector parameter $\mathbf{b}^-$ are defined as

$$\begin{cases}\mathbf{b}^+=\left[a_1^+\ldots a_n^+\,b_1^+\ldots b_n^+\,c_1^+\ldots c_n^+\right]\\ \mathbf{b}^-=\left[a_1^-\ldots a_m^-\,b_1^-\ldots b_m^-\,c_1^-\ldots c_m^-\right]\end{cases},\tag{40}$$

and where it is assumed that a path C exists which goes from $\sigma-j\infty$ to $\sigma+j\infty$, (σ real) so that all the points $\{a_r^+\}_{r=1}^n$ lie to the left of C and the points $\left\{a_q^-\right\}_{q=1}^m$ lie to the right of C. When $m=0$ (the set $\left\{a_q^-\right\}_{q=1}^m$ is empty) the function $\mathcal{B}$ is denoted as $\mathcal{B}^+\left(x;\beta,\mathbf{b}^+\right)$ and termed right-unilateral generalized Lauricella function $\mathcal{B}^+$; otherwise, when $n=0$ (the set $\{a_r^+\}_{r=1}^n$ is empty) the function $\mathcal{B}$ is denoted as $\mathcal{B}^-\left(x;\beta,\mathbf{b}^-\right)$ and termed left-unilateral generalized Lauricella function $\mathcal{B}^-$.

4.2. Convergence

Sufficient conditions for the convergence of the previously formally defined functions are given below.

Lemma 3 *If* $\operatorname{Re}\alpha > 0$ *and* $\{\operatorname{Re}[a_{r,n}]\}_{r=1}^{m} > 0$ *then the value of the incomplete generalized Lauricella function* $\bar{\mathcal{A}}$ *is finite for any value of* x *and* y.

Proof The statement follows from the fact that the only singulary in the integrand is due to the factor $t^{\alpha-1}$ when $t \to 0$. □

Lemma 4 *If the hypothesis of previous Lemma holds and the following extra hypothesis is added:* $\max\{\operatorname{Re}[a_{r,n+1}], \ldots, \operatorname{Re}[a_{r,2n}]\}_{r=1}^{m} \leqslant 0$ *; then, the value of the generalized Lauricella function* $\mathcal{A}$ *is finite for any value of* $x \in (0, \infty)$.

Proof We only need to prove the convergence in the integral which defines $\mathcal{A}$ when $t \to \infty$. From (16) and the extra hypothesis it follows that for sufficiently large t

$$\left|\prod_{r=1}^{m} \Phi_3^{(n)}\left[a_{r,1}, \ldots, a_{r,n-1}; a_{r,n}; a_{r,n+1}t, \ldots, a_{r,2n}t\right]\right| \leqslant \frac{K e^{\varepsilon m t}}{t^{-m+\sum_{r=1}^{m} a_{r,n}}},$$

where $\epsilon > 0$ and K is independent of t and ϵ. Taking the limit when $\epsilon \to 0$ yields

$$\left|\prod_{r=1}^{m} \Phi_3^{(n)}\left[a_{r,1}, \ldots, a_{r,n-1}; a_{r,n}; a_{r,n+1}t, \ldots, a_{r,2n}t\right]\right| \leqslant \frac{K}{t^{-m+\sum_{r=1}^{m} a_{r,n}}}.$$

Thus, there exist real and positive numbers C and t_0 such that

$$|\mathcal{A}(x; \alpha, \mathbf{A})| \leqslant C + K \int_{t_0}^{\infty} e^{-xt} t^{\operatorname{Re}\left[\alpha-1+m-\sum_{r=1}^{m} a_{r,n}\right]} dt < \infty$$

for any value of $x \in (0, \infty)$, which completes the proof. □

Lemma 5 *If* $\operatorname{Re}\beta > 1$ *then the generalized Lauricella functions* $\mathcal{B}$, $\mathcal{B}^{+}$ *and* $\mathcal{B}^{-}$ *give finite values for* $x \in [0, \infty)$.

Proof Let us consider the path $s = jt$ and $|t| \to \infty$ in the integral of (39). The result follows after checking that the module of the integrand behaves as $|t|^{-\operatorname{Re}\beta}$ when $|t| \to \infty$ and $x \in [0, \infty)$. □

4.3. Connections with Classical Lauricella Functions

The connections of the previously defined functions and the classical Lauricella functions are revealed in the next formulae.

The functions F_A, F_D and Φ_1 are particularizations of the $\mathcal{A}$-type functions. First note that

$$\begin{cases} \Phi_3^{(2)}\left[a_1; a_2; a_3 t, 0\right] = {}_1F_1\left[a_1; a_2; a_3 t\right], \\ \Phi_3^{(n+1)}\left[a_1, \ldots, a_n; a_{n+1}; a_{n+2}t, \ldots, a_{2n+1}t, 0\right] = \\ \qquad\qquad \Phi_2^{(n)}\left[a_1, \ldots, a_n; a_{n+1}; a_{n+2}t, \ldots, a_{2n+1}t\right]. \end{cases} \tag{41}$$

Then, comparison of (10) and Definition 3 yields

$$F_A^{(n)}\left[a, b_1, \ldots, b_n; c_1, \ldots, c_n; x_1, \ldots, x_n\right] = \mathcal{A}\left(1; a, \begin{bmatrix} b_1 & c_1 & x_1 & 0 \\ \vdots & \vdots & \vdots & \vdots \\ b_n & c_n & x_n & 0 \end{bmatrix}\right), \tag{42}$$
$$\mathrm{Re}\left\{x_1 + \ldots + x_n\right\} < 1,\ \mathrm{Re}\left\{a\right\} > 0.$$

Comparison of (12) and Definition 3 yields

$$F_D^{(n)}\left[a, b_1, \ldots, b_n; c; \frac{x_1}{s}, \ldots, \frac{x_n}{s}\right] = s^a \mathcal{A}\left(s; a, \left[b_1 \ldots b_n \, c \, x_1 \ldots x_n \, 0\right]\right),$$
$$\mathrm{Re}\left\{a\right\} > 0,\ \mathrm{Re}\left\{s\right\} > \max\left\{0, \mathrm{Re}\left\{x_1\right\}, \ldots, \mathrm{Re}\left\{x_n\right\}\right\}. \tag{43}$$

Finally, comparison of (15) and Definition 3 yields

$$\Phi_1^{(n)}\left[a, b_1, \ldots, b_{n-1}; c; \frac{x_1}{s}, \ldots, \frac{x_n}{s}\right] = s^a \mathcal{A}\left(s; a, \left[b_1 \ldots b_{n-1} \, c \, x_1 \ldots x_n\right]\right),$$
$$\mathrm{Re}\left\{a\right\} > 0,\ \mathrm{Re}\left\{s\right\} > \max\left\{0, \mathrm{Re}\left\{x_1\right\}, \ldots, \mathrm{Re}\left\{x_{n-1}\right\}\right\}. \tag{44}$$

Otherwise, the functions Φ_2 and Φ_3 are particularizations of the $\mathcal{B}$-type functions. Comparison of (14) and Definition 4 yields ($t > 0$)

$$\Phi_2^{(n)}\left[b_1, \ldots, b_n; a; x_1 t, \ldots, x_n t\right] = \frac{1}{t^{a-1}} \mathcal{B}^+\left(t; a, \left[x_1 \ldots x_n \, b_1 \ldots b_n \, \underbrace{0 \ldots 0}_{n}\right]\right), \tag{45}$$
$$\mathrm{Re}\left\{a\right\} > 0.$$

Finally, comparison of (16) and Definition 4 yields ($t > 0$)

$$\Phi_3^{(n)}\left[b_1, \ldots, b_{n-1}; a; x_1 t, \ldots, x_n t\right] = \frac{1}{e} \frac{e^{-x_n t}}{t^{a-1}} \times$$
$$\mathcal{B}^+\left(t; a, \left[(x_1 + x_n) \ldots (x_{n-1} + x_n) \, x_n \, b_1 \ldots b_{n-1} \left(a - \sum_{r=1}^{n-1} b_r\right) \underbrace{0 \ldots 0}_{n-1} \, 1\right]\right),$$
$$\mathrm{Re}\left\{a\right\} > 0. \tag{46}$$

5. Applications of the Functions $\mathcal{A}$, $\bar{\mathcal{A}}$, $\mathcal{B}$, $\mathcal{B}^+$ and $\mathcal{B}^-$ to Wireless Communications Theory

In this Section it is shown how this new family of special functions allow us to express in exact closed-form important analytically open problems in the field of wireless communications.

5.1. Application Example of the $\mathcal{A}$-type Functions

Let us consider a diversity system for Rician shadowed fading channels with adaptive modulation and selection combining with L branches. To obtain the system error probability for a set of modulation formats including BPSK and M-QAM it is necessary to compute integrals of the form

$$\mathcal{I} = \int_0^y Q\left(\sqrt{p\gamma}\right) f_{\max\{\gamma_k\}}(\gamma)\, d\gamma, \tag{47}$$

where p is a modulation-dependent parameter and $\max\{\gamma_k\}$ is the equivalent SNR at the combiner output. The CDF of the SNR at the k-th branch (the L branches are considered mutually independent) is given by [22]

$$F_{\gamma_k}(\gamma) = \frac{1+\kappa_k}{\bar{\gamma}_k\left(1+\kappa_k/m_k\right)^{m_k}}\gamma\Phi_2^{(2)}\left(1-m_k, m_k; 2; -(1+\kappa_k)\frac{\gamma}{\bar{\gamma}_k}, -\frac{1+\kappa_k}{1+\kappa_k/m_k}\frac{\gamma}{\bar{\gamma}_k}\right). \tag{48}$$

Then, after integration by parts $\mathcal{I}$ can be expressed in terms of the incomplete generalized Lauricella function $\bar{\mathcal{A}}$ as follows

$$\begin{aligned}
\mathcal{I} = {} & Q\left(\sqrt{py}\right)\prod_{k=1}^{L} F_{\gamma_k}(y) - \int_0^y \frac{\sqrt{p}}{2\sqrt{2\pi}} e^{-\frac{p\gamma}{2}}\gamma^{-1/2}\prod_{k=1}^{L} F_{\gamma_k}(\gamma) d\gamma = \\
& Q\left(\sqrt{py}\right)\prod_{k=1}^{L} F_{\gamma_k}(y) - \frac{\sqrt{p}}{2\sqrt{2\pi}}\left(\prod_{k=1}^{L}\frac{1+\kappa_k}{\bar{\gamma}_k\left(1+\kappa_k/m_k\right)^{m_k}}\right)\times \\
& \int_0^y e^{-\frac{p\gamma}{2}}\gamma^{L-1/2}\prod_{k=1}^{L}\Phi_2^{(2)}\left(1-m_k, m_k; 2; -(1+\kappa_k)\frac{\gamma}{\bar{\gamma}_k}, -\frac{1+\kappa_k}{1+\kappa_k/m_k}\frac{\gamma}{\bar{\gamma}_k}\right) d\gamma = \\
& Q\left(\sqrt{py}\right)\prod_{k=1}^{L} F_{\gamma_k}(y) - \frac{\sqrt{p}\,\Gamma(L+1/2)}{2\sqrt{2\pi}}\left(\prod_{k=1}^{L}\frac{1+\kappa_k}{\bar{\gamma}_k\left(1+\kappa_k/m_k\right)^{m_k}}\right)\times \\
& \bar{\mathcal{A}}\left(\frac{p}{2}, y; L+1/2, \begin{bmatrix} 1-m_1 & m_1 & 2 & -(1+\kappa_1)\frac{1}{\bar{\gamma}_1} & -\frac{1+\kappa_1}{1+\kappa_1/m_1}\frac{1}{\bar{\gamma}_1} & 0 \\ \vdots & \vdots & \vdots & \vdots & \vdots & \vdots \\ 1-m_L & m_L & 2 & -(1+\kappa_L)\frac{1}{\bar{\gamma}_L} & -\frac{1+\kappa_L}{1+\kappa_L/m_L}\frac{1}{\bar{\gamma}_L} & 0 \end{bmatrix}\right).
\end{aligned} \tag{49}$$

5.2. Application Example of the $\mathcal{B}$-type Functions

The η-μ fading distribution has been proposed to model a general Non-Line-Of-Sight (NLOS) propagation scenario. By two shape parameters η and μ, this model includes some classical fading distributions as particular cases, e.g. Nakagami-q (Hoyt), one-sided Gaussian, Rayleigh and Nakagami-m. A detailed description of the η-μ fading model can be found in [17].

Recently, the outage probability for η-μ/η-μ interference-limited scenarios has been obtained in [23] under certain restrictions of the μ parameters of the involved fading distributions. Obtaining the outage probability without parameter restrictions is an open problem of great interest in wireless communications. Let us focus on a simplified case in which there is one Signal-Of-Interest (SOI) and one interferer.

We are interested in the following outage probability

$$P_o = \Pr\left[\frac{X}{Y} \leqslant \eta\right], \tag{50}$$

where X represents the power of the SOI, Y represents the power of the interferer and η is a predefined threshold. Both X and Y are η-μ independent squared random variables with MGF given respectively by [24]

$$\begin{cases} M_X(s) = \left(\frac{a_{X,1}a_{X,2}}{s^2}\right)^{\mu_X}\left(1-\frac{a_{X,1}}{s}\right)^{-\mu_X}\left(1-\frac{a_{X,2}}{s}\right)^{-\mu_X}, \\ a_{X,1} = \frac{h_X - H_X}{2\Omega_X\mu_X h_X}, a_{X,2} = \frac{h_X + H_X}{2\Omega_X\mu_X h_X}, \end{cases} \tag{51}$$

and

$$\begin{cases} M_Y(s) = \left(\frac{a_{Y,1}a_{Y,2}}{s^2}\right)^{\mu_Y}\left(1-\frac{a_{Y,1}}{s}\right)^{-\mu_Y}\left(1-\frac{a_{Y,2}}{s}\right)^{-\mu_Y}, \\ a_{Y,1} = \frac{h_Y - H_Y}{2\Omega_Y\mu_Y h_Y}, a_{Y,2} = \frac{h_Y + H_Y}{2\Omega_Y\mu_Y h_Y} \end{cases} ; \tag{52}$$

where the meaning of the statistical parameters is detailed in [17] and [24]. Using the MGF-based method derived in [23] to compute the outage probability P_o, the following expression involving the bilateral generalized Lauricella function $\mathcal{B}$ is obtained

$$\begin{aligned} P_o = {} & \frac{1}{2\pi j}\int\limits_{c-i\infty}^{c+i\infty}\frac{1}{s}\mathcal{M}_X\left(-\frac{s}{\eta}\right)\mathcal{M}_Y(s)ds = \\ & \frac{\left(\eta^2 a_{X,1}a_{X,2}\right)^{\mu_X}\left(a_{Y,1}a_{Y,2}\right)^{\mu_Y}}{\Gamma\left(2\left(\mu_X+\mu_Y\right)+1\right)}\frac{\Gamma\left(2\left(\mu_X+\mu_Y\right)+1\right)}{2\pi j}\int\limits_{c-i\infty}^{c+i\infty}\frac{1}{s^{2(\mu_X+\mu_Y)+1}}\times \\ & \left(1+\frac{\eta a_{X,1}}{s}\right)^{-\mu_X}\left(1+\frac{\eta a_{X,2}}{s}\right)^{-\mu_X}\left(1-\frac{a_{Y,1}}{s}\right)^{-\mu_Y}\left(1-\frac{a_{Y,2}}{s}\right)^{-\mu_Y}ds = \\ & \frac{\left(\eta^2 a_{X,1}a_{X,2}\right)^{\mu_X}\left(a_{Y,1}a_{Y,2}\right)^{\mu_Y}}{\Gamma\left(2\left(\mu_X+\mu_Y\right)+1\right)}\times \\ & \mathcal{B}\left(0; 2\left(\mu_X+\mu_Y\right)+1, \left[-\eta a_{X,1}\ -\eta a_{X,1}\ \mu_X\ \mu_X\ 0\ 0\right], \left[a_{Y,1}\ a_{Y,1}\ \mu_Y\ \mu_Y\ 0\ 0\right]\right). \end{aligned} \tag{53}$$

6. Conclusion

Along this chapter general results on the classical Lauricella functions F_A, F_D, Φ_1, Φ_2 and Φ_3 and their applications to the performance analysis of wireless communication systems are obtained. Then, new special functions which include these classical Lauricella functions as particular cases are proposed. This new type of functions, named $\mathcal{A}$, $\bar{\mathcal{A}}$, $\mathcal{B}$, $\mathcal{B}^+$ and $\mathcal{B}^-$, allow us to express in exact closed-form a variety of important analytically open problems in wireless communications.

7. Acknowledgments

This work was partially supported by the Spain Government and FEDER project TEC2011-25473.

References

[1] M. K. Simon and M-S. Alouini, *Digital Communications over Fading Channels* (2nd edn), New Jersey: Wiley, 2005.

[2] H. Exton, *Multiple Hypergeometric Functions and Applications*, Wiley–Halsted, 1976.

[3] A. M. Mathai and R. K. Saxena, *The H-function with Applications in Statistics*, New York: Wiley–Halsted, 1978.

[4] M. A. Chaudhry and S. M. Zubair, *On a Class of Incomplete Gamma Functions with Applications*, Boca Raton: Chapman and Hall/CRC, 2001.

[5] F. Yilmaz and M-S. Alouini, 'Outage Capacity for Multicarrier Systems', *IEEE 17th International Conference on Telecommunications (ICT)*, 2010, pp. 260–265, April 2010.

[6] F. J. López-Martínez, E. Martos-Naya, J. F. Paris and U. Fernández Plazaola, 'Generalized BER Analysis of QAM and its Application to MRC under Imperfect CSI and Interference in Ricean Fading Channels', *IEEE Transactions on Vehicular Technology*, pp. 2598-2604, June 2010.

[7] H. M. Srivastava and P. W. Karlsson, *Multiple Gaussian Hypergeometric Series*, Wiley–Halsted, 1985.

[8] A. M. Cohen, *Numerical Methods for Laplace Transform Inversion*, Springer, 2007.

[9] I. S. Gradshteyn and I. M. Ryzhik, *Table of Integrals, Series, and Products* (7nd edn), Academic Press: San Diego, 2007.

[10] J. M. Romero-Jerez and A. J. Goldsmith, 'Performance of Multichannel Reception with Transmit Antenna Selection in Arbitrarily Distributed Nagakami Fading Channels', *IEEE Transactions on Wireless Communications*, vol. 8, no.4, pp. 2006–2013, April 2009.

[11] A. M. Tulino and S. Verdú, *Random matrix theory and wireless communication,* Now Publishers Inc., 2003.

[12] J. F. Paris, 'Unified Analysis of the Average Gaussian Error Probability for a Class of Fading Channels', http://arxiv.org/abs/1103.0502, March 2011.

[13] L. Rayleigh, 'On the Resultant of a Large Number of Vibrations of the Same Pitch and of Arbitrary Phase', *Philosophy Magazine*, vol. 27, no 6, 460-469, 1889.

[14] R. S. Hoyt, 'Probability Functions for the Modulus and Angle of the Normal Complex Variate. *Bell Systems Technology Journal*', vol. 26, no. 4, pp. 318–359, 1947.

[15] M. Nakagami, 'The m-Distribution- A General Formula of Intensity Distribution of Rapid Fading', in *Statistical Methods in Radio Wave Propagation.*, Pergamon Press: Oxford, pp. 3–36, 1960.

[16] A. Abdi, C. Lau, M-S. Alouini and M. Kaveh, 'A New Simple Model for Land Mobile Satellite Channels: First- and Second-order Statistics', *IEEE Transactions on Wireless Communications*, vol. 2, no. 5, pp. 611-615, 2003.

[17] M. D. Yacoub, 'The κ-μ and the η-μ Distribution', *IEEE Antennas and Propagation Magazine*, vol.49, no. 2, pp. 68–81, 2007.

[18] X. Lei and P. Fan, 'On the Error Performance of M-ary Modulation Schemes on RicianNakagami Fading Channels', *Wireless Personal Communication*, vol. 53, no. 4, pp 591–602, June 2010.

[19] D. Morales-Jiménez and J. F. Paris, 'Outage Probability Analysis for η-μ Fading Channels', *IEEE Communications Letters*, pp. 521-523, June 2010.

[20] J. F. Paris, 'Closed-form Expressions for the Rician Shadowed Cumulative Distribution Function', *Electronics Letters*, vol. 46, no. 13, pp. 952-953, June 2010.

[21] N. Y. Ermolova, 'Useful Integrals for Performance Evaluation of Communication Systems in Generalised η-μ and κ-μ fading channels', *IET Communications*, vol. 3, no 2, pp 303-308, 2009

[22] M.C. Clemente, F. Ruiz-Vega, P. Otero and J.F. Paris, 'Closed-Form Analysis of Adaptive Coded Modulation over Rician Shadowed Fading Channels', *Electronics Letters*, vol. 47, no. 3, pp. 217-218, February 2011

[23] J. F. Paris, 'Outage Probability in η-μ/η-μ and κ-μ/η-μ Interference-limited Scenarios', *IEEE Transactions on Communications*, to appear.

[24] N. Y. Ermolova, 'Moment Generating Functions of the Generalized η-μ and η-μ Distributions and Their Applications to Performance Evaluations of Communication Systems', *IEEE Communications Letters*, vol. 12, no. 7, pp. 502-504, July 2008.

In: Communication Systems: New Research
Editor: Vyacheslav Tuzlukov
ISBN: 978-1-62618-654-5

Chapter 12

A SYSTEMS APPROACH TO MODELLING COMMUNICATIVE PROCESSES

Deirdre Pratt*
Durban University of Technology, Durban,
KwaZulu-Natal, South Africa

ABSTRACT

This chapter deals with a new systems approach to the modelling of communicative processes. Systemic modelling makes it possible to arrive at generalizations about social phenomena which have the force of the law-like propositions of the natural sciences. In the process documented in Robert Franck's seminal work on systemic modelling, which uses a classical induction method similar to reverse engineering, the necessary functions required to effect a social phenomena or process are first identified. These constitute the formal or theoretical aspect of the social mechanism involved. A practical (or "empirical") model is then formulated to describe how the functions are carried out in actual social contexts. Once this model has been validated by its match with observable events, the theoretical model, or "architecture of functions", can then be validated against the practical model. The theoretical aspects of the model make it possible to generalize about the same social process in very different contexts, as well as the possibly of its being transferable to other areas or fields. This chapter will show how data from over 54 video protocol analyses gathered to formulate a model of written communication were used to establish the architecture of functions underpinning communicative interactions. The system of functions thus identified was considered to constitute a generalizable principle underpinning design and inquiry processes. It also offered insights into the functions language is thought to perform, suggesting that these are in fact communicative functions rather than linguistic per se. This account offers an exemplar of a systems approach to modelling in the social sciences, while at the same time developing a deep structure which might be used to explain the varied and highly idiosyncratic surface manifestations of spoken, written and computer-mediated interactions.

* Durban University of Technology, Box 1334, Durban, KwaZulu-Natal, South Africa. Email: deep@dut.ac.za.

Keywords: Communication, written composition, systemic modelling, classical induction, interaction

INTRODUCTION

Researchers in the Information Age (one is tempted to say: "Information *Overload* Age") now have the possibility of cross-correlating models, methods and data over a wide range of fields and disciplines to come up with new cross-field and interdisciplinary syntheses. One of the problems of modelling social processes, however, is that modelling itself is by no means an agreed on phenomenon, and the term "model" can be (and frequently is) used in various disciplines to apply to exemplars, types, data sheets, processes, case-studies and anecdotal accounts, the latter being more in the nature of "metaphors" (Makhubu, 2011, p. 58). As Franck (2002, p. 2) points out, the different types of modelling, in particular, the disparate elements in various social science models, make it difficult to compare models in different areas. To remedy this "malaise", Franck suggests a modelling process which identifies the functions necessary for the social process to occur, which, he suggests, will make it possible to compare models in different fields. It has been the author's experience that using Franck's modelling process has not only made it possible to view writing as a communication mode sharing functions in common with other modes, but that the "architecture of functions" underpinning written communication might be viewed as a principle informing other types of interaction, notably those in design, curriculum, e-learning and research processes (Pratt, 2011b).

While the research documented in this chapter goes beyond written composition in arriving at a model of communicative interactions, it will be shown how the more generalized model was formulated from the specifics of research into written composition. While the communicative functions are masked in writing because of its nature as a delayed interaction, ironically writing revealed the communicative functions more clearly than would have been possible in oral mode, as the delay in writing etiolates the process into a series of stages, and the functions which are usually carried out simultaneously in speech are in writing observable in a series of clearly demarcated phases. This chapter will first look at the modelling process used to develop both theoretical and applied models, then at the video protocol method which was used to capture the data. Next the results of the modelling process will be given, involving the model of communicative functions and two applied models of composing. The chapter will conclude after giving an account of the various practical applications of the models: it will show how the generalizable principle developed in the modelling process can be used to explain language functions, communication modes and genres, and how the modelling process was used to inform the design of composition software.

RESEARCH ORIENTATION

Critical realism was adopted as the research orientation from which the final results of the modelling were presented as (1) it was congruent with the author's belief and value system, (2) a typical preoccupation of critical realism is the modelling of social processes, and (3) it reflected the interplay of theory and practice which had occurred throughout the investigation. It also emphasised the need for application of theory in practice (Bhaskar, 1986, pp. 170-172), and it was not just the nature of written communication which was being investigated, but also ways of improving academic writing practice, first at school, later at university level. Apart from the fact that Bhaskar's tripartite ontology (Bhaskar, 1978, p. 56) was congruent with my own view of reality, his description of causal mechanisms fitted my belief that social processes, like natural processes, were not entirely random, ad hoc or idiosyncratic but followed a deep-structure systemic operation:

> There is nothing esoteric or mysterious about the concept of the generative mechanisms of nature, which provide the real basis of causal laws. *For a generative mechanism is nothing other than a way of acting of a thing*. It endures, and under appropriate circumstances is exercised, as long as the properties that account for it persist. Laws then are neither empirical statements (statements about experiences) nor statements about events. Rather they are statements about the ways of acting of independently existing and transfactually active things (Bhaskar, 1978, pp. 51-52, my emphasis).

This suggests that generative mechanisms go beyond the concept of mechanical causality to the idea that natural and social processes have an "essence", or DNA-like quality which has a real existence (i.e. in the Domain of real). The existence of these "inner structures" can be apprehended hypothetically only, in terms of their "fit" with events as experienced by humans: "Theory is not an elliptical way of referring to experience, but a way of referring to hypothesized inner structures of the world, which experience can … confirm or falsify" (Bhaskar, 1978, p. 158).

The key mechanisms involved in social processes are not easily identified, as these are not directly observable, and it is difficult to detect significant patterns in the surface impressions of complex, layered, dynamic social processes (Bhaskar, 1979, p. 62). Writing is one such activity where the researcher can easily become lost in the detail and miss the underlying systemic relations; there is also a tendency for researchers to focus on *what* writers write as opposed to what they do as they write, and the majority of research has tended to be form-based (Raimes, 1991, p. 409) and is still largely text-focused. The predominant postmodern influence in written composition has also tended to focus on discourse as text; the term "discourse" itself - often poorly defined, if at all - tends to end up meaning "text" by default (see Van Dijk, 1997, p. 3).

Critical realism holds that the reality of a social phenomenon can be grasped in transcendental "leaps", much in the way that Kuhn (1962) describes the advent of new scientific discoveries. This requires the type of modelling which includes a theoretical component. As Judd suggests:

> Theory mediates between the real world and our empirical experience of it. Being that we do not typically have immediate access to generative mechanisms or social structures, *theory is our provisional method for making our way in the world* (2003, p. 59, my emphasis).

The relationship between theory and practice in critical realism is complex, and there is not necessarily a one-to-one relationship between surface manifestations of a social phenomenon and its deep-level structure as represented in theory. Apart from suggesting that inquiry involves layers of stratification (Bhaskar, 1978, pp. 154-175), with rule-of-thumb precepts at the surface level and complex theories at deeper levels (1978, p. 160), Bhaskar does not specify any particular methodology for arriving at actual descriptions or models of social mechanisms. He does make the point that models do not always contain a theoretical component (1978, p. 192); a model of a social mechanism would require a theoretical aspect to explain its causal functioning. Franck's modelling process, below, clarifies the theoretical aspect of social mechanisms in showing how modelling involves the interplay between formal (i.e. theoretical) and applied (i.e. practical) aspects of a mechanism.

METHODS

Franck's Modelling Process

As Meehan points out: "A system explanation can be viewed as a formal pattern, a map, that can be imposed or overlaid on the empirical world. If the pattern fits the empirical data, it serves as an explanation or guide to the empirical events that fall within the pattern (1968, p. 63).[1] Using insights from studies of modelling by 13 researchers in 9 disciplines, Franck summarizes the modelling process which provides a "system explanation" thus:

> (1) Beginning with the systematic observation of certain properties of a given social system, (2) we infer the formal (conceptual) structure which is implied by those properties. (3) This formal structure, in turn, guides our study of the social mechanism which generates the observed properties. (4) The mechanism, once identified, either confirms the advanced formal structure, or indicates that we need to revise it (2002, p. 295).

Franck's modelling process complements Bhaskar's critical realist philosophy, showing how models of social mechanisms arrived at by this process (classical induction) have both a formal and applied aspect. The formal aspect is the system of functions "without which" the social process cannot take place; he terms this a *theoretical model*, or "architecture of functions" (2002, pp. 88, also pp. 229-238). The applied aspect of the social mechanism, that is the way in which it operates in actual social contexts, he terms an *empirical model*.[2] Franck's modelling process can be broken down into a series of stages, as follows:

1. The *properties* of a social system are carefully observed and defined.
2. A *theoretical model* is formulated on the basis of the functions needed to achieve the above properties: the model consists of an "architecture of functions".
3. The *mechanism* which achieves the system's properties is inferred from applying the theoretical model to real-life situations or data.
4. An *empirical model* is formulated, depicting the operating of the mechanism in a real-life situation.

---

[1] Note that Meehan's "empirical" is the equivalent of Bhaskar's "domain of real", and not his "domain of empirical".

[2] Bhaskar and Franck use the term "empirical" with completely different force and meaning: Bhaskar's "domain of empirical" refers to the dimension of thought, while Franck's "empirical model" refers to the practical application of a social process ion a specific context.

5. The empirical model is then *tested* in a real-life situation or against data, to see whether it actually generates the properties of the system.
6. The theoretical model is *validated* by being tested against the empirical model, to see whether the theoretical model needs adjustment to fit the real-life functioning of the social system.
7. If the theoretical model, that is the system of functions, is seen to be generalizable so as to explain the properties of a process in another discipline or field, it can be said to have the force of a *principle* (Pratt, 2011a, pp. 46-47).

According to Franck, arriving at the formulation of generalizable principles is desirable in social science research, as generalizability allows researchers the opportunity to consolidate their knowledge base by testing the same principle in a different discipline, area or field (2002, p. 297). As the principle involves the system of functions required to effect a process, and does not describe the actual process itself, the principles derived (i.e. in social science modelling) may be applicable even to phenomena in physical science and vice versa (Franck, 2002, pp. 297-298; see also Peeters, 2002, pp. 158-160).

There can be considerable overlap and recursion in actual instances of modelling, particularly in steps 1., 2., 3. and 4. of the modelling process. According to Franck, it is not unusual for the empirical model implicitly to pre-empt the deep-level system of functions comprising the theoretical model:

> In truth, when we construct a causal model, we quite naturally take advantage of the resources of the functional approach, without taking the trouble to model the combination of functions of a system. It is no accident that we select, within our causal model, certain variables in preference to others: we choose those variables because we think they are covering this or that social function (2002, p. 96).[3]

The systemic modelling process can also take place in recurring cycles: three such cycles were identified in the course of my inquiry, involving (1) my masters research, (2) the discovery of the system of functions and resultant refining of the empirical model, and (3) the development of composition software based on the models (developing the program menu and submenus involved re-conceptualizing the application of the theoretical model, and the software itself could be seen as yet another empirical model).

Modelling as Retrospective Process

While the stages described by Franck were carried out over a 20-year period, it was only by hindsight that the stages could be identified clearly to produce a coherent account of the modelling. A retrospective application of Franck's theoretical overview of modelling then not only "made sense" of the inquiry processes but also authenticated Franck's theory, as contained in his model of the modelling process (2002, p. 295). The modelling process started as an attempt by the author to develop a practical method for improving writing competences at secondary school level. I was at the time registered for a masters in Applied Linguistics,

[3] Note that Franck uses the term *causal model* and *empirical model* interchangeably.

and was reading up on the processes involving in composing, an approach which was then identified as the "process approach" to written composition.

A study of the literature as well as my own experiences in teaching composition and grappling with composing in a new "more scientific" genre (i.e. Applied Linguistics) revealed the possible existence of distinct stages in composing, without revealing precisely what these might constitute. Conferencing (i.e. a type of interactive interview) with my class of freshman students required the speedy formulation of a practical model of composing in stages, which took the form of a "user's model" of writing (Pratt, 1987).

The modelling process, then, was carried out in stages in the following projects /discoveries:

- 1986/7: Master's research, formulating and testing out of an applied (or "empirical") model, the user's model of writing on secondary school pupils, mother-tongue (MT) English speakers.
- 1991: Pilot project involving 5 English second language (ESL) University students.
- 1993: Project involving 18 technikon mother tongue (MT) English and ESL students writing revision assignments in three academic subjects (Education, Social Science and Political Science).
- 1995: Project involving 5 technikon students writing revision assignments in Economics.
- 1999: Project involving 11 fourth semester Electrical Engineering technikon students at various stages of writing their Electronic Design Reports.
- 2000: Identification, from the user's model, of key "aspects" of writing, which were later found to be the communicative functions, or theoretical model of writing.
- 2000/6: Developing composition software based on the models.
- 2003: Identifying the research orientation as post-positivist (i.e. critical realist).
- 2004: Obtaining Franck's (2002) model of the modelling process "formalised" the methodology to be used in my doctoral research.
- 2006/8: Doctoral research, developing a theoretical model of writing, refining the applied model formulated in 1986/7, and validating the models using the composing of 13 Town and Regional Planning students as well as the body of data gathered in previous projects.

The doctoral research (Pratt, 2007b) can be seen to be the culmination of following up at a deeper theoretical level the modelling process started in masters research.

The above list might suggest that the modelling occurred in linear progression, but this was by no means the case. For example, the formulation and testing out of the applied model usually occurs in the later stages of Franck's modelling process: my master's study was already testing out an empirical (i.e. practical) model which had yet to be explained (i.e. by formulating a theoretical model) and validated (i.e. by further data). The latter processes, however, could not be satisfactorily accomplished until an appropriate research orientation (critical realism) and methodology (Franck's modelling process) had been found.

Video Protocol Analysis

Video protocol analysis is a method which was developed in the course of the modelling for eliciting data on composing by recording live composing sessions on videotape (later DVD) and using playback to reconstruct composing processes together with the writer (see Pratt, 2010). Video protocol analysis is better known in the context of testing audience response to software (Macleod & Rengger, 1993) as well as for gathering data on interactions between people and computers (Cockburn & Dale, 1997; Pratt & Gutteridge, 2006), where data can be measured and analysed digitally to obtain quantitative feedback on user data. Its use in the social sciences is not so common, and is rare in written composition; although it is mentioned by Jones (1982), Matsuhashi (1982) and Pianko(1979) (1979) the author was not able to obtain any details of the method of analysis used beyond the fact that they video-recorded their subjects while composing, without requiring them to verbalise their composing aloud. "After each session, they review the tapes with the writers, probing for information concerning what went on in their heads at crucial points" (Candlin, 1981, p. 5). In the data gathering and analysis carried out by the author the video protocol method was preferred to "think aloud" protocols, where data are captured on audio-recorder (Raimes, 1985, p. 234), as it requires no coaching in advance and does not require writers to verbalise their composing processes while composing, which might interfere with nonverbal processes. In particular, visual cues and body language enable writers to recall detailed cognitive processes for some time after the composing session. Being filmed does not appear to cause participants undue stress or discomfort, and most writers seem to forget the cameras once they focus on the writing task. To make the process less mysterious (i.e. in terms of what data are being captured) the camera image/s being recorded should visible to participants (i.e. on a monitor screen), which offers them the opportunity of ensuring that they stay within camera range, thus affording them the status of co-researchers rather than passive "subjects".

Both the technology and the method of reconstruction were refined in the course of the long-term modelling over a number of projects. Initially there was so much surface data available that checklists were used in an attempt to focus on key composing activities. Later as the "fit" of the user's model of writing (or initial practical model) with actual instances of composing was established, the number of repetitions of discrete acts was found not to be as important as the writer's focus at any given moment during composing. This is because it was found that the same activity (e.g. writing on paper with a pen) might have a completely different focus and purpose depending on whether the writer was taking notes, generating new ideas, structuring for the reader, preparing the piece of writing for final submission, or reflecting on progress by jotting down thoughts. In fact the user's model made the video protocol method feasible, as it suggested the areas of focus which might be involved in composing; the fact that these could be reconstructed with the help of writers from the recordings in turn authenticated the stages of the model, each with its different area of focus. This does not imply that the research was theory driven (i.e. positivist): hypothesizing about the nature of the "real" by using models provided a way of making sense of multiple sense-impressions (Bhaskar, 1979, p. 62) in the "transcendental leaps" which are a feature of critical realist methodology, and observation which is "theory *laden*" is not necessarily "theory-*determined*" (Sayer, 1992, p. 73).

STAGES OF THE WRITING PROCESS

1	Prewriting -	Consider purpose and reader gather data, let it mull round. - DATA GATHERING	R
2	Draft writing -	Suggest structures or outlines, jot down ideas or fragments, write larger pieces. - IDEA GENERATION	E C U
3	Major editing -	Reread and structure for reader, order, add, delete (go back to 2 if necessary). - IDEA ORGANIZATION/STRUCTURING	R S
4	Minor editing - and polishing	Check for correctness, check format and minor editing conventions. - EDITING	I V
5	Evaluation -	Assessment (by writer and others) in terms of purpose. - EVALUATING	E

Figure 1. The user's model of composing developed in master's research (Pratt, 1987, p. 35).

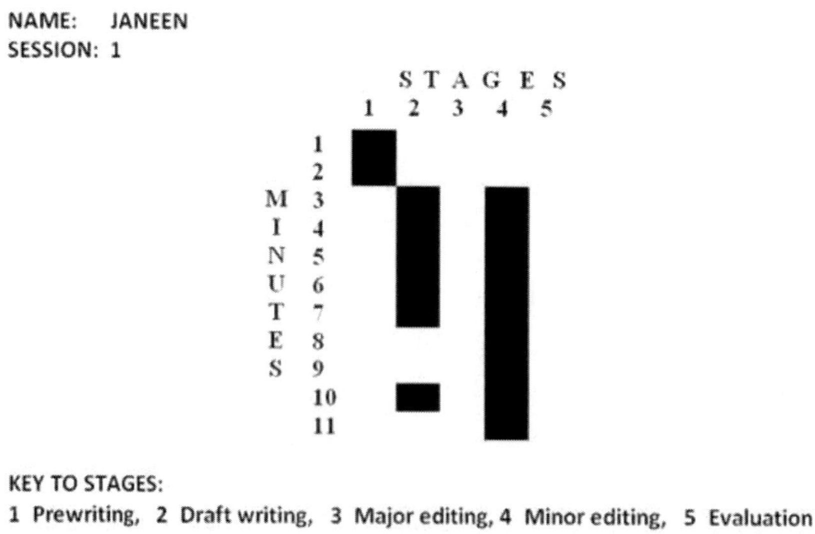

Figure 2. Example of graph showing writing profile (from Pratt, 1987, p. 59).

For the first set of video protocols, borrowed equipment was used, with a single camera (on tripod) focused over the writer's shoulder on to the emerging text as she composed. Facial expressions, though limited in side view, proved so useful in providing cues for reconstructing composing behaviour that in subsequent protocols two cameras were used in split screen mode, one focusing on the emerging text, the other on the writer's facial and body movements. Initially the recordings were captured on videotape, later on DVD; an MX10 mixer was used to capture recordings in split screen mode. Computer composing was captured using an ATI card to feed the screen display to the external mixer; this limited the "body" camera to head and shoulders view only so as not to clutter the screen display and obscure text. While theoretically the whole computer composition process could have been recorded on computer using a web camera and screen recordings, the poor quality computer equipment available to staff and students at the university where the research was carried out, as well as the expense involved in acquiring editing software, made this option not feasible; it would also have meant that any glitches in recording (an inevitable part of video protocol analysis) would have interfered with the writer's use of the computer while commencing composing, when the writer's concentration is most vulnerable and most easily compromised.

Data obtained from the video protocol analyses were analysed using the five stages identified in the user's model (the first empirical model) the existence of which in actual instances of composing were validated in data obtained from all 54 composing sessions, but which was analysed in detail in the 1986, 1993 and 2006 sessions (the 1999 data were drawn from partial and not completed instances of composing). The user's model comprised five stages, with practical advice given at each stage to assist learners to emulate the behaviours identified in the literature as that of competent writers (Pratt, 2011a, p. 68). Even at the applied model stage it can be seen that the user's model is an algorithm allowing for infinite variety of writing performance. But what the user's model did not do was explain *why* these behaviours were "good" or how contingent factors might affect "good" composing behaviour (perhaps even make it "bad"). Both the implied value system ("good" vs. "bad") and the omission to account for instances such as examination writing (inevitably "bad") as well as highly idiosyncratic writing behaviour were serious flaws in the model from a theoretical viewpoint. It was anticipated that investigating the deep structure of composing might find solutions to these problems.

Establishing the focus of the writer during the course of a composing session enabled the construction of not only narrative writing profiles (Pratt, 1987, p. 52) but also graphs which revealed the systemic nature of composing as a series of stages. The graph in Figure 2 not only shows the learner writer's focus in the course of composing, but also illustrates visually how she is blocking her composing by focusing on corrections (a case of "spelling anxiety") rather than idea generation. For a full two minutes Janeen stopped focusing on her essay content to rehearse the spelling of "crawling" in the margin. No time was given to structuring or evaluating her text for its general effect, and, losing confidence in her ability to compose, she finished in 11 minutes. After a short intervention where I explained a simplified version of the user's model[4], her composing profile graph (Figure 3) showed that she had composed over a longer period (15 minutes), had made attempts at structuring for the reader, had left corrections mainly until the end, and had attempted to evaluate her work (i.e. the "5 Cs" had modelled "good" writing behaviour for her and she had based her subsequent composing behaviour on that of competent writers). Her written texts after the intervention with the user's model comprised rough drafts as well as neat drafts, and showed signs of reworking for

[4] The "five Cs" model for school pupils, with stages simplified to: 1. Consider your thoughts, 2. Create, 3. Chop 'n change, 4. Check and 5. Criticize.

the reader and more syntactic complexity as a result of these attempts. Interviews established that her concept of writing had changed to viewing it as a series of stages (i.e. those shown in the model). In a later follow-up interview she revealed that she had made a truly brilliant conceptual leap in successfully applying the model to the task of answering comprehension questions. Janeen, whose rating as a writer, as she said, "would be pretty low because my spelling's atrocious"(Pratt, 1987, p. 82), *had perceived that the model dealt with communicative interactions other than essay writing*, which means that she must have intuitively grasped the deep structure of the user's model, which was later identified as an interactive principle.

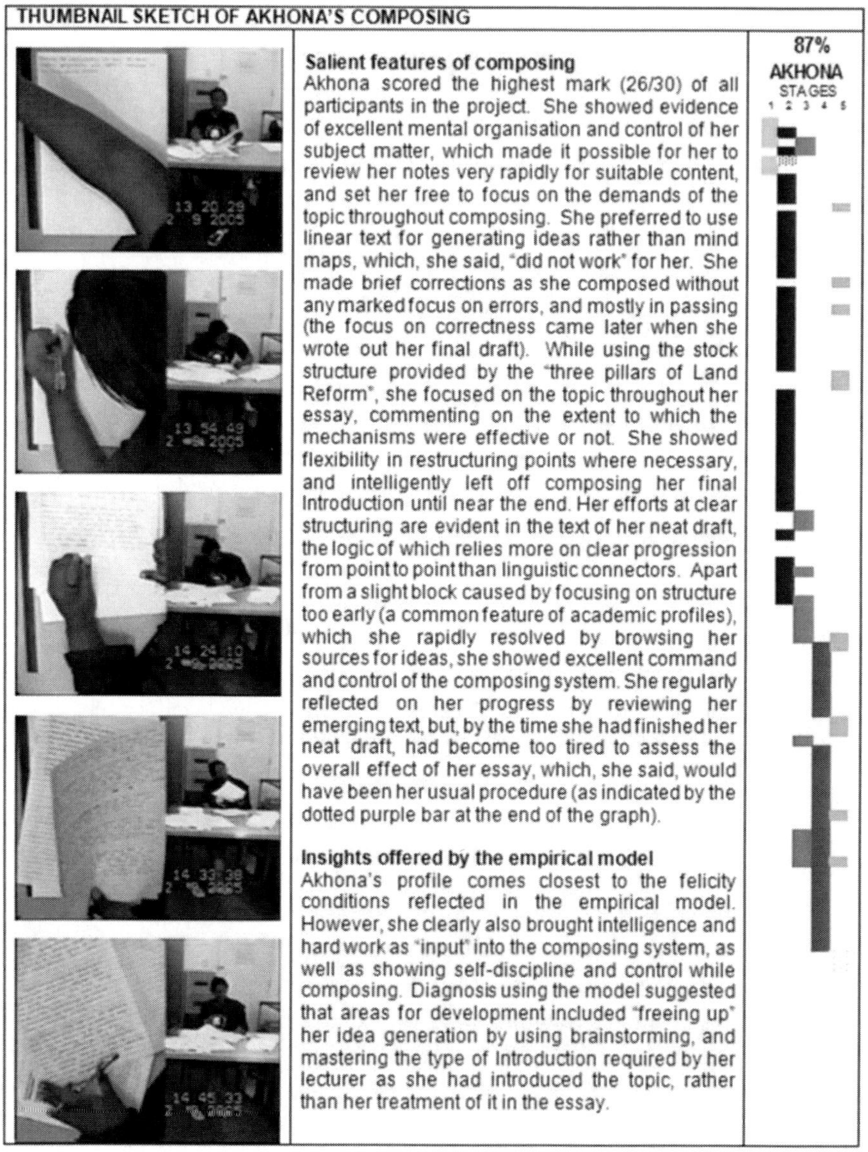

Figure 3. Example of a "thumbnail sketch" of student composing.

The video protocol analyses conducted later at tertiary level involved composing sessions up to two to three hours in duration, which resulted in more complex graphs with more evidence of recursion; the more advanced students reworked their drafts much in the manner of academic staff members reworking papers and journal articles. Colour coding was using to provide a clearer contrast between stages of composing, which mean that graphs could be shrunk in size and still maintain their distinctive profiles, enabling comparisons between not just writing profiles obtained in "before" and "after" composing sessions but between the profiles of writers in larger student groups (seePratt, 2011a, pp. 156-157). Figure 3 shows how, by the end of the modelling process, "thumbnail sketches" could be drawn up of student writers. But this is to pre-empt the development of the theoretical model of written composition and the revised empirical model, a composing algorithm based on the user's model but with features which accounted for how contextual factors impacted on the composing process. This was necessary, as the user's model did not explain how some writers with "good" profiles did not score so well in composition assignments as others whose composing did not follow the profile associated with the habits of competent writers. It was not until Franck's modelling process had been applied (in retrospect) to the modelling so far and had been used to develop - and validate - further models of composing that this could be done.

RESULTS

This section will deal with the models developed in the course of working through Franck's process retrospectively from 2004 backwards to identify the essential functions in composing and formulate the theoretical model; thence moving onwards to the last series of protocols in 2006 which involved the refining of the empirical model into a composing algorithm.

Model of Communicative Functions

1. The Properties of a Social System are Carefully Observed and Defined.

The properties of composing had been "carefully observed and defined" in video protocols and augmented by the author's teaching, lecturing and personal experience over a period of twenty years, so that a detailed picture had been drawn up. It was the categorisation of the social system as "composing" which proved to be problematic, until the social system was reformulated as "communication in written mode". Composing is in fact more properly the *mechanism* (see 3. below) whereby the social system, communication in written mode, is achieved.

2. A Theoretical Model is Formulated on The Basis of the Functions Needed to Achieve the Above Properties: The Model Consists of an "Architecture Of Functions".

Identification of the functions required to achieve the properties of written communication proved difficult. Firstly, composing is a highly idiosyncratic form of communication owing to the fact that it is a form of distanced communication split into two interactions by proxy, namely writing and reading, therefore one is dealing with a proxy system, the full communicative nature of which is somewhat masked (Pratt, 2011a, p. 47). Next, more than one level of systems is involved, in the sense that writing is one of several different communication *modes*; this led to the postulation of a primary communicative system which manifests differently in the different modes, whereupon it takes on the modal characteristics of each mode (see Pratt, 2011a, pp. 120-121). The primary communicative system would then constitute the "system of essential functions", that is the functions which would need to be carried out to effect communication in a given mode. It followed from this premise that the theoretical model of composing would constitute a system of communicative functions, or functions "without which" communication could not occur (or could not occur successfully).

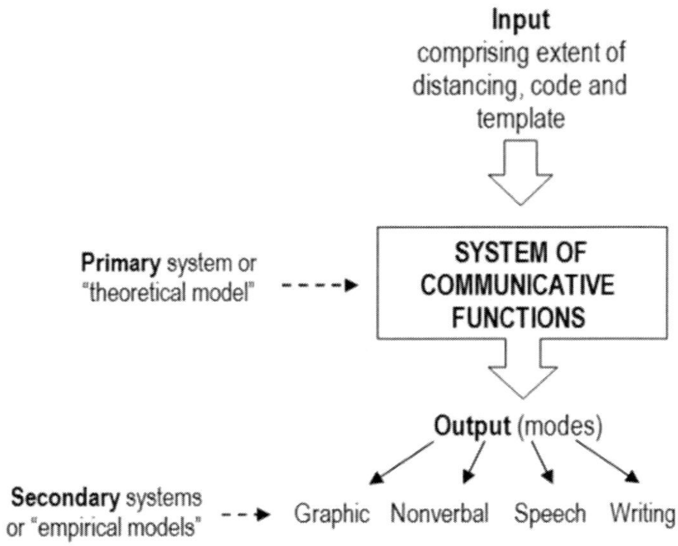

Figure 4. Primary communicative system underpinning modes at secondary level.

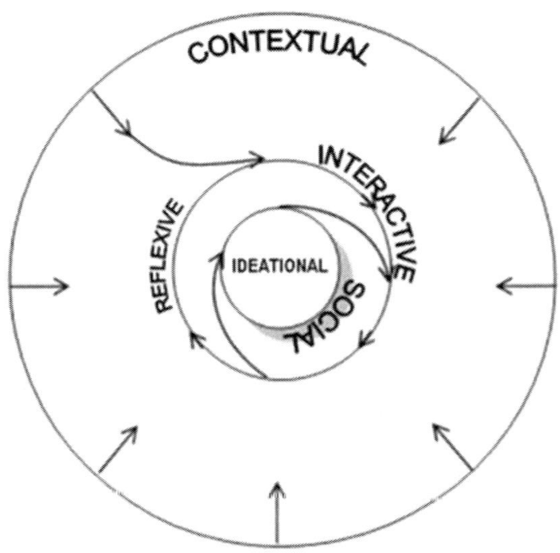

Figure 5. The system of communicative functions (from Figure 7.3 in Pratt, 2007b, p. 170).

Somewhat earlier on in the process (in 2000) key communicative aspects of writing had been tentatively identified, namely the contextual, ideational, interactive, social and reflexive aspects of writing, which could be related to each stage of the user's model, although stage 1 presented problems (dealt with in Pratt, 2011a, pp. 138-141). Applying Franck's description of the modelling process suggested that these "aspects" comprised the system of communicative functions which were involved not only in written communication, but also in other communication modes. This led to re-conceptualizing composing as a mechanism for effecting "communication in written mode", and attempting to sketch the systemic relationships between the functions, as shown in Figure 5.

The system of communicative functions is in fact the "theoretical model of composing", although these functions are masked as follows in written mode:

Contextual:	The contextual function was the most difficult function to identify, as it is masked by the need to gather data in advance for the proxy interaction involved in writing. This means that "data gathering" is moved up into the preliminary area. Initially it was also difficult to distinguish between social issues in the *context* and social issues in the *written text* (this difficulty was compounded by the fact that, in writing, the social context is established largely by means of the written text.
Ideational:	The ideational function is the least masked function in composing, as it is achieved by generating ideas, mainly achieved in composing by rapidly jotting down rough drafts; but accommodating the reader's needs also generates further ideational content, as shown by the "vortex" between stages 2 and 3 in both empirical models (see Figures 1 and 6).
Interactive:	The interactive function was identified relatively easily, as it is known that turn-taking in conversation is replaced in writing mainly by structuring for the reader (Condon & Cech, 1999), represented by "reader accommodation" in the refined empirical model.
Social:	The social function (orthographic correctness) was also a difficult function to identify, partly because of the fact that the contextual function includes issues such as social context. However, it is accepted by linguists that precise articulation in speech and orthographic correctness in writing are social, rather than semantic, issues (Palmer, 1971). The social function in writing is realised mainly by achieving a level of orthographic correctness *appropriate to the social mores* governing that genre as well as the specific audience and context, and the function is carried out mainly in editing (*minor* editing, as opposed to restructuring).
Reflexive:	The reflexive function refers to evaluation of progress by the writer or others in terms of the context and intended audience. Evaluation includes evaluation of the effectiveness of the composing strategies used, as well as an appraisal of the emerging (or final) drafts.

To sum up the relationship between the functions as shown in Figure 4: the communicative interaction is driven by the context (i.e. in terms of its providing the motive for communicating), which prompts an interaction generating message (or idea) content, which in turn needs to be governed by social appropriateness, the whole process being regulated by feedback. This systemic functioning is thought to apply to all communicative interactions, and not just writing, but in delayed interactions such as writing the operation of the functions may be masked or adapted so that the communicative functions are not so easily distinguishable. In some modes, for example, speech, the functions may be carried out near simultaneously; in others, for example, writing, the functions are stretched out into separate stages. It is the latter effect which actually made identification of the functions possible, even though they are masked in writing. It would not have been possible to detect these functions from detailed observation of spoken interactions, but in retrospect they can be seen to be the functions "without which" communication could not take place with any degree of success.

3. The *mechanism* which achieves the system's properties is inferred from applying the theoretical model to real-life situations or data.

While this part of the modelling process is expressed as two stages (3. and 4.), it must be noted that a mechanism has two aspects, formal and applied, so that both the empirical and theoretical models are aspects of the same mechanism. "Composing" (as mentioned above) is then the mechanism which achieves "communication in written mode" and composing has a formal aspect, the system of functions which need to be performed in writing, and an applied aspect, the composing algorithm whereby this performance is achieved.

4. An *empirical model* is formulated, depicting the operating of the mechanism in a real-life situation.

One empirical model, the user's model, had already been formulated right at the beginning of the inquiry process, but, as was stated earlier, this model did not explain why the strategies it contained were effective, nor how local contingent factors might affect composing. The video protocol analyses had revealed instances where students following the model profile of competent writers did not necessarily achieve good composition scores, and also suggested that contingent factors were a key issue, particularly with regard to differing assessment criteria. The user's model might be retained for pedagogical purposes, but it did not depict "the operating of the mechanism in a real-life situation" satisfactorily. A more "analytical" empirical model was therefore formulated and tested out (and in the process validated) in the last series of video protocol analyses. This comprised a composing algorithm, slightly modified to show how the communicative functions are effected in written communication. Because it is an algorithm, the processes it describes are still fairly generalised (see Figure 6). However, the inclusion of an input option (i.e. of *contingent factors*) into the composing system offers insights into how the processes are actually carried out in real world situations. The notion of *conceptual mechanisms* impacting on composing was also introduced, as "the writer's concept of composing can act as a conceptual mechanism in guiding composing practice" (Pratt, 2011a, p. 117). The user's model and the composing algorithm are in fact both conceptual mechanisms guiding writing practice. If the writer's concept does not "fit" with what is known about competent writing practice, however, it may impact negatively on composing procedure, as did Janeen's near obsessive focus on spellings, which interfered with other composing processes until she was offered a concept of composing (the simplified user's model) in which a focus on correctness was delayed until after rough drafts had been generated and the content re-worked for the intended reader.

The Systemic Model of Composing

5. The empirical model is then *tested* in a real-life situation or against data, to see whether it actually generates the properties of the system.

These issues hinged around the two main developments in the model, the inclusion of contingent factors as input into the composing system, and the notion that the carrying out of each stage might not just pre-empt but actually affect the carrying out of subsequent stages. The refined empirical model, now in the form of the composing algorithm shown in Figure 6 was tested out in the 13 more video protocol analyses, and also against the data gathered in all of the previous protocols. Because the refined model explained both the extra- and intra-systemic operation of composing, it provided a more detailed and satisfactory explanation (i.e. a better fit) of the properties of communication in written mode:

> The revised model provided a far more cogent explanation of actual instances of student composing, without the anomalies or exceptions resulting from analysis with the first empirical model. In particular, it explained how internal variation within the composing system could be triggered by external contingent factors, as well as by the results of internal processing. Most importantly, the model helped to explain the degree of success (i.e. in terms of scores given by an independent assessor) achieved by each student writer, and made it possible to diagnose areas for learner development with specific reference to the observed composing procedure of that learner (Pratt, 2011a, p. 147).

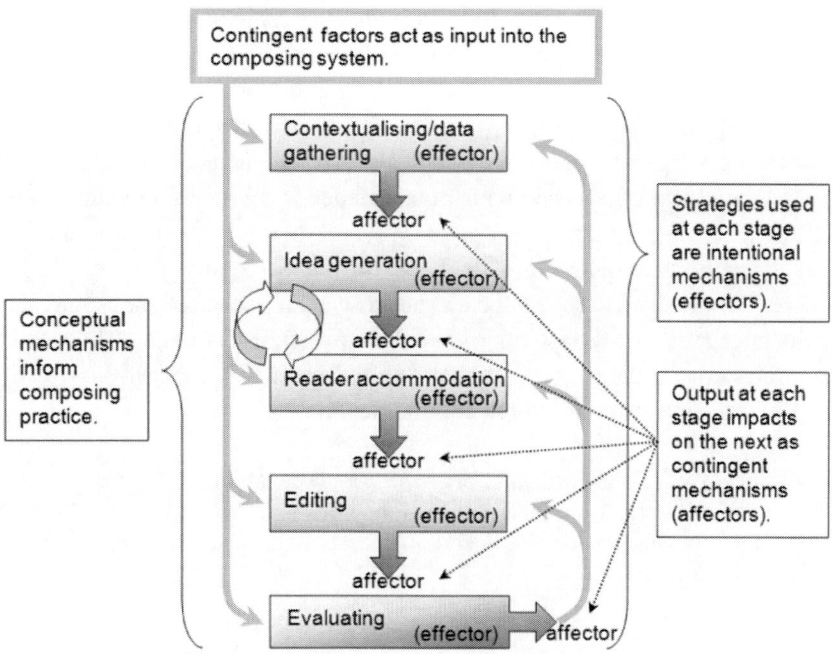

Figure 6. The composing algorithm.

Validation of the Models

The "better fit" and diagnostic potential of the refined empirical model in multiple actual instances of written communication then constituted validation of the second empirical model in terms of the purpose of the modelling, namely, to be the basis for composition instruction which might offer students insight into their composing performance and identify specific areas for development. While only 13 instances of composing were analysed after the empirical model had been refined, the composing algorithm also made sense of the until now unexplained factors in the previous 41 instances.

The data obtained in the 13 video protocols showed that the revised empirical model went some way towards explaining the effect contingent factors had on the composing system, as well as how the carrying out of the function involved in one stage might affect the performance of the function of a later stage. The protocols also offered insights into how contingent factors and systemic operation might affect the shape composing takes in specific instances, cause writer's block, and/or prompt recursion. This is a significant advance on the first empirical model (the user's model), which could not explain how apparently "good" profiles could result in poor composition scores. The input option also helped to explain why the function of each stage was performed effectively (or not); the model showed that the effectiveness of the writer's performance had implications for his/her performance in other stages. This was now a more significant issue than how often redrafting might occur, or even the exact order in which stages might occur. On the other hand, the input option had revealed that recursion could now be seen as being prompted by contingent factors, and not just as the result of the function of a previous stage not having been performed satisfactorily. Much of the intra-systemic variation, then, could be seen as the result of input into the system, with contingent factors contributing to the systemic concatenation of events. The only true systemic variation which could be established at this stage was that which was prompted by the writer's self-assessment of progress, with assessment being a common recurring systemic function. This is possibly because the feedback loop is an integral part of the system in terms of regulating and adjusting performance. In spite of the importance of extra-systemic factors, the video protocols suggested that composing should be viewed primarily as a systemic process, and that key local criteria and constraints needed the writer's attention at the stage of contextualising the written interaction (stage 1). Failure to do so meant that composing expertise could not be marshalled effectively to fulfil the requirements of the given task. However, while the precise nature of local criteria was an aspect of input into the system, *considering* local criteria was a systemic issue. To sum up, knowledge of how to engage in the composing system appeared to be the key to success in student writing, justifying the focus on the systemic operation of written communication:

> This study suggests that, as with all forms of communication, while local contextual issues may materially affect the outcome, it is the core systemic functions which play a critical part in satisfying the demands and/or overcoming the constraints of any given context (Pratt, 2011a, p. 158).

6. The theoretical model is *validated* by being tested against the empirical model, to see whether the theoretical model needs adjustment to fit the real-life functioning of the social system.

The correspondences between the theoretical model and the two empirical models are shown in Table 1. The functions in the theoretical model were derived from reflecting on the user's model, which pre-empted it, which, as Franck comments, sometimes occurs (2002, p. 96). There was thus already a match between the system of communicative functions and the stages of the first empirical model. The terms used for the stages of the refined empirical model (the composing algorithm), while based on those in the first empirical model, can be seen to have changed slightly to show how the communicative functions are effected in writing: they are in fact adaptations of the communicative functions to suit the modal idiosyncrasies of writing as a temporally and spatially distanced form of communication. In written communication the reader is not physically present during the interaction (or, if present, not available), so that the writer has to accommodate the intended reader's needs in advance (i.e. by structuring clearly and elaborating or editorialising where necessary); as the written text is a social artefact reflecting (and establishing) the writer's social status, the social function is carried out by editing the text so that it presents in a manner appropriate to the writer's actual - or imagined - social status. The "fit" between stages 1, 2 and 4 of the algorithm with the corresponding communicative function is self-apparent.

Table 1. Correspondences between theoretical and empirical models

1st Empirical Model	Theoretical Model	2nd Empirical Model
USER'S MODEL	COMMUNICATIVE FUNCTIONS	COMPOSING ALGORITHM
1 Prewriting	*CONTEXTUAL*	Contextualizing/data gathering
2 Draft writing	*IDEATIONAL*	Idea generation
3 Major editing	*INTERACTIVE*	Reader accommodation
4 Minor editing and polishing	*SOCIAL*	Editing
5 Evaluation	*REFLEXIVE*	Evaluating

The inclusion of data-gathering in stage 1 (where it is clearly situated when instances of composing are observed) requires some explanation, however. As the text in writing, unlike the text in a conversation (unrecorded, that is), is captured in a more permanent form, it takes on the aspect of a social artefact which requires a fair amount of crafting. The crafting process, combined with the absence of the an interlocutor to assist with negotiating the social aspects of the written interaction, requires much more attention to ideational content than that required when initiating a conversation, where the ideational content is generated in turn-taking and is not usually known in advance. As a result of these factors, "slippage" takes place in that a consideration of *potential* ideational content takes place even at the contextualising stage (see Pratt, 2011a, pp. 138-141 for a more detailed account).

It must be emphasized that there is no one method of validating human performance models, that different types of models require very different validation techniques, and that the validity of a model needs to be assessed in terms of its purpose. It was not the position of this study that the theoretical model could be validated in the sense of its being able to generate law-like empirical regularities (see Franck, 2002, pp. 297, also 232-234). The data was obtained from video protocol analyses of the composing of 54 students over a period of 20 years, but the empirical models produced in this research were not tested out statistically in large data samples in an attempt to show them to be reliably predictive, as the outcomes of open-ended social processes cannot be predicted reliably. Bunge points out that not all determination is causal, and categorizes "interaction" as a form of determination distinct from causal (1979, pp. 18-20, 30). The focus in this study was on identifying the constants and variables of an interactive social process, thereby giving participants more control over the process. According to the principle set forth by Franck, empirical models are validated in terms of their explanatory force. The empirical models developed should be evaluated in terms of their intended practical application. While in this case, it was to form the basis for composition software, the empirical models have application for composition instruction in general, with the user's model having direct pedagogical application, and the composing algorithm providing a useful tool to study students' composing procedures with a view to diagnosing areas for improvement. The composing algorithm provides a "better" explanation of composing than any offered previously in distinguishing between its extra- and intra-systemic operation, thus accounting not only for the commonalities and variables in composing, but also the infinite variety found in actual composing behaviour, which until now has defied description.

Generalizable Principle

7. If the theoretical model, that is the system of functions, is seen to be generalizable so as to explain the properties of a process in another discipline or field, it can be said to have the force of a *principle*.

As well as having relevance for communication in different modes (see below), the model of communicative functions offers a generalizable interactive principle for use in other areas or fields of social science. It has been used to explain curriculum design in both classroom-based and e-learning courses (Pratt, 2011b). Learning, like composing, can be viewed as an interactive process which must be contextualised, generate knowledge content, be governed by local criteria, and offer opportunities for feedback, whether by means of self-, peer- or instructor feedback, or formal testing (Pratt, 2011b, pp. 10-11). The system of communicative functions also provided the basis for an applied (i.e. empirical) model of blended learning (Gutteridge, 2009; Pratt & Gutteridge, 2006). Piloting mixed-mode research modules designed to build research capacity has suggested that the communicative functions may constitute a principle underpinning inquiry, as applied to an empirical model of research processes which forms the basis of the modules (Pratt, 2009). The principle might be applied as a theoretical framework in disciplines involving design, whether artistic or industrial, and has been used by a Television student at BTech level to provide a framework for analysing features of successful film documentaries (Van Heerden 2008). The author has used the principle to investigate the nature of hypermedia communication, suggesting that the type and extent of distancing involved are more significant than synchronicity or asyncronicity per se (Pratt, 2007a, pp. 710-711).

Application of the Models

Apart from the applications mentioned above, the models offered insights into the language functions as described Halliday's systemic functional linguistics (Halliday, 1985; Halliday & Hasan, 1989), as well as providing a tentative rationale for the formation of the different communication modes and genres. It must be emphasized that these comments are speculative, and have not yet been followed up in formal inquiry. Another application which will be described here is the software based on the modelling, which has not been dealt with in detail in this account so as to focus more specifically on the modelling process. The software did, however, provide the motivation for embarking on Franck's modelling process and framing it within the critical realist orientation for doctoral research. Moreover, adapting the contents - and force - of the empirical models for composition instruction in a different medium (i.e. digital) added another dimension which assisted in formulating the theoretical model.

Language Functions

There are elements in the communicative functions of the theoretical model which correspond roughly with Halliday's language functions (Matthiessen & Halliday, 1997), although the communicative functions were derived from a consideration of the stages of the user's model, the correspondences with the language functions being noted only later. It must be stressed, however, that the functions in the theoretical model are those which need to be carried out for effective *communication* to *occur*, while Halliday's functions are those which *language carries out*. Viewing language as one of the mechanisms for effecting communication would account for the correspondences (i.e. language, as mechanism, would presumably be performing the communicative functions). The relevant language functions are termed the "textual", "ideational" and "interpersonal" functions respectively (Halliday & Hasan, 1989, pp. 44-49). The ideational function, according to Halliday and Hasan, "is the learning or thinking function" which is performed by language (1989, p. 44), which is usually identified with the propositional content of an utterance:

> (ii) The **ideational** metafunction is concerned with 'ideation' ? grammatical resources for ***construing*** our experience of the world around us and inside us. One of its major grammatical systems is TRANSITIVITY, the resource for construing our experience the flux of 'goings-on', as structural configurations; each consisting of a process, the participants involved in the process, and circumstances attendant on it. For example: [Location:] *in the open glade* [Actor:] *the wild rabbits* [Process:] *danced* [Accompaniment:] *with their shadows* (Matthiessen & Halliday, 1997, p. 10).

In this chapter, the term "ideational" refers to a function which needs to be performed for *communication* to take place, and refers to "propositional content" or "message content", which, however, is not to be equated with "meaning", which is the desired outcome of the whole communicative interaction. If, as Mathiessen and Halliday state, language performs an ideational function, language might be viewed as a *mechanism* carrying out the ideational function in communication. However, it is not the only mechanism to do so, as the ideational function may be performed by codes other than verbal language, and by other factors, such as actions, events, objects and placements. The context itself often communicates propositional content, for example, a woman alone in a bar; certain aspects of interaction also communicate, for example, writing to end a relationship instead of doing so in a face-to-face interaction. It must also be noted that, while Halliday and Hasan connect the ideational with "learning or thinking", the ideational function in the communicative system of functions is not limited to generating propositional content in the form of abstract ideas or information, but can refer to social message content.

The next language function mentioned by Halliday which might be viewed as corresponding with one of the communicative function is "interpersonal". According to Kern and Warschauer, by interpersonal Halliday meant "the use of language to maintain social relations" (2000, p. 5). However, it is used with an interactive force, as Halliday contends that language used to encode "interpersonal meaning" fulfils an interactive function: "The sentence [i.e. the one just discussed] is not only a representation of reality; it is also a piece of *interaction* between speaker and listener" (Halliday & Hasan, 1989, p. 20, my emphasis). In the theoretical model, the term "interactive" is used rather than "interpersonal", and it is not just seen as signalling interpersonal relationships but also as having generative power in producing ideational content.

According to Kern and Warschauer, Halliday's "textual" function refers to the production of "situationally relevant discourse" (2000, p. 5). As Halliday and Hasan explain: "All use of language has a context. The 'textual' features enable the discourse to cohere not only with itself but also with its context of situation" (1989, p. 44). They add: "The context of situation, as defined in these terms, is the immediate environment in which a text is actually functioning" (1989, pp. 44-46). The term "situationally relevant discourse" suggests social appropriateness, which is usually signalled by use of socially appropriate language forms or genres, but is also indicated by choice of content. Halliday appears to have conflated the two notions into one function, possible reasons being (1) that social appropriateness is usually defined with reference to a specific social context, and (2) because it would be near-impossible to separate the two functions in the language of written verbal texts (where the only way in which context can be indicated *is* by means of text). In the system of communicative functions the two notions contained in Halliday's "textual" function are separated into a contextual function and a social function (note that there are social aspects of context). The term "social", although it refers to the social loading of text, was used rather than "textual" in the system of functions as text is considered to be a *mechanism* for effecting distanced communication, rather than a function. Finally, while two communicative functions can be linked to Halliday's "textual" function, the concept of the "social" is actually realised at three different levels in the composing algorithm (Figure 5), as follows.

Contingent social factors [INPUT INTO THE SYSTEM] are considered when the writer contextualises [CONTEXTUAL FUNCTION] the writing task in stage 1, which influences the editing of the text in Stage 3 so that it has an appropriate social loading [SOCIAL FUNCTION]. Thus the social context comprises extra-systemic input into the system (1st level), is processed by the writer at the stage of contextualising the task (2nd level), which is an intra-systemic issue, and, as the contextual elements drive the interaction, they also prompt editing of text so that socially appropriate language is used (3rd level) at the editing stage, which is also an intra-systemic activity. Composition approaches which focus on "appropriate discourse" (i.e. text) at the outset tend to overlook the complexity of social aspects of composing, and are thus not very helpful in assisting learner writers to achieve socially appropriate text.

The insistence by Halliday on the inherent functionality of language (1989, p. 17) supports the position that language is one of the mechanisms involved in carrying out the five communicative functions identified as being prerequisites for communication. However, while there is a connection between Halliday's language functions and the communicative functions, Halliday is referring mainly to written texts, which means that the functions may have become obscured by their role in textual encoding. Many of the details of the functional features which Halliday identifies as "language functions" are rather those required by the need to encode verbal language in written texts, and those dealing with textual relations in distanced interactions, rather than any functions which can be seen to be integral to language. In Halliday's account, then, it is difficult to separate the functions played by language in communication from the functions language plays in dealing with the *distanced aspect* of communication (i.e. in terms of all of the communicative functions needing to be encoded in text). He has, however, provided enough correspondences with the communicative functions to indicate that language is a key mechanism in carrying out some of the functions which are prerequisites for *communication* to occur. It must be emphasized that language is not the only mechanism, however, and that its functional role is exaggerated out of all proportion when written verbal texts only are considered.

Modes and Genres

The identification of the system of communicative functions as a primary system with the modes as secondary systems (see Figure 4) suggested how input into the primary system might explain the various communication modes. In other words, modes are not just arbitrary choices, but are dictated by factors impacting on the communicative system, or "input" into the system. It must be noted that communication is a daunting system to investigate, as so many levels of mechanisms can be seen to be operating. This is the "Chinese box" syndrome of systemic investigation mentioned by Bunge (1979, p. 245), and which realist inquiry seems to encourage (i.e. in its emphasis on "stratification"). At the *macro* level, communication itself can be viewed as a mechanism creating cohesion in society; at a lower level, writing is one of the mechanisms effecting distanced communication; lower still, the strategies of the stages of the writing process could be seen as mechanisms carrying out composing; at the *micro* level, various language repertoires constitute the mechanisms which generate text on a page (although one could go to even lower levels, for example, physio-algorithmic functioning, neuronic functioning, and so on). In this chapter composing has been identified as the mechanism used for effecting written communication, but that is for the purpose of framing the study only.

Input into the system of functions at a *macro* level can be seen to determine the form taken by output, that is the form taken by the communication modes. Table 2 suggests that input in terms of the degree of distancing (as opposed to immediacy), the code, and the template can be seen as determining the resulting communication mode. Thus written communication is the modal option used if the input to the system is *temporal* distancing, *verbal script* is used for code, and the result is recorded on a *two-dimensional* template. The various genres operating within these modes cab be explained if "medium" is added to the input variables. Thus input with regard to distancing, code, external template and medium could be seen to form the sequence of causal factors which determine the genre options available (see Table 3). It is local circumstances, often beyond the interlocutors' control, which determine distancing, code and the necessity for a template, and these are factors with contingent determination. It must be noted that interactants who are geographically close to their intended audience may deliberately choose a certain mode (or genre) to introduce (artificially) an element of distancing (for example, the "Dear John" letter). Intentional determination could be said to be operating in this case, although contingent determination might influence the actual route used (for example, a hard copy letter rather than SMS or email). Distancing is not always a choice dictated by coldness or aversion: in mass examinations writing (i.e. temporally distanced communication, as the examiner and assessor may be in the same room) is the mode of choice because mass oral examinations are not feasible.

Table 2. Input into the communicative system determining mode

INPUT			OUTPUT
Distancing	Code	Template	Mode
(Immediate)	Spoken language	(memory)	Speech
Temporal	Verbal script	Two-dimensional	Writing
(Immediate)	Body language	(the human body)	Nonverbal communication
Temporal	Graphic signs	Two-dimensional	Graphic communication

Table 3. Input into the communicative system determining genre

INPUT				OUTPUT
Distancing	Code	Template	Medium	Genre
(Immediate)	Spoken language	(none)	Face-to-face	Conversation
Valence	Spoken language	(none)	Face-to-face	Formal speech
Spatial	Spoken language	(none)	Mobile or telephone	Phone conversation
Spatial	Spoken language	(none)	Computer + modem	Internet conference
Spatial	Verbal script	(none)	Computer + modem	Chat room conversation
Temporal/spatial	Verbal script	Electronic lattice	Computer + modem	Email 402
Temporal	Verbal script	Two-dimensional template	Pen and paper	Handwritten script
Temporal	Graphics	Two-dimensional template	Pencil and paper	Sketch
Temporal/valence[5]	Graphics	Two-dimensional template	Oils and brushes	Painting
Temporal	Spoken language	Temporal sequential template	Tape (or digital) recorder	Sound recording
Temporal	Spoken language, movement, sound, NVC, graphics.	Temporal sequential template	Video recorder	Video recording
Temporal/valence	Spoken language, movement, sound, NVC, graphics.	Temporal sequential template	Video camera/ film camera	Film

[5] The term "valence" is used for the type of distancing which occurs when the interaction is value-loaded, thus "distanced" in terms of being rehearsed/replayed by proxy. It is used in Vroom's expectancy model (in Fielding, 1993, pp. 39-40) to signify the expected value of work outcomes.

It is thought that a study of the effects of distancing, code, external template and medium in different forms of electronic communication might make it easier to distinguish between the effects caused by mode and medium in areas such as distance education over the Internet. An investigation into to how the effects of an electronic template (i.e. two-dimensional monitor screen) template differs from that provided by paper, and the reduction of temporal distancing in electronic communication (i.e. which can be near-instantaneous) could go some way to distinguishing between the effects of the electronic medium and the mode (i.e. writing) involved. Condon and Cech (1999) report that they were unable to resolve this issue when identifying the characteristics of discourse strategy management in a digital environment.

Software Design

The application of the modelling process dealt with in my doctoral research was a process-based writing tutor program, informed mainly by the user's model (which had been designed for pedagogical use), but with an input option as suggested by the second empirical model. The pedagogical model comprised a composing algorithm on which the computer program could be based, and in itself constituted a "social algorithm", or type of conceptual mechanism informing social practice (Blunt Bugental, 2000). *NEWT* (the New Electronic Writing Tutor) could be used by learners on a home computer or in a computer laboratory while composing on word processor. The basic specifications for use were that:

- the tutor program should be accessible to individual learners (i.e. be able to be installed on home or institutional computers);
- the program display should remain on the screen while the user was working on a word processor;
- the program should be able to be customized to suit local academic requirements;
- user input could be saved, retrieved and displayed on-screen.

The completed program ran from a floating menu bar which hung superimposed over a Microsoft Word page, but which remained on view when typing occurred, unlike most Microsoft help menus.

Figure 7. The floating menu bar of the writing tutor program.

The algorithm provided by the recursive stages of the user's model formed the core of the writing tutor program, and it was the focus on translating the stages and strategies of the user's model into menu and submenu items which triggered discovery of the "aspects of composing" which were later found to the functions forming the theoretical model. The input option was added later, after the second empirical model had been formulated; programming led to re-conceptualizing the input in terms of "teacher advice" (in keeping with the pedagogical focus of the software), which section was, however, for input by the student user.

Table 4. Stages of composing as reflected in writing tutor menu items

Theoretical Model	1st Empirical Model	2nd Empirical Model	Software
SYSTEM OF FUNCTIONS	USER'S MODEL	COMPOSING ALGORITHM	NEWT Menu Items
CONTEXTUAL	1 Prewriting	Contextualizing/data gathering	Preparing to write
IDEATIONAL	2 Draft writing	Idea generation	Writing rough drafts
INTERACTIVE	3 Major editing	Reader accommodation	Revising for your reader
SOCIAL	4 Minor editing and polishing	Editing	Editing and proof-reading
REFLEXIVE	5 Evaluation	Evaluating	Evaluating your writing

The stages of the user's model (together with consideration of the insights provided by the other two models were represented in the writing tutor menu as shown in Table 4. The wording of menu items needed to be clear but as brief as possible, and the terms used had to make sense to users who had not studied composition theory, that is, they had to be framed in practical terms. On the other hand, the stages of the composing algorithm needed to be clearly differentiated, so that users would be aware of the change of focus in each stage. Table 4 shows how the NEWT menu items were underpinned theoretically by the theoretical model, but drew on elements of both empirical models.

Far more in the way of guidelines and strategies could be added than could be contained in the two-dimensional surface of a diagram of the model, as well as the "teacher's advice" input option, which was stand-alone, but had links to each stage section. The problem now was which advice could be taken as generic, and which aspects needed to be omitted or left for the input option, as they depended on context. These were the same issues which were being played out in parallel in the modelling process, and which forced decisions on both sides. Table 5 shows the full main menu of the writing tutor program which could be collapsed into a reduced menu (by clicking on HELP WITH COMPOSING) showing only the stages of composing; this helped to reduce screen clutter as well as focusing the writer on the stages while composing.

There is a tendency in composition software design to focus on what computers can do (for example, databases, help menus, text analyzers, formatting) rather than on instructional needs. The focus in designing the writing tutor program was to translate into digital form the insights gained from the modelling, as well as the instructional materials and strategies which had been used in combination with the user's model. Nothing was included because it was a feature which programming could offer; on the other hand, various programming options were used to transform instructional routines and make them more interesting and accessible.

The writing tutor program fits into the category of a process-based tutor, but was designed to be a learning experience, not a teaching programme. It is not a simulation in the sense of a simulation model (Burch, 2002), yet is very much like the kind of simulation users experiences in *Flight Simulator*, *Age of Empires* or *SimCity*, where expertise is developed as result of the user's working through self-selected routines at various levels. It is then a type of procedural simulation (Alessi & Trollip, 1991), but more open-ended than most. The writing tutor program can be used for developing composing expertise by individual students, by the teacher as an enhancement for classes of students, or could form the basis for a composition teaching programme.

Table 5. Expanded and reduced menus of the writing tutor program

NEWT - x
Menu +/-
Program overview
Tutorial
About composing
Assess your writing expertise
Teacher's advice
HELP WITH COMPOSING
Preparing to write >
Writing rough drafts >
Revising for your reader >
Editing and proof-reading >
Evaluating your writing >
Mind map
Flow chart
Help with writer's block
Inner dialogues
Composing on computer
Ideas database
Readings database
Working notes

NEWT - x
Menu +/-
Tutorial
Preparing to write >
Writing rough drafts >
Revising for your reader >
Editing and proof-reading >
Evaluating your writing >
EXPAND

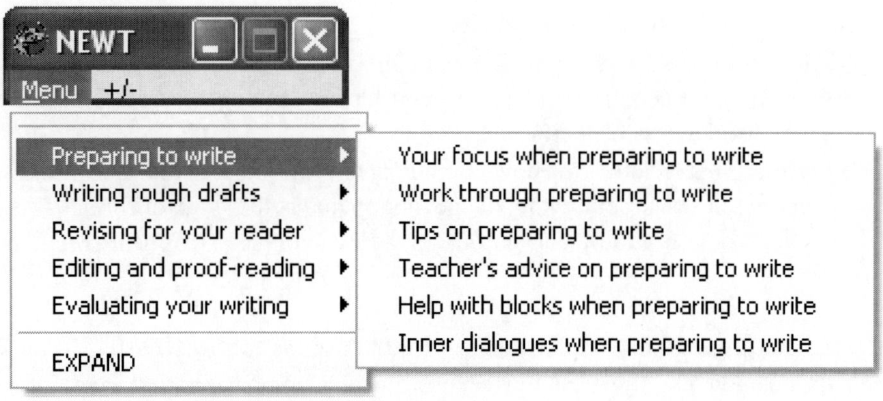

Figure 8. Submenu items running off the main menu stages.

It has conferencing-type elements, but the five stages which are the core of the program constitute an algorithm rather than a heuristic (Pratt 2005a). The program contains text directions for heuristic and invention strategies. It also contains revision strategies and checklists, and acts partly as an organiser in prompting students to take notes and to save references in a database. These features were added as the result of the theory underpinning the writing tutor program, and not derived from the example of other process-based tutor programs.

An advantage of the help menu structure is the way it steps information in series of recursive hyperlinked sections for the user. Each stage is divided into six sub-stages (see Figure 8). The standard items include *Focus*, *Tips*, *Teacher's advice*, *Help with blocks* and *Inner dialogue* (*Blocks* and *Inner dialogues* can also be from the main menu). As each stage of composing has specific needs related to its function, non-standard items (i.e. related to each stage) were *Work through preparing to write*, *Creative idea-generating strategies*, *Some structures you could use*, *Editing checklist* and *Evaluation checklist. Work through preparing to write* leads the writer through the process of considering audience, purpose and genre. *Creative idea-generating strategies* suggests a number of strategies of generating ideas. *Some structures you could use* suggests stock school composition structures. When any of the six sub-stages of the five-stage menu is accessed, a prompt appears at the bottom of the screen giving the focus for that stage, but users can remove the prompts easily by mouse-clicking on them.

As well as being based on the theoretical and empirical and models of composing, the writing tutor was informed by the modelling in another sense, in following the educational design principle suggested by the communicative functions:

- **Contextual:** This function relates to the social context in which knowledge is constructed, and requires the course designer to decide how learning is to be contextualised.
- **Ideational:** This function relates to the source of the knowledge to be constructed, or the process whereby knowledge actually comes into being (it also raises the question of course content).
- **Interactive:** As knowledge is constructed in learning interactions (including interactions with resources), the course designer needs to anticipate how participants will interact in constructing knowledge.
- **Social:** The social parameters, conventions or constraints operating in a given learning situation need to be identified and made explicit to learners, particularly in respect of local assessment criteria.
- **Reflexive:** This relates to how participants will reflect on and assess their performance in constructing knowledge, and includes the issue of formal assessment (if any) and how it will be carried out, as well as course assessment (Pratt, 2005, p. 138).

The writing tutor program is contextualised for the user as follows. Firstly, It remains on screen and its position is controlled by the user, which emphasized its accessibility as a resource. Next, the "help menu" look of the program contextualises it as potentially helpful, and the *NEWT* acronym and logo suggests a friendly helper. The program overview, text description of the user's model and tutorial explain its use. Finally, the users themselves provide the context by adding input in the *Teacher's advice* section.

Figure 9. The *NEWT* "friendly lizard" desktop icon.

In the writing tutor program **ideational** content ranges from the menu items and instructions to actual lesson content and examples. There was a tension in designing the writing tutor program in terms of adding my own composition instructional materials or leaving the users more room for input. The user's model already had a body of fairly generalized (i.e. not context-specific) instructional materials associated with it, geared to each stage. These needed only to be collated, sized to fit the tutor menu, and adapted to suit general use for learners from 12 to 18 years of age. The writing tutor program could then be seen as a virtual extension of myself and my experience of teaching written composition. However, it is not a highly idiosyncratic personal account, as it based on a theoretical model of written composition which establishes the commonalities in composing and shows how local variables may impact in specific instances of composing. User input, though limited, makes an important contribution to the ideational content of the writing tutor program when in use, and also makes the program more interactive.

The writing tutor was initially envisaged as being much more *interactive* in the sense of having a number of routines in which the user could join as in a "live" lesson. However, a help menu is also a highly interactive option, one which is controlled by the user according to need (Pellone, 1995, p. 8). A computerised tutor, while it is less versatile than a human tutor, is always "there" for individual learners, no matter what time of day and night. The program contains a few interactive routines and prompts, such as *Assess your writing expertise* and *Work through preparing to write*, and the prompts which appear as help is accessed for each stage. An interactive feel is also given by the *Blocks* and *Inner dialogues* hyperlinks.

Social aspects include local requirements, in this case catered for by the *Teacher's advice* input and display sections. Initially two programs were designed, one for junior school and one for high school/university use. It was later decided to conflate these into one version, suitable for users from 12 to 18 years of age. This was a challenge in terms of expressing the content in suitable language (a social issue) so as not to baffle the juniors or bore the seniors. The program has a "teacher-like" tone, as composing is learned in a school context. The "friendly lizard" NEWT logo adds a sense of fun, and the tutorial, mind map and flow chart applications give the sense of computer games, offsetting the pedagogical tone.

The tutor has features which encourage *reflexive* habits in the user by including the *Working notes* menu option. *Assess your writing expertise* (an automatically scored self-test) can also be used as a basis for reflection. There is not much opportunity for user feedback on the program itself, although students are encouraged to communicate suggestions to a NEWT website (which is how most course feedback is obtained).

Conclusion

Writing is a mode of communication which it is essential for students in formal education to master in order to progress academically, ostensibly because it is currently the mode most associated with intellectual development. This was not so in the past, when poets who could not compose in their heads - and remember - 800 lines of oral verse were considered uneducated; it may not be so in the future, when verbal language might well be replaced by holograms or pictorial glyphs (the younger generation already seem to be shedding their verbal skills in anticipation of "picting" with their tablets). But it is the case now, and there is little agreement amongst researchers and teachers as to what writing is, let alone how it should best be learned. Written composition was a contentious, highly divided field when the author commenced the research described in this chapter (North, 1987, p. iii), and has, if possible, become even more contentious and divided. In the last 30 years form-based approaches were superseded by process approaches, then diversified into postmodern approaches which focused mainly on "discourse" (i.e. language), including critical (i.e. Marxist) and constructivist (or construct*ionist*) approaches, as well as the New Literacy approaches focusing on "situated practice". One of the reasons interchanges between composition researchers can become so acrimonious[6] is because "views of writing - and how it should be learned - inevitably reflect views as to what constitutes reality as well as knowledge, and cut to - or rather, cut into - one's most cherished beliefs and values" (Pratt, 2011a, p. ix). Accounts of writing are not innocent, which is why the author's research orientation was specified near the beginning of this chapter; readers may not agree with the opinions given here, but they cannot claim that they were presented from a neutral position.

The hypotheses presented in the theoretical model of communicative functions and associated empirical models are not meant to challenge the diversity of views in the field or to stifle healthy debate; they can in fact be seen to provide a rationale for current teaching approaches to written composition, and in that sense support certain of the elements on which these approaches focus. The *process approach* to written composition emphasises the *ideational* and *interactive* functions in composing, the expressive and cognitive schools (see Reid, 1993, pp. 3-4) focusing on the ideational; practitioners who emphasise interactive methods of responding to students' texts, such as conferencing, focus on the expressive (Graves, 1978; Zamel, 1985). The *conferencing* method favoured in process pedagogy emphasises that composing is part of an *interaction*, and conferencing could also be viewed as a mechanism for effecting the *reflexive* function. *Genre-based approaches* focus on the *contextual* and *social* functions of composing (but mainly in the written text, see Cazden et al., 1996; Cope & Kalantzis, 1993; Swales, 1990), as do *social constructionism* (Bartholomae, 1985; Bizzell, 1992; Bruffee, 1984, 1986; R.M. Coe, 1986; R.M. Coe, 1987) and *critical approaches* (Berlin, 1988; Clark, 1992; Faigley, 1986; Ivanic & Simpson, 1992). Critical approaches also include *reflective* studies where the of the writer's "voice" is handled with individual student writers in retrospective fashion. The focus on situated practice of the **New Literacy** approach is with the extra- and intra-systemic functioning of the social element in composing, although they would not probably not express it in those terms (see Cazden et al., 1996, pp. 85-86; Paltridge, 2004; Street, 2003).

[6] The 1993 interchange between Canagarajah (emancipatory interest) and Raimes (hermeneutic) is fairly acrimonious, but that between Hairston (1982), Blom (1984) and Parsons (1985) becomes positively vitriolic.

It is interesting to see that Cazden et al. acknowledge the limitations of immersion in a social practice, as it can obscure the systemic aspects of literacy (Cazden et al., 1996, p. 85; see also Street, 2003, pp. 3-4); interesting, because, if there is any one premise on which composition theorists would agree unanimously, it would probably be rejection of the notion that composing comprises a system. The process approach was criticised for representing composing as a "universal process" (Faigley, 1986, p. 40; Kostelnick, 1989, p. 278; Krapels, 1990, p. 53; Lynn, 1987, p. 908; Young, 1978). Even process practitioners themselves cautioned that writing is too complex and too idiosyncratic for researchers to be able to identify one common writing procedure (Raimes 1985, Spack 1984). It is not, of course, the author's position that composing *is* "one common procedure": the modelling process suggests a far more complex systemic operation. It goes without saying that the research in this chapter was not exactly received with cries of joy by the composition fraternity. However, I was encouraged by the comments of one of the blind-reviewers of my book (Pratt, 2011a), whom I suspect, was a critical realist, as s/he clearly understood the thrust and purpose of the work:

> Pratt is able to provide a sophisticated model of writing that lends itself 1) to pedagogical application in the classroom or writing center and 2) to the development of a computer program that prompts writers to attend to the five communicative functions, not in a lockstep fashion but in a process that allows for the recursive nature of writing (Deelen-Mans, 2009)[7].

The reviewer added: "I am already planning to use the composing model in my Advanced Composition course next semester." Professor Robert Franck was also very complimentary in the Preface to the book, but modesty prevents me from citing excerpts here.

In conclusion, even though my field is Applied Linguistics, and my passion - and art - is in crafting the written text, in my opinion composition research has focused too much on language and the written text, and too little on the interactive, communicative aspects of writing. It is hoped that this chapter goes some way towards remedying this imbalance.

REFERENCES

Alessi, S. M., & Trollip, S. R. (1991). *Computer-based instruction: methods and development* (2nd ed.). London: Englewood Cliffs.

Bartholomae, D. (1985). Inventing the university. In M. Rose (Ed.), *When a writer can't write*. New York: Guilford Press.

Berlin, J. A. (1988). Rhetoric and ideology in the writing class. *College English, 50*(5), 477-494.

Bhaskar, R. (1978). *A realist theory of science*. Hassocks: Harvester Press.

Bhaskar, R. (1979). *The possibility of naturalism: a philosophical critique of the contemporary human sciences*. Brighton: Harvester Press.

Bhaskar, R. (1986). *Scientific realism and human emancipation*. London: Verso.

Bizzell, P. (1992). *Academic discourse and critical consciousness*. Pittsburgh: University of Pittsburgh Press.

[7] Ms Bernadette Deelen-Mans is the Springer representative who forwarded the blind peer review reports.

Blom, T. E. (1984). Counterstatement: response to Maxine Hairston, “The winds of change: Thomas Kuhn and the revolution in the teaching of writing” (with a Reply by Maxine Hairston). *College Composition and Communication 35*(4), 489-494.

Blunt Bugental, D. (2000). Acquisition of the algorithms of social life: a domain-based approach. *Psychological Bulletin 126*(2), 187 -219.

Bruffee, K. A. (1984). Collaborative learning and the "conversion of mankind". *College English, 46*(7), 635-652.

Bruffee, K. A. (1986). Social construction, language and the authority of knowledge; a bibliographical essay. *College English, 48*(8), 773-790.

Bunge, M. (1979). *A world of systems. Treatise on basic philosophy, Ontology 2*. Dordrecht: Reidel Publishing Company.

Burch, T. K. (2002). Computer modelling of theory, explanation for the 21st century. In R. Franck (Ed.), *The explanatory power of models: bridging the gap between empirical and theoretical research in the social sciences*. Norwell, MA: Kluwer Academic Publishers.

Candlin, C. N. (Ed.). (1981). *Learning to write: first language/second language*. Essex: Longman.

Cazden, C., Cope, B., Fairclough, N., Gee, J., Kalantzis, M., Kress, G., Luke, A., Luke, C., Michaels, S. & Nakata, M. (1996). A pedagogy of multiliteracies: designing social futures. *Harvard Educational Review, 66*(6), 60-92.

Clark, R. (1992). Principles and practice of CLA in the classroom. In N. Fairclough (Ed.), *Critical language awareness*. Essex: Longman.

Cockburn, A., & Dale, T. (1997, 11-19 November 1997). *CEVA: a tool for collaborative video analysis*. Paper presented at the International ACM SIGGROUP Conference on Supporting Group Work 1997, Phoenix, Arizona, USA.

Coe, R. M. (1986). Teaching writing: the process approach, humanism, and the context of “crisis”. In S. de Castell, A. Luke & K. Egan (Eds.), *Literacy, society and schooling*. Cambridge: Cambridge University Press.

Coe, R. M. (1987). An apology for form: or, who took the form out of the process? *College English 49*(1), 13-28.

Condon, S. L., & Cech, C. G. (1999). Discourse management in three modalities. In S. Herring (Ed.), *Computer-mediated conversation*. Cresskill, NJ: Hampton Press.

Cope, B., & Kalantzis, M. (1993). *The powers of literacy. A genre approach to teaching writing*. London and Washington: Falmer Press.

Deelen-Mans, B. (2009). Blinded review reports. Email of 26 November 4:45 p.m. to Dr D.D. Pratt, deep@intekom.co.za.

Faigley, L. (1986). Competing theories of process: a critique and a proposal. *College English, 48*(6), 527-542.

Fielding, M. (1993). *Effective communication in organizations*. Ndabeni, Cape: The Rustica Press.

Franck, R. (Ed.). (2002). *The explanatory power of models: bridging the gap between empirical and theoretical research in the social sciences*. Norwell, MA: Kluwer Academic Publishers.

Graves, D. H. (1978). *Balance the basics: let them write. Paper on research about learning.* New York: Ford Foundation.

Gutteridge, R. (2009). *The impact of socio-cultural factors on blended learning in the development of academic literacy in a tertiary vocational context.* MTech in Education, Durban University of Technology, Durban.

Hairston, M. (1982). Thomas Kuhn and the revolution in the teaching of writing. *College Composition and Communication, 33*(1), 76-88.

Halliday, M. A. K. (1985). *Spoken and written language*. Oxford: Oxford University Press.

Halliday, M. A. K., & Hasan, R. (1989). *Language, context, and text: aspects of language in a social-semiotic perspective* (2nd ed.). Hong Kong: Oxford University Press.

Ivanic, R., & Simpson, J. (1992). Who's who in academic writing. In N. Fairclough (Ed.), *Critical Language Awareness*. Essex: Longman.

Jones, C. S. (1982). *Composing in a second language: a process study*. Paper presented at the 16th Annual TESOL Convention, Honolulu.

Judd, D. (2003). *Critical realism and composition theory*. London: Routledge.

Kern, R., & Warschauer, M. (2000). Theory and practice of network-based language learning. In M. Warschauer & R. Kern (Eds.), *Network-based language teaching: concepts and practice*. Cambridge: Cambridge University Press.

Kostelnick, C. (1989). Process paradigms in design and composition: affinities and directions. *College Composition and Communication, 40*(3), 267-281.

Krapels, A. (1990). An overview of second language writing process research. In B. Kroll (Ed.), *Second language writing: research insights for the classroom*. New York: Longman.

Kuhn, T. (1962). *The structure of scientific revolutions* (2nd ed.). Chicago: University of Chicago Press.

Lynn, S. (1987). Reading the writing process: toward a theory of current pedagogies. *College English, 49*(8), 902-910.

Macleod, M., & Rengger, R. (1993). *The development of DRUM: a software tool for video-assisted usability evaluation.* Paper presented at the HCI'93 [online]. Available: www.usabilitynet.org/papers/drum93.pdf (3 January 2006).

Makhubu, R. L. (2011). *Development of an interpreting services model for the Durban University of Technology.* Unpublished doctoral thesis, Durban University of Technology.

Matsuhashi, A. (1982). Explorations in the real-time production of written discourse. In N. Nystrand (Ed.), *What writers know: the language, process, and structure of written discourse*. New York: Academic Press.

Matthiessen, C., & Halliday, M. A. K. (1997). Systemic functional grammar: a first step into the theory. In M. A. K. Halliday & R. P. Fawcett (Eds.), *New developments in systemic linguistics* [online]. Available: http://minerva.ling.mq.edu.au/resource/VirtuallLibrary/Publications/sfg_firststep/SFG%20intro%20New.html (17 February 2006).

Meehan, E. J. (1968). *Explanation in social science: a system paradigm*. Homewood, Illinois: The Dorsey Press.

North, S. (1987). *The making of knowledge in composition studies*. Upper Montclair, N.J.: Boynton/Cook.

Palmer, F. (1971). *Grammar*. London: Penguin.

Paltridge, B. (2004). *Approaches to Teaching Second Language Writing*. Paper presented at the 17th English Australia Conference, Adelaide.

Parsons, G. M. (1985). *Constraints on the multidisciplinary approach: uses and misuses of the Kuhnian paradigm*. Paper presented at the Annual Meeting of the Conference on College Composition and Communication (36th, Minneapolis, MN, March 21-25, 1985).

Peeters, D. (2002). On modelling in human Geography. In R. Franck (Ed.), *The explanatory power of models: bridging the gap between empirical and theoretical research in the social sciences*. Norwell, MA: Kluwer Academic Publishers.

Pellone, G. (1995). Educational software design: a literature review. *Australian Journal of Educational Technology, 11*(1), 64-84. Retrieved from http://www. ascilite.org.au/ ajet/ajet11/pellone.html

Pianko, S. (1979). A description of the composing processes of college freshman writers. *Research in the Teaching of English, 13*(1), 5-22.

Pratt, D. D. (1987). *A process approach: the formulation of a simplified conceptual framework showing the stages of the writing process, and an investigation into the effects on writing behaviour of communicating this framework directly to the learner*. Masters in Applied Linguistics, University of Natal, Durban.

Pratt, D. D. (2005, 4-6 July 2005). *An analysis of the design features of three mixed-mode courses in a master's degree programme*. Paper presented at the IASTED International Conference on Education and Technology (ICET 2005), Alberta, Canada.

Pratt, D. D. (2007a). No middle ground, but many mansions: design features of effective mixed mode courses. *SAJHE, 21(6), Special Edition from NADEOSA 2006, 21*(6).

Pratt, D. D. (2007b). *A realist approach to writing: developing a theoretical model of written composition to inform a computer mediated learning application*. DTech in Language Practice, Durban University of Technology, Durban. Retrieved from http://ir.dut. ac.za:8080/jspui/handle/10321/332.

Pratt, D. D. (2009, 10–13 July 2009). *A systems approach to the teaching/research nexus*. Paper presented at the 7th International Conference on Education and Information Systems, Technologies and Applications (EISTA 2009), Orlando, FL, USA.

Pratt, D. D. (2010). *Protocols and processes: a case book of video protocol analyses in written composition*. Saarbrücken: Lambert Academic Publishers.

Pratt, D. D. (2011a). *Modelling written communication: a new systems approach to modelling in the social sciences*. Dordrecht: Springer, Methodos Series.

Pratt, D. D. (2011b). *Scenarios for learning: an integrated approach*. Saarbrücken: Lambert Academic Publishing.

Pratt, D. D., & Gutteridge, R. G. (2006, 6-8 September). *The role of the social mechanism in social transformation: a critical realist approach to blended learning*. Paper presented at the 8th Annual Conference on World Wide Web Applications, Bloemfontein.

Raimes, A. (1985). What unskilled ESL students do as they write: a classroom study of composing. *TESOL Quarterly, 19*(2), 229-256.

Raimes, A. (1991). Out of the woods: emerging traditions in the teaching of writing. *TESOL Quarterly, 25*(3), 407-430.

Reid, J. M. (1993). *Teaching ESL writing.* Englewood Cliffs, NJ: Regents/Prentice Hall.

Sayer, R. A. (1992). *Method in social science: a realist approach* (Vol. Routledge): London.

Street, B. (2003). What's new in New Literacy studies? Critical approaches to literacy in theory and practice. *Current Issues in Comparative Education, 5*(2), 1-14.

Swales, J. M. (1990). *Genre analysis: English in academic and research settings.* Cambridge: Cambridge University Press.

Van Dijk, T. A. (1997). The study of discourse. In T. A. van Dijk (Ed.), *Discourse studies: a multidisciplinary introduction: Vol. 1: Discourse as structure and process.* London: Sage Publications.

Young, R. (1978). Paradigms and problems: needed research in rhetorical invention. In C. Cooper & L. Odell (Eds.), *Research in composing.* Urbana, Illinois: National Council of Teachers of English.

Zamel, V. (1985). Responding to student writing. *TESOL Quarterly, 19*(1), 79-101.

INDEX

D

M

N

O

S

V

W

Y